Transfer Function
Techniques for Control
Engineers

# Transfer Function Techniques for Control Engineers

D.R. Towill, M.Sc., B.Sc. (Eng.), C.Eng., M.I.Mech.E., M.I. Prod.E.

London Iliffe Books Ltd

ILIFFE BOOKS LTD
42 Russell Square London, W.C.1.



First published in 1970 by ILIFFE BOOKS LTD

592 01911 5

Printed in Hungary by Franklin Printing House

# Contents

# Preface

This book specifically aims at presenting control engineering principles, by means of transfer functions, in a way which will make the subject physically obvious to students, designers and users alike. In attempting to put old heads on young shoulders, the importance of *system* performance is continually stressed because it leads to a uniform approach inwhich behaviour in both time and frequency domains is implicitly available.

The book was suggested by Mr Frank Jones, then of Wayne Kerr Laboratories, who considered such a book necessary to speed progress in the application of analogue modelling methods to feedback control systems. The author wishes to express his gratitude to Mr Jones, and also to Mr J. Enwright of the same company, for their active encouragement in this venture.

In 1965, when approached by Mr Jones, the author was lecturing to postgraduate students at the Royal Military College of Science, and much of the book is based either on material used in those courses, or on the research into control system design which the author was undertaking at that time. The chapters are self-contained, and consequently there is some repetition. This is partly a result of the various courses at the Royal Military College of Science requiring different levels of specialisation in control engineering, and partly because of the broad divisions to be found in the application of transfer function techniques in the field.

Many practising control engineers should find the techniques described of direct interest in their work. A significant number of users and designers, particularly of military equipment, are already using these techniques which they have acquired at special lectures given by the author in various parts of the United Kingdom. As a student textbook, Chapter 4 can be used as the basis of a second year undergraduate course introducing the concepts of control. Chapters 1, 2 and 3 can form the basis of a final year undergraduate text, whilst Chapters 5, 6 and 7 can form the basis of postgraduate courses in design.

The author has been considerably stimulated by the texts of Porter, Chestnut and Mayer, Lawden, Thaler and Brown, Truxal, and Horowitz, to mention but a few. He is also aware of the contributions to his education made by Professors C. Holt Smith, and E. B. Pearson of the Royal Military College of Science, and Professor N. A. Dudley of the University of Birmingham. It is sincerely hoped that this book reflects the spirit of their teaching. The Royal Military College of Science offers unusual opportunities to develop the art of lecturing in control engineering; the many students who contributed to making the lectures so enjoyable have considerably influenced the method of presentation herein.

Many of the charts are taken, with the kind permission of the Dean, from Royal Military College of Science Technical Notes. The excellence of these charts is due in no small measure to the skill of the draughtsmen and technical illustrators employed at the college. Thanks are also due to Mr G. Simpson, who was responsible for the computer programming of many of the results, and to Mrs L. Woodrough, who typed many of the Technical Notes.

Cardiff D. R. T.

1968

CHAPTER 1

# Introduction to Control Engineering Techniques

## 1.1 The Objectives of Control Engineering

It is helpful at the start of a book on control engineering to attempt to crystallise the subject in such a way that the reader may have the objectives clearly in view. Control engineering deals with the automatic following of a command signal. The load which is being controlled can take many physical forms, and the control engineer is normally given the load characteristics and asked to design a suitable control system around these. In practice it is often difficult to determine the characteristics of the expected command signal, and at both design and testing stages it is common to assume signal characterisation which can be conveniently handled mathematically. The desired behaviour of the system is then translated into performance characteristics related to the assumed input.

The design of a control system is thus a case of choosing components suitable for controlling the load in a satisfactory manner. In most applications a feedback control system is used; thus information relating to the instantaneous load behaviour is fed back and used to modify the signal controlling the load. Feedback control systems can exhibit a phenomenon known as instability which results in the load oscillating in a violent and unsatisfactory manner, even when the command signal is zero. The faster the control system acts, the more likely is there to be unstable behaviour, and so the designer must build adequate compensation, or protection, into the system.

In addition to the command signal, spurious and therefore unwanted signals will appear at various points in the system, and one of the designer's tasks is to make sure that the system can discriminate between command signals and these spurious signals, i.e. to follow the former and reject the latter. Geometrical and commercial considerations are also highly important. The designer is often restricted by cost, weight, space, and delivery date considerations and, more often than is admitted, by lack of information on load and input signal characteristics.

## 1.2 Historical Review

The Second World War gave tremendous impetus to the discipline of control engineering. Through sheer necessity a vast number of scientists and engineers of widely varying backgrounds were brought together to solve fire control problems. In order to achieve a high kill probability from any weapon system, it is necessary to track the target accurately, and to predict the subsequent movement of the target. This problem is worsened if either the tracker, whether optical or radar, or the weapon launcher, or both of them, is located on a ship or a tank which is itself subject to some complicated motion. A feedback control system is then necessary to stabilise the firing platform, so as to attenuate the short-term disturbances due to the rolling and pitching of the ship or tank. Further feedback control systems are to be found positioning the tracker and gun in response to some command signal.

The advent of guided missiles vastly increased the use of feedback control systems. For example, the missile autopilot is a form of feedback control system, as is the actuator loop for positioning the missile's flight control surfaces. In a sophisticated missile there may be dozens of feedback control systems performing a wide variety of tasks. Space vehicle applications have been a natural extension of this work. Fixed-wing aircraft have been equipped with autopilots for over two decades, and the autopilot reliability has now reached the stage when regular blind-landing of passenger aircraft is expected by 1970.

Since the Second World War the use of feedback control techniques has been extended to the control of machine tools and to the general automation of industrial processes, whilst chemical processes have been automated using feedback control techniques for well over three decades, and a vast technology has been built up in this field. Speed control has been a problem for many years, particularly in the generation of electrical power supplies. The use of a feedback control system together with a governor for measuring the actual engine speed, is probably the oldest known application of the feedback technique. This particular application is termed a regulator, since the function of the system is to 'regulate' or keep the speed constant in the face of fluctuating external demands for electrical power.

In electronic engineering the science of feedback is well established for obtaining signal amplification. The theory of feedback amplifiers was well understood in the early 1930s, and it was natural that many new problems arising in other feedback control fields should be solved using the techniques originally developed by Nyquist and Bode for the design of feedback amplifiers. This approach to feedback control system design based mainly on the assessment of system stability dominated the design process during the Second World War. Since then many alternative techniques have been developed, allowing the control engineer to choose the design technique best suited to his particular application. Control engineering tends to be mathematical in nature, since the behaviour of systems is inevitably studied by using differential equations. Transient behaviour is studied by using the Laplace transform method of solution, the classical method of solution or, in large design offices, by the use of analogue or digital computers for direct solution. Steady state behaviour is normally studied by vector methods. It is possible to present the solution to many of these equations in graphical form, and this does allow very many control systems to be designed using simple graphical techniques. This book concentrates on simple algebraic and graphical techniques, and in particular those techniques which relate to the system transfer function.

## 1.3 Feedback Control Systems

We may wish to control such diverse physical quantities as the position of a massive object, the temperature of a furnace, or the voltage in a particular part of a circuit. There are two distinct ways in which these quantities may be controlled. We may firstly determine the performance of the system under consideration and derive the law connecting the controlled variable (e.g. furnace temperature) and the power source (e.g. electrical current), and thereafter assume that setting a given current will result in the desired furnace temperature. This so called 'open loop' or 'non-feedback' control method suffers from many disadvantages, the most obvious being that changes in furnace conditions will result in a different law connecting the controlled variable and the power source. Thus a given current will not necessarily correspond to a required furnace temperature. Since no information about load conditions is available to modify the supply from the power source, no corrective action based on load behaviour can be taken.

Secondly, we may use the feedback control principle, measure the controlled variable, and feedback this information to modify the supply from the power source. Thus any variation in the controlled variable is automatically measured, and suitable compensation of power supply is made to reduce the difference between the desired and actual value of the controlled variable.

This book is mainly concerned with servomechanisms, which are feedback control systems used for controlling the speed or position of inertia loads. The formal definition of a servomechanism will assist in emphasising the feedback control principle. There are four defining characteristics for the servomechanism as follows:

a. *It is error actuated*, and hence there must be a constant comparison between the desired and actual load behaviour. The instantaneous difference between desired and actual load behaviour is termed the error. It is often necessary to use special devices known as transducers to measure desired and actual load behaviour and convert the information to an electrical signal, as this is often the most convenient form for data transmission and data processing.
b. *There is normally power amplification*. This implies that a low-power error signal, often measured in watts, is used to control a high-power source, such as some form of motor, often of many horsepower. Usually the error signal requires pre-amplification before it can usefully control the motor, and this may be achieved electrically, mechanically, pneumatically, or hydraulically, using a signal amplifier.
c. *Operation is completely automatic*. Operation can be quite remote in character, as, for example, in a nuclear reactor, where the input signal may be some distance from the load.
d. *There is a mechanism, or mechanical drive, driving the load*. Due to the normal high cruise speed of most motors it is necessary to incorporate some form of gear reduction between the motor and load. This gear ratio can often be chosen to achieve the best balance between the motor inertia and the load inertia.

It is only the last characteristic that differentiates between the feedback control system and the servomechanism.

Control engineering lends itself to the representation of various components in block diagram form, with signal paths shown

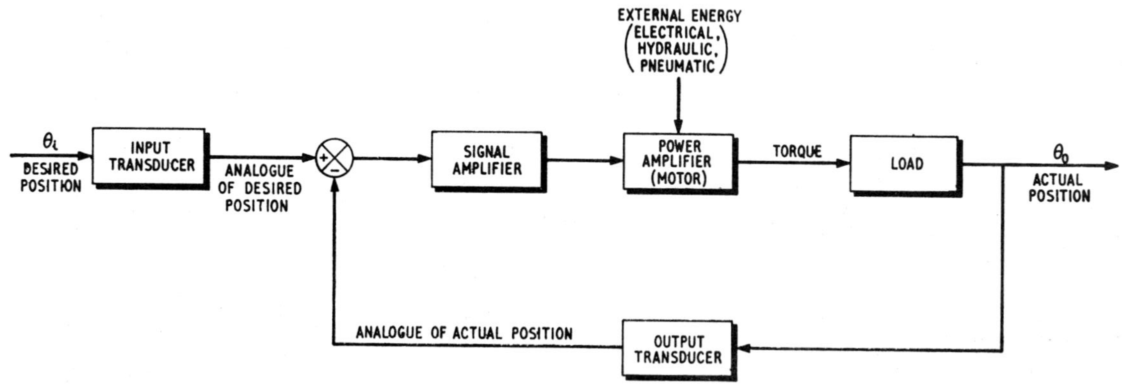


*Fig. 1.1. Block diagram of basic angular position control servomechanism*

as lines joining the appropriate components. Fig. 1.1 shows the block diagram of a basic angular position control system, and the basic components already described are clearly shown. Since the servomechanism shown is controlling angular position, the load will be a moment of inertia rather than an inertia. Unless dealing with specific systems we shall represent the command signal, which in this case is the desired position, as $\theta_i$, and the controlled variable, in this case the load position, as $\theta_0$. We shall see later that a servomechanism which contained only the components shown in Fig. 1.1 would probably be unsatisfactory in operation, and it would be necessary to introduce other components to shape the signals in various parts of the servo.

As self-explanatory examples of simple servos, Fig. 1.2 shows a gun control servomechanism and Fig. 1.3 shows a turbogenerator

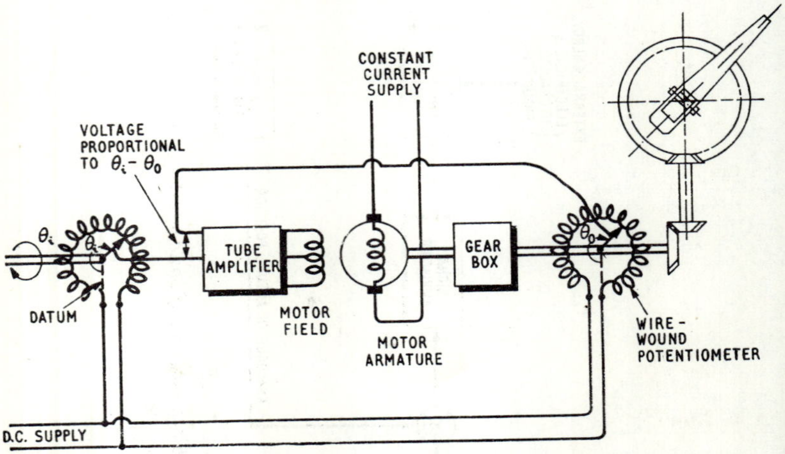


*Fig. 1.2. A simple d.c. gun control servomechanism*

speed control system. Both servomechanisms would, of course, require a little more sophistication than has been shown in these schematic drawings. The speed control system acts as a regulator, since we would set a desired speed by adjusting the governor and valve settings; the system would then attempt to keep the speed constant in the face of fluctuating electrical power demand. The governor acts as a transducer and converts a change in speed to a displacement of the linkage, which in turn adjusts the position of a control valve controlling the pressure and hence the flow through the pneumatically operated steam control valve. An increase in steam flow increases the turbine speed and vice versa.

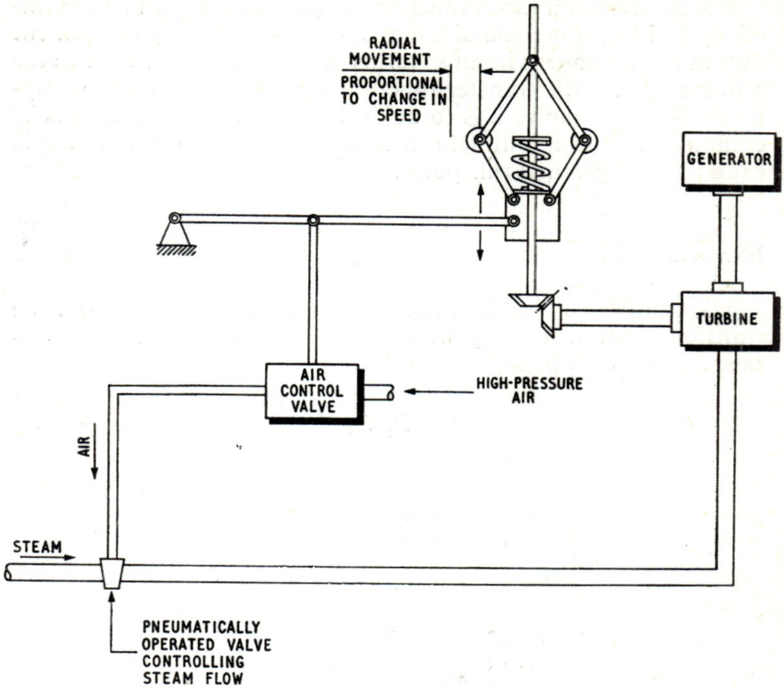


*Fig. 1.3. Turbogenerator speed control system*

## 1.4 Block Diagram and Transfer Function Manipulation

Each of the individual components shown in Figs. 1.1, 1.2 and 1.3 will have physical laws describing their behaviour. For example, the field-controlled d.c. motor shown in Fig. 1.2 will have a definite relationship between the field current exciting the motor, and the motor output torque. It is very convenient to define this relationship as the transfer function of the component, and for linear systems it is normally expressed in terms of the operator $s = d/dt$. Thus if displacement is $x$, then velocity is $dx/dt = sx$ and acceleration is $d^2x/dt^2 = s^2x$, and so on. For systems where the behaviour obeys the law of superposition, the operator s may be treated as an algebraic quantity, which can be manipulated in accordance with normal algebraic principles. If the law of superposition is not obeyed, the system will be 'non-linear', and the transfer function is best interpreted graphically except in special circumstances.

The theoretical transfer function of a component can be obtained by writing down the differential equation relating component output to component input in terms of the operator s, whence for a linear system the transfer function will be a ratio of two polynomials in s. Chapter 2 is to be devoted to transfer functions of components, but it will help to clarify discussion if a few simple examples are given at this point.

EXAMPLE 1.4.1

Fig. 1.4 shows an electrical network, and it is desired to find the transfer function relating the output voltage $E_2$ to the input voltage $E_1$. Now, applying Ohm's Law,

$$E_1 = (R_1 + R_2)I + \int \frac{I}{C_2}\,dt$$

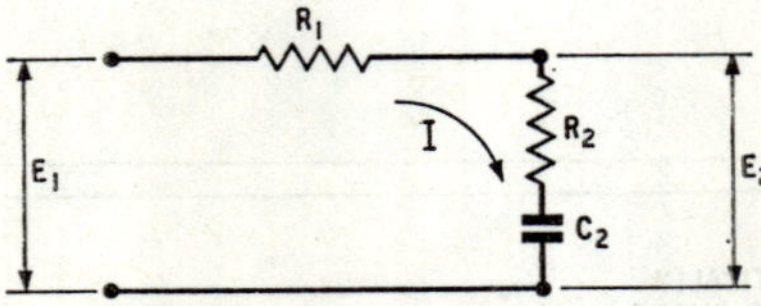


$$\frac{E_2}{E_1}(s) = \frac{1 + T_1 s}{1 + T_2 s} \quad \text{WHERE} \quad T_1 = R_2 C_2, \quad T_2 = (R_1 + R_2) C_2$$

*Fig. 1.4. Transfer function for a simple electrical network*

and, in operator form,

$$E_1 = \left[(R_1 + R_2) + \frac{1}{C_2 s}\right] I$$

since

$$\int = \frac{1}{d/dt} = \frac{1}{s}$$

The output voltage is

$$E_2 = \left(R_2 + \int \frac{1}{C_2}\,dt\right) I$$

and, in operator form,

$$E_2 = \left(R_2 + \frac{1}{C_2 s}\right) I$$

Hence

$$\frac{E_2}{E_1}(s) = \frac{R_2+1/C_2s}{(R_1+R_2)+1/C_2s}$$

which may be simplified to

$$\frac{E_2}{E_1}(s) = \frac{1+T_1s}{1+T_2s} \tag{1.1}$$

where $T_1 = R_2C_2$ and $T_2 = (R_1+R_2)C_2$. The significance of writing $E_2/E_1(s)$ rather than just $E_2/E_1$ is that we have written the relationship between $E_2$ and $E_1$ as a ratio of polynomials in s.

Example 1.4.2

Fig. 1.5 shows a spring-mass system commonly met in many engineering applications. Applying Newton's Law to the mass $M$, we obtain

$$\frac{M\mathrm{d}^2y}{\mathrm{d}t^2} = k(x-y)+C\left(\frac{\mathrm{d}x}{\mathrm{d}t}-\frac{\mathrm{d}y}{\mathrm{d}t}\right)$$

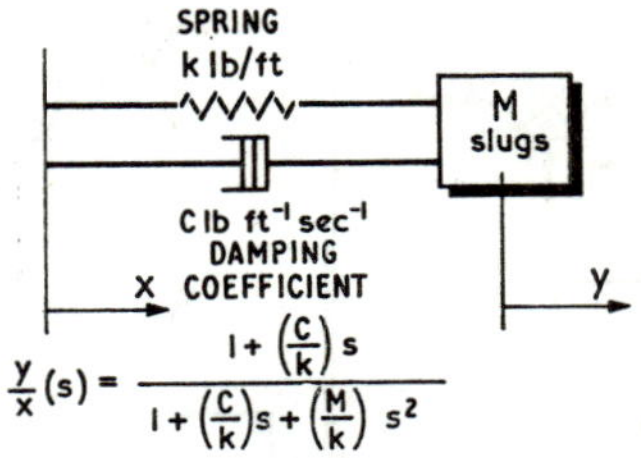


*Fig. 1.5. Transfer function for a simple spring-mass system*

and, rearranging and putting in operator form,

$$\left[\left(\frac{M}{k}\right)s^2+\left(\frac{C}{k}\right)s+1\right]y = \left[\left(\frac{C}{k}\right)s+1\right]x$$

yielding the transfer function

$$\frac{y}{x}(s) = \frac{1+\left(\frac{C}{k}\right)s}{1+\left(\frac{C}{k}\right)s+\left(\frac{M}{k}\right)s^2} \tag{1.2}$$

The denominator of Equation 1.2 is of second order in s whilst the denominator of Equation 1.1 is of only first order in s.

These two examples should serve to show the reader the kind of expression we shall meet frequently in subsequent chapters. We shall now discuss the manipulation of transfer functions; this procedure is necessary in order to simplify the algebra which is inevitably found in dealing with feedback control systems. We shall adopt the general terminology of $G(s)$ for a transfer function which occurs in the feedforward path of a system, and $H(s)$ for a transfer function which occurs in a feedback path. Fig. 1.6 shows

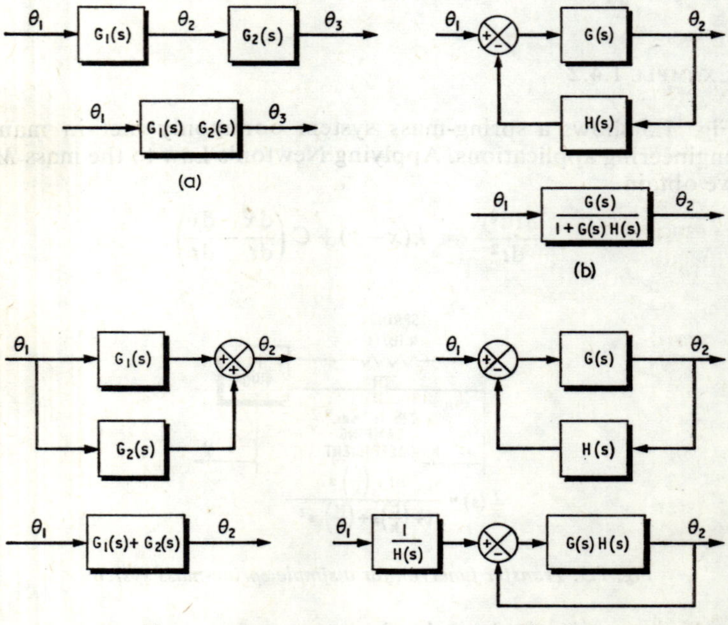


*Fig. 1.6. Block diagram algebra — some important transformation theorems applicable to linear transfer functions: (a) multiplication theorem; (b) negative feedback loop removal theorem; (c) feedforward loop removal theorem; (d) feedback block removal theorem*

the four most important theorems relating to block diagram and transfer function manipulation. These theorems, in general, only apply to systems for which the principle of superposition holds.

The multiplication theorem allows us to replace two transfer functions $G_1(s)$ and $G_2(s)$ in series by a single transfer function

$G_1(s)G_2(s)$. In order to justify the negative feedback loop theorem, let us consider the behaviour of the loop. The signal entering the feedforward element $G(s)$ of Fig. 1.6(b) is $\theta_1 - H(s)\theta_2$, and the signal leaving the feedforward element is $\theta_2 = G(s)[\theta_1 - H(s)\theta_2]$. Thus $\theta_2 = G(s)\theta_1 - G(s)H(s)\theta_2$ which may be arranged to give

$$\frac{\theta_2}{\theta_1}(s) = \frac{G(s)}{1+G(s)H(s)} \tag{1.3}$$

Equation 1.3 may be easily remembered as

$$\frac{\theta_2}{\theta_1}(s) = \frac{\text{feedforward T.F.}}{1+\text{feedforward T.F.}\times\text{feedback T.F.}}$$

Where two feedforward paths $G_1(s)$ and $G_2(s)$ exist in parallel they may be combined as shown in Fig. 1.6(c) into a single feedforward block with transfer function $G_1(s)+G_2(s)$. Occasionally we may wish to remove a block in a feedback path in order to manipulate a system with unity feedback path, since some specialised design techniques are applicable only to unity feedback systems. The procedure is to multiply the feedforward path inside the loop by $H(s)$, and divide the feedforward path before the comparator by $H(s)$. Proof that the systems are identical dynamically is straightforward, since

$$\frac{\theta_2}{\theta_1}(s) = \frac{1}{H(s)}\left[\frac{G(s)H(s)}{1+G(s)H(s)}\right]$$

leaving us with the standard form of Equation 1.3. Provided the system does obey the law of superposition, the order in which the transfer functions are written does not matter. Thus $H(s)G(s) = G(s)H(s)$.

## 1.5 Reduction of Multi-loop Systems

Many feedback control systems consist of a single feedback loop controlling the variable under consideration, whilst at the other end of the scale a complex chemical process control system can have many feedback loops all contributing to the control of the main process. An example of a multi-loop system is shown in Fig. 1.7(a), and we shall apply the negative feedback loop removal theorem and the multiplication theorem to obtain $\theta_0/\theta_i\,(s)$ in terms of the component transfer functions. Applying the negative feedback loop removal theorem to the minor loop, we obtain the transfer function shown in Fig. 1.7(b). Further use of the negative

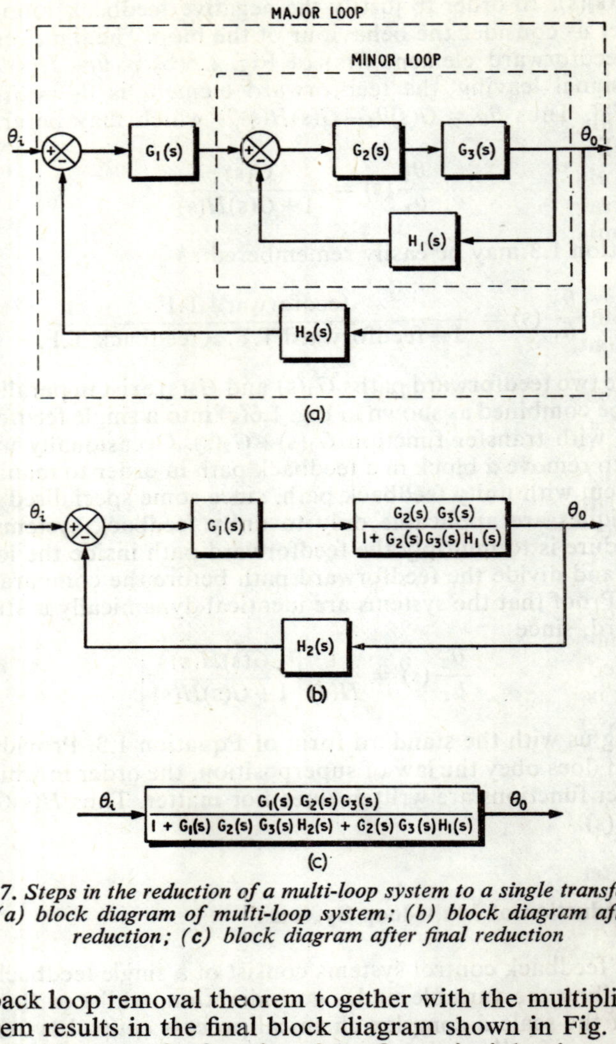


*Fig. 1.7. Steps in the reduction of a multi-loop system to a single transfer function: (a) block diagram of multi-loop system; (b) block diagram after first reduction; (c) block diagram after final reduction*

feedback loop removal theorem together with the multiplication theorem results in the final block diagram shown in Fig. 1.7(c). The resulting transfer function, from first principles, is

$$\frac{\theta_0}{\theta_i}(s) = \frac{G_1(s)\left[\dfrac{G_2(s)G_3(s)}{1+G_2(s)G_3(s)H_1(s)}\right]}{1+H_2(s)\,G_1(s)\left[\dfrac{G_2(s)G_3(s)}{1+G_2(s)G_3(s)H_1(s)}\right]}$$

which simplifies to

$$\frac{\theta_o}{\theta_i}(s) = \frac{G_1(s)G_2(s)G_3(s)}{1+G_1(s)G_2(s)G_3(s)H_2(s)+G_2(s)G_3(s)H_1(s)} \tag{1.4}$$

Equation 1.4 may appear quite formidable to the lay reader, but he can rest assured that it is normally advantageous to manipulate during the symbolic stage and then to substitute component transfer functions into the final equation, rather than to manipulate the component transfer functions directly.

## 1.6 The General Control Problem and the Advantages of Feedback Control

The command signal, conveying as it does the desired load position in a servomechanism, is not the only signal appearing in the system which will affect the load position. Often there will be load disturbances such as windage forces which tend to move the load from the desired position. There may also be spurious signals, termed 'noise', which may generate internally or externally to the control system. Many electronic devices generate such spurious signals, and noise may also arise from such sources as valve pistons sticking and structural vibrations set up by oscillating masses. Not only does the noise signal cause the load to deviate from the desired position, but it may also cause excessive power consumption through the motor attempting to follow the noise signal. The function of the control system is to minimise the error arising from all three sources.

Advantages of feedback control over the alternative 'non-feedback' or 'open loop' control mode have been succintly summarised by Horowitz[1] as:

a. Better rejection of noise and disturbance signals.
b. Improvement in system linearity.
c. Decrease in the effect of variations in load characteristics on system performance.

Feedback control systems are generally more complex and more expensive than their non-feedback counterparts. One major point, which has been made by Truxal[2] but which is not universally known, is that for a load invariant linear system, without noise or disturbance signals, the command signal following requirement can be achieved more simply by a non-feedback type of control system with a suitably designed pre-filter. However, it is not possible to guarantee that noise signals will not be generated somewhere in the system, although in some applications the noise level will be more significant than in others. Load characteristics are often only scantily known, and it is necessary to impose very close tolerances in manufacture of all components in order to obtain high performance with the non-feedback type of control system. With feedback control much less precision is required in manufacture, and lack of load data is not such an embarrassment at the design stage. For many systems such as governors, radar servos, and process plants, the disturbance rejection capability is of prime importance and therefore feedback control is obligatory.

Despite the fact that the advantages of feedback control are mainly in noise and disturbance rejection, the reader will find that the majority of the items included in performance specifications relate to following the command signal; this is not unnatural, since we require the command signal following function to be achieved despite the presence of contaminating signals and, as we shall see later, having accepted the advantages of feedback control, it is possible to write specifications based on the command signal which themselves are influenced by signal rejection requirements. For example, system bandwidth refers to the ability of the system to follow the command signal, but inherently the bandwidth also affects the filtering of high-frequency noise and therefore the bandwidth is chosen with both these facets in mind.

## 1.7 The Three General Sources of Error

Fig. 1.8 shows the block diagram for the general control problem. It has been assumed that the noise *N* and the load disturbance *U* enter at different points in the feedforward path. The total

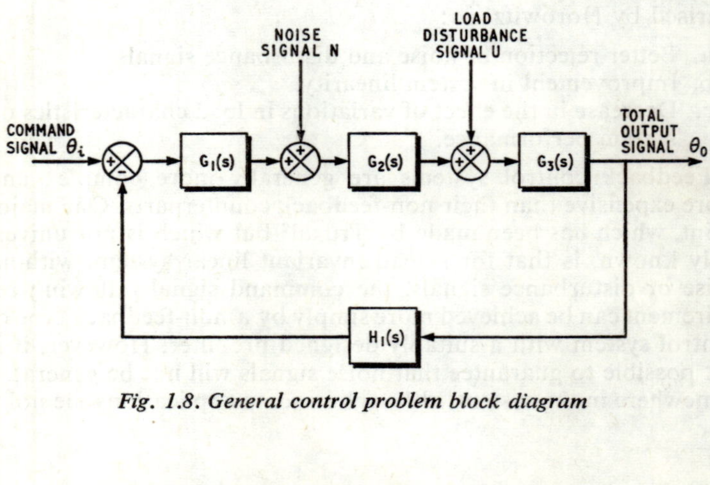


*Fig. 1.8. General control problem block diagram*

output signal $\theta_0$ is now due to some combination of $\theta_i$, $N$, and $U$, which arises from the efforts of the system to follow all three signals. Clearly we would like to ensure that the system follows only the command signal $\theta_i$ and at the same time ignores, or 'filters', the noise signal $N$ and the load disturbance signal $U$. If we assume that the three signals are statistically independent of each other, we may obtain the transfer functions $\theta_0/\theta_i(s)$, $\theta_0/N(s)$ and $\theta_0/U(s)$ by considering each input in turn with the

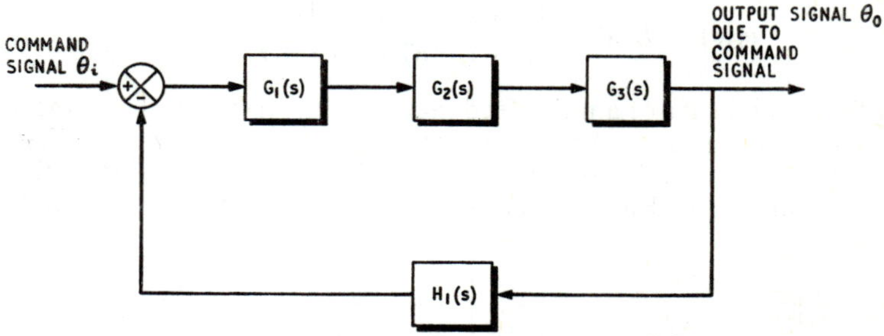


*Fig 1.9. Block diagram relating system output to command signal input*

others zero. Fig. 1.9 shows the resultant block diagram for determining the transfer function $\theta_0/\theta_i(s)$.

Applying the theorems of Section 1.4, we may write

$$\frac{\theta_0}{\theta_i}(s) = \frac{G_1(s)G_2(s)G_3(s)}{1+G_1(s)G_2(s)G_3(s)H_1(s)} \qquad (1.5)$$

The error in following this command signal is $\theta = \theta_i - \theta_0$, and to clarify the principle involved Fig. 1.10 shows the general block diagram for obtaining the error signal. For the purposes of this

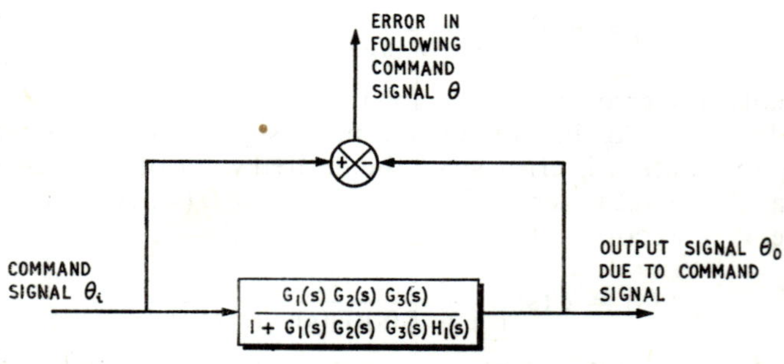


*Fig. 1.10. Block diagram for deriving the error arising in following the command signal*

section we shall define the error in following the command signal as $\theta = e_c$, thus we may write

$$e_c(s) = \theta_i(s) - \theta_0(s)$$

Hence, substituting from Equation 1.5,

$$e_c(s) = \theta_i(s)\left[1 - \frac{G_1(s)G_2(s)G_3(s)}{1+G_1(s)G_2(s)G_3(s)H_1(s)}\right]$$

and therefore

$$e_c(s) = \theta_i(s)\left\{\frac{1+G_1(s)G_2(s)G_3(s)[H_1(s)-1]}{1+G_1(s)G_2(s)G_3(s)H_1(s)}\right\} \tag{1.6}$$

In the special case of a unity feedback system $H(s) = 1$, the numerator of Equation 1.6 becomes unity, thus considerably simplifying the expression. To obtain the error due to the noise signal, the block diagram of Fig. 1.11 is used. Note that all the noise which

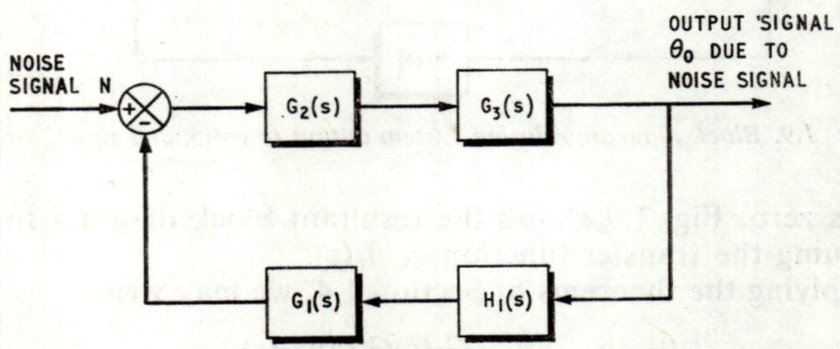


*Fig. 1.11. Block diagram relating system output to noise signal input*

appears at the system output is, in fact, error, since the desired value of noise at this point is zero; thus $e_N/N(s) = \theta_0/N(s)$, and, applying the theorems of Section 1.4, we obtain

$$e_N(s) = N(s)\left[\frac{G_2(s)G_3(s)}{1+G_1(s)G_2(s)G_3(s)H_1(s)}\right] \tag{1.7}$$

Finally, the error due to the load disturbance can be obtained from Fig. 1.12. Again all the disturbance signal which appears at the system output is error since the desired value of load disturbance at this point is zero. Thus $e_u/U(s) = \theta_0/U(s)$. Applying the theorems of Section 1.4 to Fig. 1.12, we obtain

$$e_u(s) = U(s)\left[\frac{G_3(s)}{1+G_1(s)G_2(s)G_3(s)H_1(s)}\right] \tag{1.8}$$

It will be noted that the Equations 1.6, 1.7 and 1.8 will have the same denominator. This denominator is in fact the system characteristic equation which we shall be meeting on many occasions throughout the book. Before discussing the mathematical significance of the system characteristic equation, let us first examine the significance of the error transfer functions which arise from

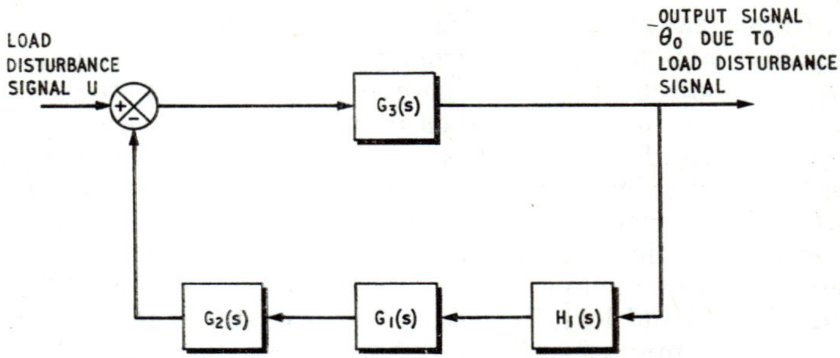


*Fig. 1.12. Block diagram relating system output to load disturbance signal*

the various sources. All three error transfer functions may be expressed in terms of the system transfer function $\theta_0/\theta_i(s)$ as follows:

$$\frac{e_c}{\theta_i}(s) = 1 - \frac{\theta_0}{\theta_i}(s) \tag{1.9}$$

$$\frac{e_N}{N}(s) = \frac{\theta_0}{\theta_i}(s)\left[\frac{1}{G_1(s)}\right] \tag{1.10}$$

and

$$\frac{e_u}{U}(s) = \frac{\theta_0}{\theta_i}(s)\left[\frac{1}{G_1(s)G_2(s)}\right] \tag{1.11}$$

Now the system which made all three errors simultaneously equal to zero for all time would be absolutely perfect in performance. It would also probably have wrecked itself at the start of its working life due to the infinitely high accelerations required to achieve this performance, and the design time would be prohibitively large even if the design itself were an engineering proposition, which is not possible anyway. We must therefore modify our concept of the perfect feedback control system to take account of these practical limitations, and define a good system as a system which minimises some function of the total error. For example, we might say that the best system (performancewise) is that which minimises the expression $\lambda_1\overline{e_c^2(t)}+\lambda_2\overline{e_N^2(t)}+\lambda_3\overline{e_u^2(t)}$, where the errors quoted are mean square values, and each $\lambda$ is a weighting multiplier depending on the application (they could all be unity in the absence of further knowledge). We are therefore arbitrarily saying that the best system is that system which minimises the total mean square error, since the mean square values are additive if the signals $\theta_i$, $U$, and $N$ are statistically independent.

If the system under consideration is a governor or a regulator, then the command signal is normally zero, and hence the requirement of Equation 1.9 may be ignored. Similarly, it is often found that small instrument-type servos are not subjected to disturbance torques, etc., and hence for these servos the requirement of Equation 1.11 may be ignored. For applications where significant noise and/or disturbance signals are anticipated in addition to a command signal following requirement, the dilemma of the control engineer is now obvious from a qualitative inspection of Equations 1.9, 1.10 and 1.11. To reduce the errors due to the noise and disturbance sources involves a reduction in $\theta_0/\theta_i(s)$, but this action increases the error in following the command signal, and therefore a compromise is indicated on mathematical as well as on engineering grounds. Detailed examples of the best compromise based on the minimum total mean square error criterion have been reported by Towill,[3] and the interested reader is referred to the original paper for information on how the total error is affected by changes in the system transfer function.

Very few feedback control systems can economically justify the detailed consideration of all three errors at the design stage. It is part of the art of control system design to translate the working environment into a performance specification which is mathematically tangible and physically measurable, and which will permit a realistic design to be achieved within a commercial timescale. For certain applications where a 1% gain in performance of a feedback control system can significantly improve weapon kill lethality or process plant yield, the involved and costly calculations required to minimise total error are justified. Charts which assist in design for total error minimisation have been developed by Coales and Lawrence,[4] and are applicable for particular forms of command and noise signals. For further general information in this field, and especially for a detailed account of the total mean square error criterion, the reader is referred to exhaustive studies by Norbert Wiener.[5] This book will, however, concentrate on the very wide class of control systems which may be designed primarily on the following of the command signal requirement but with due intuitive regard paid to the problems of noise and disturbance signal rejection. Section 1.19 will enlarge on the problem of writing the specification in such a manner as to provide realistic design objectives.

## 1.8 System Stability

A feedback control system is defined as linear if the principle of superposition holds, i.e. the behaviour of the system in following a signal of magnitude $X+Y$ is the sum of the behaviour in following signals of magnitude $X$ and $Y$ applied separately. The non-dimensional response of the system $\theta_0/\theta_i(s)$ is therefore independent of the magnitude of the signal applied to the system. An alternative definition of linearity is that a system is termed linear if the behaviour of the system can be adequately described, for all values of time, by a *single* linear differential equation with constant coefficients. The differential equation will appear in the following form:

$$a_n s^n \theta_0 + a_{n-1} s^{n-1} \theta_0 + \dots a_0 \theta_0 = b_q s^q \theta_i + b_{q-1} s^{q-1} \theta_i + \dots b_0 \theta_i \tag{1.12}$$

and thus a unique transfer function relating output to input will exist as

$$\frac{\theta_0}{\theta_i}(s) = \frac{b_0 + b_1 s + \dots b_q s^q}{a_0 + a_1 s + \dots a_n s^n} \tag{1.13}$$

Normally, for servomechanisms $b_0 = a_0 = 1$ and $n > q$. For systems where $a_0 \neq b_0$ the reader is referred to Chestnut and Mayer[6] for possible implications. The values of s which make the numerator of Equation 1.13 zero are termed system zeros, whilst those values of s which make the denominator of Equation 1.13 zero are termed system poles. Such a system will therefore have $q$ system zeros and $n$ system poles and hereafter will be defined as a type $(q, n)$ system.

In return for the advantages of feedback control previously discussed, feedback control systems can sometimes exhibit the phenomenon of instability. The type of instability which can result depends entirely on whether or not the system is linear. For a one-to-one linear system the definition of a stable system is that following some transient (short-term) disturbances $\lim_{t \to \infty} \theta_0 = \theta_i$. If a linear system is unstable, then $\lim_{t \to \infty} \theta_0 = \infty$, as shown in Fig. 1.13. Generally the unstable system will oscillate to infinity as shown, although theoretically it is possible to follow a smooth exponential path to infinity.

A non-linear system is merely a system to which the principle of superposition cannot be applied. The non-linear system can become unstable in the same way as a linear system, but in addition it is possible for a non-linear system to oscillate continuously about the desired output position as shown in Fig. 1.14. This is termed a limit cycle, and is acceptable for certain control systems provided that the limit cycle frequency and amplitude are acceptable in value. Thus instability in the limit cycle mode is a matter of defining whether a limit cycle is acceptable and, if so, what are reasonable values for the amplitude and frequency of the limit cycle.

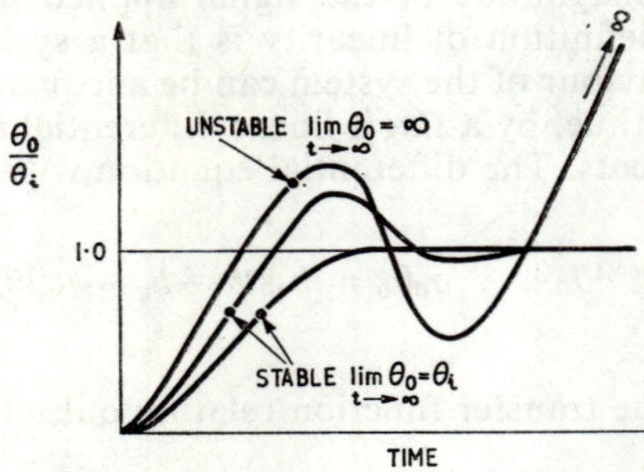


*Fig. 1.13. Stable and unstable response modes possible in a linear system*

In general, it is not possible to describe the behaviour of a non-linear system exactly by a transfer function of the type shown in Equation 1.13. Under these circumstances the transfer function may be interpreted as the graphical relationship between output and input for the particular type and magnitude of input under consideration. Derivation of these graphical relationships for certain non-linear systems will be fully discussed in Chapter 7. The response of a stable system is measured by considering, for example, the time taken for the system to satisfactorily follow the command signal. Thus a system which follows from point $A$ to

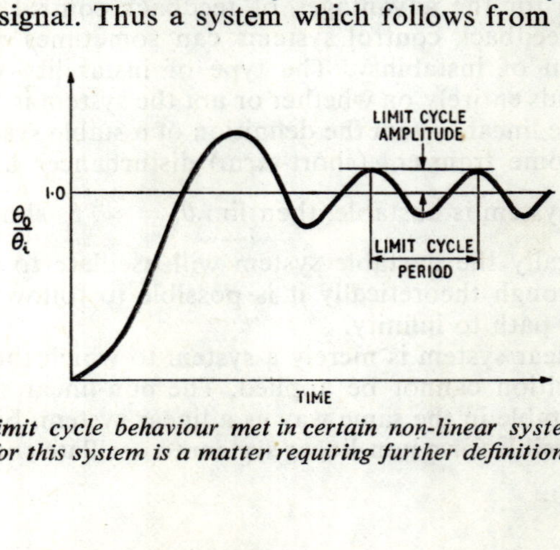


*Fig. 1.14. Limit cycle behaviour met in certain non-linear systems (stability for this system is a matter requiring further definition)*

point $B$ in 0·5 sec is a faster system than one which requires 5 sec for the same movement. It is worth noting at this stage that if 5 sec is a satisfactory response time, then a system following in 0·5 sec is probably too expensive to consider. Response and stability are irrevocably connected, and the control system designer's problem is often to make sure the system is fast enough without tending to become unstable.

### 1.9 The Characteristic Equation

We have already mentioned the system characteristic equation, which we shall now formally define from Equation 1.12 as

$$(a_0+a_1s+a_2s^2+a_3s^3+\ldots a_ns^n)\theta_0 = 0 \qquad (1.14)$$

Absolute stability of the system depends only on the roots of the characterisation equation, i.e. the poles of the system transfer function. As an example of how the system characteristic equation is derived from the system block diagram, let us consider the missile roll autopilot shown in Fig. 1.15.[7] The individual component transfer functions will be derived in Chapter 2; for the moment it is only necessary to state that the actual control surface deflection $\xi_0$ is obtained from an electro-hydraulic servo, and that in the steady state a given control surface deflection $\xi_0$ results in a proportional *roll rate* of the missile. The feedforward transfer function is

$$G(s) = \frac{K_gK_a(1+T_es)}{(1+T_cs)(1+T_as)s}$$

and the feedback transfer function is

$$H(s) = 1$$

Using the theorem for obtaining the system transfer function for a feedback control system discussed in Section 1.4, we obtain

$$\frac{\phi_0}{\phi_D} = \frac{1+T_es}{1+\left(T_e+\dfrac{1}{K_gK_a}\right)s+\left(\dfrac{T_a+T_c}{K_gK_a}\right)s^2+\left(\dfrac{T_aT_c}{K_gK_a}\right)s^3} \qquad (1.15)$$

and the characteristic equation is thus

$$\left[1+\left(T_e+\frac{1}{K_gK_a}\right)s+\left(\frac{T_a+T_c}{K_gK_a}\right)s^2+\left(\frac{T_aT_c}{K_gK_a}\right)s^3\right]\phi_0 = 0 \qquad (1.16)$$

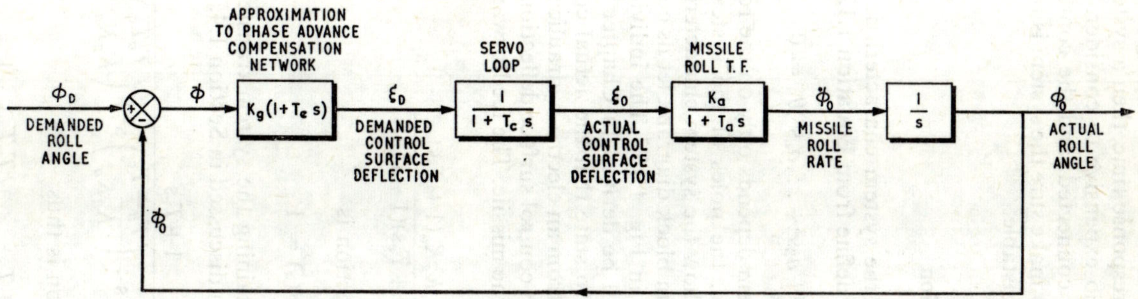


$$G(s) = \frac{K_g K_a (1 + T_e s)}{(1 + T_c s)(1 + T_a s) s}$$

$$H(s) = 1$$

$$\frac{\phi_0}{\phi_D} = \frac{G(s)}{1 + G(s)H(s)}$$

THEREFORE

$$\frac{\phi_0}{\phi_D} = \frac{1 + T_e s}{1 + \left(T_e + \frac{1}{K_g K_a}\right) s + \left(\frac{T_a + T_c}{K_g K_a}\right) s^2 + \left(\frac{T_a T_c}{K_g K_a}\right) s^3}$$

THE CHARACTERISTIC EQUATION IS THUS

$$\left[1 + \left(T_e + \frac{1}{K_g K_a}\right) s + \left(\frac{T_a + T_c}{K_g K_a}\right) s^2 + \left(\frac{T_a T_c}{K_g K_a}\right) s^3\right] \phi_0 = 0$$

*Fig. 1.15. Example of the derivation of the system characteristic equation from the system block diagram*

Comparing Equations 1.16 and 1.13, the polynomial coefficients are

$$a_0 = b_0 = 1{\cdot}0 \quad b_1 = T_e \quad a_1 = T_e + \frac{1}{K_g K_a}$$

$$a_2 = \frac{T_a + T_c}{K_g K_a} \quad a_3 = \frac{T_a T_c}{K_g K_a}$$

## 1.10 System Poles and Zeros

It is a fundamental property of polynomial series that both the numerator and denominator of Equation 1.13 can always be factorised into first and second order terms. For example, a cubic polynomial will have either three first order terms or one first order

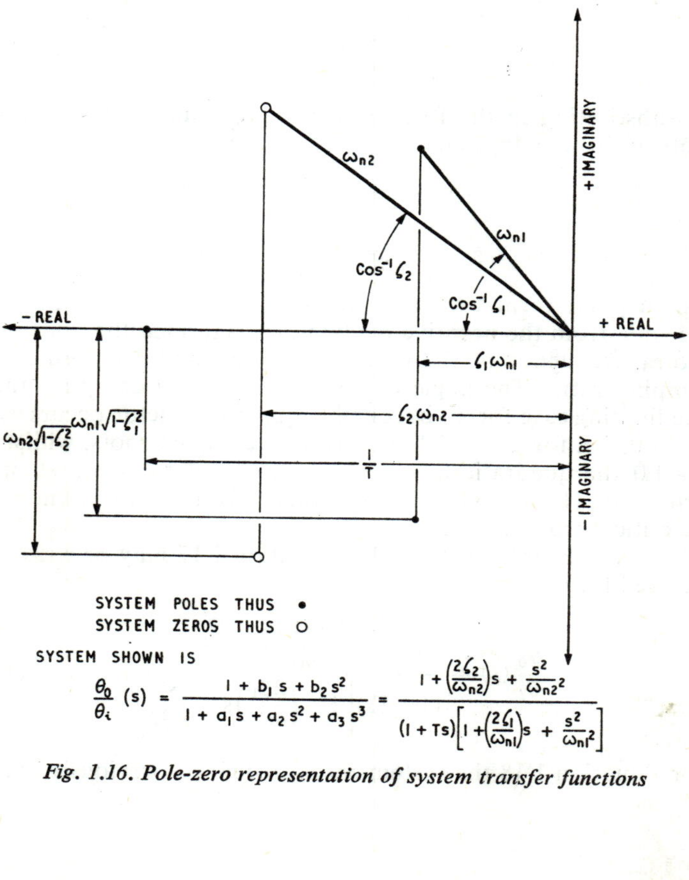


*Fig. 1.16. Pole-zero representation of system transfer functions*

term and one second order term. Fig. 1.16 shows a transfer function,

$$\frac{\theta_0}{\theta_i}(s) = \frac{1+b_1 s+b_2 s^2}{1+a_1 s+a_2 s^2+a_3 s^3} \tag{1.17}$$

where the numerator cannot be factorised into first order terms, and where the denominator may be factorised by synthetic division, Newton's method, etc., into a first order term and a second order term. It will be remembered that the roots of a quadratic equation with imaginary parts can be written, in standard form, and in schoolboy notation, as

$$x = \frac{-b \pm j\sqrt{4ac-b^2}}{2a}$$

It is particularly convenient to write an irreducible quadratic factor in the form

$$\frac{s^2}{\omega_n^2}+\left(\frac{2\zeta}{\omega_n}\right)s+1 = 0$$

By substituting in the formula for $x$ given above, the complex roots of this quadratic factor, in Cartesian form, are

$$s_1 = -\zeta\omega_n+j\omega_n\sqrt{1-\zeta^2}$$

and

$$s_2 = -\zeta\omega_n-j\omega_n\sqrt{1-\zeta^2}$$

and in polar form the roots are at a radius $\omega_n$ and an angle $\pm\cos^{-1}\zeta$ from the negative real axis. $\omega_n$ is termed the undamped natural frequency of the quadratic factor, and $\zeta$ is termed the damping ratio. The angle $\cos^{-1}\zeta$ is termed the damping angle. The limiting case for an irreducible quadratic factor occurs when $\zeta = 1{\cdot}0$, i.e. for $\zeta < 1{\cdot}0$ there are two complex roots, whilst for $\zeta > 1{\cdot}0$ the quadratic factor can be reduced to two first order terms, say, $(1+T_1 s)(1+T_2 s)$. For this reason $\zeta = 1{\cdot}0$ is known as the critical damping ratio.

For the example of Fig. 1.16, Equation 1.17 may be written in factored form

$$\frac{\theta_0}{\theta_i}(s) = \frac{1+\left(\frac{2\zeta_2}{\omega_{n2}}\right)s+\frac{s^2}{\omega_{n2}^2}}{(1+Ts)\left[1+\left(\frac{2\zeta_1}{\omega_{n1}}\right)s+\frac{s^2}{\omega_{n1}^2}\right]} \tag{1.18}$$

For Equation 1.18 the system zeros as defined previously are the roots of

$$1+\left(\frac{2\zeta_2}{\omega_{n2}}\right)s+\frac{s^2}{\omega_{n2}^2} = 0$$

i.e.

$$z_1 = -\zeta_2\omega_{n2}+j\omega_{n2}\sqrt{1-\zeta_2^2}$$

and

$$z_2 = -\zeta_2\omega_{n2}-j\omega_{n2}\sqrt{1-\zeta_2^2}$$

Similarly the system poles are the roots of

$$(1+Ts)\left[1+\left(\frac{2\zeta_1}{\omega_{n1}}\right)s+\frac{s^2}{\omega_{n1}^2}\right] = 0$$

i.e.

$$s_1 = -\frac{1}{T}$$

$$s_2 = -\zeta_1\omega_{n1}+j\omega_{n1}\sqrt{1-\zeta_1^2}$$

and

$$s_3 = -\zeta_1\omega_{n1}-j\omega_{n1}\sqrt{1-\zeta_1^2}$$

These system poles and zeros are shown in Fig. 1.16; the symbol for a system pole is ●, and the symbol for a system zero is ○. From a knowledge of the system poles the absolute stability of the system can be readily determined by the following simple rule: 'All system poles must be real and negative, or if forming a complex pair must have negative real parts.' Thus terms of the form $s_1 = -1/T$, $s_2 = -\zeta_1\omega_{n1}+j\omega_{n1}\sqrt{1-\zeta_1^2}$ are stable. If we refer to Fig. 1.16, this may be expressed as requiring all poles to be in the left-hand plane.

## 1.11 The Effect of Pole Position on Modal Behaviour

If the system is subject to a disturbance, the behaviour of the system is determined by solving the differential equation using classical or Laplace transform methods, and may be written in the form

$$\theta_0 = A_0+A_1e^{s_1t}+A_2e^{s_2t}+A_3e^{s_3t}+\ldots \quad (1.19)$$

Complex pairs such as $s_2$ and $s_3$ may be combined, and, for example, the response of the system of Fig. 1.16 may be rewritten

$$\theta_0 = A_0+A_1e^{-t/T}+A_4e^{-\zeta_1\omega_{n1}t}\cos\left(\omega_{n1}\sqrt{1-\zeta_1^2}\,t+\nu\right) \quad (1.20)$$

Note that we have replaced the arbitrary constants $A_2$ and $A_3$ by a more convenient pair of arbitrary constants $A_4$ and $\nu$. Terms such as $A_1e^{-t/T}$ and $A_4e^{-\zeta_1\omega_{n1}t}\cos(\omega_{n1}\sqrt{1-\zeta_1^2}\,t+\nu)$ are termed 'response modes'. The arbitrary constants $A_1$, $A_4$ and $\nu$ are dependent on the type of input disturbance, the system initial conditions, i.e. whether the system is at rest or not, and on the position of any zeros such as $z_1$ and $z_2$. In Chapter 3 we shall show how these arbitrary constants may be determined from the system pole-zero array.

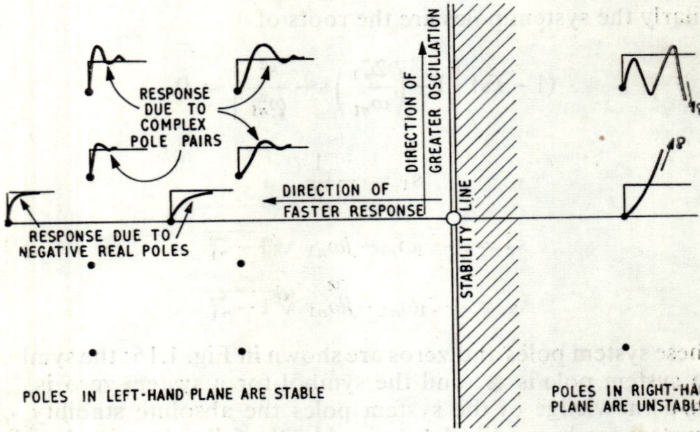


*Fig. 1.17. Effect of pole position on the response mode due to that pole*

The behaviour of each response mode depends on the position of that particular pole or pole pair in the root plane. Fig. 1.17 shows how the behaviour is affected by a movement of the pole. A movement to the left, away from the origin, results in a faster mode, whilst a movement of complex poles away from the negative real axis results in a more oscillatory mode. For convenience these modes are shown as following a sudden demand signal but the modes are recognisable for a wide variety of input signals. We must emphasise that the presence of zeros will considerably affect the response for a given pole position, and might make the system far too lively even if all the system poles are real; on the other hand a zero can be used to eliminate the effect of a sluggish pole.

A commonly taken measure of absolute stability is the damping ratio $\zeta$ associated with a quadratic response mode. The angle $\cos^{-1}\zeta$ is termed the damping angle, and a commonly accepted maximum value for this angle is 60°, but this will depend very much on the application.

## 1.12 Routh Hurwitz Criterion for Determining System Absolute Stability

In Chapter 3 we shall discuss the graphical solution of the characteristic equation. The reader is no doubt aware of the labour involved in solving high order polynomials, and it is necessary to establish whether a proposed design is stable at an early stage in the proceedings before wasting time analysing an obviously unstable system. A simple numerical method which does not involve formal solution of the differential equation is the Routh Hurwitz stability criterion.[8] The first step in applying the test is to arrange the Routh Hurwitz array. Using the characteristic equation of Equation 1.14 and, for illustration, assuming $n = 4$, then the array is

$$
\begin{array}{c|lll}
s^4 & a_4 & a_2 & a_0 \\
s^3 & a_3 & a_1 & 0 \\
s^2 & \alpha_4 = \dfrac{a_3 a_2 - a_4 a_1}{a_3} & a_2 = \dfrac{a_3 a_0 - a_4 \times 0}{a_3} & \\
s^1 & \beta_4 = \dfrac{\alpha_4 a_1 - a_3 \alpha_2}{\alpha_4} & \beta_2 = 0 & \\
s^0 & \gamma_4 = \dfrac{\beta_4 \alpha_2 - \alpha_4 \times 0}{\beta_4} & &
\end{array}
$$

The sufficient and necessary conditions for system stability are satisfied if, and only if:

a. The coefficients of the characteristic equation are all positive and non-zero.
b. The numbers in the left-hand column, $a_4, a_3, \alpha_4, \beta_4, \gamma_4$, are all positive.

If the system is unstable, the number of sign changes of the left-hand column indicates the number of roots of the characteristic equation with positive real parts.

EXAMPLE 1.12.1

Characteristic equation is $s^4+2s^3+10s^2+2s+1 = 0$.

$$
\begin{array}{c|lll}
s^4 & 1 & 10 & 1 \\
s^3 & 2 & 2 & \\
s^2 & 9 & 1 & \\
s^1 & 16/9 & & \\
s^0 & 1 & &
\end{array}
$$

No negative numbers in the left-hand column, hence the system is stable and there are no roots with positive real parts.

EXAMPLE 1.12.2

Characteristic equation is $s^4+2s^3+2s^2+10s+1 = 0$.

| | | | |
|---|---|---|---|
| $s^4$ | 1 | 2 | 1 |
| $s^3$ | 2 | 10 | |
| $s^2$ | $-3$ | 1 | |
| $s^1$ | 32/3 | | |
| $s^0$ | 1 | | |

System is unstable since there is a negative number in the left-hand column. Since there are two sign changes in the left-hand column, there are two roots with positive real parts.

There are two special cases which arise from the existence of noughts in the array.[9] Firstly, when a nought appears in the left-hand column, the procedure is to reverse the order of the coefficients of the characteristic equation. The new equation is then studied in the same way as the original equation, and this new equation will have the same number of roots with positive real parts as the original equation. Secondly, should a row, say $s^r$, possess only noughts, then the last non-zero row $s^{r+1}$ is differentiated with respect to s, and the resulting coefficients are used in the $s^{r+1}$ row.

The Routh Hurwitz array is also useful in determining limiting component values for system stability.

EXAMPLE 1.12.3

The characteristic equation of Equation 1.16 is

$$\left(\frac{T_aT_c}{K_gK_a}\right)s^3+\left(\frac{T_a+T_c}{K_gK_a}\right)s^2+\left(T_e+\frac{1}{K_gK_a}\right)s+1 = 0$$

and the Routh Hurwitz array is

$$\begin{array}{l|ll}
s^3 & \dfrac{T_aT_c}{K_gK_a} & T_e+\dfrac{1}{K_gK_a} \\[2ex]
s^2 & \dfrac{T_a+T_c}{K_gK_a} & 1 \\[2ex]
s^1 & \dfrac{(T_a+T_c)\left(T_e+\dfrac{1}{K_gK_a}\right)-T_aT_c}{T_a+T_c} & \\[2ex]
s^0 & 1 &
\end{array}$$

For limiting stability, by inspection we see that the $s^1$ term in the left-hand column must be zero. Hence

$$T_e + \frac{1}{K_g K_a} = \frac{T_a T_c}{T_a + T_c}$$

Suppose $T_e$ is the component value under consideration, then the limiting value of $T_e$ is given by

$$T_e = \frac{T_a T_c}{T_a + T_c} - \frac{1}{K_g K_a} \tag{1.21}$$

These examples show the simplicity of determining absolute stability from the system characteristic equation using the Routh Hurwitz criterion. However, the method, with one exception to be discussed later, does not yield much information about the nearness of a system to instability. Nevertheless, the test is so simple that it should be applied to any design to check that an error has not been made in the design calculations.

## 1.13 Compensation Methods in Feedback Control System Design

We have yet to discuss the detailed behaviour of feedback control systems, but our elementary study of stability and measures of stability, such as the damping angle and damping ratio, has already indicated that suitable compensation must be built into the system. For example, in Fig. 1.15, $1+T_e s$ can be thought of as a network used for such compensation, and we can see by inspecting Equation 1.21 that $T_e$ does contribute to the stability of the system since, with $T_e$ present, we may write

$$\frac{1}{K_g K_a} \geqslant \frac{T_a T_c}{T_a + T_c} - T_e \tag{1.22}$$

Thus the greater is $T_e$, the less likelihood is there of the system being unstable.

There are many compensation techniques available of building adequate stability into systems with fixed or semi-fixed load characteristics. The two most common involve the introduction of a single compensation network into the system as shown in Figs. 1.18 and 1.19, where it can be seen that the network may be inserted before the fixed or semi-fixed transfer function $G_2(s)$, or in some feedback path which may or may not form a minor feedback loop. The system of Fig. 1.15 is series compensated by the compensation transfer function $G_c(s) = 1+T_e s$.

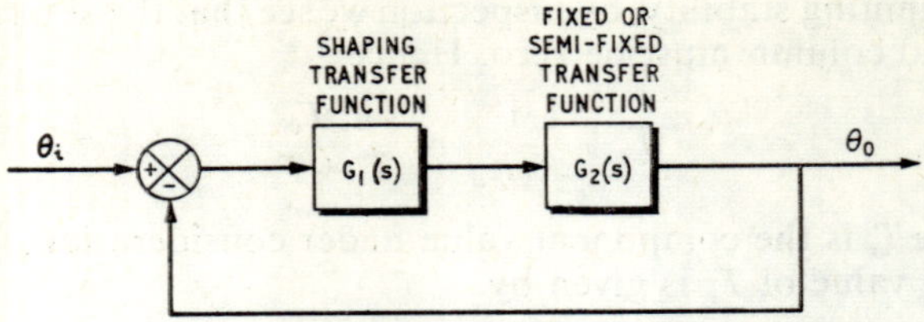


*Fig. 1.18. Series or cascade compensation method of obtaining a satisfactory design*

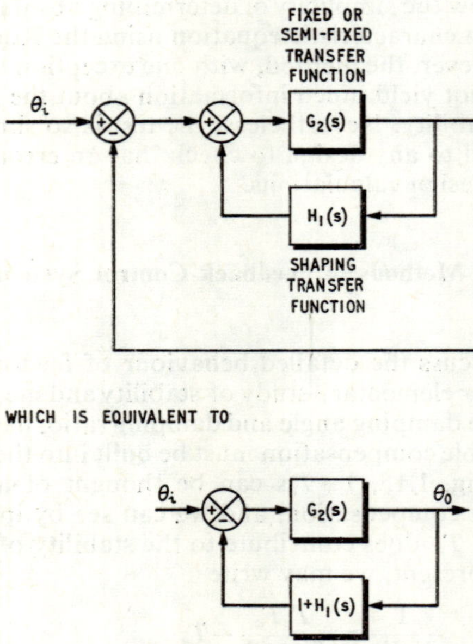


*Fig. 1.19. Feedback compensation method of obtaining a satisfactory design*

The choice between series or feedback methods of compensation, or a combination of these methods, lies with the designer. Generally speaking, feedforward compensation can be achieved using simple networks, whereas often more complex and heavier equipment, such as rate gyroscopes and tachogenerators, need to be used in feedback compensation.

If noise signal rejection considerations are important, then a preference for feedback compensation is usually, but not always, indicated.

## 1.14 Control System Inputs

The inputs likely to excite a control system vary enormously with individual applications. Some systems may expect to receive smooth, slowly changing signals, whilst other systems may expect to receive completely random inputs. Fig. 1.20(a) shows a sample

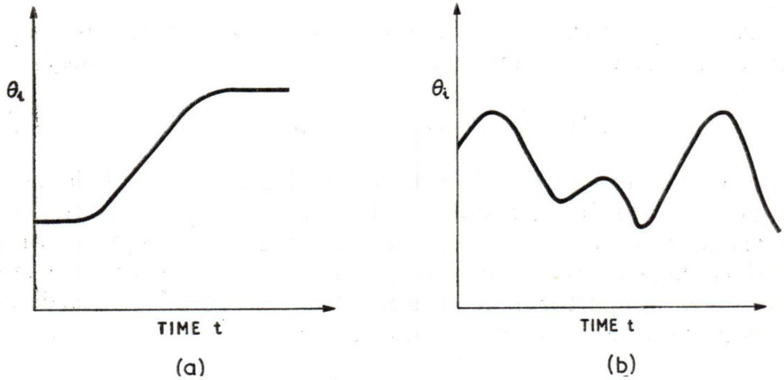


*Fig. 1.20. Typical control system inputs: (a) sample input to a machine tool servo; (b) sample input to a tank gun control servo*

input to a machine tool control servo which is cutting a straight side with radii at each end, and there are no violent changes in input signal, whilst Fig. 1.20(b) shows a sample input to a tank gun control servo used for controlling the tank gun elevation. Due to the relative motion of the gun and target the input is far from smooth. Often with such inputs it is found necessary to smooth the signal before it reaches the control system in order to remove the short-term components caused by the tank crossing uneven terrain. The smoothing network is designed to generate the best possible predicted signal, and the gun servo responds to this predicted signal.

The degree to which the input signal, whether command, noise, or disturbance stimulus, is known depends on the application. When the application is at the boundary of existing knowledge, very little may be known about the input signal characteristics, and the control system designer may be working very much in the dark at this stage. Once a prototype is built and environmentally tested, it is then possible to use not only newly acquired performance data but also any further data on system inputs which may become available at the test stage. For example, the disturbance loadings on aircraft power controls can only be determined with reasonable accuracy once the prototype aircraft has actually flown, although it is naturally to be hoped that the aircraft designers' prediction of such loadings would certainly be in the right parish.

When the input is reasonably smooth, it is possible to represent the input characteristics in one of two simple ways. Firstly, we may represent the input in the time domain as a polynomial in terms of the time $t$, thus

$$\theta_i(t) = r_0+r_1t+r_2t^2 \tag{1.23}$$

and secondly, in the frequency domain we may represent the input as a Fourier Series in terms of the fundamental frequency $\omega$

$$\theta_i(t) = \sum_{n=0}^{\infty}(F_n \sin n\omega t+Q_n \cos n\omega t) \tag{1.24}$$

Both Equation 1.23 and Equation 1.24 are deterministic, since if $\theta_i$ is known at time $t$ the use of the appropriate equation completely defines $\theta_i$ at time $t+\Delta t$. Deterministic input signals can be readily expressed mathematically, and many control system design techniques and performance specifications are based on such deterministic inputs; these we shall now review in detail.

## 1.15 Deterministic Inputs Used in the Assessment of Control System Performance

Equation 1.23 leads naturally to the consideration of the three individual terms as assumed control system inputs. As shown in Fig. 1.21, the $r_0$ term is a displacement step function, often assumed to be of unity magnitude in linear systems, the $r_1t$ term is a velocity step function $\Omega t$, and the $r_2t^2$ term is an acceleration step function $ft^2/2$. The mathematical definition of these inputs is as follows.

| | | |
|---|---|---|
| Displacement step function | $\theta_i = 0$ | $t = 0$ |
| | $\theta_i = 1{\cdot}0$ | $t > 0$ |
| Velocity step function | $\theta_i = 0$ | $t = 0$ |
| | $\theta_i = \Omega t$ | $t > 0$ |
| Acceleration step function | $\theta_i = 0$ | $t = 0$ |
| | $\theta_i = \dfrac{ft^2}{2}$ | $t > 0$ |

Their Laplace transforms are $1/s$, $\Omega/s^2$, and $f/s^3$ respectively. In the classical solution of differential equations these inputs govern the steady state solution and also the coefficients $A_1$, $A_2$, etc. of the complete solution as, for example, in Equation 1.20.

A very considerable advantage of the frequency form shown in Equation 1.24 is that the steady state solution to such an input can be calculated easily by substituting $j\omega$ for s in the system transfer function and evaluating the resulting vector as a magnitude and phase angle. The deterministic sine wave used to assess performance is shown in Fig. 1.21(d). The time taken for one complete oscillation of this sinusoidal input is $2\pi/\omega$ sec, and the peak value is $|\theta_i|$.

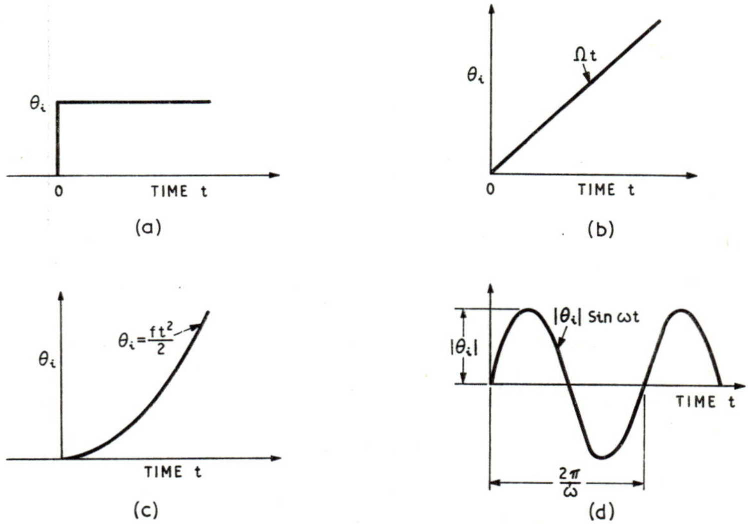


*Fig. 1.21. Deterministic inputs used in the assessment of control system performance; (a) displacement step function: (b) velocity step function; (c) step acceleration input; (d) sinusoidal input*

For linear systems it is convenient to deal with these deterministic inputs singly, and then, if necessary, which is rare, to superimpose several inputs; this is achieved by superimposing the solution to the individual inputs. In many instances it is possible to specify the performance of a system perfectly adequately by choosing several points on the response to one of these deterministic inputs, such as several points on the step function response, or by choosing unique singular points on the individual responses to several widely varying deterministic inputs. For example, we might specify system performance by choosing one point on the displacement step function response, one point on the velocity step function response and one point on the acceleration step function response.

## 1.16 Performance Criteria Based on Deterministic Inputs

We shall now review some of the more important performance criteria associated with deterministic inputs. Fig. 1.22 shows a typical step function response for a servomechanism. A number of the more common performance criteria based on this type of input is also shown. The theoretical responses can be obtained using Laplace transform or classical methods for solving the system

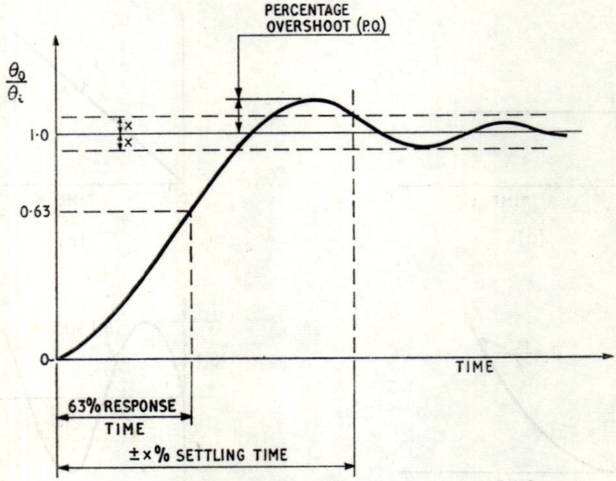


*Fig. 1.22. Some performance criteria based on the displacement step function response*

differential equation. An analytical/graphical method for the determination of the step function response for known system pole-zero arrays will be discussed in Chapter 3. Actual experimental responses are recorded on a cathode-ray oscilloscope, XY recorder, etc. The percentage overshoot (P.O.) can be a very useful and self-evident criterion for certain applications. There are many possible criteria to apply along the time scale of the step function response; 63% response time is a useful measure of behaviour during the early stages of response, and $\pm x\%$ settling time, which is the time at which the system error is less than, and for ever after less than $x\%$, is a useful measure of the later stages of response.

Fig. 1.23 defines performance criteria based on step velocity and step acceleration inputs. The criteria based on these inputs are particularly useful if the true system inputs can be reasonably approximated by Equation 1.23. In particular, in practice the steady state velocity lag is specified for a large percentage of applications.

'Steady state velocity lag' is the steady state displacement error in following a step velocity input, and 'steady state acceleration lag' is the steady state displacement error in following a step acceleration input.

Let us develop the expressions for steady state velocity and acceleration lags from first principles, using the particular integral method of the classical theory of differential equations. Let the

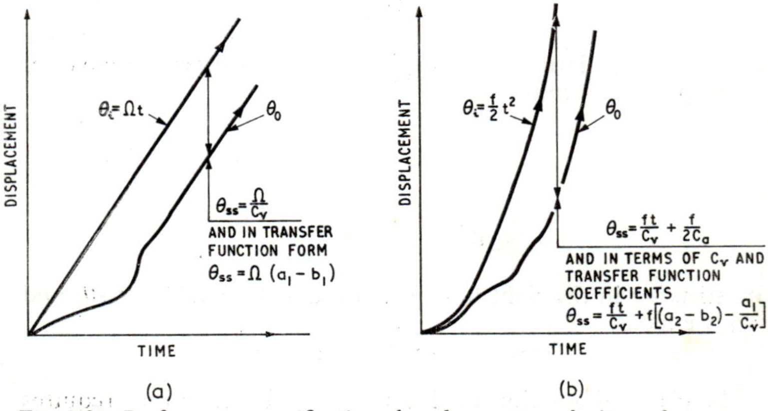


*Fig. 1.23. Performance specifications based on step velocity and step acceleration inputs: (a) steady state velocity lag for a system with zero steady state displacement lag; (b) steady state acceleration lag for a system with zero steady state displacement and velocity lags*

system transfer function be

$$\frac{\theta_0}{\theta_i}(s) = \frac{1+b_1s+b_2s^2+\ldots b_qs^q}{1+a_1s+a_2s^2+\ldots a_ns^n}$$

then

$$\frac{\theta}{\theta_i}(s) = 1-\frac{\theta_0}{\theta_i}(s)$$

Hence

$$\theta = \theta_i\left(1-\frac{1+b_1s+b_2s^2\ldots b_qs^q}{1+a_1s+a_2s^2\ldots a_ns^n}\right)$$

Therefore

$$\theta = \theta_i\left[\frac{(a_1-b_1)s+(a_2-b_2)s^2+\ldots}{1+a_1s+a_2s^2+\ldots}\right] \quad (1.25)$$

and the general steady state solution we require will be

$$(1+a_1s+a_2s^2)\theta = [(a_1-b_1)s+(a_2-b_2)s^2]\theta_i \quad (1.26)$$

Since the step acceleration input is itself only of second order it is not necessary to consider terms in the transfer function higher than $a_2$ and $b_2$.

### CASE 1. STEP VELOCITY INPUT

For the step velocity input, by definition,

$$\theta_i(t) = \Omega t$$

Hence, differentiating the required number of times, we have for the step velocity input

$$\frac{d\theta_i}{dt} = s\theta_i = \Omega$$

and

$$\frac{d^2\theta_i}{dt^2} = s^2\theta_i = 0$$

Substituting in the right-hand side of Equation 1.25 for $s\theta_i$ and $s^2\theta_i$, we have

$$(1+a_1s+a_2s^2)\theta = (a_1-b_1)\Omega$$

The classical solution of linear differential equations requires that a suitable particular integral be assumed for the steady state value of $\theta$. Let us assume a solution of the form

$$\theta_{ss} = A+Bt$$

Then

$$s\theta_{ss} = B$$

and

$$s^2\theta_{ss} = 0$$

Substituting in the left-hand side of Equation 1.26 to evaluate $A$ and $B$, we have

$$A+Bt+a_1B = (a_1-b_1)\Omega$$

Clearly, by inspection, $B = 0$, leaving the particular integral as

$$\theta_{ss} = (a_1-b_1)\Omega \tag{1.27}$$

where $\theta_{ss}$ positive indicates a lag of the system output behind the input. Often it is convenient to specify the velocity constant $C_v$ instead of the steady state velocity lag $\theta_{ss}$. $C_v$ is defined by the relationship $\theta_{ss} = \Omega/C_v$ whence, from Equation 1.27,

$$C_v = \frac{1}{a_1-b_1} \tag{1.28}$$

CASE 2. STEP ACCELERATION INPUT

For the step acceleration input, by definition,

$$\theta_i(t) = \frac{ft^2}{2}$$

Hence, differentiating the required number of times, we have for the step acceleration input

$$\frac{d\theta_i}{dt} = s\theta_i = ft$$

and

$$\frac{d^2\theta_i}{dt^2} = s^2\theta_i = f$$

Substituting in the right-hand side of Equation 1.26, we have

$$(1+a_1s+a_2s^2)\theta = (a_1-b_1)ft+(a_2-b_2)f$$

Again we have to assume a suitable particular integral for the steady state value of $\theta$. If we assume a solution of the form

$$\theta_{ss} = A+Bt$$

then

$$s\theta_{ss} = B$$

and

$$s^2\theta_{ss} = 0$$

Substituting these values in the left-hand side of Equation 1.26, we obtain

$$A+Bt+a_1B = (a_1-b_1)ft+(a_2-b_2)f$$

Hence

$$B = (a_1-b_1)f$$

and

$$A = (a_2-b_2)f-a_1(a_1-b_1)f$$

Substituting back the values of $A$ and $B$ in the expression for $\theta_{ss}$, the final solution is

$$\theta_{ss} = (a_1-b_1)ft+(a_2-b_2)f-a_1(a_1-b_1)f \tag{1.29}$$

Equation 1.29 contains three terms. The first term is a lag which increases linearly with time if $a_1 \neq b_1$, i.e. the system is not a zero velocity lag system. The second and third terms are both constant, and the third term will be zero if $a_1 = b_1$. Thus if the system is a zero velocity lag system, the steady state error reduces to

$$\theta_{ss} = (a_2-b_2)f \tag{1.30}$$

The acceleration constant, $C_a$, is defined from Equation 1.29, by ignoring the term which increases with time, as $\theta_{ss} = f/2C_a$. Thus the full expression for $C_a$ becomes

$$\frac{1}{2C_a} = (a_2-b_2)-a_1(a_1-b_1) \tag{1.31}$$

or alternatively, in terms of $C_v$,

$$\frac{1}{2C_a} = (a_2-b_2)-\frac{a_1}{C_v}$$

If the system is a zero velocity lag system, then the expression for $C_a$ simplifies to

$$\frac{1}{2C_a} = a_2-b_2 \tag{1.32}$$

The reader should note from Equations 1.27 to 1.32 that the error constants $C_a$ and $C_v$ are directly related to the coefficients of the system transfer function. It may be argued that $C_a$ is of little use as a specification in a system where $C_v \neq \infty$, since the steady state lag will increase linearly with time. Indeed, it is possible to have a system with $C_a = \infty$, but $C_v \neq \infty$, thus again resulting in a system with a steadily increasing steady state lag. For example, consider a simple second order system with transfer function

$$\frac{\theta_0}{\theta_i}(\mathrm{s}) = \frac{1}{1+\left(\dfrac{2\zeta}{\omega_n}\right)\mathrm{s}+\dfrac{\mathrm{s}^2}{\omega_n^2}}$$

Thus the polynomial coefficients are:

$$\begin{aligned} b_0 &= 1 \qquad & a_0 &= 1 \\ b_1 &= 0 \qquad & a_1 &= 2\zeta/\omega_n \\ b_2 &= 0 \qquad & a_2 &= 1/\omega_n^2 \end{aligned}$$

From Equation 1.31 $C_a = \infty$ when $a_2-b_2 = a_1(a_1-b_1)$. Hence, substituting, we have

$$\frac{1}{\omega_n^2} = \left(\frac{2\zeta}{\omega_n}\right)^2$$

Hence

$$\zeta = \tfrac{1}{2}$$

Thus a simple second order system with $\zeta = \frac{1}{2}$ will have $C_a = \infty$, but $C_v = 1/a_1 = \omega_n$ and is therefore not infinite. However, even if both $C_v$ and $C_a$ are not infinite, their inclusion in a specification is sometimes justified as their presence does afford some control over the lower transfer function coefficients. If the reader is in any doubt on the merit of using the error coefficients then it is always advisable to work from first principles from Equation 1.26, using numerical values where appropriate. In practice, $C_v$ is very frequently specified, and the design methods to be developed in subsequent chapters will often assume that a $C_v$ requirement is to be met. If the system does not possess at least one *pure* integrator in the loop, then a steady state error will exist when following a step function input. Under these circumstances it is useful to define performance in terms of the position constant as well as the velocity and acceleration constants. For further details the reader is referred to Chestnut and Mayer[6].

To summarise, for a servomechanism to have zero velocity lag $C_v = \infty$, and for a servomechanism to have zero acceleration lag in the true sense, i.e. no displacement lag whether time-dependent or not, $C_v = C_a = \infty$. Furthermore, in terms of transfer function coefficients, the corresponding sufficient and necessary conditions are that $a_1 = b_1$ for $C_v = \infty$, and both $a_1 = b_1$ and $a_2 = b_2$ for $C_a = C_v = \infty$.

If a linear system is excited by a harmonic signal $\theta_i = |\theta_i| \sin \omega t$ where $\theta_i$ is the peak value, the output of the linear system will, in the steady state, be a sine wave of the same frequency, but generally lagging behind the input with a waveform $\theta_0 = |\theta_0| \sin (\omega t - \psi)$ as shown in Fig. 1.24(a). If the system transfer function is known, $|\theta_0/\theta_i|$ and $\psi$ can be easily calculated for a given value of $\omega$ from the system transfer function by substituting $j\omega$ for s and evaluating the resulting vector. Thus substituting $j\omega$ in Equation 1.13, we obtain

$$\frac{\theta_0}{\theta_i}(j\omega) = \frac{1 + jb_1\omega - b_2\omega^2 - jb_3\omega^3 + \cdots j^q b_q \omega^q}{1 + ja_1\omega - a_2\omega^2 - ja_3\omega^3 + \cdots j^n a_n \omega^n} \tag{1.33}$$

Equation 1.33 may be summated as the ratio of complex numbers, thus

$$\frac{\theta_0}{\theta_i}(j\omega) = \frac{A + jB}{C + jD}$$

and using complex number theory, we may write

$$\left|\frac{\theta_0}{\theta_i}\right|_{\omega} = \sqrt{\frac{A^2 + B^2}{C^2 + D^2}} \tag{1.34}$$

and

$$\psi = \tan^{-1}\frac{D}{C} - \tan^{-1}\frac{B}{A} \tag{1.35}$$

Fig. 1.24(b) shows $|\theta_0/\theta_i|$ and $\psi$ plotted against $\omega$ for a typical servomechanism. Common criteria are $M_p$ the maximum amplitude ratio, $\omega_p$ the frequency at which $M_p$ occurs, $\omega_B$ the bandwidth, which is the highest excitation frequency at which $|\theta_0/\theta_i| = 1/\sqrt{2}$, and finally $\psi_B$ the phase lag at bandwidth. If the system

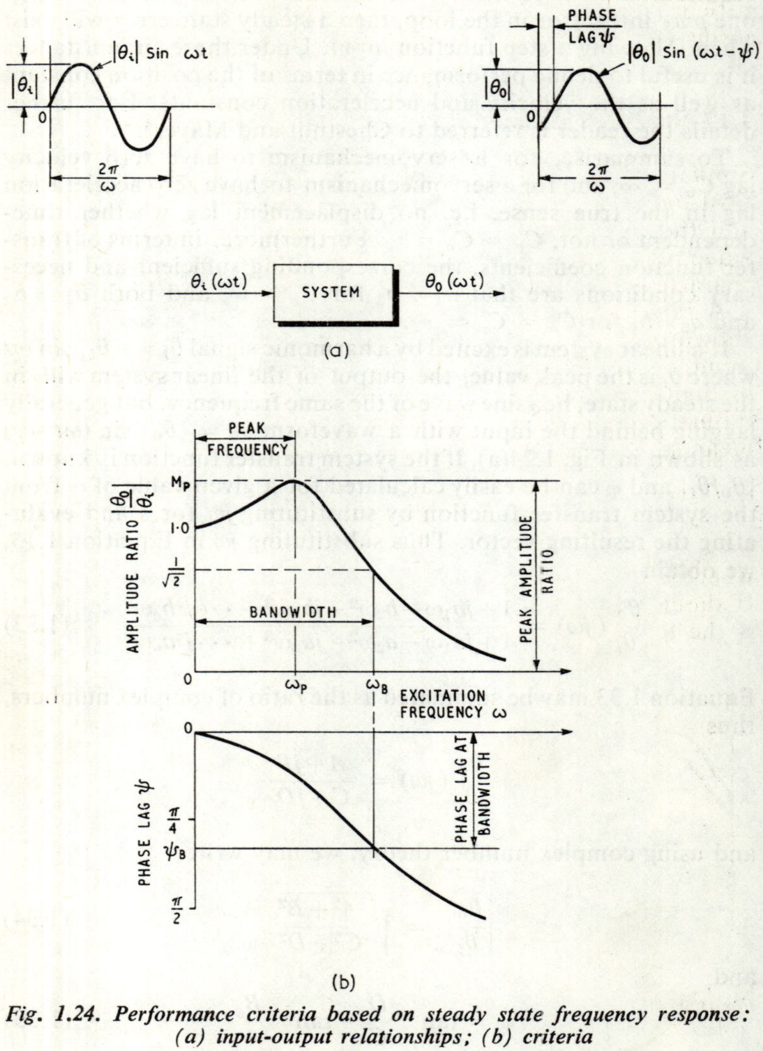


*Fig. 1.24. Performance criteria based on steady state frequency response: (a) input-output relationships; (b) criteria*

input signal can be expressed harmonically, then bandwidth is a very useful measure of the ability of the system to follow this type of input. Phase lag at bandwidth is useful if the system is a minor loop, whilst $\omega_p$ may be of interest if there are several resonant systems in series with resonant peaks $M_p$. The definition of bandwidth as the highest frequency at which the amplitude ratio is $1/\sqrt{2}$ stems from electrical engineering, since in electrical terms it is the half-power frequency.

## 1.17 Statistical Inputs

Similarly, it is possible to apply performance criteria to statistical inputs. A statistical input is completely random in nature and the value of $\theta_i$ at time $t+\Delta t$ cannot be predicted from the value of $\theta_i$ at time $t$. Nevertheless it is possible to assign certain characteristics, such as average values, to statistical inputs.

A random input $\theta_i$, with the corresponding system output is shown in Fig. 1.25. The random signal can be the actual system input, or can be generated for assessment purposes in an analogue or digital fashion. The most common performance criterion for estimating system performance following a random signal is the mean square error criterion, defined as

$$\text{MSE} = \frac{1}{\tau}\int_0^{\tau} \theta^2 \mathrm{d}t \tag{1.36}$$

where $\tau$ is the sample time and should be as long as possible. It should be noted that the square root of the mean square error is the RMS error and can be registered on an RMS voltmeter.

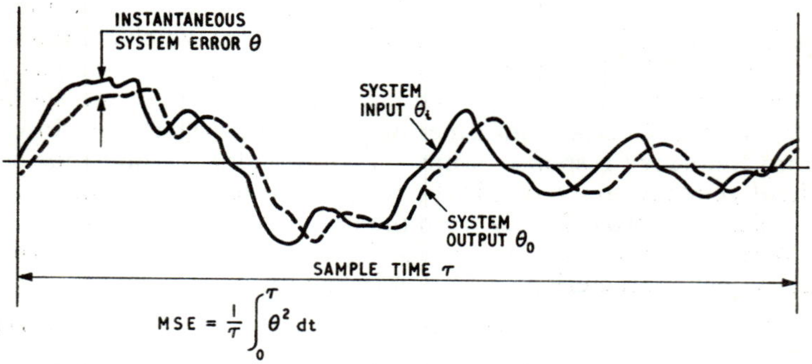


*Fig. 1.25. Mean square error criterion for estimating system performance in following a random input*

If the input signal is known as a power spectrum or in a suitable alternative form, it is possible to determine a system transfer function which will minimise the MSE. Such a system would be defined as the MSE optimum system for the particular application. A detailed discussion on the MSE criterion is included in the text of Newton, Gould and Kaiser,[10] to which the interested reader is referred, as space limitations will restrict our own studies to systems defined in terms of deterministic performance criteria.

## 1.18 Performance Criteria based on Broken Loop Methods (Design in the Frequency Domain)

System stability and system response are interdependent to some extent, particularly if the system response is dominated by a few system poles and zeros. For linear systems the position of system poles and zeros in the s plane determine the system response uniquely whatever the form of the input signal, and so far we have arbitrarily chosen damping ratio or damping angles of quadratic poles as a measure of system stability. It must be emphasised, however, that the presence of a suitably placed system zero can force $M_p$ to be so large that even for a system with a large damping ratio for the dominant complex poles, for practical purposes the system will be unstable. This is because in engineering terms $M_p = 3{\cdot}0$ is as bad as $M_p = \infty$.

Solution of the system characteristic equation in a manner suitable for system synthesis can be achieved by the use of broken loop methods. Before the advent of analogue and digital computers, and the derivation of root locus method by Evans, the open loop methods were the only practical way of designing a vast number of feedback control systems. The broken loop methods, in basic form, define stability measures which are found on the basis of experience to be generally suitable for the majority of commonly found control systems. Developments of the methods can yield certain information about the system (or closed loop) frequency response, whilst one particular method (due to Chen[11]) endeavours to obtain the system transfer function in terms of poles and zeros directly from the open loop plot.

When using broken loop methods, the first step is to rewrite the system characteristic equation in the form

$$G(s)H(s) = -1 \tag{1.37}$$

If we consider sinusoidal excitation only, we may substitute $j\omega$ for s. $G(j\omega)\ H(j\omega)$ will then be available as a vector quantity with specific magnitude and phase angle for a particular value of frequency. Because the terms $G(s)\,H(s)$ are composed of component transfer functions which are available in factored form, it follows that $G(j\omega)\,H(j\omega)$ can in general be evaluated as the product of a number of simple individual vectors, which themselves may be evaluated graphically as will be shown in Chapter 3. If the final vector, $|G(j\omega)\,H(j\omega)|\ \underline{/G(j\omega)\,H(j\omega)}$, is plotted in the complex plane as a function of frequency $\omega$, then a locus will be constructed in the frequency domain as $\omega$ varies from zero to infinity. The $\underline{/G(j\omega)\,H(j\omega)}$ phase angle is normally a phase lag which is conventionally shown clockwise. Using regeneration theory, it may be shown that, subject to restrictions on the form of system considered, in so far as each individual element in the system must itself be stable, the system will be critically stable if the locus thus drawn passes through the $(-1, j0)$ point, which in polar form has coordinates $1\ \underline{/-180°}$. Furthermore, if the locus does not enclose the $(-1, j0)$ point the system will be stable, and if it does enclose the $(-1, j0)$ point the system will be unstable. *This is the simplified Nyquist criterion which applies only if each individual element in the broken loop equation is itself stable.* For the generalised Nyquist criterion which must be applied if an individual element is itself unstable, the reader is referred to the standard description of James, Nichols and Phillips.[12] The $(-1, j0)$ point is fundamental to broken loop methods, and is conventionally referred to as the Nyquist point in deference to the originator of broken loop techniques. The Nyquist plot passing through the $(-1, j0)$ point represents a critically stable system, which in the s plane would be represented by one or two poles on the $j\omega$ axis. It follows, therefore, that the $(-1, j0)$ point is the Nyquist transform of the s plane $j\omega$ axis.

Fig. 1.26(a) shows a system with characteristic equation $1+G(s)\,H(s) = 0$. If we break the loop physically at the comparator, inject a sinusoidal signal $\bar{\theta}(j\omega)$ in the feedforward path and measure the amplitude ratio and phase angle of the feedback signal $\bar{\theta}_0(j\omega)$, we shall be able to obtain the vector quantity needed to plot the open loop response shown in Fig. 1.26(b). The interpretation of Equation 1.37 is therefore the behaviour which arises in theory or practice when the control loop is physically broken at a salient point in the loop.

It is now necessary to introduce the two common measures of stability known as the 'gain margin' (at 180° phase lag), and the 'phase margin' (at unity broken loop gain) shown in Fig. 1.26(b). Commonly accepted minimum values are gain margin = 0·6 and phase margin = 45°. Since the locus is calibrated in frequency, it is possible under some circumstances to obtain certain information which has regard to the system characteristics directly from the Nyquist plot. As we shall see in Chapter 4, it is often possible to have several systems with an identical *system* transfer function but with somewhat different Nyquist plots. Furthermore, by physical

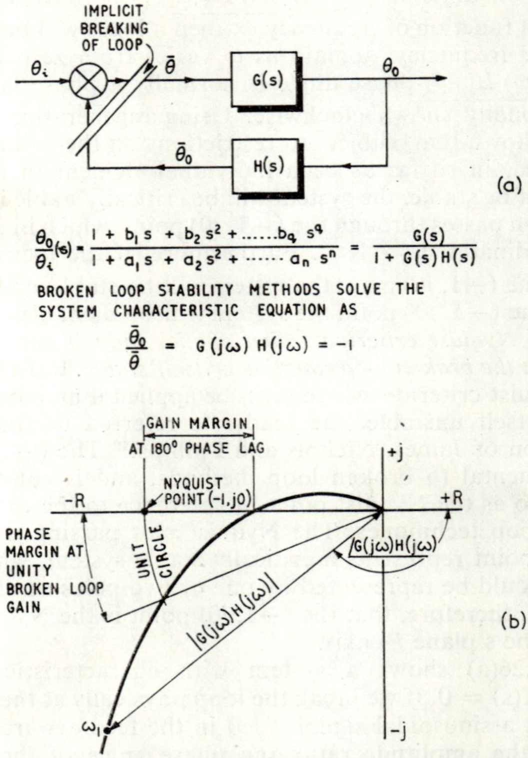


*Fig. 1.26. Stability criteria based on broken loop response: (a) block diagram illustrating broken loop approach; (b) definition of broken loop stability margin shown on Nyquist plot*

or mathematical manipulation of feedback signals, several broken loop equations can be obtained for the same system, and the only point which is exactly common to all arrangements is the critical locus as it passes through the Nyquist point. Provided the true characteristic equation is solved in broken loop form for all these arrangements, then each of these Nyquist plots will indicate absolute stability or otherwise, but the gain and phase margins will depend upon the particular arrangement of the feedback signals. It is therefore extremely useful to include a sketchof the system block diagram on every broken loop plot, showing quite clearly which feedback signals actually pass through the point at which the loop is broken.

If the system can be written in unity feedback form, and the Nyquist plot corresponding to this form is available, the system frequency response can be evaluated directly by superimposing on the Nyquist plot contours of constant system amplitude ratio and constant system phase angle, and interpolating to obtain individual points on the system frequency response as shown, for example, in Shinners[13] (Section 5.5). Because the system is unity feedback, the closed loop response is, in effect, obtained very simply by considering the vector relationship at the comparator; since $G(j\omega)$ is known, it follows that $G(j\omega)/[1+G(j\omega)]$ is also known. In fact, for unity feedback path systems, any point on the Nyquist plot can be quickly converted to the corresponding point on the system frequency response without reference to these contours, which are, however, very useful for synthesis purposes when several Nyquist plots are studied simultaneously. Because under these circumstances the system frequency response is relatively easy to obtain, broken loop techniques are often referred to as frequency domain design techniques. Provided the system is unity feedback and is dominated by a second order pole pair, then the dominant system poles can be approximated directly from the Nyquist plot using simple graphical techniques such as those outlined by Porter[14] and Meadows.[15]

The reader should take particular note that introducing a particular compensation transfer function alternately in the system feedforward path and system feedback path will result in identical Nyquist plots, but when the loop is closed the system transfer functions may be vastly different. For example, if the compensation transfer function may be approximately written $G(s) = 1+Ts$, then, when inserted in the feedforward path, the compensator will generate a system zero, whereas when the same transfer function is inserted in the feedback path it does not generate a system zero. In both cases the system poles are identical.

The Bode plot is an alternative method of plotting the broken loop response. At any value of frequency the amplitude ratio and phase lag components of $|\bar{\theta}_0/\bar{\theta}|$ are separated. The amplitude ratio is not plotted in scalar form, but in db form, i.e. $20\log_{10}|\bar{\theta}_0/\bar{\theta}|$, and both amplitude ratio and phase angle are plotted against log frequency. Plotting the amplitude ratio in logarithmic form has two outstanding advantages compared with the Nyquist form:

a. By using the logarithmic scale, vector contributions of amplitude ratios from individual components are *added* and not *multiplied*.

b. Contributions from such terms as $1+Ts$ and $\omega_n^2+2\zeta\omega_n s+s^2$ can be approximated by straight lines for high and low values of frequency.

For initial synthesis the straight line approximations can be used for intermediate values of frequency as well, and a possible solution corrected by using standard templates which are available for the various component transfer functions. In theory it is still desirable to compute the phase angles for the individual elements, or read them from standard tables, although Douce[16] has published simplified formulae for obtaining approximations to phase angle contributions. Fig. 1.27 shows a typical Bode plot, and the phase margin and gain margin are again indicated. The phase margin is measured where the amplitude ratio is 0 db, and the gain margin is measured where the phase lag is 180°. Using the straight line approximations shown, the maximum discrepancy arising from one term $1+Ts$ is 3 db, and this occurs at a frequency $\omega = 1/T$, known as the break frequency. Second order terms with low damping ratio can have very large discrepancies between the actual and straight line approximations to the logarithmic form of frequency response as will be shown in Chapter 3.

For each term involving s, there will be a slope of 20 db/decade as $\omega \rightarrow \infty$, thus if the broken loop transfer function is a second order divided by a fourth order, then the ultimate slope will be $(2\times20)-(4\times20) = -40$ db/decade. Component transfer functions in Bode form will be derived in Section 2.19. Now for a linear system, amplitude ratio and phase are uniquely related (provided the elements are non-minimum phase), and therefore it should be possible to examine system stability based on broken loop amplitude ratio only. This approach considerably speeds up system synthesis, since it is obtaining the phase data which is so time-consuming. A simple rule, due to Bode, is that the system will be stable if the slope of the Bode plot in the region of 0 db is $-20$ db/decade, and this slope is maintained constant for a region $\pm\frac{1}{2}$ decade about the crossing point. This simple rule is sufficient for approximately fixing some of the parameters. Inevitably, such a rule can only be approximate, and indeed tends to be conservative. For example, the system shown in Fig. 1.27 has a slope of $-40$ db/decade at the crossing point, and yet the system is stable, but has a phase margin of only 27°, which is considerably less than the usually recommended value of 45°.

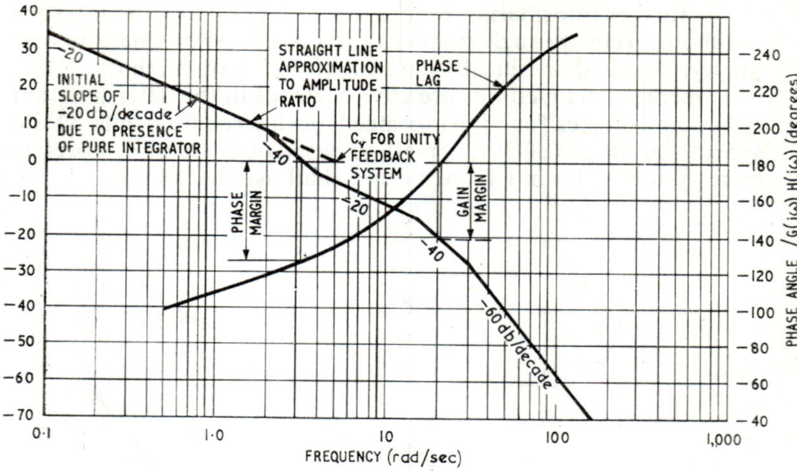


*Fig. 1.27. Bode plot of broken loop response showing phase margin and gain margin stability criteria*

Using the amplitude ratio information of the Bode plot only, Chen[11] has developed an empirical method for the determination of system poles directly. The method can be applied to non-unity feedback systems, but is particularly useful for unity feedback systems because if there is a basic integrator in the loop, $C_v$ is simply the intersection of the initial $-20$ db/decade slope (extended if necessary) to cut the 0 db line, as shown in Fig. 1.27. It is therefore possible to synthesise a variety of systems very quickly knowing full well that the $C_v$ requirement is met in each case by keeping the $-20$ db/decade initial slope intersection point at the required value. Chen's mehod will be described in detail in Section 6.13.

Finally, in this very brief review of broken loop methods, we conclude by reference to the Nichols plot. Here the amplitude ratio in decibels is plotted directly against scalar phase angle as a function of excitation frequency. Both stable and unstable systems are shown in Nichols chart form in Fig. 1.28. Gain and phase margins for the stable system are clearly indicated. Standard templates are available to reduce the labour involved in plotting Nichols charts. The logarithmic form of amplitude ratio plotting does allow the addition of component contributions in exactly the same way as the Bode plot. On the assumption $H(s) = 1{\cdot}0$, system magnitude and phase contours for $G(j\omega)/[1+G(j\omega)]$ may be superimposed on the basic Nichols plot, thus facilitating the estimation of closed loop frequency response.

For further information on broken loop design methods, readers should refer to such standard texts as Porter or Chestnut and Mayer. We shall specifically develop broken loop methods in this text wherever the methods are advantageous in the determination of system transfer functions to meet specific requirements. Quite apart from the use of Chen's method based on Bode plots to determine system poles in Chapter 6, we shall use the Nichols plot in

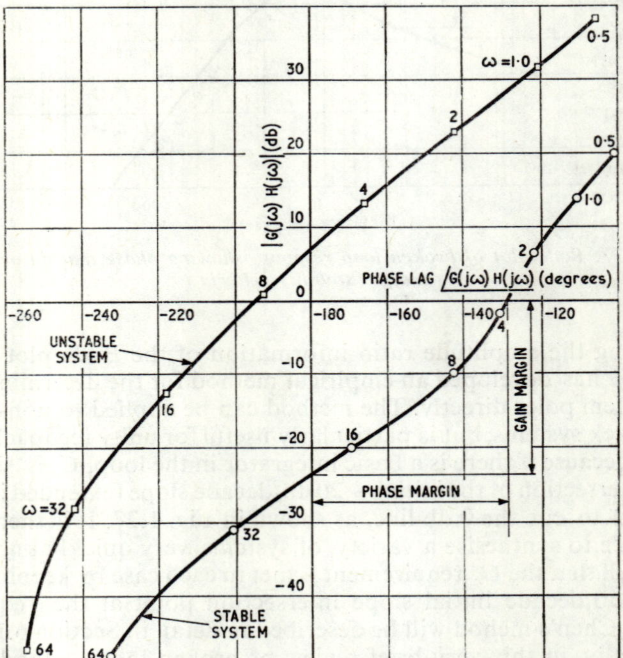


*Fig. 1.28. Nichols plot of broken loop response showing phase margin and gain margin stability criteria*

Chapter 7 to determine the frequency response of non-linear feedback control systems. Nichols plots are probably the most popular method of tackling non-linear problems at the present time, since experimental results can be easily incorporated. Broken loop tests for novel designs do tend to prevent damage of systems which are unstable, since the system may be adjusted before the loop is closed and regeneration, i.e. instability, commences. Maladjustments of feedforward gain can also be detected easily using open loop testing methods, since pure gain change will merely result in a straightforward amplification or attenuation of the broken loop amplitude ratio.

## 1.19 Discussion on the Choice of Performance Criteria for a Given Application

The choice of performance criteria for a given application is often a matter of some argument, since clearly we do not want to specify too many facets of performance as this may increase cost and complexity. The problem is therefore to decide which, and how many, of the performance criteria we have dicussed are really relevant in any particular application. Now, it is often possible to express desired performance characteristics in terms of three or four performance criteria, and the three criteria which are generally accepted as most common are bandwidth, percentage overshoot, and velocity constant.[17]

We have already stated in Section 1.7 that a certain art is involved in writing the specification so as to save unnecessary waste at the design stage. Thus we should choose a bandwidth such that most of the command signal is transmitted with little loss in amplitude ratio, and yet the bandwidth should be sufficiently low for the system to effectively filter out any high-frequency noise which may be present. In this way we are seeking, in terms of Equations 1.9, 1.10 and 1.11 to minimise $e_c$ and $e_N$ simultaneously by specifying a suitable bandwidth. Generally, power consumption increases with bandwidth, particularly with hydraulic servos, which is a further reason for minimising the bandwidth requirement and impressing upon designers the fact that too high a bandwidth can be as undesirable as too low a bandwidth. Percentage overshoot is a measure of transient response which is particularly useful for systems with a dominant pole pair, as an unambiguous understanding of the response is obtained very easily. For systems where more than two poles dominate the response, it may be necessary to specify the form of the transient response also, as will be discussed in the next section. $C_v$ is a genuine measure of the ability to follow a ramp function input, and for many applications it is possible to specify the minimum value of $C_v$ fairly accurately. An error allowable at a particular low frequency in following an input $\theta_i$ can also be interpreted as a $C_v$ requirement. Let the allowable waveform peak error be $E$ at the low frequency $\omega$, then from Equation 1.25 we may write

$$\frac{\theta}{\theta_i}(s) = \frac{(a_1-b_1)s+(a_2-b_2)s^2+\dots}{1+a_1s+a_2s^2+a_3s^3+\dots} \qquad (1.38)$$

Substituting $j\omega$ for s, because $\omega$ is small we may neglect $s^2$ and above, and furthermore $a_1\omega$ will be small compared with unity. Hence the amplitude ratio relating error to input at the low frequency is

$$\left|\frac{E}{\theta_i}\right|_{\omega} \simeq (a_1 - b_1)\omega = \frac{\omega}{C_v} \tag{1.39}$$

The specification of $C_v$ therefore automatically defines the allowable low frequency error for such a system. We shall often assume that velocity constant, percentage overshoot and bandwidth requirements are to be satisfied simultaneously, and in particular, in Chapter 4, we shall develop a method for uniquely defining two-pole one-zero systems which uniquely satisfy such a tripartite specification.

In Section 1.7 the effect of disturbance torques on system behaviour was discussed. Since it is often necessary at the design stage to assume that load disturbances are static, it is relatively straightforward to show that the effect of these disturbances are minimised by making the gain in the feedback path of the $e_u/U(s)$ transfer function as high as is practicable, consistent with system stability. In practice this is achieved by making the feedforward gain between the comparator and the effective point of entry of the disturbance as high as practicable. Examples 5.17.1, 6.12.1, and 6.12.2 are devoted to systems wherein a load disturbance requirement is written into the performance specification. In practice, many systems are non-linear for very small signals because of backlash, dead zone, etc., and it is then useful to specify the system threshold, i.e. the smallest input signal the system is capable of following. Threshold is affected by the dynamic characteristics of the input signal itself. Some threshold problems have been outlined recently by Carrington and Martin,[18] evidence indeed that much more research work is still required in the small-error operating mode. We can, however, say that the threshold level will be reduced by increasing the feedforward gain between the comparator and the non-linear element itself. Thus the requirements of threshold and system stability act in opposition. Threshold level can always be reduced by a higher standard of engineering, but this can result in an excessively expensive servo.

To summarise, a typical specification for a rectilinear position control servomechanism would then appear as follows:

| | |
|---|---|
| Percentage overshoot | 15% |
| Velocity constant | $\geqslant$ 30 $\text{sec}^{-1}$ |
| Bandwidth | $\leqslant$15 rad/sec |
| Threshold (static) | $10^{-4}$ in |

## 1.20 Performance Indices

The percentage overshoot specified for a given application determines to some extent the form of the transient response, but intuition tells us by inspection that some forms of response are

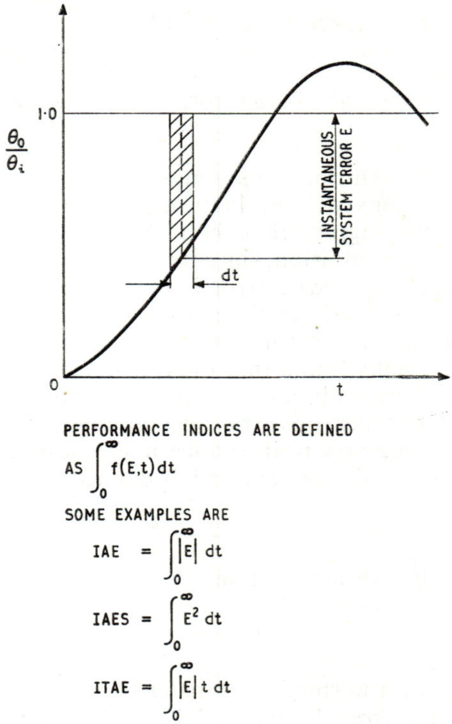


*Fig. 1.29. Definition of performance indices referred to step function response*

much better than others. Hence there must be some system transfer functions and resultant system pole-zero arrays which are better than others. Contributions in this field were made as early as 1947 by Whiteley,[19] who listed standard forms of transfer functions meeting a 10% overshoot requirement. Nowadays this intuition is crystallised by the use of performance indices, which are shown referred to the step function response in Fig. 1.29;, they are formally defined as

$$I = \int_0^\infty f(E, t)\mathrm{d}t \tag{1.40}$$

where $E$ is the system error $(\theta_i - \theta_0)$ at time $t$. The most common performance indices are:

$$\text{Integral of absolute error, IAE} = \int_0^{\infty} |E|\,\mathrm{d}t \tag{1.41}$$

$$\text{Integral of absolute error squared, IAES} = \int_0^{\infty} E^2\,\mathrm{d}t \tag{1.42}$$

$$\text{Integral of time} \times \text{absolute error, ITAE} = \int_0^{\infty} t\,|E|\,\mathrm{d}t \tag{1.43}$$

Graham and Lathrop,[20, 21] amongst others, have made fundamental contributions to the evaluation of such performance indices, and it is generally agreed that the ITAE index most nearly crystallises engineering intuition, since it may be argued that in following a step function a large error is inevitable early on, whilst a smaller error at some later time should be heavily penalised by suitably weighting the error in the time domain. For any order system it is normally found that the transfer function minimising the ITAE performance index will have an acceptable form of transient response. In order to portray and subsequently compare these responses it is necessary to introduce a simple normalising technique. If the normalised operator S is defined as follows

$$\mathrm{S}^n = a_n \mathrm{s}^n \tag{1.44}$$

or, alternatively, we may define

$$\omega_0 \mathrm{S} = \mathrm{s} \tag{1.45}$$

where $\omega_0^n = 1/a_n$, a normalised transfer function in terms of S can then be obtained from Equation 1.13 with a new set of coefficients:

$$\frac{\theta_0}{\theta_i}(\mathrm{s}) = \frac{1 + b_1'\mathrm{S} + b_2'\mathrm{S}^2 + \cdots b_q'\,\mathrm{S}^q}{1 + a_1'\mathrm{S} + a_2'\mathrm{S}^2 + \cdots a_n'\,\mathrm{S}^n} \tag{1.46}$$

where $a_r' = a_r\omega_0^r$ and $b_r' = b_r\omega_0^r$. Thus, for example, $a_2' = \omega_0^2 a_2$.

Thus the pole-zero array can be normalised, and for a given order of system a preferred form of transfer function can be evaluated. As examples, the ITAE optimum transfer function coefficients for unity numerator systems are shown in Table 1.20.1. Note that these coefficients only apply to unity numerator systems and would be different for systems with numerator terms even when following the same input.

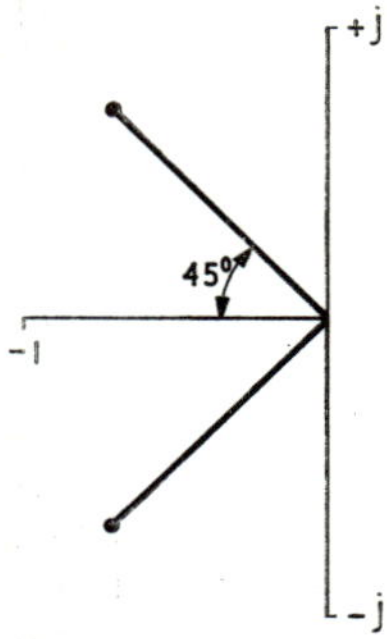


$$\frac{\theta_o}{\theta_i}(s) = \frac{1}{1 + 1{\cdot}4s + s^2} = \frac{1}{1 + (2 \times 0{\cdot}7)s + s^2}$$

SECOND ORDER

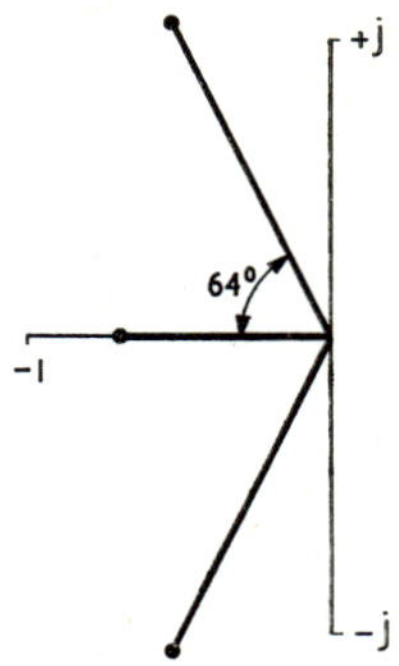


$$\frac{\theta_o}{\theta_i}(s) = \frac{1}{1 + 2{\cdot}15s + 1{\cdot}75s^2 + s^3} = \frac{1}{[s + 0{\cdot}71][s^2 + (2 \times 0{\cdot}44 \times 1{\cdot}18)s + 1{\cdot}18^2]}$$

THIRD ORDER

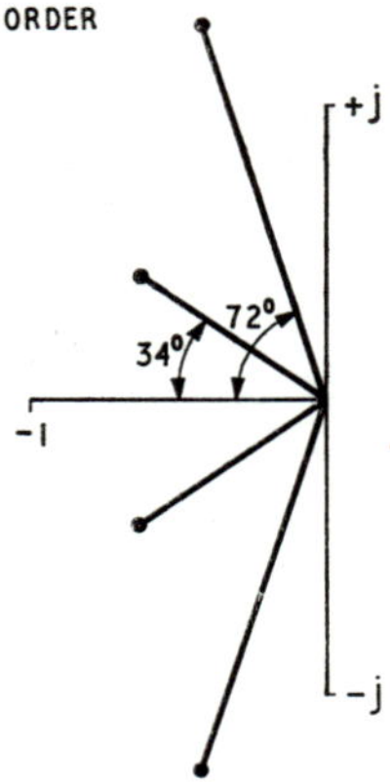


$$\frac{\theta_o}{\theta_i}(s) = \frac{1}{1 + 2{\cdot}7s + 3{\cdot}4s^2 + 2{\cdot}1s^3 + s^4} = \frac{1}{[s^2 + (2 \times 0{\cdot}83 \times 0{\cdot}75)s + 0{\cdot}75^2][s^2 + (2 \times 0{\cdot}31 \times 1{\cdot}33)s + 1{\cdot}33^2]}$$

FOURTH ORDER

*Fig. 1.30. Pole positions for normalised ITAE optimum control systems, second to fourth orders*

**Table 1.20.1**

DENOMINATOR OF NORMALISED ITAE OPTIMUM UNITY NUMERATOR SYSTEMS

| *Order* | *Denominator* |
|---|---|
| 1 | $1+S$ |
| 2 | $1+1{\cdot}4S+S^2$ |
| 3 | $1+2{\cdot}15S+1{\cdot}75S^2+S^3$ |
| 4 | $1+2{\cdot}7S+3{\cdot}4S^2+2{\cdot}1S^3+S^4$ |

The normalised pole arrays are shown in Fig. 1.30. For any application the true pole arrays are obtained by multiplying the scales by $\omega_0$. Fig. 1.31 shows the normalised step function response, and Fig. 1.32 shows the normalised frequency response for systems of second to fourth order. It cannot be too strongly emphasised that the ITAE index defines a form of response, the true response scale being determined by $\omega_0$. Thus, if we have a free choice of pole and zero positions, the ITAE index indicates one suitable array.

The systems in Table 1.20.1 are optimised by minimising the ITAE index applied to the system when following a displacement

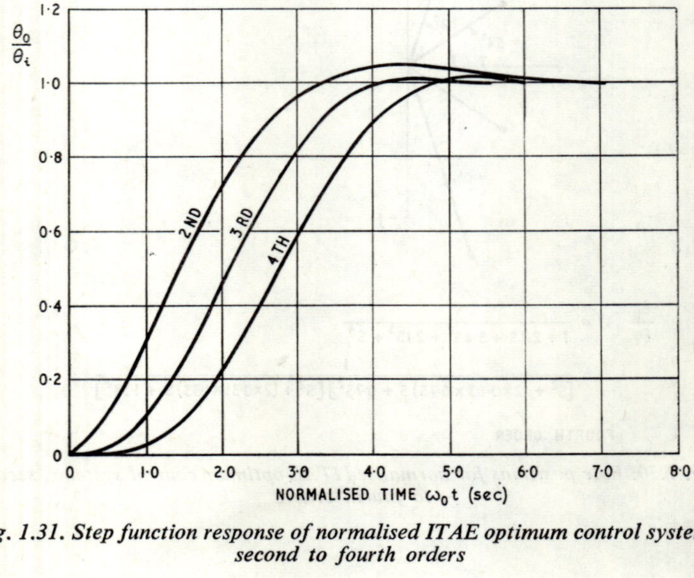


*Fig. 1.31. Step function response of normalised ITAE optimum control systems, second to fourth orders*

step function input. If the system is specifically designed to follow a velocity step function, it would be more reasonable to minimise the ITAE index following this particular input. This will be illustrated in Chapter 4, when second order systems are discussed in detail. Further examples of ITAE optimum pole-zero arrays will be given in Chapter 6.

The IAES following a displacement step function input can be readily calculated using standard integral tables, whilst except for those systems studied by Graham and Lathrop and others it is necessary to use an analogue computer to determine the ITAE for

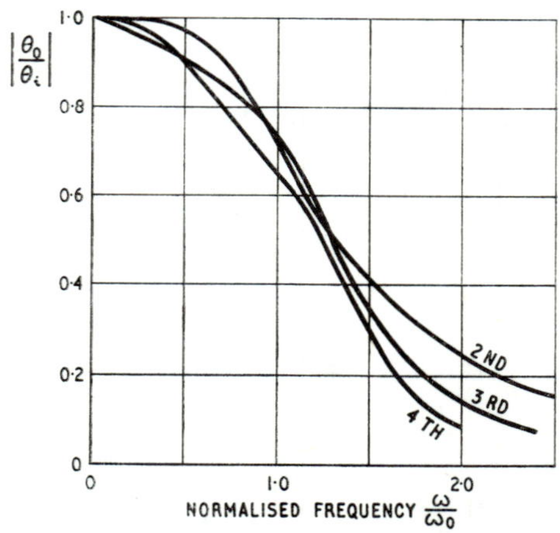


*Fig. 1.32. Frequency response of normalised ITAE optimum control systems, second to fourth orders*

a specific transfer function, and this process is repeated for different transfer function coefficients until the minimum is located. It is unfortunate that the response of the system transfer function minimising the IAES index is normally too oscillatory for many applications, since the relative ease of computation of the index does offer distinct possibilities in the algebraic study of high order systems. An outstanding advantage of the IAES index occurs when only one component value has to be determined, as it is then possible to evaluate the IAES algebraically, differentiate with respect to the component in question, and thus find the component value to minimise the IAES index. This is one of the analytic design methods favoured by Newton, Gould and Kaiser[10] for the determination of compensation terms.

## 1.21 The Butterworth Standard Forms for Unity Numerator Systems

The ITAE optimum system is obviously not the only system with a satisfactory response. For a given order of system there are many pole-zero arrays which will give a satisfactory response subject only to restrictions, such as percentage overshoot, velocity constant, etc., that we may care to impose. An early attempt to

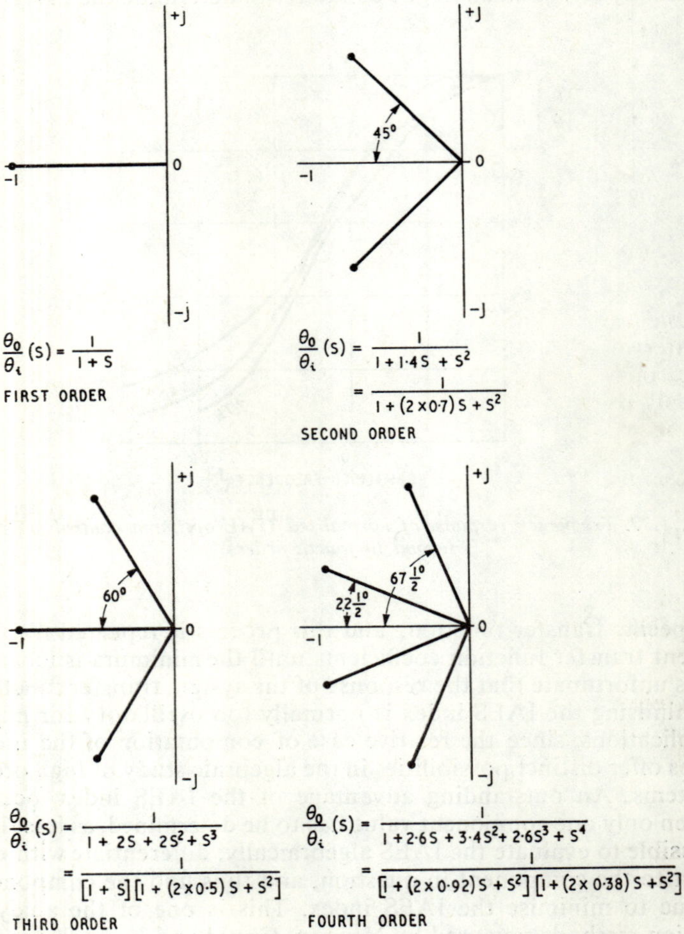


*Fig. 1.33. Normalised Butterworth filter pole positions*

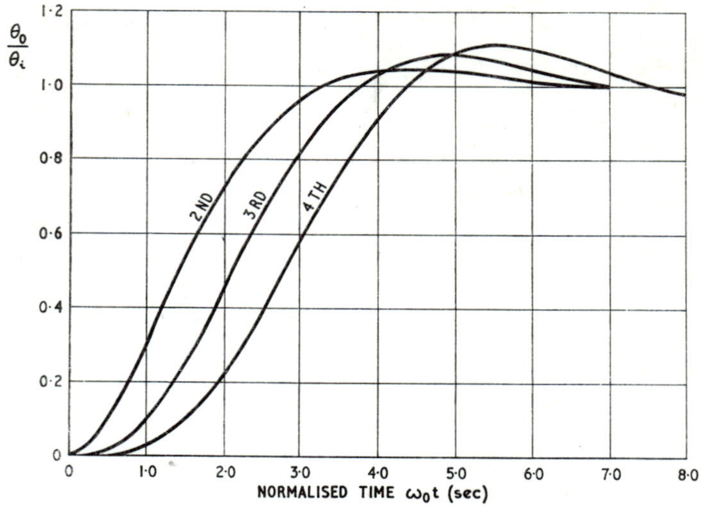


*Fig. 1.34. Step function response of normalised Butterworth filters, second to fourth orders*

define a logical pole array for unity numerator systems is due to Butterworth and was originally based on a particular pulse amplifier application. The rule is very simple: the normalised poles are evenly distributed on the unit circle in the left half of the pole plane, as shown in Fig. 1.33. Normalised transfer function coefficients for the Butterworth filters up to fourth order are given in Table 1.21.1. Fig. 1.34 shows the corresponding transient response, and Fig. 1.35 shows the corresponding frequency responses. It will be noted that the coefficients of the denominator are symmetrical, and this is a property of any standard form which positions poles on the unit circle. Although not as selective as the ITAE standard forms in the sense of picking the best intuitive response, the

**Table 1.21.1**

DENOMINATOR OF NORMALISED BUTTERWORTH STANDARD FORMS FOR UNITY NUMERATOR SYSTEMS

| *Order* | *Denominator* |
|---|---|
| 1 | $1+S$ |
| 2 | $1+1{\cdot}4S+S^2$ |
| 3 | $1+2S+2S^2+S^3$ |
| 4 | $1+2{\cdot}6S+3{\cdot}4S^2+2{\cdot}6S^3+S^4$ |

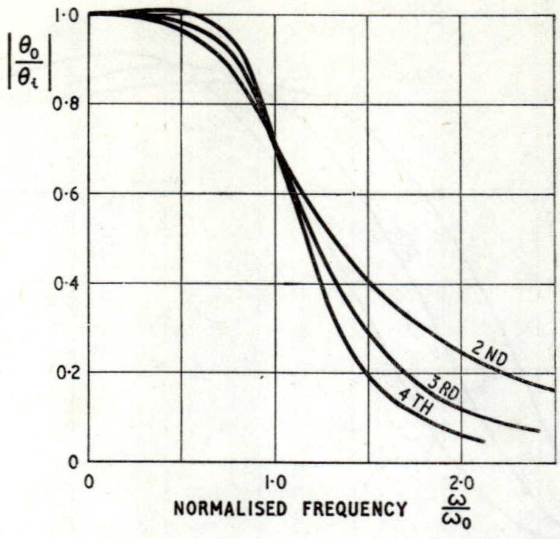


*Fig. 1.35. Frequency response of normalised Butterworth filters, second to fourth orders*

Butterworth standard forms certainly give a useful form of response. Graham and Lathrop actually used the Butterworth pole positions as starting points when iterating on an analogue computer to determine the ITAE optimum pole positions. The simplicity of the pole array for the Butterworth standard form is particularly appealing, and the Butterworth pole positions can easily be remembered.

## 1.22 Dominant Mode Approach

Standard preferred forms for system transfer functions such as the ITAE optimum forms are useful in indicating *some* of the *many* pole arrays that may be satisfactory in a given application. We might, of course, specify so many desired performance characteristics that the pole-zero array is absolutely defined by the performance specification. For example, a second order system with one zero and two poles, can, in general, satisfy required values of bandwidth, velocity constant, and percentage overshoot by one specific pole-zero array only, as we shall subsequently establish in Chapter 4.

A further important point often overlooked is that if the system is of $n^{th}$ order, in order that the $n$ poles are forced into the desired positions defined by a standard form, it is necessary for the system to possess $n$ independent adjustable parameters such as network time constants, etc. For high order systems this is often impracticable, and it is generally reckoned to provide two or three adjustable parameters only, thus placing limitations on possible pole configurations. This restriction is not terribly important, since we are rarely after absolutely the best control system possible for a given application. A control system which adequately meets the specification in every respect is all that is required, thus the pole positions indicated by a standard form are only a rough guide. In the dominant mode design technique, the available degrees of design freedom are used to force a few poles and zeros into desired positions, the other poles and zeros being driven to infinity or to cancellation.

If there are system zeros present, the ITAE optimum standard forms will change radically from the arrays shown in this chapter. Graham and Lathrop have determined standard forms for certain systems with one system zero present; generally, systems with more than one zero present would have to be studied using an analogue computer to determine the effect of variations in transfer function on the ITAE. Subsequent iteration would then determine the transfer function, and hence the pole-zero array corresponding to the minimisation of the integral. ITAE optimum standard forms for systems with $C_v = \infty$ (one system zero), and $C_v = C_a = \infty$ (two system zeros) will be examined in Chapter 6. Unless the correct number of design adjustments are available in the system, it is not possible to obtain the ITAE optimum form exactly. However, this is unimportant, since we shall see in Chapters 4, 5 and 6 that the optimum forms are only guides which should be used in particular circumstances. It must be emphasised that the designer with three degrees of freedom available will be able, in most cases, to arrive at either a second order or a third order dominant array according to his whim. Generally, the third order array will be found to be a better engineering proposition, but this will depend on the particular performance criteria of interest.

For high order systems it is often possible to inspect the system pole-zero array and make certain assumptions to reduce the order of the system, thus greatly simplifying the initial design calculations. Fig. 1.36 shows a sixth order system pole-zero array. There are three system zeros. Some rules for reducing the order of the system are as follows:

a. If a pole is close to a zero, both may be ignored since their effects tend to cancel out. The nearer the pole and zero are to the origin, the closer together must they be for their effects to cancel out.
b. Poles or zeros far to the left of the origin may be neglected.

What is meant by 'close' and 'far' is somewhat arguable. The zeros in Fig. 1.36 are certainly close enough to poles to cancel out the effects of these poles, leaving a dominantly third order system which might approximate to the ITAE form. Interpretation of 'close' and 'far' can only be based on experience, but a rough rule is that the effect of a pole or zero on the system performance is negligible if it is five times further from the origin than the poles or zeros under consideration, although some control system designers only ignore poles and zeros about ten times further from the origin. In this context, distance to the left must be measured along the negative real axis. Thus complex poles and zeros are measured by their negative real parts.

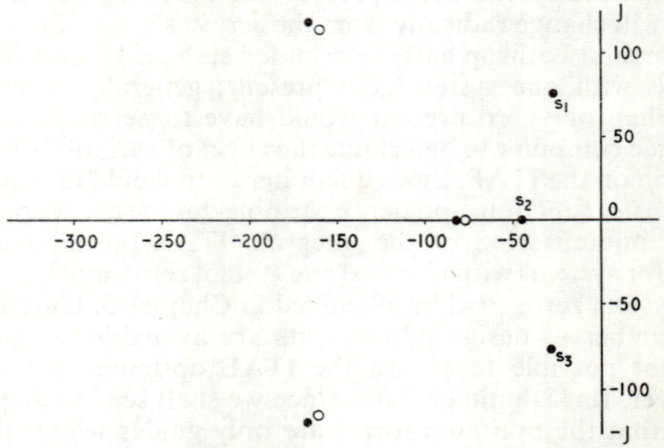


*Fig. 1.36. Pole-zero array of sixth order system with three dominant poles*

The poles and zeros which remain after this elimination process has been carried out are then called the dominant poles and zeros, since these poles and zeros tend to dominate the system response. As discussed in Section 1.11, each negative real pole and each complex pole pair will give rise to a response mode: modes due to dominant poles are termed dominant modes. In present-day literature, considerable emphasis is placed on the viability of a second order dominant response. These modes will, of course, be shaped by the dominant zeros via the coefficients of the differential equation solution. Using the dominant mode approach, it is possible to compute, for example, bandwidth to an accuracy of 5%, compared with the true value obtained by taking all the poles and zeros into account, and thus it is possible to determine very quickly whether the system is likely to meet a specified value of bandwidth. Judicious use of the dominant mode approach can be combined with the standard form approach if desired, as might be indicated from Fig. 1.36. Some specific examples of the dominant mode approach will be dealt with in detail in Chapters 3, 5 and 6, and these examples will assist the reader in interpreting 'close' and 'far' in his own particular problem.

## 1.23 Concluding Remarks

This chapter has given the reader an insight into the objectives of control engineering, and in particular an insight into the function of feedback control systems. The significance of the system transfer function and of the related poles and zeros has been briefly examined. Methods of specifying system performance have been described in some detail, and some preferred forms of transfer function based on the ITAE index and on the Butterworth filter discussed, so that the reader can appreciate the necessity for clear thinking at the specification stage. In Chapter 2 we shall derive the transfer functions for individual components, whilst in Chapter 3 we shall firstly show how system performance can be determined from the system pole-zero array, and secondly we shall examine how the system pole-zero array can be determined from a knowledge of the individual component transfer functions using the root locus method due to Evans.

Subsequent chapters will deal with the analysis and synthesis of a variety of feedback control systems. Considerable emphasis will be placed on the use of graphical techniques to evaluate system transfer functions and system responses.

CHAPTER 2

# Derivation of Transfer Functions for Components and Simple Control Systems

## 2.1 Introduction

There is nothing magic about the concept of a transfer function, which is simply the relationship between the output and input to a black box. For linear systems, the relationship will be independent of the magnitude of the input signal, and it is often particularly convenient to express the cause and effect relationship as the ratio of two polynomials in the operator $s = d/dt$. The coefficients of these two polynomials uniquely define the response of the element to any input stimulus. These coefficients are normally obtained by writing down the differential equations governing dynamic behaviour, as was seen in Examples 1.4.1 and 1.4.2 of Chapter 1. Having determined the transfer function in terms of s, it is normal practice to represent this transfer function by poles and zeros in the s plane. When the black box is a complete system, then these poles and zeros are the system poles and zeros as defined in Fig. 1.16, whilst if the black box is a motor or some such item, then the poles and zeros are component poles and zeros, and their symbolic s plane representation will be somewhat different, as we shall see in Chapter 3.

A given transfer function written in terms of the operator s will have a unique response to a particular input, and this will normally be expressed graphically, in the form, for example, of a step function or a harmonic response. This unique property applies in reverse. Provided the system or component is linear, then a given response curve can only be exactly fitted at all points by one unique transfer function written in terms of the operator $d/dt$. Sometimes it is convenient to leave the transfer function in this graphical form, and in this chapter we shall examine the logarithmic representation of the frequency response of transfer functions in some detail. For non-linear systems the graphical representation of transfer functions is often very helpful, as the dynamic performance can be crystallised as a set of response curves, where input amplitude, by definition, is the salient variable. A necessary preliminary to the study of system transfer functions and the derivation of system transfer functions which are mutually compatible with both performance specifications and with load characteristics, is an understanding of component transfer functions, and how individual components are combined to form a feedback control system. We shall now derive the transfer functions of a wide variety of elements, including load elements, power elements, and compensation elements. Control engineering is now so catholic that many basic disciplines form an integral part of transfer function derivation, and we shall see that a knowledge of aerodynamics, mechanics, fluid dynamics, and electronics is necessary for even the relatively few transfer functions that we shall have space to include here. The great advantage of the transfer function concept is that provided the concept is properly understood, then the control engineer merely specifies acceptable transfer functions for particular components, and need not necessarily be involved in detailed derivation. For a more extensive review of transfer function derivation the reader is referred to such reference works as the *Control Engineers' Handbook*.[22] The examples chosen here give a reasonable coverage of the important physical principles involved, and are also relevant to the detailed design examples to be considered in later chapters.

## 2.2 Transfer Function for an Inertia Load with Viscous Damping

In servomechanisms the load to be controlled is normally a mass of metal, and there will often be some form of viscous damping characteristic inherent in the load which will provide a resistive torque-opposing motion proportional, not to displacement, but to velocity. The drive to the load may be rotary or straight line, depending on the motor and the particular application. Fig. 2.1(a) shows a rotary inertia load with rotary viscous damping. Let the torque applied to the load be $T$ lb ft, the moment of inertia be $J$ slugs ft$^2$, and the viscous damping be $f$ lb ft rad$^{-1}$ sec, whilst the load displacement is measured in radians. The transfer function required is the relationship between load displacement and applied torque, and this is obtained by using Newton's Law. Let:

Load displacement $= \theta_0$

Inertia torque $= J\ddot{\theta}_0 = Js^2\theta_0$

Damping torque $= f\dot{\theta}_0 = fs\theta_0$

Applied torque $= T$

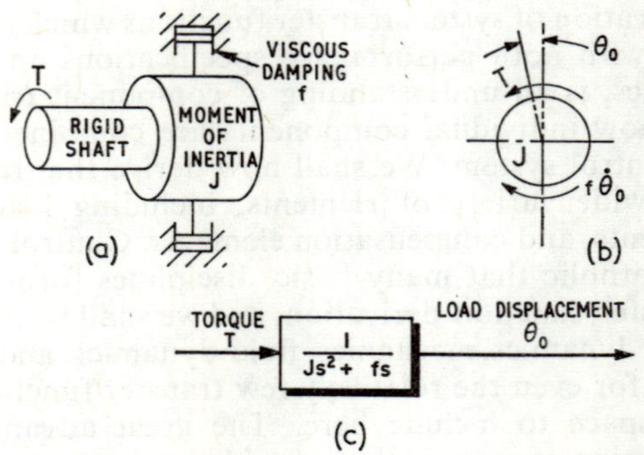


*Fig. 2.1. Rotary inertia load with viscous damping: (a) schematic diagram; (b) free-body diagram; (c) block diagram form of transfer function*

From Newton's Law the inertia torque is equal to the nett torque acting on the body, hence from the free-body diagram of Fig. 2.1(b) we may write

$$Js^2\theta_0 = T - fs\theta_0$$

and, transposing,

$$T = Js^2\theta_0 + fs\theta_0 \tag{2.1}$$

and, in transfer function form,

$$\frac{\theta_0}{T} = \frac{1}{Js^2+fs} \tag{2.2}$$

Thus the diagrammatic representation of the load shown in Fig. 2.1(a) is at all times equivalent to the block diagram form of the transfer function shown in Fig. 2.1(c).

The rectilinear drive equivalent of Fig. 2.1(a) is shown in Fig. 2.2(a). For an applied force $F$ lb, load inertia $M$ slugs, load displacement $x_0$ ft, and a damping force of $f$ lb ft$^{-1}$ sec, we may write:

Load displacement $= x_0$

Inertia force $= M\ddot{x}_0 = Ms^2x_0$

Damping force $= f\dot{x}_0 = fsx_0$

Applied force $= F$

Applying Newton's Law we may write the expression for inertia force from the free-body diagram of Fig. 2.2(b) as

$$Ms^2x_0 = F - fsx_0$$

and, transposing,

$$F = Ms^2x_0 + fsx_0 \tag{2.3}$$

and, in transfer function form,

$$\frac{x_0}{F} = \frac{1}{Ms^2+fs} \tag{2.4}$$

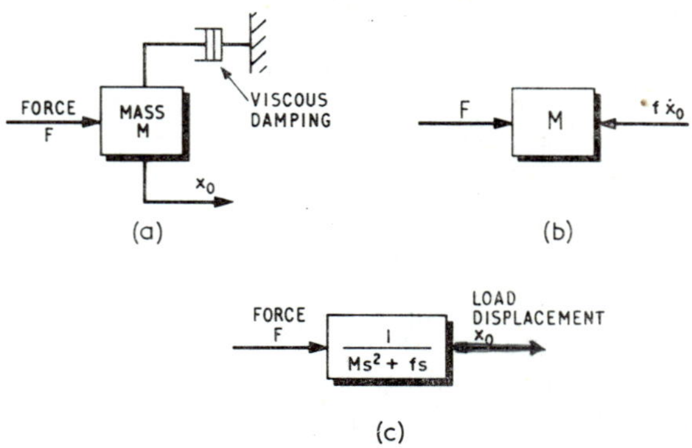


*Fig. 2.2. Rectilinear inertia load with viscous damping: (a) schematic diagram; (b) free-body diagram; (c) block diagram form of transfer function*

The block diagram equivalent is shown in Fig. 2.2(c). The force-length-time system of dimensional units (slugs, lb, ft, and slugs ft$^2$, lb ft, rad) has been chosen for convenience, but any consistent set of units may be applied. From the similarity of Equations 2.2 and 2.4 we see that the load movements for both rotary and rectilinear motion are analogous, and in general it is possible, as we shall see shortly, to derive electric circuits with the same transfer functions as those mechanical elements. The physical quantities involved may be vastly different, but the transfer functions will be similar. In later sections we shall derive transfer functions for a large number of rectilinear mechanical networks, but for all examples given an equivalent rotary network can be devised.

## 2.3 Total Inertia for a System with Gearing

In many servomechanisms the power amplifier, or motor, is a device with a high cruise speed and, for a variety of reasons which will become apparent later, it is pertinent to introduce a gear reduction unit between the power element and the load. For rotary output servos a conventional gearbox may be used, whilst for rectilinear output servos a linkage may be used to achieve the speed reduction. A hybrid system is also possible with a rotary motor and gearbox whose output is translated into linear motion of the load by using, say, a rack and pinion or a leadscrew.

A typical gear-driven load is shown in Fig. 2.3(a). Initially we shall assume that the gears themselves are inertialess. If $\theta_m$ is the motor shaft displacement, and $\theta_0$ is the load displacement, then the gear ratio $N$ is defined by $\theta_m/\theta_0$, where $N$ is normally $\gg 1{\cdot}0$ for servomechanisms with electric motor power amplifiers. We can refer the inertia to the motor shaft or to the output shaft, whichever is desired; the following analysis refers the load characteristics to the motor shaft. The angular relationships may be written as

$$\left.\begin{aligned} \theta_m &= N\theta_0 \\ \mathrm{s}\theta_m &= N\mathrm{s}\theta_0 \\ \mathrm{s}^2\theta_m &= N\mathrm{s}^2\theta_0 \end{aligned}\right\} \tag{2.5}$$

For a given motor acceleration the load acceleration is reduced by the factor $N$. The torque required to accelerate the load and overcome viscous friction at the load is

$$T_{\text{load shaft}} = J_L\mathrm{s}^2\theta_0 + f_L\mathrm{s}\theta_0$$

where $J_L$ is the load inertia and $f_L$ is the load viscous damping. Substituting for $\mathrm{s}^2\theta_0$ and $\mathrm{s}\theta_0$ from Equation 2.5, we have

$$T_{\text{load shaft}} = \left(\frac{J_L}{N}\right)\mathrm{s}^2\theta_m + \left(\frac{f_L}{N}\right)\mathrm{s}\theta_m \tag{2.6}$$

To obtain a given value of torque at the load shaft, consider the gearbox as a pair of simple spur gears as shown in Fig. 2.3(b). The common element is, of course, the force $F$ at the contact point. If $r_1$ is the pitch radius of the load shaft gear, and $r_2$ is the pitch radius of the motor shaft gear, where, in general, $r_1 > r_2$, then the torque at the load shaft is $r_1F$ and at the motor shaft

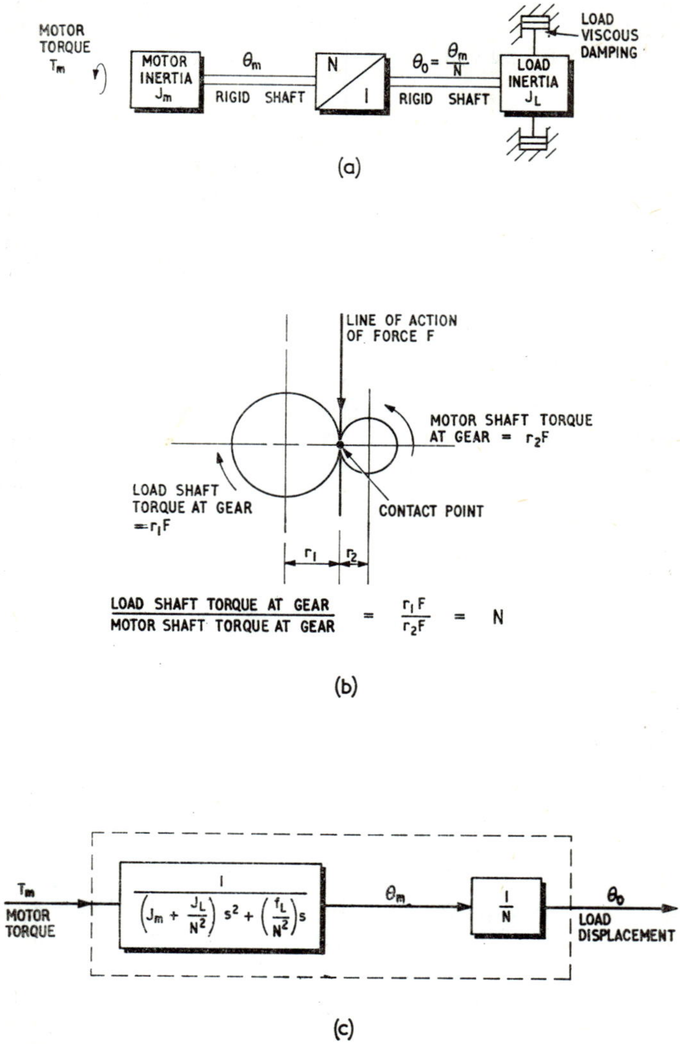


*Fig. 2.3. Total inertia for a motor and load with gear ratio $\theta_M/\theta_0 = N$: (a) schematic diagram; (b) representation of the gearbox by a pair of spur gears in order to evaluate the torque relationship; (c) block diagram of the transfer function*

is $r_2F$. Clearly the ratio of the motor shaft torque to load shaft torque is $r_2/r_1$, i.e. to obtain a particular torque $T_L$ at the load shaft, a torque $(r_2/r_1)T_L$ is required at the motor shaft. But $r_1/r_2$ is, by definition, the gear ratio $N$. Thus we may write

$$T_{\text{motor shaft}} = \frac{T_{\text{load shaft}}}{N} \tag{2.7}$$

and hence the torque at the motor required to overcome load dynamics is obtained by substituting Equation 2.7 into Equation 2.6, leaving the simple result

$$\underset{\text{To overcome load dynamics}}{T_{\text{motor shaft}}} = \left(\frac{J_L}{N^2}\right) s^2\theta_m + \left(\frac{f_L}{N^2}\right) s\theta_m \tag{2.8}$$

We must now add the torque required to overcome the inertia of the motor itself, and this torque will be $J_m s^2\theta_m$. The motor torque/motor shaft transfer function can therefore be derived from the differential equation

$$T_m = J_m s^2\theta_m + \left(\frac{J_L}{N^2}\right) s^2\theta_m + \left(\frac{f_L}{N^2}\right) s\theta_m$$

to yield

$$\frac{\theta_m}{T_m} = \frac{1}{\left(\frac{f_L}{N^2}\right) s + \left(J_m + \frac{J_L}{N^2}\right) s^2} \tag{2.9}$$

with the appropriate block diagram representation shown in Fig. 2.3(c).

Although the absolute values of load and motor inertias may be very different, their individual contributions to the system dynamics may be of the same magnitude. For example, for a motor moment of inertia of 0·01 slugs ft², a gear ratio of 100 : 1, and a load moment of inertia of 100 slugs ft², then $J_m$ and $J_L/N^2$ are exactly equal. By making $N$ very large, it can be seen from Equation 2.9 that the effect will be to diminish the effect of the load on the motor/load transfer function, leaving substantially the motor characteristics only. This may be one reason for a judicious choice of gear ratio; another may arise because of the desire to optimise system response for a particular type of input, such as a constant acceleration demand, or alternatively it may be necessary to choose the value of gear ratio which will minimise the motor torque required for given load velocity and acceleration requirements.

EXAMPLE 2.3.1

A motor with moment of inertia $J_m$ and viscous damping coefficient $f_m$, both referred to the motor shaft, is used to control, through gearing, an inertia load with moment of inertia $J_L$ and viscous damping coefficient $f_L$, both referred to the load shaft. Determine the value of gear ratio $N$ which will minimise the motor torque needed to meet a specification which requires that the load is to accelerate at $\ddot{\theta}_L$ when traversing at a velocity $\dot{\theta}_L$.

The motor torque equation referred to the motor shaft is

$$T_m = f_m s\theta_m + J_L s^2\theta_m + \left(\frac{f_L}{N^2}\right) s\theta_m + \left(\frac{J_L}{N^2}\right) s^2\theta_m \qquad (2.10)$$

Since the specification is written in terms of the load behaviour, we must rewrite the right-hand side of Equation 2.10 in terms of load velocity and load acceleration. Thus, substituting $s\theta_m = N\dot{\theta}_L$ and $s^2\theta_m = N\ddot{\theta}_L$, Equation 2.10 becomes

$$T_m = f_m N\dot{\theta}_L + J_m N\ddot{\theta}_L + \frac{f_L\dot{\theta}_L}{N} + \frac{J_L\ddot{\theta}_L}{N} \qquad (2.11)$$

where $\dot{\theta}_L$ and $\ddot{\theta}_L$ are the specified values of load velocity and load acceleration required simultaneously from the performance specification. Differentiating Equation 2.11 with respect to the gear ratio $N$, we obtain

$$\frac{\partial T_m}{\partial N} = f_m\dot{\theta}_L + J_m\ddot{\theta}_L - \frac{f_L\dot{\theta}_L}{N^2} - \frac{J_L\ddot{\theta}_L}{N^2} \qquad (2.12)$$

and equating to zero we obtain the optimum value of $N$ which is defined by

$$N = \sqrt{\frac{f_L\dot{\theta}_L + J_L\ddot{\theta}_L}{f_m\dot{\theta}_L + J_m\ddot{\theta}_L}} \qquad (2.13)$$

Assuming the following values

$J_L = 100$ slugs ft$^2$

$J_m = 0{\cdot}01$ slugs ft$^2$

$\dot{\theta}_L = 1$ rad/sec

$\ddot{\theta}_L = 20$ rad/sec$^2$

$f_L = 500$ lb ft rad$^{-1}$ sec

$f_m = 0{\cdot}2$ lb ft rad$^{-1}$ sec

then

$$N = \sqrt{\frac{(500\times 1)+(100\times 20)}{(0\cdot 2\times 1)+(0\cdot 01\times 20)}}$$

Therefore

$$N = \sqrt{\frac{2{,}500}{0\cdot 40}} = 79\cdot 1 \qquad (2.14)$$

The motor speed at which the specification is met is $N\dot{\theta}_L = 79\cdot 1\times 1 = 79\cdot 1$ rad/sec $= 760$ rev/min, and the motor torque required at this speed is obtained from Equation 2.11. Hence

$$T_m = \underbrace{(0\cdot 2\times 79\cdot 1\times 1)}_{\text{Motor damping term}}+\underbrace{(0\cdot 01\times 79\cdot 1\times 20)}_{\text{Motor inertia term}}+\underbrace{\frac{500\times 1}{79\cdot 1}}_{\text{Load damping term}}+\underbrace{\frac{100\times 20}{79\cdot 1}}_{\text{Load acceleration term}}$$

$$= 15\cdot 9+15\cdot 9+63\cdot 0+252\cdot 0$$

$$= 346\cdot 8 \text{ lb ft}$$

Therefore

$$\text{motor horsepower} = \frac{T_m\theta_m}{550}$$

$$= \frac{79\cdot 1\times 346\cdot 8}{550}$$

$$= 50\cdot 3 \text{ hp}$$

About 90% of this horsepower is used to overcome load inertia and load damping under these optimum conditions. The distribution of horsepower taken by load and motor in this example is far from equal; in automobile transmission applications, equality of distribution of power between load and motor is quite common.

Alternatively, we could have imposed a constraint on the motor speed or on the motor horsepower instead of on the motor torque, but Example 2.3.1 is useful in illustrating how a motor can be chosen to fulfil a given specification. The choice of motor is obviously affected by the performance requirements of the load given in the form of Example 2.3.1, which in turn would be broken down from the *system* performance specification. As we have already seen in Chapter 1, the system performance may be specified in terms of percentage overshoot, velocity error constant, and system bandwidth; since the velocity error constant is defined in terms of an acceptable error for a given system input velocity, $\dot{\theta}_L$ can be assumed to be equal to this given input velocity. A value for $\ddot{\theta}_L$ can be estimated from a knowledge of input conditions for the worst expected demand signal. Alternatively, $\dot{\theta}_L$ and $\ddot{\theta}_L$ can be chosen from an examination of the required system frequency response characteristics, and evaluated as a function of $\omega$. For further information on the choice of optimum gear ratios the reader is referred to Section 13, by Davis and Ledgerwood, in the *Control Engineers' Handbook*. It should be realised that the assumption of maximum load velocity occurring simultaneously with the maximum load acceleration may be pessimistic and may lead to the choice of a motor which, strictly speaking, is larger than is really required. However, in many instances, load data is scanty and imprecise anyway, and it is necessary to apply a suitable factor of safety to motor power requirements.

## 2.4 Voltage-smoothing Circuit

As the next element in our *pot-pourri* review of transfer function derivation, let us consider the voltage-smoothing circuit shown in Fig. 2.4(a). The function of this circuit is to accept a.c. and d.c. electrical signals but to filter out the a.c. components, leaving

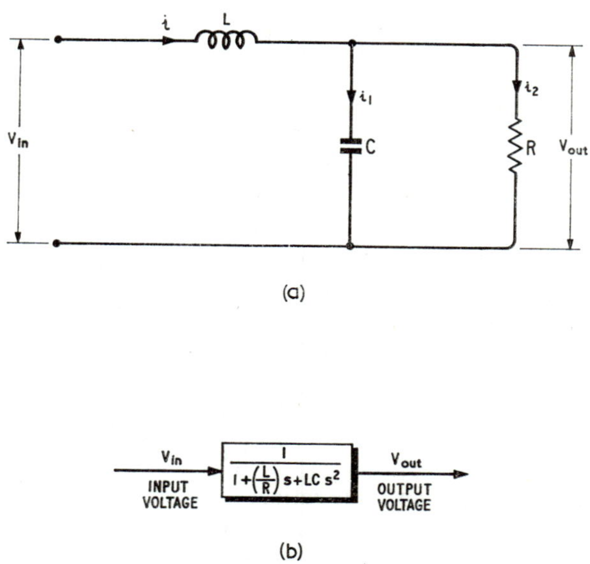


*Fig. 2.4. Voltage-smoothing circuit: (a) circuit diagram; (b) block diagram of transfer function*

essentially a d.c. output. In the analysis of electrical circuits, Ohm's Law and Kirchoff's Laws replace Newton's Law in the derivation of the component transfer function. For the voltage-smoothing circuit the transfer function required is that relating output voltage to applied voltage.

Applying Ohm's Law to the circuit we derive the voltages

$$V_{\text{in}} = Lsi + V_{\text{out}} \tag{2.15}$$

$$V_{\text{out}} = Ri_2 = \frac{i_1}{Cs} \tag{2.16}$$

and

$$i = i_1 + i_2 \tag{2.17}$$

$i$, $i_1$, and $i_2$ are eliminated from Equations 2.15, 2.16 and 2.17 as follows. From Equations 2.15 and 2.17,

$$V_{\text{in}} = Ls(i_1 + i_2) + V_{\text{out}}$$

and, from Equation 2.16,

$$i_2 = \frac{V_{\text{out}}}{R}$$

and

$$i_1 = V_{\text{out}}Cs$$

Hence

$$V_{\text{in}} = Ls\left(\frac{V_{\text{out}}}{R} + V_{\text{out}}Cs\right) + V_{\text{out}}$$

and, in transfer function form,

$$\frac{V_{\text{out}}}{V_{\text{in}}}(s) = \frac{1}{1 + \left(\frac{L}{R}\right)s + LCs^2} \tag{2.18}$$

which is shown in block diagram form in Fig. 2.4(b).

From Equation 2.18 it is apparent that at d.c. frequency ($\omega = 0$), $V_{\text{out}}/V_{\text{in}} = 1{\cdot}0$. To study the smoothing action at high frequency, substitute $j\omega$ in Equation 2.18. This gives

$$\frac{V_{\text{out}}}{V_{\text{in}}}(j\omega) = \frac{1}{1 + j\omega\left(\frac{L}{R}\right) - LC\omega^2}$$

and evaluating the amplitude ratio,

$$\left(\frac{V_{\text{out}}}{V_{\text{in}}}\right)_{\omega} = \frac{1}{\sqrt{(1 - LC\omega^2)^2 + \frac{\omega^2L^2}{R^2}}} \tag{2.19}$$

If both $L$ and $C$ are relatively large, then the amplitude ratio may be approximated by $1/LC\omega^2$, and the filtering effect, being proportional to $\omega^2$, is then immediately apparent.

## 2.5 Transformer Circuit

It is required to determine the transfer function relating secondary current to applied voltage for the transformer circuit shown in Fig. 2.5(a), where the two individual circuits are coupled with a mutual inductance $M$.

For the primary circuit side we may write the voltage equation

$$V_{in} = R_1 i_1 + L_1 s i_1 + M s i_2 \tag{2.20}$$

and for the secondary circuit side, assuming no load effects,

$$0 = R_2 i_2 + L_2 s i_2 + M s i_1 \tag{2.21}$$

The primary current $i_1$ may be eliminated between Equations 2.20 and 2.21 by writing

$$i_1 = -\frac{R_2 i_2 + L_2 s i_2}{M s}$$

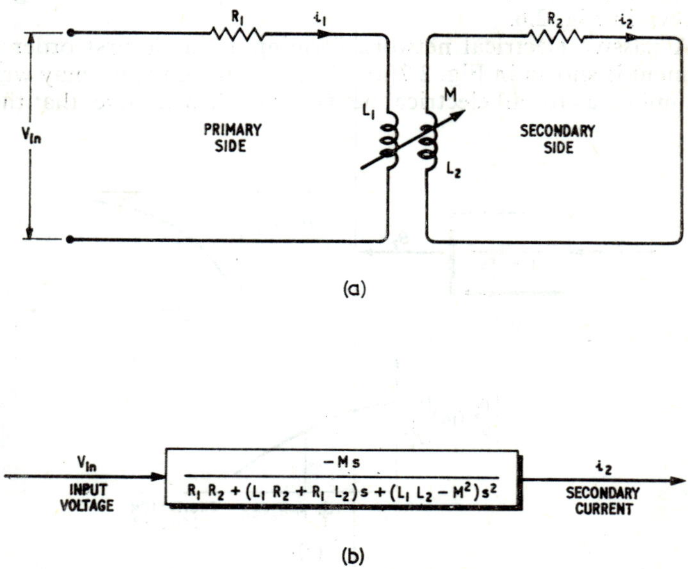


*Fig. 2.5. Transformer circuit; (a) circuit diagram; (b) block diagram of the transfer function*

giving

$$V_{in} = Msi_2 - \left(\frac{R_2+L_2s}{Ms}\right)(R_1+L_1s)i_2$$

and the transformer circuit transfer function is then

$$\frac{i_2}{V_{in}}(s) = \frac{-Ms}{R_1R_2+(L_1R_2+R_1L_2)s+(L_1L_2-M^2)s^2} \tag{2.22}$$

The block diagram form of Equation 2.22 is shown in Fig. 2.5(b).

## 2.6 Simple First Order Lag Transfer Function

Transfer functions of the form $\theta_2/\theta_1(s) = 1/(1+Ts)$ are of common occurrence in control engineering and are termed simple first order lags. The interpretation of the transfer function in terms of common performance criteria is particularly straightforward. In response to a step function input, the element described by the simple lag transfer function will reach 63% of the final value at a time $T$ sec after the application of the input, whilst the element will have a bandwidth for harmonic response of $1/T$ rad/sec. These characteristics of the simple first order lag are shown in Fig. 2.6.

A passive electrical network forming a simple first order lag element is shown in Fig. 2.7(a). Using Ohm's Law, we may write, assuming, as for all electrical networks dealt with here, that there

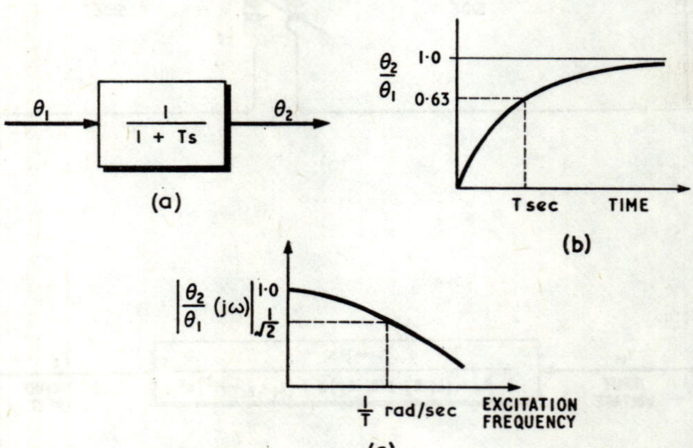


*Fig. 2.6. The first order lag: (a) block diagram representation of transfer function; (b) transient response; (c) frequency response*

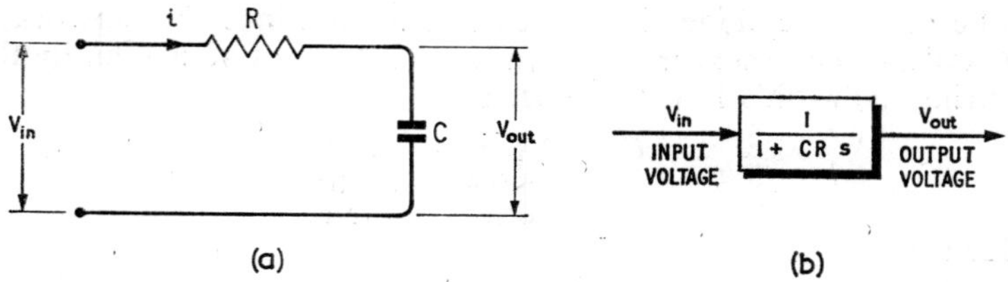


*Fig. 2.7. RC passive network first order lag: (a) circuit diagram; (b) block diagram representation of transfer function*

are no load effects,

$$V_{in} = Ri + \frac{i}{Cs} \tag{2.23}$$

and

$$V_{out} = \frac{i}{Cs} \tag{2.24}$$

and, eliminating $i$ and multiplying the top and bottom by $Cs$, we obtain the transfer function

$$\frac{V_{out}}{V_{in}}(s) = \frac{1}{1+CRs} \tag{2.25}$$

i.e. $T = CR$. Fig. 2.7(b) shows the block diagram equivalent of Equation 2.25.

A mechanical equivalent network is shown in Fig. 2.8(a). A spring of stiffness $k$ is mounted in series with a dashpot of viscous damping coefficient $f$. The input displacement is applied to the spring, and the output displacement is measured as the displacement of the dashpot piston (or any point on the rigid member connecting dashpot and the lower end of the spring).

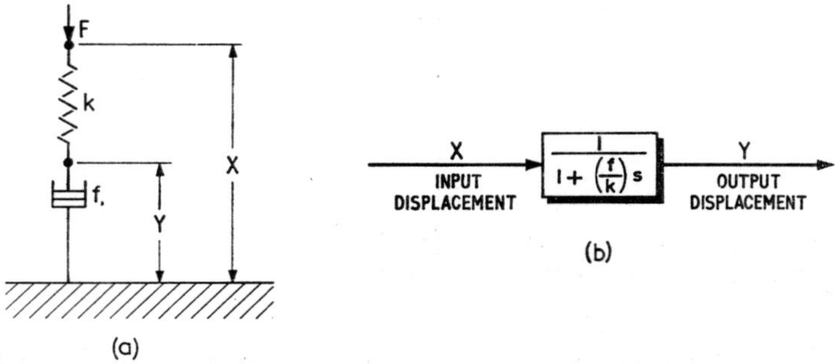


*Fig. 2.8. Mechanical network first order lag: (a) schematic diagram; (b) block diagram representation of transfer function*

The dashpot cylinder is earthed as shown. Since the network is massless, the same force is seen by the dashpot as is seen by the spring. Newton's Law therefore states

$$F = \underbrace{k(X-Y)}_{\text{Spring force}} = \underbrace{fsY}_{\text{Dashpot force}} \tag{2.26}$$

Hence

$$kX = (k+fs)Y$$

and the mechanical network transfer function is therefore

$$\frac{Y}{X}(s) = \frac{1}{1+\left(\frac{f}{k}\right)s} \tag{2.27}$$

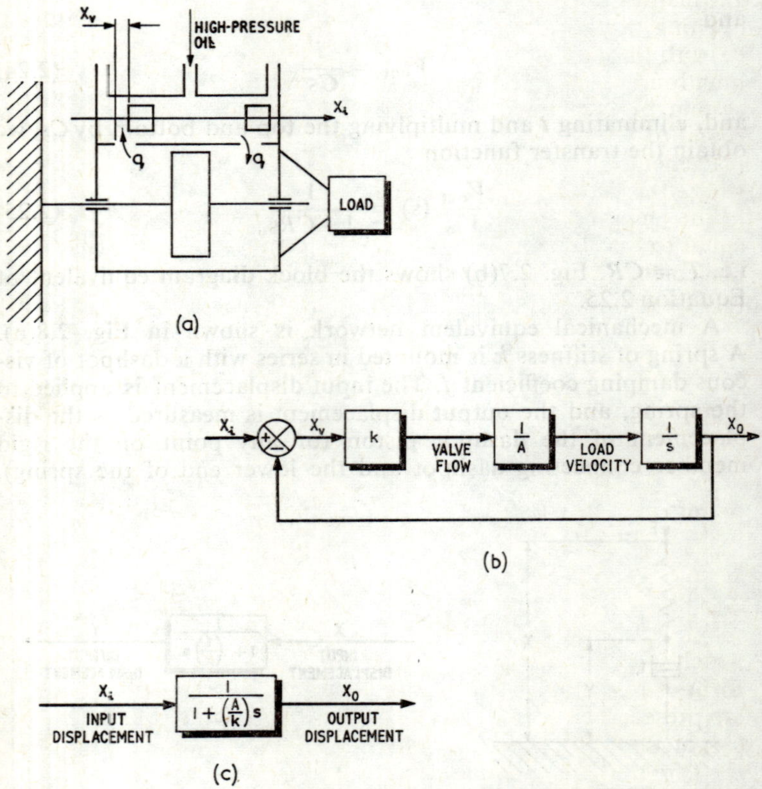


*Fig. 2.9. Simple arrangement of a mechanical-hydraulic servo giving a first order lag: (a) schematic diagram; (b) operational diagram; (c) block diagram of transfer function*

where $T = f/k$. Fig. 2.8(b) shows the block diagram representation of Equation 2.27.

Hydraulic and pneumatic valve-controlled jack-type motors have a *flow* rate which may be assumed proportional to the valve opening for small displacements of the valve and for negligible load and compressibility effects. The valve opening and load movement are therefore related by the equation

$$q = kX_v = AsX_0$$

where $q$ is the valve flow, $k$ is the valve gain (flow per unit opening), $X_v$ is the valve opening, $A$ is the jack cross-sectional area, and $X_0$ is the load displacement. It can be seen that the relationship between load motion and valve opening is the basic integrator property, since the load speed is proportional to the valve opening. If the load displacement is measured and then compared with the desired load displacement, the basic valve-jack combination may be turned into a displacement-displacement control system by demanding a control-valve opening which is proportional to the difference between the desired and actual load displacements. One method of achieving this is shown in Fig. 2.9(a), where the jack piston is earthed and the load is rigidly attached to the jack body. Using this method, the valve opening is made equal to the difference between the desired and actual load displacements. Many other variations are possible. Treating the system as a feedback control system in its own right, we may write

$$\frac{X_0}{X_i}(s) = \frac{G(s)}{1+G(s)H(s)}$$

and

$$H(s) = 1{\cdot}0 \qquad G(s) = \frac{X_0}{X_v} = \frac{k}{As}$$

giving

$$\frac{X_0}{X_i}(s) = \frac{1}{1+\left(\dfrac{A}{k}\right)s} \tag{2.28}$$

which is again a first order lag with $T = A/k$. The minor loop form of block diagram is shown in Fig. 2.9(b), and the first order lag presentation is shown in Fig. 2.9(c). A detailed non-linear analysis of this type of hydraulic servomechanism is included in Chapter 7; in the simple unloaded case the first order lag can often be a reasonable approximation to actual behaviour. There are, of course, a multitude of other elements and systems which can be adequately described by simple first order lags. Space limitations unfortunately preclude the detailed description of any other elements, but the examples chosen should convince the reader that many apparently different systems can be described by similar transfer functions.

## 2.7 Low-frequency Differentiating Networks

A passive *RC* network capable of generating an output signal approximately proportional to the rate of change of the input signal, for slowly changing input signals, is shown in Fig. 2.10(a).

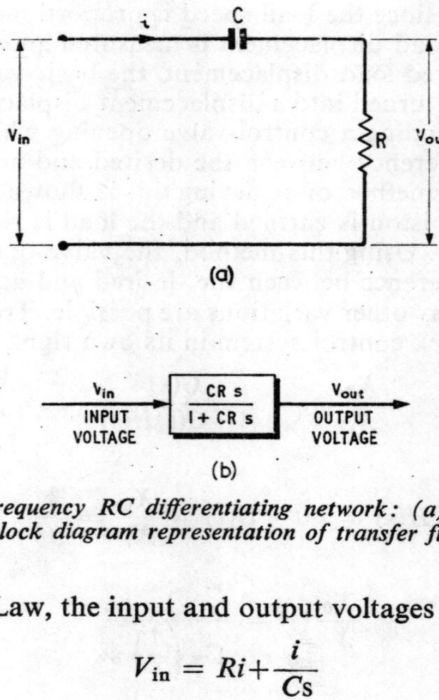


*Fig. 2.10. Low-frequency RC differentiating network: (a) circuit diagram; (b) block diagram representation of transfer function*

Using Ohm's Law, the input and output voltages are

$$V_{\text{in}} = Ri + \frac{i}{Cs} \tag{2.29}$$

and

$$V_{\text{out}} = Ri \tag{2.30}$$

Eliminating the current *i*, and multiplying top and bottom by *Cs*, the transfer function is

$$\frac{V_{\text{out}}}{V_{\text{in}}}(\text{s}) = \frac{CRs}{1+CRs} = \frac{Ts}{1+Ts} \tag{2.31}$$

where $T = CR$. The transfer function block diagram is shown in Fig. 2.10(b). From Equation 2.31 we see that at low frequency ($s \approx 0$), the network transfer function is approximately $V_{out}/V_{in} = Ts$, i.e. the output voltage is approximately the derivative of the input voltage for slowly changing input signals.

The mechanical equivalent network is shown in Fig. 2.11(a). Dashpot and spring are in series, the input displacement is applied

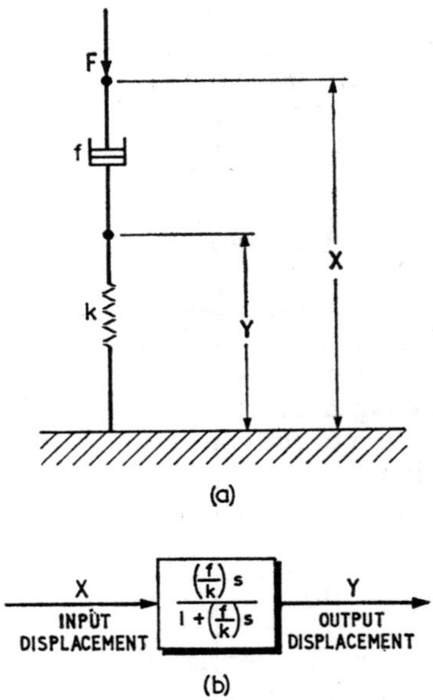


*Fig. 2.11. Low-frequency mechanical differentiating network: (a) schematic diagram; (b) block diagram representation of transfer function*

to the dashpot piston, and the output displacement is obtained from the spring movement. Since the dashpot and spring are in series, the applied force is completely seen by both the dashpot and by the spring. Newton's Law therefore allows us to write

$$F = fs(X - Y) = kY \tag{2.32}$$

and hence, in transfer function form,

$$\frac{Y}{X}(s) = \frac{\left(\frac{f}{k}\right)s}{1 + \left(\frac{f}{k}\right)s} \tag{2.33}$$

i.e. $T = f/k$. A block diagram representation of Equation 2.33 is shown in Fig. 2.11(b), and by inspection of Equation 2.33 we again see that at low frequencies a signal proportional to the input derivative is generated. Both electrical and mechanical low-frequency differentiating networks can be used to shape the signals in various parts of feedback control systems in order to provide the necessary compensation for meeting system performance specifications. Discussions on the use of these and other networks in the design process will be the subject of Chapters 4, 5 and 6, especially Chapter 6. Once the compensation network transfer function required has been determined, the choice of the mechanism for generating the compensation network, i.e. electrical, mechanical, hydraulic, etc., will be a matter of convenience.

## 2.8 Conventional Phase Lag Networks

Two very common compensation networks often used in the series compensation mode are conventional phase lag and conventional phase lead networks. The detailed application of the conventional phase lag network is dealt with in Section 6.9, and the detailed application of the conventional phase lead network is dealt with in Section 6.10. We shall define the conventional lag network to have the transfer function

$$\frac{\theta_2}{\theta_1}(s) = \frac{1+Ts}{1+\alpha Ts} \tag{2.34}$$

where $\alpha > 1{\cdot}0$. At low frequencies the steady state amplitude ratio is unity, whilst at high frequencies the steady state amplitude ratio is $1/\alpha$. At very low frequencies and at very high frequencies the phase lag of the output waveform behind the input waveform will be zero, but at intermediate frequencies the output waveform will lag considerably behind the input waveform, hence the expression 'lag network'. Other transfer functions with different lag characteristics do exist, but the form of Equation 2.34 is the most commonly used lag network.

Fig. 2.12(a) shows an *RC* lag network. Using Ohm's Law, the differential equations describing behaviour are

$$V_{\text{in}} = R_1 i + R_2 i + \frac{i}{C_2 s} \tag{2.35}$$

and

$$V_{\text{out}} = R_2 i + \frac{i}{C_2 s} \tag{2.36}$$

Eliminating $i$ and multiplying the top and bottom by $C_2s$, we have the transfer function

$$\frac{V_{out}}{V_{in}}(s) = \frac{1+R_2C_2s}{1+(R_1+R_2)C_2s} \tag{2.37}$$

Hence $T = R_2C_2$ and $\alpha = (R_1+R_2)/R_2$. Fig. 2.12(b) shows the transfer function block diagram representation.

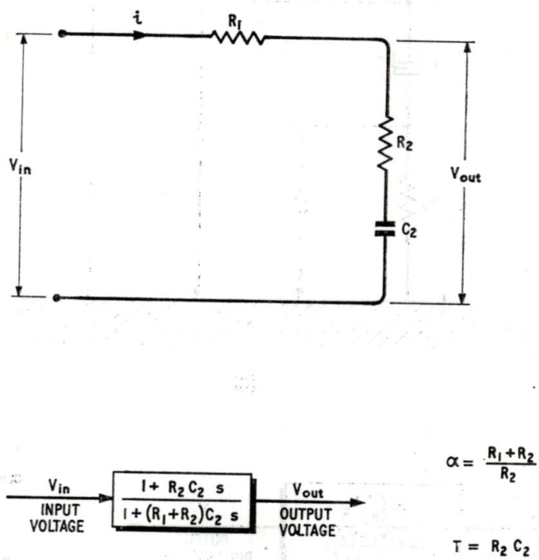


*Fig. 2.12. Conventional phase lag RC network: (a) circuit diagram; (b) block diagram of transfer function*

A mechanical lag network is shown in Fig. 2.13(a). There are two springs and a dashpot in series, the dashpot being situated between the two springs as shown. The input signal is applied to the first spring, and the output signal is the displacement of the lower end of the same spring. Since the springs and dashpot are in series, the same force will be transmitted through each. It is necessary to consider the displacement at the dashpot body. Let this be $X'$, then from Newton's Law we may write

$$F = k_1(X-Y) = fs(Y-X') = k_2X' \tag{2.38}$$

Hence

$$k_1(X-Y) = fs\left[Y-\frac{k_1(X-Y)}{k_2}\right]$$

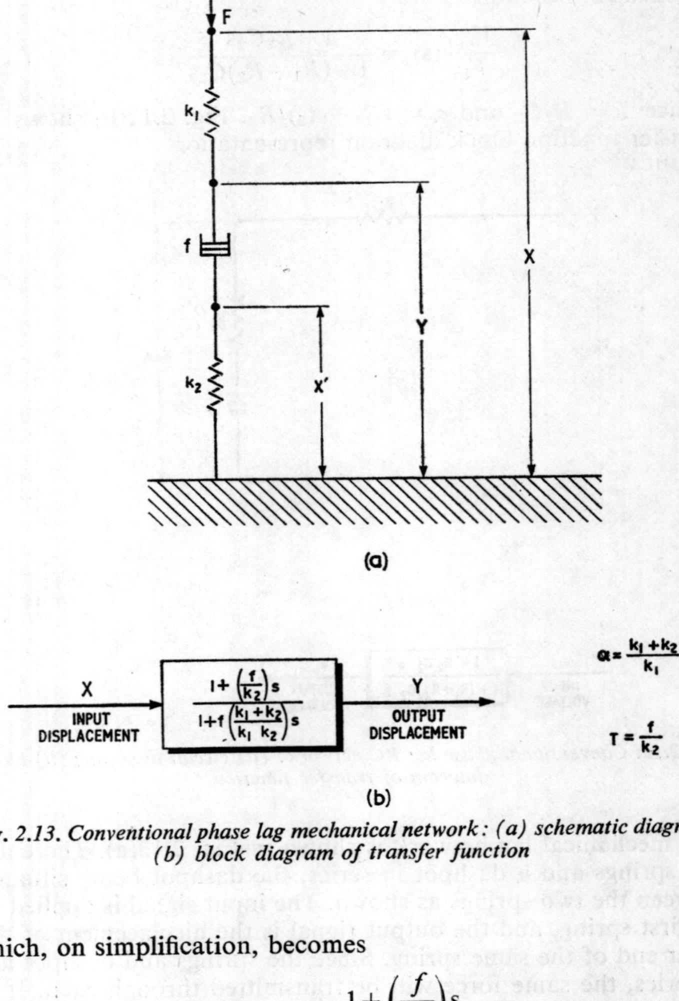


*Fig. 2.13. Conventional phase lag mechanical network: (a) schematic diagram; (b) block diagram of transfer function*

which, on simplification, becomes

$$\frac{Y}{X}(s) = \frac{1+\left(\dfrac{f}{k_2}\right)s}{1+f\left(\dfrac{k_1+k_2}{k_1k_2}\right)s} \tag{2.39}$$

i.e. $T = f/k_2$, and $\alpha = (k_1+k_2)/k_1$. The block diagram representation is shown in Fig. 2.13(b).

## 2.9 Conventional Phase Lead Network

The conventional phase lead network transfer function is defined by

$$\frac{\theta_2}{\theta_1}(s) = \frac{1}{\alpha}\left(\frac{1+\alpha Ts}{1+Ts}\right) \tag{2.40}$$

and it will be seen from Equation 2.40 that at low frequencies the

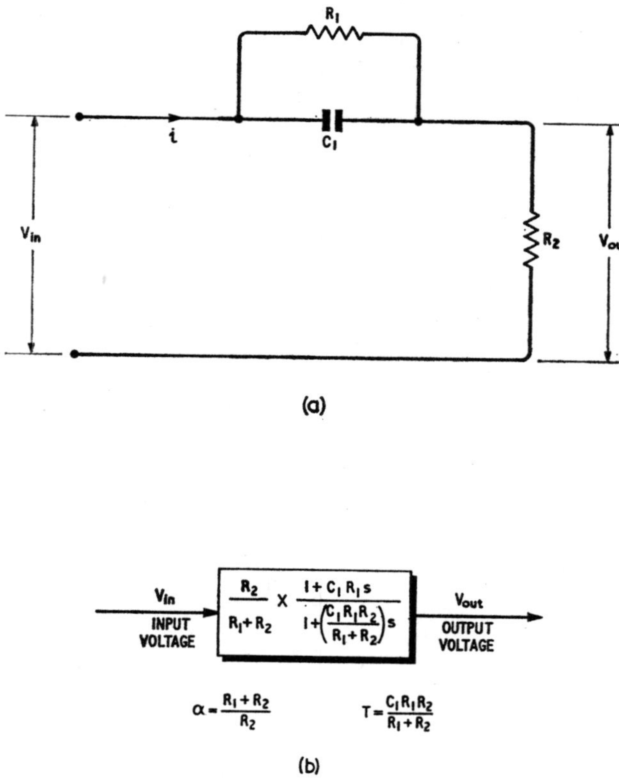


*Fig. 2.14. Conventional phase lead RC network: (a) circuit diagram; (b) block diagram representation of transfer function*

amplitude ratio will be $1/\alpha$, where $\alpha > 1{\cdot}0$, and at high frequencies the amplitude ratio will be unity. At very low frequencies and at very high frequencies the phase lag between input and output waveform will be zero, but at intermediate frequencies the output waveform will lead the input waveform, thus leading to the description of the phase lead network; although there are many networks with phase lead characteristics, the form shown in Equation 2.40 is the most common. Since the d.c. gain is $1/\alpha$, the feedforward gain in a control system should be increased by a factor $\alpha$ to compensate for the introduction of the conventional phase lead network.

Fig. 2.14(a) shows a passive *RC* phase lead network. Using Ohm's Law, the output and input voltages may be derived as

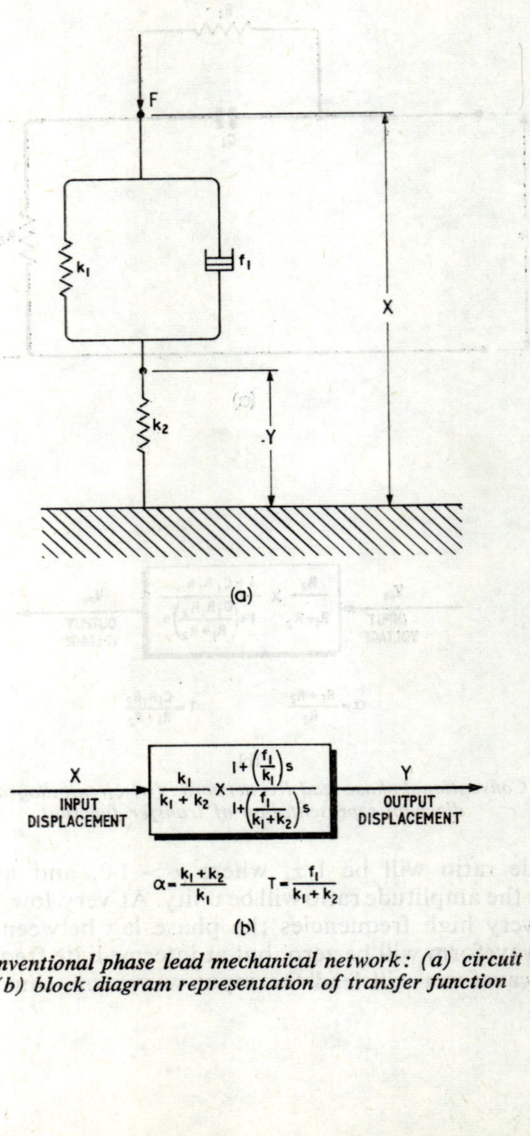


$$\frac{k_1}{k_1 + k_2} \times \frac{1 + \left(\frac{f_1}{k_1}\right)s}{1 + \left(\frac{f_1}{k_1 + k_2}\right)s}$$

$$\alpha = \frac{k_1 + k_2}{k_1} \qquad T = \frac{f_1}{k_1 + k_2}$$

*Fig. 2.15. Conventional phase lead mechanical network: (a) circuit diagram; (b) block diagram representation of transfer function*

follows

$$V_{\text{in}} = iR_2 + \frac{i}{1/R_1 + C_1 s} \tag{2.41}$$

and

$$V_{\text{out}} = iR_2 \tag{2.42}$$

Eliminating $i$ between Equations 2.41 and 2.42, and multiplying numerator and denominator by $1/R_1 + C_1 s$, we obtain

$$\frac{V_{\text{out}}}{V_{\text{in}}}(s) = \frac{R_2/R_1 + R_2 C_1 s}{1 + R_2/R_1 + R_2 C_1 s}$$

and, on simplification,

$$\frac{V_{\text{out}}}{V_{\text{in}}}(s) = \frac{R_2}{R_1 + R_2} \times \frac{1 + C_1 R_1 s}{1 + \left(\dfrac{C_1 R_1 R_2}{R_1 + R_2}\right) s} \tag{2.43}$$

hence $\alpha = (R_1 + R_2)/R_2$, and $T = C_1 R_1 R_2/(R_1 + R_2)$. Fig. 2.14(b) shows the transfer function block diagram representation.

The equivalent mechanical phase lead network is shown in Fig. 2.15(a). A spring and dashpot are connected in parallel, and another spring joins this combination to earth. Input displacement is applied to the free end of the spring and dashpot combination, whilst the output displacement is taken off the earthed spring as shown. The applied force is shared between the first spring and dashpot, but is wholly transmitted to earth through the second spring. Applying Newton's Law we may therefore write

$$F = k_1(X - Y) + f_1 s(X - Y) = k_2 Y \tag{2.44}$$

Therefore

$$k_2 Y + k_1 Y + f_1 s Y = f_1 s X + k_1 X$$

and the required transfer function is

$$\frac{Y}{X}(s) = \frac{k_1 + f_1 s}{(k_1 + k_2) + f_1 s} = \frac{k_1}{k_1 + k_2} \times \frac{1 + \left(\dfrac{f_1}{k_1}\right) s}{1 + \left(\dfrac{f_1}{k_1 + k_2}\right) s} \tag{2.45}$$

thus giving $\alpha = (k_1 + k_2)/k_1$ and $T = f_1/(k_1 + k_2)$, with the appropriate transfer function block diagram shown in Fig. 2.15(b).

## 2.10 *RCL* Network with Complex Zeros

In Chapter 6 we shall see that, under certain circumstances, it may be required to generate a compensation network with complex zeros. When the feedforward transfer function is derived for the complete control system, the complex zeros will cancel out the effect of unwanted complex poles emanating from the load characteristics. Such compensation transfer functions with complex zeros may be obtained using either *RC* networks or *RCL* networks. *RC* networks can only be used if real network poles are acceptable, whereas, by judicious choice of component values, *RCL* networks can have complex zeros but either complex poles or real poles as desired. A detailed discussion of the synthesis of compensation networks is outside the scope of this book, and the

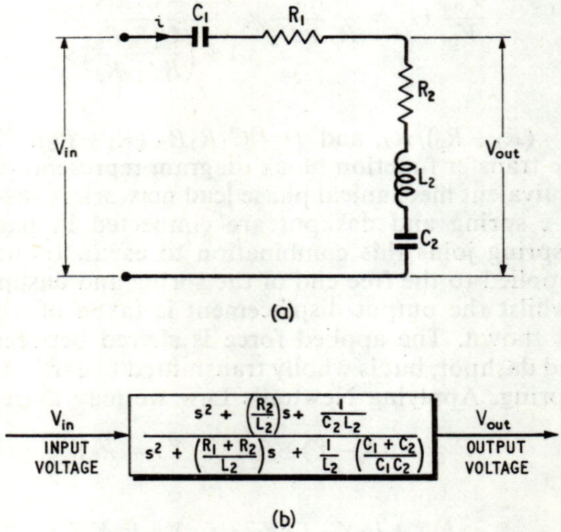


$$\frac{s^2 + \left(\frac{R_2}{L_2}\right)s + \frac{1}{C_2 L_2}}{s^2 + \left(\frac{R_1 + R_2}{L_2}\right)s + \frac{1}{L_2}\left(\frac{C_1 + C_2}{C_1 C_2}\right)}$$

*Fig. 2.16. RCL network which can generate complex network zeros as required in the cancellation compensation technique: (a) circuit diagram; (b) block diagram representation of transfer function*

interested reader is referred to Truxal[23] for detailed accounts of the use of Guillemin's procedure for synthesising *RC* ladder networks in parallel, and Dasher's procedure for synthesising *RC* connected bridge twin-T structures. Both procedures lead to the design of *RC* networks with complex zeros.

As an example of the analysis of *RCL* networks, consider the network shown in Fig. 2.16(a). Applying Ohm's Law, we have the input and output voltages defined by

$$V_{\text{in}} = \frac{i}{C_1 s} + R_1 i + V_{\text{out}} \tag{2.46}$$

and

$$V_{\text{out}} = R_2 i + L_2 s i + \frac{i}{C_2 s} \tag{2.47}$$

Hence

$$\begin{aligned} V_{\text{in}} &= \frac{i}{C_1 s} + R_1 i + R_2 i + L_2 s i + \frac{i}{C_2 s} \\ &= (R_1 + R_2) i + L_2 s i + \frac{i}{s}\left(\frac{1}{C_1} + \frac{1}{C_2}\right) \end{aligned}$$

Therefore

$$\frac{V_{\text{out}}}{V_{\text{in}}}(s) = \frac{R_2 + L_2 s + \dfrac{1}{C_2 s}}{(R_1 + R_2) + L_2 s + \dfrac{1}{s}\left(\dfrac{1}{C_1} + \dfrac{1}{C_2}\right)}$$

and, in simplified form,

$$\frac{V_{\text{out}}}{V_{\text{in}}}(s) = \frac{s^2 + \left(\dfrac{R_2}{L_2}\right) s + \dfrac{1}{C_2 L_2}}{s^2 + \left(\dfrac{R_1 + R_2}{L_2}\right) s + \dfrac{1}{L_2}\left(\dfrac{C_1 + C_2}{C_1 C_2}\right)} \tag{2.48}$$

The block diagram form of Equation 2.48 is shown in Fig. 2.16(b). In standard form, we may write

$$\frac{V_{\text{out}}}{V_{\text{in}}}(s) = \frac{s^2 + 2\zeta_1 \omega_{n1} s + \omega_{n1}^2}{s^2 + 2\zeta_2 \omega_{n2} s + \omega_{n2}^2} \tag{2.49}$$

where

$$\omega_{n1} = \sqrt{\frac{1}{C_2 L_2}} \qquad \omega_{n2} = \sqrt{\frac{1}{L_2}\left(\frac{C_1 + C_2}{C_1 C_2}\right)}$$

$$\zeta_1 = \frac{R_2}{2 L_2 \omega_{n1}} \qquad \zeta_2 = \frac{R_1 + R_2}{2 L_2 \omega_{n2}}$$

By judicious choice of component values, $\zeta_1$ and $\zeta_2$ may be real or complex. In an example quoted by Truxal, the following values are used: $C_1 = \frac{1}{399}$ F, $C_2 = \frac{1}{2}$ F, $R_1 = 38\ \Omega$, $R_2 = 2\ \Omega$ and $L_2 = 1$ H. Equation 2.49 then becomes

$$\frac{V_{\text{out}}}{V_{\text{in}}}(s) = \frac{s^2 + 2s + 2}{s^2 + 40s + 401} \tag{2.50}$$

giving

$$\omega_{n1} = 1{\cdot}414 \qquad \omega_{n2} \simeq 20$$

$$\zeta_1 = 0{\cdot}707 \qquad \zeta_2 \simeq 1{\cdot}0$$

With these component values we see that the compensation network zeros are complex, since $\zeta_1 < 1{\cdot}0$, whilst the compensation network poles are real, since $\zeta_2 \simeq 1{\cdot}0$. If $R_1$ is now reduced from 38 Ω to 18 Ω, the compensation network transfer function becomes

$$\frac{V_{\text{out}}}{V_{\text{in}}}(\text{s}) = \frac{\text{s}^2+2\text{s}+2}{\text{s}^2+20\text{s}+401} \tag{2.51}$$

Then

$$\omega_{n1} = 1{\cdot}414 \qquad \omega_{n2} \simeq 20$$

$$\zeta_1 = 0{\cdot}707 \qquad \zeta_2 \simeq 0{\cdot}50$$

thus resulting in a compensation network with complex zeros and complex poles which might be considered useful in specific circumstances. In both Equations 2.50 and 2.51 the d.c. gain is 2/401, and thus the feedforward gain of the control system would have to be increased by this amount to compensate for the introduction of the compensation network.

## 2.11 D.C. Generators

D.C. generators are frequently found in control systems, and are used as power amplifiers and for measurement of velocity for use in feedback compensation methods. A typical d.c. generator is shown in Fig. 2.17(a). The voltage $V_g$ generated by the generator is proportional to the speed of rotation $n$ of the armature and to the flux density $\phi$; thus we may write

$$V_g(\text{s}) = K_1 n\phi \tag{2.52}$$

The flux density is proportional, in the linear operating region, to the field current $i_f$; hence

$$\phi = K_2 i_f \tag{2.53}$$

Combining Equations 2.52 and 2.53, the generator voltage may be written

$$V_g(\text{s}) = K_g i_f \tag{2.54}$$

where $K_g = K_1K_2n$. At this stage we assume that the armature speed $n$ is constant. Applying Ohm's Law to the field circuit,

$$V_f(\text{s}) = R_f i_f + L_f \text{s} i_f$$

Therefore

$$V_f(\text{s}) = i_f(R_f + L_f\text{s}) \tag{2.55}$$

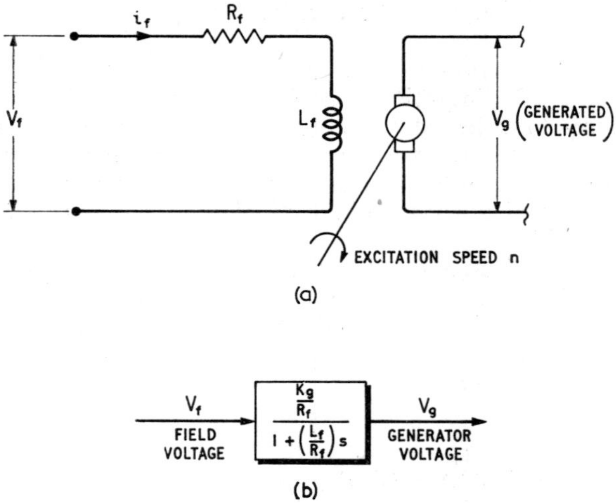


*Fig. 2.17. D.C. generator: (a) circuit diagram; (b) block diagram representation of transfer function*

Eliminating field current $i_f$ between Equations 2.54 and 2.55 we obtain the transfer function relating generated voltage to applied voltage,

$$\frac{V_g}{V_f}(\mathrm{s}) = \frac{K_g/R_f}{1+\left(\frac{L_f}{R_f}\right)\mathrm{s}} \tag{2.56}$$

i.e. the transfer function is a simple first order lag multiplied by a linear gain term. Note that $V_g$ is not the voltage available to drive the load, as there will be a voltage drop between generator and load. Considerable power amplification is possible using the d.c. generator since the field power input is small. When used as an amplifier, the d.c. generator is termed a rotating amplifier, and there are several commercial variations, particularly the Amplidyne and the Ward Leonard drive. A block diagram of Equation 2.56 is shown in Fig. 2.17(b).

If the field voltage is kept constant, then the generator voltage is defined by Equation 2.52, where $\phi$ is now a constant, and $n$, the shaft speed, will be $\mathrm{d}\theta_0/\mathrm{d}t$. We may therefore write

$$\frac{V_g}{\mathrm{s}\theta_0}(\mathrm{s}) = K_3 \tag{2.57}$$

where $K_3 = K_1\phi$. When used in this fashion the d.c. generator is termed a tachogenerator.

## 2.12 Armature-controlled D.C. Servomotor

Of the many types of a.c. and d.c. motors found in control systems, space limitations restrict us to an examination of the armature-controlled d.c. servomotor only. Further information can be obtained from a variety of sources, one comprehensive review of d.c. motors, a.c. motors, and tachogenerators is given by Davis and Ledgerwood in Section 12 of the *Control Engineers' Handbook*.[22]

The armature-controlled d.c. servomotor is shown in Fig. 2.18(a). Field current is constant, and the torque developed by the motor is proportional to the armature current. The total moment of inertia referred to the motor is $J$, and there is viscous damping $f$ also referred to the motor. Back e.m.f. is induced in

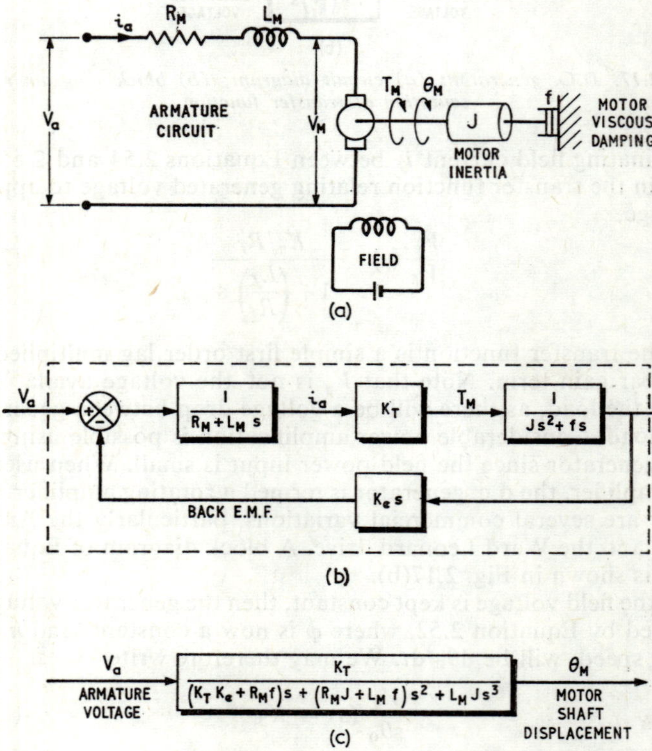


*Fig. 2.18. Armature-controlled d.c. servomotor: (a) circuit diagram; (b) function block diagram; (c) block diagram representation of transfer function*

the armature circuit with sign opposite to the control voltage and magnitude proportional to the speed of rotation of the armature. The required transfer function is that relating motor shaft position to applied armature voltage. Back e.m.f. is then

$$V_M = K_e s\theta_M \tag{2.58}$$

the motor torque equation is

$$T_M = K_T i_a = Js^2\theta_M + fs\theta_M \tag{2.59}$$

and, applying Ohm's Law to the armature circuit,

$$V_a = V_M + R_M i_a + L_M s i_a \tag{2.60}$$

If Equation 2.59 is rewritten as

$$\frac{T_M}{i_a} = K_T \qquad \frac{\theta_M}{T_M} = \frac{1}{Js^2+fs}$$

and Equation 2.60 is rewritten as

$$\frac{i_a}{V_a - V_M} = \frac{1}{R_M + L_M s}$$

then the simple block diagram of Fig. 2.18(b) may be drawn for the armature-controlled d.c. motor.

The required transfer function may now be written down from the relationship

$$\frac{\theta_M}{V_a}(s) = \frac{\text{feedforward T.F.}}{1+\text{feedforward T.F.}\times\text{feedback T.F.}}$$

$$= \frac{\dfrac{K_T}{(R_M+L_M s)(Js^2+fs)}}{1+\dfrac{K_T K_e s}{(R_M+L_M s)(Js^2+fs)}}$$

which in simplified form is

$$\frac{\theta_M}{V_a}(s) = \frac{K_T}{(K_T K_e + R_M f)s + (R_M J + L_M f)s^2 + L_M J s^3} \tag{2.61}$$

and the block diagram representation of the transfer function of Equation 2.61 is shown in Fig. 2.18(c).

## 2.13 Ward Leonard System

The Ward Leonard system is often used as the basis of an all-electric d.c. control system.[24] It consists of a d.c. generator and an armature-controlled d.c. motor, the generator voltage being applied to the armature circuit of the motor. In this context the d.c. generator acts as a rotating power amplifier which supplies the power which, in turn, drives the prime mover and hence the load. Fig. 2.19(a) shows the Ward Leonard system. Since we have already derived the individual transfer functions, we may derive the

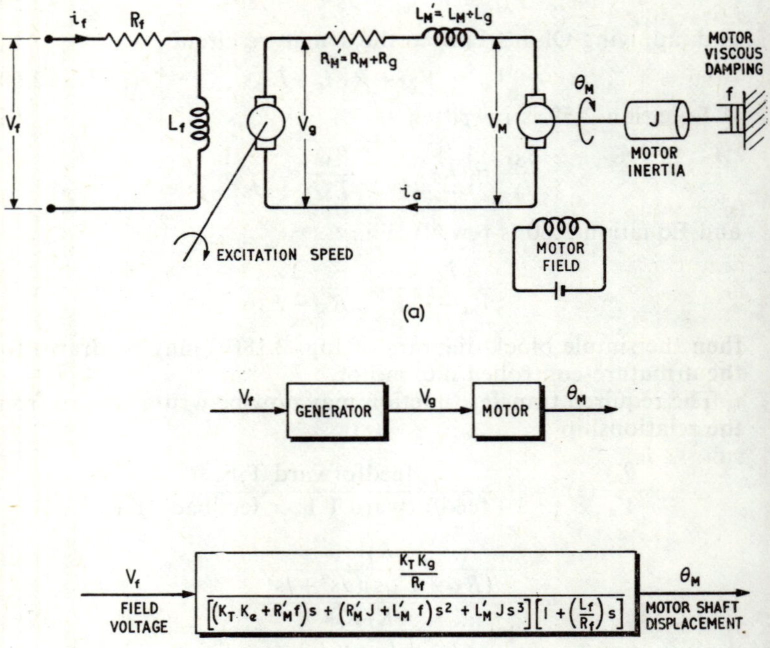


*Fig. 2.19. Ward Leonard system: (a) circuit diagram; (b) block diagram representation of transfer function*

overall transfer function relating motor displacement to the generator control voltage $V_f$ by multiplying the two transfer functions together. The armature inductance and resistance of the d.c. motor should then be modified to include any inductance or resistance due to the generator, but on the motor side of the generator. The overall transfer function is

$$\frac{\theta_M}{V_f}(s) = \frac{\theta_M}{V_g}(s) \times \frac{V_g}{V_f}(s)$$

Therefore

$$\frac{\theta_M}{V_f}(s) = \frac{K_T K_g / R_f}{[(K_T K_e + R'_M f)s + (R'_M J + L'_M f)s^2 + L'_M J s^3]\left[1 + \left(\frac{L_f}{R_f}\right)s\right]} \tag{2.62}$$

where $R'_M$ and $L'_M$ indicate that the motor armature resistance and inductance have been modified to allow for the effect of the motor side of the generator. The block diagram representation of Equation 2.62 is shown in Fig. 2.19(b).

## 2.14 Electronic Amplifier

In most control systems it is necessary to amplify the feedforward path signal before the signal is capable of controlling the motor, or main power amplifier. There are many types of amplifiers, including thermionic, solid state, pneumatic and hydraulic. The choice of a basic form of amplifier is to some extent a matter of convenience, depending often on the availability of stabilised power supplies. As a typical example, consider the thermionic triode amplifier shown in Fig. 2.20(a), and the associated small-signal equivalent circuit shown in Fig. 2.20(b). From the equivalent circuit we may use Ohm's Law to write down the input and output

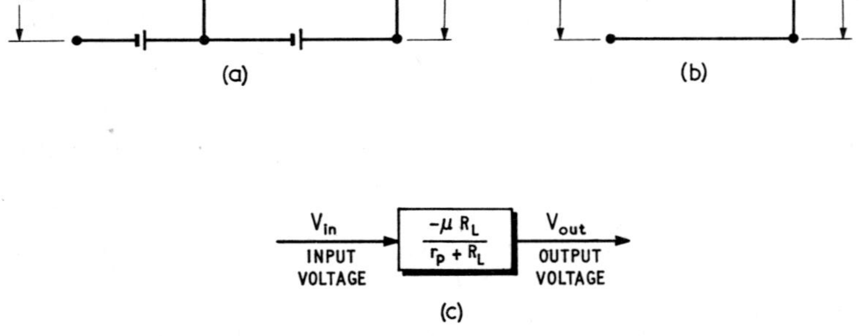


*Fig. 2.20. Simple electronic amplifiers: (a) circuit diagram; (b) equivalen circuit; (c) block diagram representation of transfer function*

voltages:

$$-\mu V_{in} = i(r_p+R_L) \tag{2.63}$$

and

$$V_{out} = iR_L \tag{2.64}$$

where $\mu$ is termed the amplification factor, $r_p$ is the dynamic plate resistance of the thermionic tube, and $R_L$ is the load resistance, assumed to be ohmic.

Eliminating $i$ from Equations 2.63 and 2.64, we obtain the amplifier transfer function

$$\frac{V_{out}}{V_{in}}(s) = \frac{-\mu R_L}{r_p+R_L} \tag{2.65}$$

The minus sign of Equation 2.65 is typical of a large class of thermionic amplifiers. Block diagram representation of the amplifier transfer function is shown in Fig. 2.20(c).

## 2.15 Hydraulic Motor and Pump Unit

A constant-speed variable-delivery hydraulic pump, feeding a variable-speed hydraulic motor, which is a pump running in reverse, is often used as the drive unit of machine tools. A functional diagram of the motor-pump unit is shown in Fig. 2.21(a), and the block diagram symbolically showing the mode of operation in Fig. 2.21(b). The pump output rate may be considered proportional to the angle of the controlling swashplate, hence

$$Q_p = K_p\theta_p \tag{2.66}$$

There are three forms of flow in the motor: $Q_M$ the swept volume flow, which is proportional to the output speed, $Q_L$ the leakage flow in the motor, and $Q_c$ the compressibility flow, which arises because in a high-performance unit the oil cannot be considered incompressible. Flow considerations therefore allow us to write

$$Q_p = Q_M+Q_L+Q_c \tag{2.67}$$

and, by definition,

$$Q_M = V_M s\theta_0 \tag{2.68}$$

---

*Fig. 2.21. Hydraulic pump and motor unit for fluid power transmission: (a) functional diagram of hydraulic pump and motor power transmission unit; (b) block diagram of operation of hydraulic pump and motor power transmission unit; (c) flow diagram showing relationship between various flows and transmission output-input transfer function; (d) block diagram representation of transmission unit transfer function; (e) incorporation of hydraulic transmission unit in a position control system*

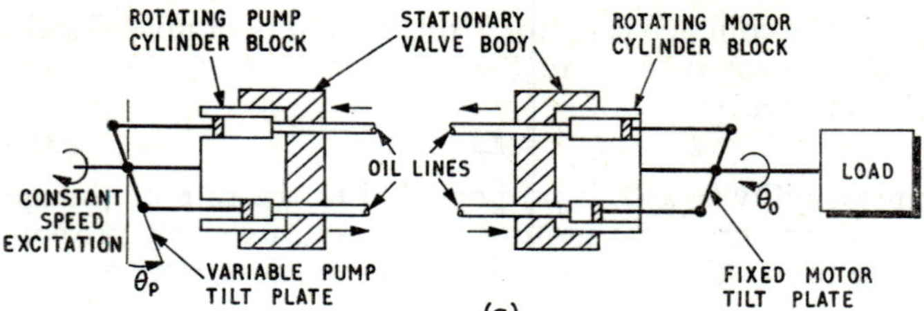
ROTATING PUMP CYLINDER BLOCK
STATIONARY VALVE BODY
ROTATING MOTOR CYLINDER BLOCK
OIL LINES
LOAD
CONSTANT SPEED EXCITATION
$\theta_P$
VARIABLE PUMP TILT PLATE
$\theta_0$
FIXED MOTOR TILT PLATE

(a)

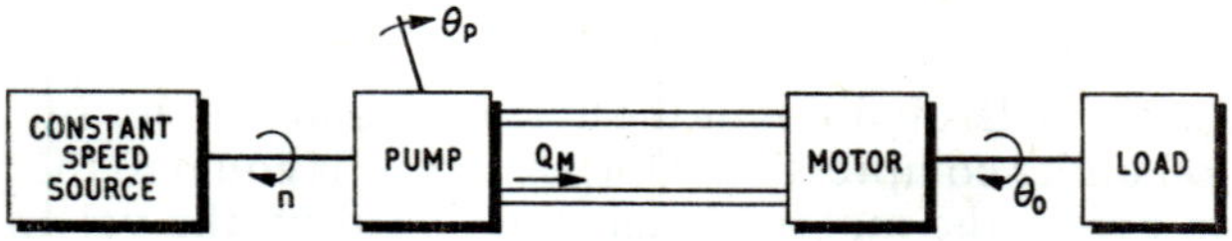
$\theta_P$
CONSTANT SPEED SOURCE
n
PUMP
$Q_M$
MOTOR
$\theta_0$
LOAD

(b)

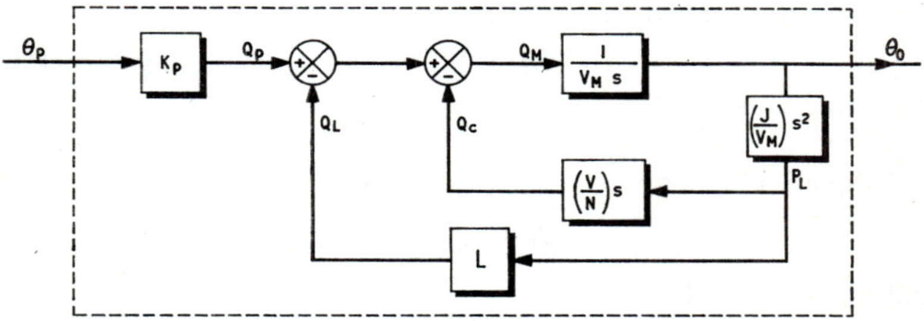
$\theta_P$
$K_P$
$Q_P$
$Q_L$
$Q_C$
$Q_M$
$\frac{1}{V_M s}$
$\theta_0$
$\left(\frac{J}{V_M}\right)s^2$
$\left(\frac{V}{N}\right)s$
$P_L$
L

(c)

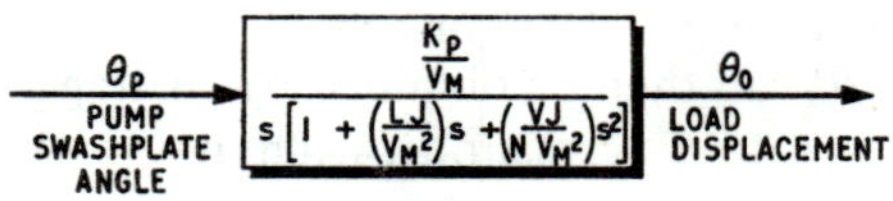
$\theta_P$
PUMP SWASHPLATE ANGLE
$\frac{\frac{K_P}{V_M}}{s\left[1+\left(\frac{LJ}{V_M^2}\right)s+\left(\frac{VJ}{N V_M^2}\right)s^2\right]}$
$\theta_0$
LOAD DISPLACEMENT

(d)

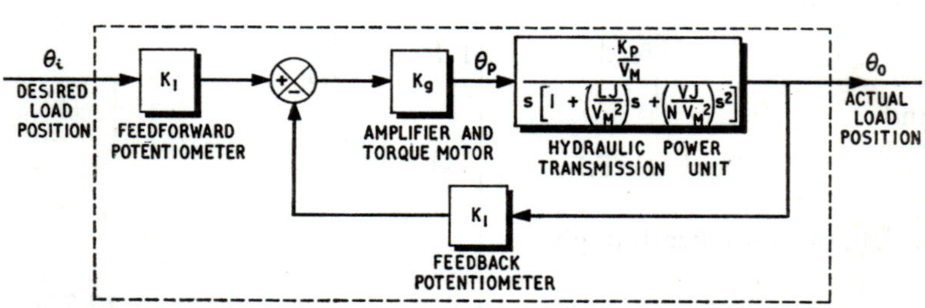
$\theta_i$
DESIRED LOAD POSITION
$K_1$
FEEDFORWARD POTENTIOMETER
$K_g$
AMPLIFIER AND TORQUE MOTOR
$\theta_P$
$\frac{\frac{K_P}{V_M}}{s\left[1+\left(\frac{LJ}{V_M^2}\right)s+\left(\frac{VJ}{N V_M^2}\right)s^2\right]}$
HYDRAULIC POWER TRANSMISSION UNIT
$\theta_0$
ACTUAL LOAD POSITION
$K_1$
FEEDBACK POTENTIOMETER

(e)

where $V_M$ is the volumetric motor displacement per radian. Leakage flow $Q_L$ is proportional to the pressure drop $P_L$ across the motor pistons, hence

$$Q_L = LP_L \tag{2.69}$$

The compressibility flow $Q_c$ is proportional to the rate of change of pressure drop across the motor pistons, and inversely proportional to the total volume of oil under compression. Hence

$$Q_c = \left(\frac{V}{N}\right) sP_L \tag{2.70}$$

where $N$ is the bulk modulus of the fluid used and $V$ is the total volume of oil under compression. Output torque is equal to the pressure drop across the motor pistons multiplied by the volumetric motor displacement per revolution, and if the load is purely inertia then the output torque will be equal to moment of inertia multiplied by the angular acceleration; thus

$$P_L V_M = Js^2\theta_0 \tag{2.71}$$

Rewriting Equation 2.67 as

$$Q_M = Q_p - Q_c - Q_L$$

and Equation 2.68 as

$$\theta_0 = \frac{Q_M}{V_M s}$$

and Equation 2.71 as

$$P_L = \left(\frac{J}{V_M}\right) s^2\theta_0$$

the flow diagram of Fig. 2.21(c) may be drawn. The relationship between $Q_p$ and $\theta_0$ can then be obtained by writing

$$\frac{\theta_0}{Q_p}(s) = \frac{\text{feedforward T.F.}}{1+\text{feedforward T.F.} \times \text{ feedback T.F.}}$$

where

$$\text{feedforward T.F.} = \frac{1}{V_M s}$$

and

$$\text{feedback T.F.} = \left(\frac{J}{V_M}\right) s^2 \left[\left(\frac{V}{N}\right) s + L\right]$$

which, on simplification, gives

$$\frac{\theta_0}{Q_p}(s) = \frac{1}{s\left[V_M + \left(\frac{LJ}{V_M}\right) s + \left(\frac{VJ}{NV_M}\right) s^2\right]} \tag{2.72}$$

Now,

$$\frac{\theta_0}{\theta_p}(s) = \frac{\theta_0}{Q_p}(s) \times \frac{Q_p}{\theta_p}(s)$$

and, in simplest form,

$$\frac{\theta_0}{\theta_p}(s) = \frac{K_p/V_M}{s\left[1+\left(\frac{LJ}{V_M^2}\right)s+\left(\frac{VJ}{NV_M^2}\right)s^2\right]} \qquad (2.73)$$

which is shown in transfer function block diagram form in Fig. 2.21(d). If compressibility and leakage may be neglected, then Equation 2.73 becomes simply

$$\frac{\theta_0}{\theta_p}(s) = \frac{K_p/V_M}{s} \qquad (2.74)$$

and the motor-pump unit is simply a basic integrator. To incorporate the motor-pump combination in a position control servomechanism, one method would be to convert the desired shaft position into a voltage using a potentiometer, measure the output displacement using a potentiometer with the same gain, measured in volts per radian, and subtract output voltage at the comparator. The error signal thus formed can be used as the basis of a proportional control system in which we move the pump swashplate a distance proportional to the error signal. A torque motor, with limited output stroke, can be used to operate the swashplate, and the error signal can be amplified to the level required to operate the torque motor using a thermionic amplifier of the type discussed in Section 2.14. If the amplifier and torque motor may be considered linear over the operating range of interest, and also to have a frequency response which is essentially flat in the operating range of interest, then we may represent amplifier and torque motor by a simple linear gain $K_g$. The block diagram of the complete position control system is shown in Fig. 2.21(e), and the system transfer function is

$$\frac{\theta_0}{\theta_i}(s) = \frac{G_p(s)G(s)}{1+G(s)H(s)}$$

where

$$G_p(s) = \text{potentiometer gain } K_1$$

$$H(s) = \text{potentiometer gain } K_1$$

and

$$G(s) = \frac{K_g K_p/V_M}{s\left[1+\left(\frac{LJ}{V_M^2}\right)s+\left(\frac{VJ}{NV_M^2}\right)s^2\right]}$$

$$= \frac{1}{1+\left(\frac{V_M}{K_g K_1 K_p}\right)s+\left(\frac{LJ}{K_g K_1 K_p V_M}\right)s^2+\left(\frac{VJ}{K_g K_1 K_p V_M}\right)s^3} \tag{2.75}$$

In the standard polynomial form of Chapter 1, we have

$$a_0 = b_0 = 1$$

$$a_1 = \frac{V_M}{K_g K_1 K_p} \qquad a_2 = \frac{LJ}{K_g K_1 K_p V_M} \qquad a_3 = \frac{VJ}{K_g K_1 K_p V_M}$$

Note that if the input potentiometer is omitted, the system will no longer display the one-to-one d.c. relationship. The negative sign which was encountered in Equation 2.65 in describing the thermionic amplifier is of no significance in this problem provided the terminals are connected correctly.

## 2.16 Accelerometer

It is often necessary to measure the acceleration of a body in space to provide either a reference signal for navigation purposes or a compensation signal for feedback compensation. Fig. 2.22(a) shows the basic linear accelerometer which is used to measure the acceleration in a particular direction. The frame is rigidly attached to the missile vehicle or machine tool table whose acceleration is to be measured, and the displacement of the mass $M$ relative to the frame is a measure of the acceleration of the body to which the

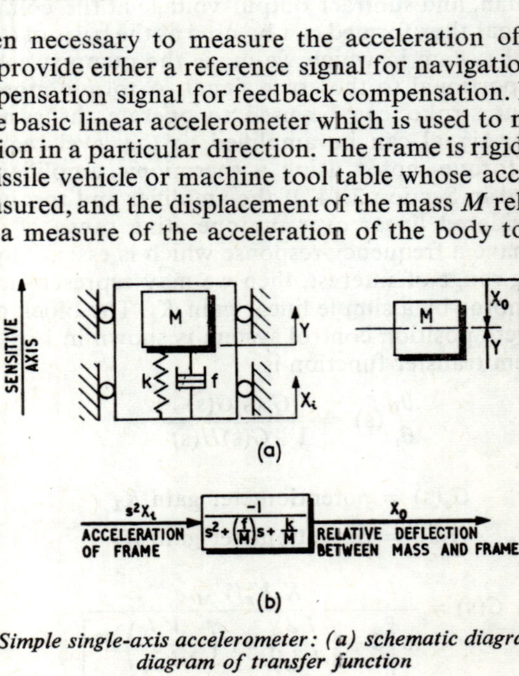


*Fig. 2.22. Simple single-axis accelerometer: (a) schematic diagram; (b) block diagram of transfer function*

frame is attached. Let $Y$ = motion of the mass $M$ in space, let $X_i$ = motion of the frame in space, and let $X_0 = Y - X_i$ = motion of the mass relative to the frame. $X_0$ is the quantity which is easily measured. The compression of the spring is $k(X_i - Y)$, and applying Newton's Law to the mass $M$, we have mass × acceleration = nett force; i.e.

$$Ms^2Y = k(X_i - Y) + fs(X_i - Y) \tag{2.76}$$

Eliminating $Y$ by using the relationship $Y = X_0 + X_i$, we obtain

$$Ms^2X_0 + Ms^2X_i = -kX_0 - fsX_0$$

and, transposing and dividing by $M$, we have the accelerometer transfer function

$$\frac{X_0}{s^2X_i}(s) = -\frac{1}{s^2 + \left(\frac{f}{M}\right)s + \frac{k}{M}} \tag{2.77}$$

i.e. at low frequencies the relative displacement $X_0$ between mass and frame is proportional to the acceleration of the frame, since we may write for $s \simeq 0$

$$X_0 = -\frac{M}{k}(s^2X_i)$$

The relative displacement is of the opposite sign to the frame acceleration, as would be expected, due to mass displacement lagging behind the frame displacement as the frame is accelerated. Equation 2.77 may be written in standard form

$$\frac{X_0}{s^2X_i}(s) = -\frac{1}{s^2 + 2\zeta\omega_n s + \omega_n^2} \tag{2.78}$$

and for accurate following of the frame acceleration, $\omega_n$ should be as high as possible, and $\zeta$ should be about 0·70. There is an engineering limit to the undamped natural frequency possible with the simple basic accelerometer of Fig. 2.22(a), and sophisticated accelerometers use the force-balance principle, where an electrical restoring force is used in place of the mechanical spring, and the current required to balance the acceleration force is used as a measure of the frame acceleration. Pendulum accelerometers may also be used to measure frame acceleration. The transfer function of the accelerometer will be of the form shown in Equation 2.78 whatever the method of providing the spring resistance. For a full review of the various types of accelerometers the reader is referred to the appropriate chapter by Canfield.[25]

## 2.17 Rate Gyroscope

The measurement of the rate of turn of a body moving in space is necessary for many missile and fire control applications. A convenient method of measuring angular rate is to use a rate gyroscope. There are many other applications of the gyroscopic principle to be found in control engineering, and these are described in Canfield; the rate gyroscope is sufficient for our purpose in introducing readers to the derivation of component transfer functions.

In Fig. 2.23(a) the gyro wheel is rotating at a constant angular velocity $\Omega$ rad/sec about the $y$ axis. Due to the moment of inertia $I$ of the gyro wheel, the wheel will possess angular momentum $I\Omega$ about the $y$ axis. If a torque $T$ is applied about the $x$ axis, then the gyro will precess about the $z$ axis at an angular rate of $w$ rad/sec, the angular rate $w$ being equal to $T/I\Omega$. Conversely, if an angular rate $w$ is applied about the $z$ axis, then a torque is induced about the $x$ axis. This property is used in the rate gyroscope, since the torque induced about the $x$ axis is proportional to the angular

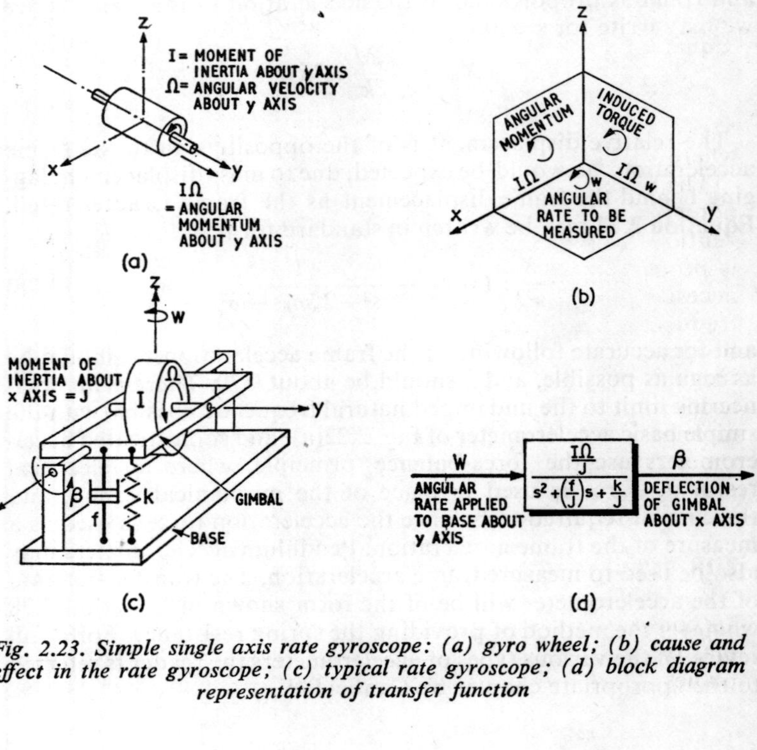


*Fig. 2.23. Simple single axis rate gyroscope: (a) gyro wheel; (b) cause and effect in the rate gyroscope; (c) typical rate gyroscope; (d) block diagram representation of transfer function*

rate applied about the $z$ axis. Cause and effect are shown diagrammatically in Fig. 2.23(b).

A rate gyroscope is shown in Fig. 2.23(c). The angular velocity to be measured rotates the case about the $z$ axis, and the resultant torque rotates the frame about the gimbal axis, which is the $x$ axis. The gyro moment of inertia about the $x$ axis is $J$, and a spring term $k$ and viscous damping term $f$ are also present. If the frame rotation about the $x$ axis is $\beta$, then the torque equation is

$$T = I\Omega w = J\ddot{\beta}+f\dot{\beta}+k\beta \tag{2.79}$$

and the gyro transfer function is

$$\frac{\beta}{w}(s) = \frac{I\Omega}{Js^2+fs+k} \tag{2.80}$$

The block diagram for this transfer function is shown in Fig. 2.23(d). For relatively slow inputs the angle of deflection about the $x$ axis is simply related to the rate of turn about the $z$ axis by the simple formula

$$\frac{\beta}{w}(s) = \frac{I\Omega}{k} \tag{2.81}$$

Equation 2.80 can be written in standard form

$$\frac{\beta}{w}(s) = \frac{I\Omega/J}{s^2+2\zeta\omega_n s+\omega_n^2} \tag{2.82}$$

where $\omega_n^2 = k/J$, and $\zeta = f/2J\omega_n$.

The angular deflection $\beta$ can be turned into a voltage using a potentiometer or a microsyn, which is a derivation from the conventional E-type pickoff. To reproduce faithfully a signal which is proportional to the rate of turn of the body of interest, it is necessary in Equation 2.82 to have as high an undamped natural frequency as possible. In practice the natural frequency tends to be rather low, and hence for many high-performance applications the rate gyroscope is replaced by an angular differentiating accelerometer (A.D.A.), which generates a signal proportional to angular rate provided that the excitation frequency is above the A.D.A. resonant frequency. The A.D.A. resonant frequency may therefore be chosen as low as 1 c/sec.

## 2.18 Missile Roll Transfer Function[26]

In controlling the motion in space of guided missiles, it is normal practice to control motion in three mutually perpendicular planes. These are, respectively, pitch (up and down), yaw (side to side), and roll (motion about the missile longitudinal axis). Guidance signals, which tell the missile what motion is required to hit the target, are therefore resolved into components in each of these three planes. In general, a control system, or autopilot, is required in each plane for the purpose of forcing the missile to follow the guidance signal. Before an autopilot can be designed it is necessary

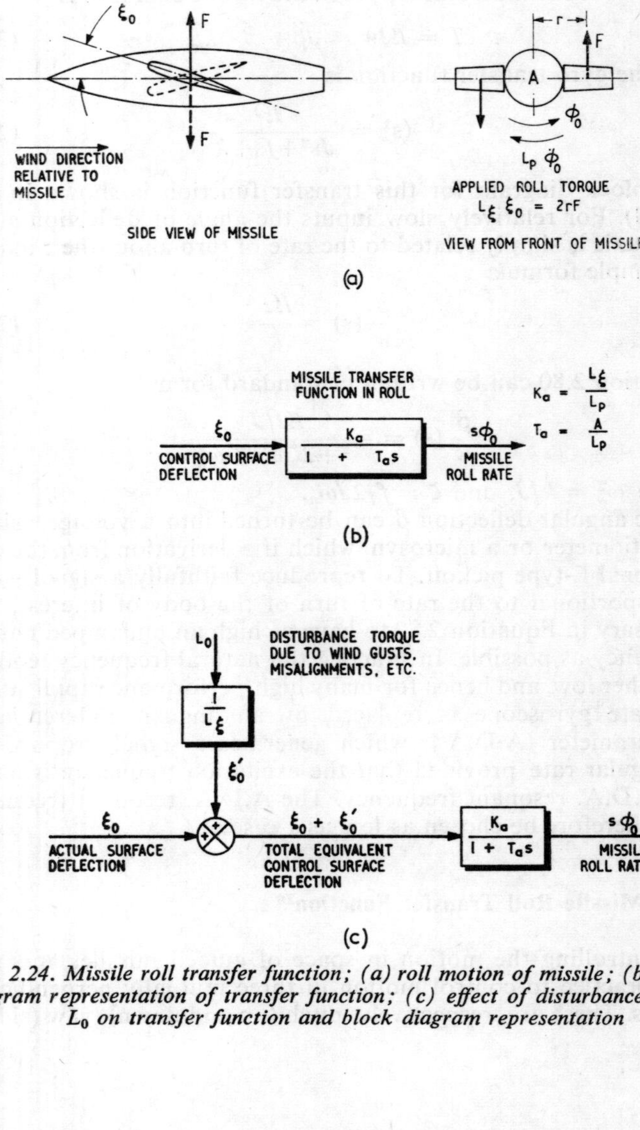


*Fig. 2.24. Missile roll transfer function; (a) roll motion of missile; (b) block diagram representation of transfer function; (c) effect of disturbance torque $L_0$ on transfer function and block diagram representation*

to know the missile transfer function, which is derived from a knowledge of aerodynamics and applied mechanics. We shall derive only the roll transfer function for the missile. This is the simplest transfer function to derive, although the pitch and yaw transfer functions can often be finally expressed as second or third order transfer functions.

Motion about the missile longitudinal axis is induced by moving a pair of wings or control surfaces differentially, as shown in Fig. 2.24(a). In the roll mode each control surface is moved differentially from the equilibrium position by an angle $\xi_0$. From aerodynamic considerations, a force proportional to $\xi_0$ is developed as a result of this movement and will act at the centre of pressure of the control surface and perpendicular to the surface. Multiplying this force by the distance from the line of action of the force to the longitudinal axis of the missile, we see that a torque is developed about this axis which will tend to rotate the missile in the direction of the applied torque. The total torque induced on the missile by the control surface deflection may be written $T_{\text{induced}} = L_\xi \xi_0$. If the missile moves four control surfaces in the roll control mode, then $L_\xi$ will be double the value for the missile with only two control surfaces operating. As the missile rotates about the longitudinal axis, the change in angle of incidence creates an opposing torque proportional to the roll rate $s\phi_0$; this has the same effect as a viscous damping torque.

If the moment of inertia of the missile is $A$ slugs ft², the viscous drag coefficient is $L_p$ lb ft rad$^{-1}$ sec, and the induced torque coefficient is $L_\xi$ lb ft/rad of control surface deflection, then Newton's Law applied to the missile roll mode yields

$$As^2\phi_0 = L_\xi \xi_0 - L_p s\phi_0 \tag{2.83}$$

The transfer function required is the roll rate $s\phi_0$ induced per unit control surface deflection $\xi_0$. Thus

$$\frac{s\phi_0}{\xi_0}(s) = \frac{L_\xi/L_p}{1+\left(\dfrac{A}{L_p}\right)s} \tag{2.84}$$

and, defining $L_\xi/L_p = K_a$ the aerodynamic gain in rad sec$^{-1}$ rad$^{-1}$ and $A/L_p = T_a$ the aerodynamic time constant, Equation 2.84 may be written in the standard form

$$\frac{s\phi_0}{\xi_0}(s) = \frac{K_a}{1+T_a s} \tag{2.85}$$

The block diagram representation of Equation 2.85 is shown in Fig. 2.24(b). If an additional torque $L_0$ acts on the missile due to wind conditions or assembly malalignments, the torque equation can be rewritten as

$$As^2\phi_0 + L_p s\phi_0 = L_\xi \xi_0 + L_0 \tag{2.86}$$

and the additional torque can therefore be looked upon as a dummy control surface deflection $\xi_0$ defined by

$$\xi_0' = \frac{L_0}{L_\xi} \tag{2.87}$$

The missile roll transfer function block diagram is then modified to the form shown in Fig. 2.24(c). The missile roll transfer function parameters vary with both missile speed and altitude, and the problem facing the autopilot designer is how to meet performance specifications when controlling a widely varying load transfer function. As far as this book is concerned we shall concentrate many of our examples on the flight case corresponding to $K_a =$ 335 rad sec$^{-1}$ rad$^{-1}$, and $T_a = 0{\cdot}179$ sec. It should be noted that the choice of a missile roll transfer function to illustrate many of the transfer function techniques outlined in this book is based on the fact that such a standard mathematical model does allow these techniques to be developed consistently from first principles, and a background of information built up which facilitates the development of the subject. In this respect the book follows a highly successful teaching pattern developed at the Royal Military College of Science,[7] but it should be clearly understood that transfer function techniques are universal in application, and that this book is not meant to teach autopilot design specifically.

## 2.19 Representation of Component Transfer Functions

There are several methods of representing component transfer functions. In this book, in general, we shall only be concerned with two forms of representation, the s plane and the logarithmic frequency response. If the component transfer function is

$$\frac{\theta_2}{\theta_1}(s) = \frac{f_1(s)}{f_2(s)} \tag{2.88}$$

then those values of s which make the numerator of s equal to zero are component zeros, marked thus ⊙ and those values of s which make the denominator of s equal to zero are component poles marked thus ×. These values can be marked in the s plane which is drawn with plus and minus real and imaginary parts. To quote

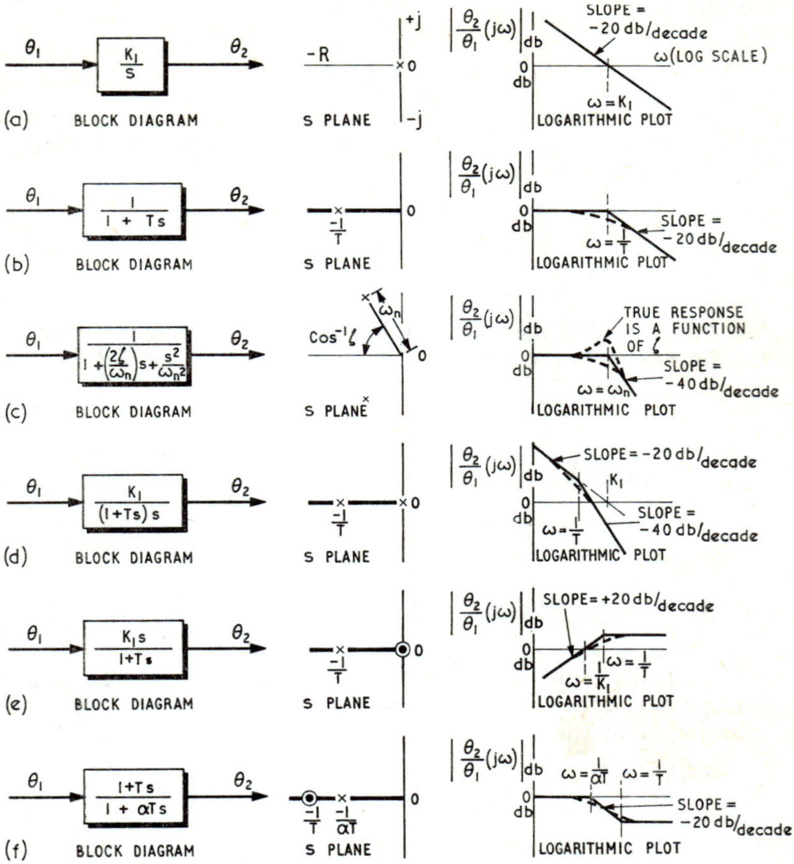


*Fig. 2.25. Block diagram,* s *plane, and asymptotic logarithmic frequency response representation of some simple transfer functions: (a) basic integrator; (b) first order lag; (c) second order lag; (d) first order lag followed by a basic integrator; (e) low-frequency differentiating network; (f) conventional lag network*

simple examples, the basic integrator may be written

$$\frac{\theta_2}{\theta_1}(s) = \frac{K_1}{s} \tag{2.89}$$

There are no component zeros, but there is a component pole at s = 0, and this is shown in Fig. 2.25(a). The first order lag transfer function is,

$$\frac{\theta_2}{\theta_1}(s) = \frac{1}{1+Ts} \tag{2.90}$$

Once again there are no component zeros, but there is a component pole at $s = -1/T$, as shown in Fig. 2.25(b). The second order lag transfer function is

$$\frac{\theta_2}{\theta_1}(s) = \frac{1}{1+\left(\frac{2\zeta}{\omega_n}\right)s+\frac{s^2}{\omega_{n^2}}} \tag{2.91}$$

Assuming $\zeta < 1{\cdot}0$, then the denominator poles are complex, and may be written in the form

$$\left.\begin{aligned} s_1 &= -\zeta\omega_n + j\omega_n\sqrt{1-\zeta^2} \\ s_2 &= -\zeta\omega_n - j\omega_n\sqrt{1-\zeta^2} \end{aligned}\right\} \tag{2.92}$$

or in the polar form

$$\left.\begin{aligned} s_1 &= \omega_n \underline{/+\cos^{-1}\zeta} \\ s_2 &= \omega_n \underline{/-\cos^{-1}\zeta} \end{aligned}\right\} \tag{2.93}$$

with the angles measured from the negative real axis as shown in Fig. 2.25(c).

The s plane representation of linear component transfer functions possesses the additive property, and hence the s plane representation of a basic integrator following a first order lag will be as shown in Fig. 2.25(d). Thus the component transfer function

$$\frac{\theta_2}{\theta_1}(s) = \frac{K_1}{(1+Ts)s} \tag{2.94}$$

will have a component pole at $s = 0$ and at $s = -1/T$. The s plane representation of the low-frequency differentiating network discussed in Section 2.7 is shown in Fig. 2.25(e), the transfer function being

$$\frac{\theta_2}{\theta_1}(s) = \frac{K_1 s}{1+Ts} \tag{2.95}$$

where for generality we have assumed $K_1 \neq T$. There is thus a component zero at $s = 0$, and a component pole at $s = -1/T$. Finally, the conventional lag network discussed in Section 2.8 has the transfer function

$$\frac{\theta_2}{\theta_1}(s) = \frac{1+Ts}{1+\alpha Ts} \tag{2.96}$$

with a zero at $s = -1/T$, and a pole at $s = -1/\alpha T$, as shown in Fig. 2.25(f). Note that the s plane representation of component transfer functions is independent of feedforward gain $K_1$. The feedforward gain $K_1$ of the components will be highly relevant when we are determining the closed loop transfer function from a knowledge of the component transfer functions.

Turning now to the logarithmic form of component frequency response (which is dealt with in great detail in Chestnut and Mayer, Vol. I), any linear transfer function which is available as a series of first and second order factors can be represented by straight-line approximations to the amplitude ratio part of the steady state frequency response. For a linear system, amplitude and phase parts of the frequency response are uniquely related, and therefore it may be argued that only the amplitude ratio is required in order to design a satisfactory feedback control system. The logarithmic representation, like the s plane, possesses the very important additive property, and thus the overall transfer function resulting from a number of components connected in series can be obtained by *adding together* the individual responses. Several basic forms of logarithmic representation exist. Consider firstly transfer functions involving only component poles or zeros at the origin. The transfer function may be written

$$\frac{\theta_2}{\theta_1}(s) = K_1 s^n \tag{2.97}$$

where $n$ is a positive or negative integer, positive integers meaning differentiation and negative integers meaning integration. Standard logarithmic form is, in decibels,

$$20 \log_{10} \left| \frac{\theta_2}{\theta_1}(j\omega) \right| = 20 \log_{10} |K_1(j\omega)^n| \tag{2.98}$$

Now in this instance $|j\omega| = \omega$, and hence we may write

$$20 \log_{10} \left| \frac{\theta_2}{\theta_1}(j\omega) \right| = 20 \log_{10} K_1 + 20\, n \log_{10} \omega \tag{2.99}$$

The first part of Equation 2.99 is a constant, and the second part is linear in $\log_{10} \omega$. If the response is plotted on a linear (decibel) versus $\log \omega$ scale, the relationship defined by Equation 2.99 will be a straight line with a slope $20n$ db/decade. If the component is a basic integrator, then $n = -1$, and there will be a slope of $-20$ db/decade, with an intersection point on the $\log \omega$ axis defined by $\omega = K_1$, as shown in Fig. 2.25(a); this is drawn alongside the s plane representation to allow visual correlation between the two approaches. If the component is a double integrator then the slope, due to $n = -2$, will be $-40$ db/decade with an intersection point on the log $\omega$ axis defined by $\omega = \sqrt{K_1}$. The intersection point is obtained by equating Equation 2.99 to zero, and paying due regard to the sign of $n$. Alternatively, the position of the line in the linear-log plane can be determined by considering the case when $\omega = 1{\cdot}0$, since the decibel amplitude ratio at this frequency is given by the left-hand term to the right of Equation 2.99 as $20 \log_{10} K_1$.

For a first order lag, the amplitude ratio is

$$20 \log_{10} \left| \frac{\theta_2}{\theta_1}(j\omega) \right| = 20 \log_{10} \frac{1}{\sqrt{1+\omega^2 T^2}} \qquad (2.100)$$

which can be rewritten as

$$20 \log_{10} \left| \frac{\theta_2}{\theta_1}(j\omega) \right| = -10 \log_{10}(1+\omega^2 T^2) \qquad (2.101)$$

Consider now the region where $\omega \ll 1/T$, then Equation 2.101 may be approximated by

$$20 \log_{10} \left| \frac{\theta_2}{\theta_1}(j\omega) \right| \simeq -10 \log_{10} 1 \simeq 0 \qquad (2.102)$$

since $\omega^2 T^2$ is then much smaller than unity. The low-frequency asymptote is then the log $\omega$ axis. For higher frequencies where $\omega \gg 1/T$, $\omega^2 T^2$ is much larger than unity and Equation 2.101 may be approximated by

$$20 \log_{10} \left| \frac{\theta_2}{\theta_1}(j\omega) \right| \simeq -10 \log_{10} \omega^2 T^2 \qquad (2.103)$$

and, on simplification,

$$20 \log_{10} \left| \frac{\theta_2}{\theta_1}(j\omega) \right| = -10 \log_{10} T^2 - 20 \log_{10} \omega \qquad (2.104)$$

The left-hand term to the right of Equation 2.104 is a constant, and the high-frequency asymptote is therefore defined by

$$20 \log_{10} \left| \frac{\theta_2}{\theta_1}(j\omega) \right| = -20 \log_{10} \omega \qquad (2.105)$$

which is a straight line of slope $-20$ db/decade when plotted on a linear (decibel) versus log $\omega$ scale. Intersection of the low-frequency asymptote defined by Equation 2.102 and the high-frequency asymptote defined by Equation 2.105 is given by $\omega = 1/T$. Fig. 2.25(b) shows the logarithmic plot of the amplitude ratio part of the steady state frequency response. The exact response is shown dotted, but will rarely be used in the synthesis methods to be described later in this book. The intersection point of the two asymptotes is referred to as the break point, and the frequency $\omega = 1/T$ is referred to as the break frequency. If the transfer function is $1+Ts$ and not $1/(1+Ts)$, then the low-frequency asymptote is still zero but the high-frequency asymptote is now +20 db/decade. Transfer functions of the form $(1+Ts)^n$ will have the same break point $\omega = 1/T$, and the high frequency asymptote is $20n$ db/decade.

Quadratic factors remain to be examined. If the component transfer function is

$$\frac{\theta_2}{\theta_1}(s) = \frac{1}{1+\left(\dfrac{2\zeta}{\omega_n}\right)s+\dfrac{s^2}{\omega_n^2}}$$

then for values of $\omega \ll \omega_n$, we may say that terms due to $j\omega$ and $(j\omega)^2$ are much smaller than unity, and hence

$$20 \log_{10}\left|\frac{\theta_2}{\theta_1}(j\omega)\right| \simeq 20 \log_{10} 1 \simeq 0 \qquad (2.106)$$

and the low-frequency asymptote is again the log $\omega$ axis. For frequencies such that $\omega \gg \omega_n$, then only the $s^2/\omega_n^2$ term is significant, and we may write the high-frequency amplitude ratio as

$$20 \log_{10}\left|\frac{\theta_2}{\theta_1}(j\omega)\right| \simeq 20 \log_{10}\frac{\omega_n^2}{\omega^2}$$

and this can be rewritten as

$$20 \log_{10}\left|\frac{\theta_2}{\theta_1}(j\omega)\right| \simeq 20 \log_{10} \omega_n^2 - 40 \log_{10} \omega \qquad (2.107)$$

The first term on the right-hand side of Equation 2.107 is a constant, hence the second term represents the high-frequency asymptote, which can therefore be written as

$$20 \log_{10}\left|\frac{\theta_2}{\theta_1}(j\omega)\right| \simeq -40 \log_{10} \omega \qquad (2.108)$$

which is a straight line of slope −40 db/decade. The intersection point of the high and low frequency asymptotes is given by $\omega = \omega_n$. The logarithmic representation using these high and low frequency asymptotes is shown in Fig. 2.25(c). Note that the representation is independent of $\zeta$. Two actual responses are sketched in Fig. 2.25(c). For values of $\zeta < 0{\cdot}70$, a peak is to be expected, whereas for values of $\zeta > 0{\cdot}70$, the true response will lie below the asymptotic approximation. The derivation from the asymptotic approximation is obviously dependent on $\zeta$, but the approximation is reasonably good for frequencies less than $\omega_n/5$ and greater than $5\omega_n$ irrespective of the value of $\zeta$. To some extent the presence of a component with complex poles or zeros which are very lightly damped ($\zeta$ small) may prevent the use of a design method based on the asymptotic approximations to amplitude ratio only, and the actual value of amplitude ratio must be used or, alternatively, the phase information must be made available.

Logarithmic approximations to transfer functions in series are shown in Fig. 2.25(d), (e) and (f). For the first order lag followed by a basic integrator, the asymptote has a slope of $-20$ db/decade until a frequency of $\omega = 1/T$ is reached, when the first order lag term is added, resulting in a subsequent slope of $-40$ db/decade. The low-frequency differentiating network has a positive slope of 20 db/decade until $\omega = 1/T$, where the lag term comes into play, generating a negative slope of $-20$ db/decade; this cancels with the positive slope giving an asymptote of zero for frequencies greater than $\omega = 1/T$. For the conventional lag network, the slope at low frequencies is zero. By definition the lag term comes into play first, and hence at $\omega = 1/\alpha T$ the slope becomes $-20$ db/decade. At frequency $\omega = 1/T$, the positive slope is significant and cancels the negative slope, leaving the nett slope zero for frequencies greater than $\omega = 1/T$.

Considerable emphasis will be placed in this book on the s plane method of representation of transfer functions. Logarithmic asymptotic approximations to frequency response amplitude ratios will be used for two distinctly separate purposes. In Chapter 3 the asymptotic approximation will be used to determine the approximate frequency response for a complete feedback control system, given the system pole-zero array, so that a rapid comparison of various proposed system pole-zero arrays can be made. But in Chapter 6, an approximate method will be described by which the system poles and zeros, and hence the system transfer function, can be obtained directly from the broken loop, frequency response asymptotes, provided the system is written in unity feedback form.

## 2.20 Building up a Feedback Control System from Basic Components

To illustrate the procedure in building up a feedback control system from basic components, suppose it is required to design an autopilot for the missile roll channel which will control the missile roll position. The first element is therefore the load, which will be the transfer function defined by Equation 2.85, which relates missile roll rate $s\phi_0$ to the control surface deflection $\xi_0$. Since a position control system is required, the roll position of the missile is measured using a gyroscope, and this will measure a voltage proportional to roll position. Thus we may write the feedback voltage

$$V_0 = K_0\phi_0 \tag{2.109}$$

and there will be an inherent integration in the feedforward path appearing after the missile because we are physically measuring missile position. If a one-to-one system is required, we must pass the demanded roll position $\phi_D$ through a suitable transducer to generate an input voltage $V_i$, such that

$$V_i = K_0\phi_D \tag{2.110}$$

By subtracting the output voltage from the input voltage, using a summing network, we can generate an error signal

$$V_e = V_i - V_0 \tag{2.111}$$

which can be used as the basis of a simple proportional control system. The missile roll movement is to be generated by movement of the control surface. As our first simple control rule, let us ask for a control surface deflection $\xi_D$ which is proportional to the error voltage $V_e$. Thus, we may write

$$\xi_D = K_e V_e \tag{2.112}$$

where $K_e$, the controller gain, may cover the gain of any amplifiers required to raise the power level of the signal demanding the control surface deflection. For a very fast servomechanism loop, the actual control surface deflection may be considered as equal to the demanded control surface deflection. As a second approximation, we may consider the servomechanism loop to take the form of Equation 2.28, with a simple first order lag with time constant $T_c$. We thus write

$$\frac{\xi_0}{\xi_D} = \frac{1}{1+T_c s} \tag{2.113}$$

The basic block diagram now takes the form shown in Fig. 2.26. The pre-loop transfer function is

$$G_p(s) = K_0$$

and the feedforward transfer function is given by

$$G(s) = \frac{K_e K_a}{(1+T_c s)(1+T_a s)s}$$

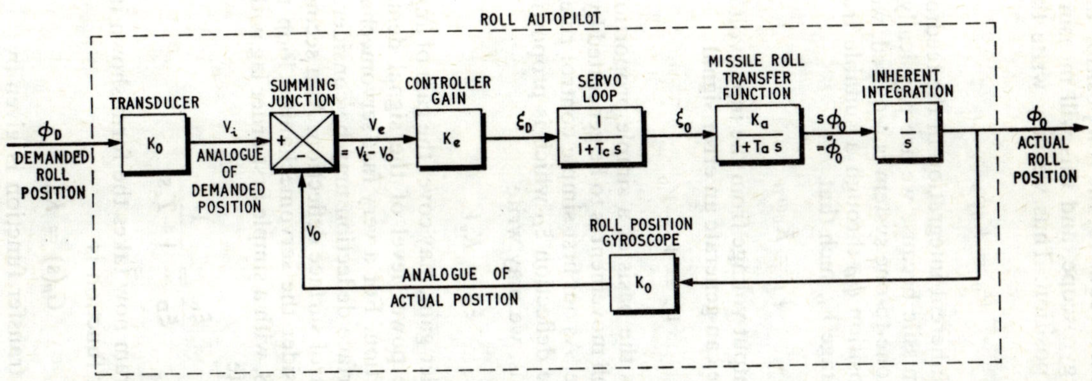


$$\frac{\phi_0}{\phi_D}(s) = \frac{1}{1 + \left(\frac{1}{K_e K_a K_0}\right)s + \left(\frac{T_a + T_c}{K_e K_a K_0}\right)s^2 + \left(\frac{T_a T_c}{K_e K_a K_0}\right)s^3}$$

*Fig. 2.26. Building up the basic missile roll autopilot*

and

$$H(s) = K_0$$

Thus the system transfer function is

$$\frac{\phi_0}{\phi_D}(s) = \frac{G_p(s)G(s)}{1+G(s)H(s)} \tag{2.114}$$

In simplest form this becomes

$$\frac{\phi_0}{\phi_D}(s) = \frac{1}{1+\left(\frac{1}{K_eK_0K_a}\right)s+\left(\frac{T_a+T_c}{K_eK_0K_a}\right)s^2+\left(\frac{T_aT_c}{K_eK_0K_a}\right)s^3} \tag{2.115}$$

Since $K_eK_0K_a$ is a common denominator for the $a_1$, $a_2$, and $a_3$ terms, it could be replaced by a single term. However, because $K_a$ is a load term and is varying in flight, for this particular example $K_a$ should be kept separate, but it is convenient for linear operation to write $K_g = K_eK_0$, and this $K_g$ term can then be looked upon as the lumped feedforward gain adjustment. In the process we have transformed the system into the unity feedback form. The new block diagram is shown in Fig. 2.27, and it can be seen that for this particular arrangement the signals entering the comparator are the demanded and actual roll positions, the analogue conversion term being submerged now in $K_g$. If further signals are now fed back to the comparator to stabilise the system or shape the dynamic response, these must now be correspondingly adjusted.

For example, the transfer function shown in Equation 2.115 is unlikely to result in acceptable performance, and one way of bettering this performance is to modify the control signal to take account of the rate at which the system output is moving. Let the missile roll rate be measured using a rate gyroscope sufficiently fast in response to allow the lags defined by Equation 2.82 to be neglected, and suppose a voltage proportional to roll rate is generated from the potentiometer attached to the rate gyroscope such that

$$V_R = K_r s\phi_0 \tag{2.116}$$

The demanded control surface deflection is now

$$\xi_D = K_e(V_i - V_0 - V_R) \tag{2.117}$$

as shown in Fig. 2.28. Expanding Equation 2.117, we may write

$$\xi_D = K_e(K_0\phi_D - K_0\phi_0 - K_r s\phi_0) \tag{2.118}$$

and, introducing $K_g = K_0K_e$, we have

$$\xi_D = K_g\phi_D - K_g\phi_0 - K_g\left(\frac{K_r}{K_0}\right)s\phi_0 \tag{2.119}$$

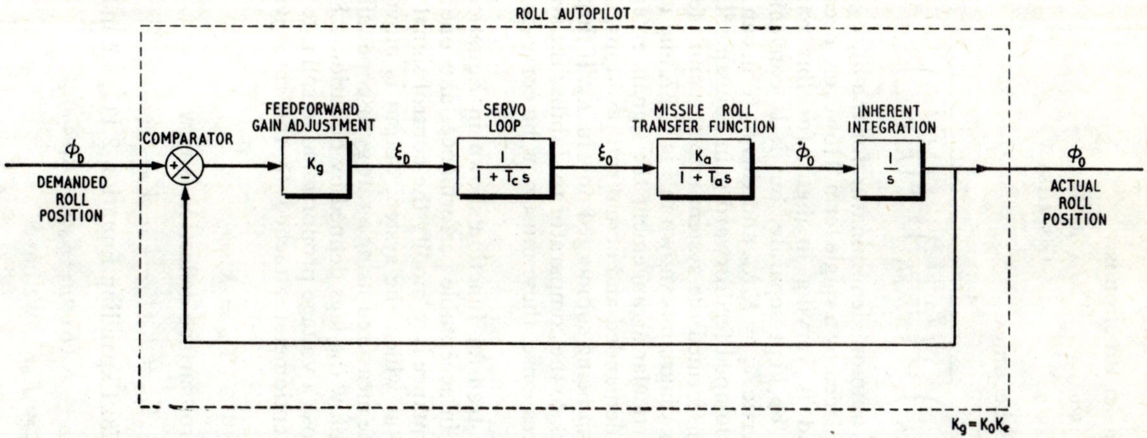


$$\frac{\phi_0}{\phi_D}(s) = \frac{1}{1 + \left(\frac{1}{K_g K_a}\right)s + \left(\frac{T_a + T_c}{K_g K_a}\right)s^2 + \left(\frac{T_a T_c}{K_g K_a}\right)s^3}$$

*Fig. 2.27. The basic missile roll autopilot of Fig. 2.26 shown with the feedback term lumped with controller gain*

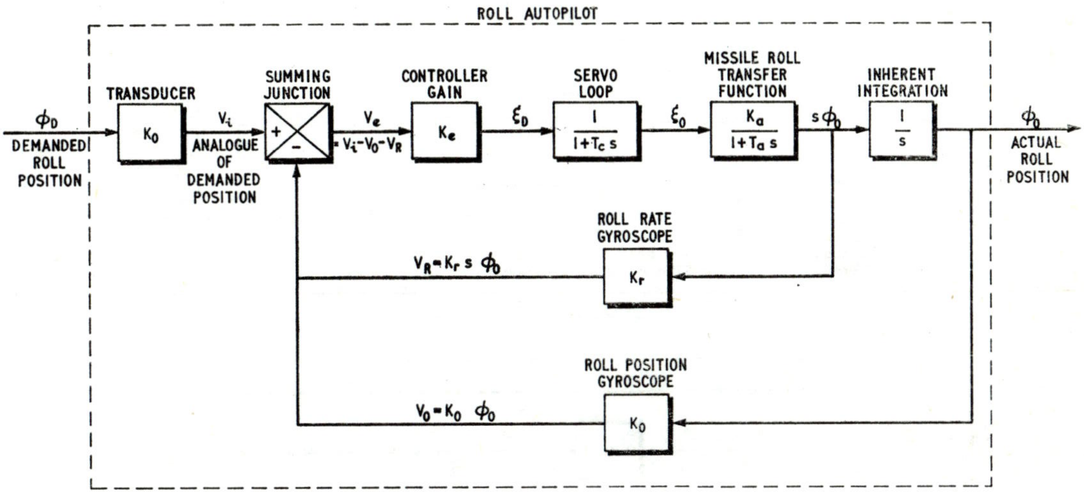


$$\frac{\Phi_0}{\Phi_D}(s) = \frac{1}{1 + \left(\frac{K_r}{K_0} + \frac{1}{K_0 K_e K_a}\right)s + \left(\frac{T_a + T_c}{K_0 K_e K_a}\right)s^2 + \left(\frac{T_a T_c}{K_0 K_e K_a}\right)s^3}$$

*Fig. 2.28. Missile roll autopilot with rate feedback compensation incorporated*

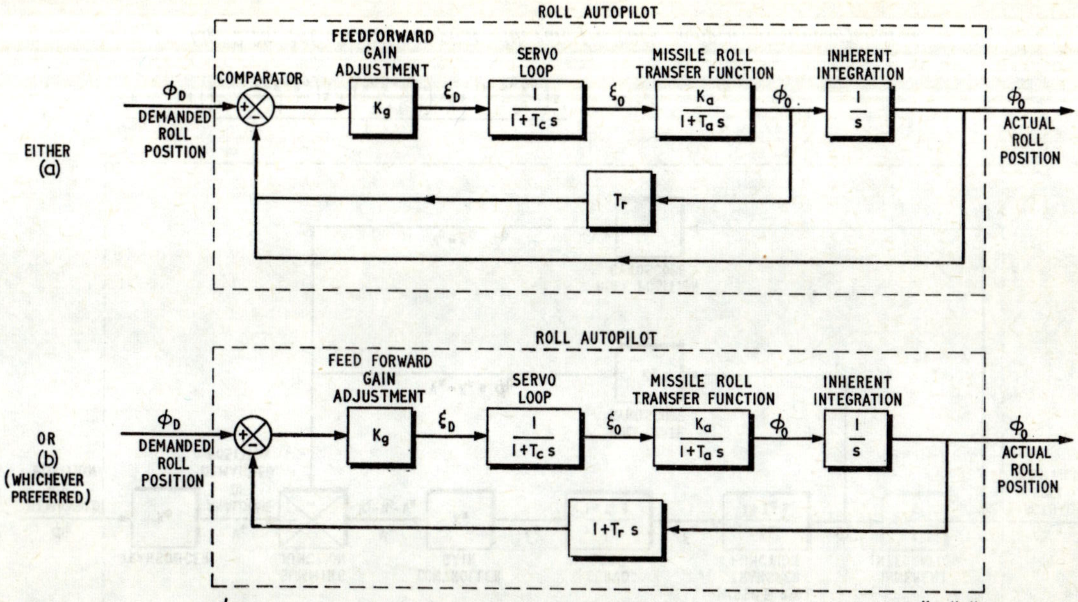


$$\frac{\phi_0}{\phi_D}(s) = \frac{1}{1 + \left(T_r + \frac{1}{K_g K_a}\right)s + \left(\frac{T_a + T_c}{K_g K_a}\right)s^2 + \left(\frac{T_a T_c}{K_g K_a}\right)s^3}$$

*Fig. 2.29. The missile roll autopilot of Fig. 2.28 shown with the feedback term lumped with controller gain — either (a) or (b), whichever preferred*

i.e. in the lumped parameter form we should write the rate feedback term at the comparator as $T_r = K_r/K_0$ to obtain true equivalence. Fig. 2.29 shows the new block diagram which lumps together all the component gains except $K_a$. For non-linear systems, where the absolute levels of signal are important, it may be advantageous to leave the block diagram in the original form.

## 2.21 Conclusions

This brief resumé of the derivation of transfer functions for specific components is included for continuity and is not intended to replace standard reference books on hardware. We are, however, now in a position to use transfer function techniques to design feedback control systems to meet particular performance requirements. Chapter 3 will show how the system characteristic equation be can solved in the s plane for known component transfer functions. This will determine the system poles as a function of a component parameter of interest.

CHAPTER 3

# Root Locus and Pole-zero Techniques

## 3.1 Introduction

If the system poles and zeros are known, then, using standard mathematical techniques, the system behaviour to any input stimulus can be determined, and the various performance criteria discussed in Chapter 1 can be evaluated. The design problem is to determine system poles and zeros which allow the system performance specification to be met. In general, there will be a variety of system pole-zero arrays which can satisfy the system performance specification; the problem is to locate a practicable pole-zero array with the minimum of design effort. For low order systems the system performance specification will uniquely define a system transfer function; for example, the specification of desired bandwidth, velocity constant, and percentage overshoot completely defines the array for a one-zero two-pole transfer function. It is therefore unneccessary to draw the root locus for many low order systems; the charts of Chapters 4 and 5 define many of the performance criteria uniquely in terms of the system transfer function.

It is difficult to define uniquely pole-zero arrays for higher order systems directly from the performance specification, and the design process becomes to some extent a matter of trial and error in order to determine suitable compensation transfer functions involving, for example, phase advance networks and/or output derivative feedback. If the design process determines the system poles and zeros directly, then system response and stability is also directly controlled. The root locus technique due to Evans[27] determines the system poles graphically in the root or s plane from a knowledge of the various component transfer functions, and, furthermore, the technique allows the locus of the poles to be traced as one of the system parameters, such as feedforward gain, is varied from zero to infinity. For a given value of this system parameter, the roots thus determined are added to the system zeros to form the system pole-zero array for this particular value of parameter. For a given value of the system parameter, the system poles and zeros are uniquely known for a linear system, and hence the system transfer function is also known uniquely. Transfer function techniques are therefore inherently related to the s plane.

During the design stage it is customary to approximate the *system frequency response* by using the Bode logarithmic approximation, thus making possible rapid correlation between system pole-zero positions and system frequency response. Once these approximate responses have indicated the best design region, more exact responses are computed from the system pole-zero array. Steady state frequency response and system velocity constant are easily determined from the system pole-zero array, and it is also similarly possible to compute the coefficients of the transient response solution using a simple vector method. If it is obvious from the system pole-zero array that only two or three poles significantly affect the system response, then the response may be obtained directly from the charts of Chapters 4 and 5.

If variations in more than one system parameter are to be studied at the design stage, the root locus plot is repeated as many times as considered necessary. For example, in designing a position servomechanism we might choose feedforward gain as the independent variable for the root locus plot, and then repeat this plot for, say, four values of motor time constant, since both feedforward gain and motor time constant may be under the designer's control. The best combination of feedforward gain and motor time constant is then found by examining all four plots, choosing certain combinations of interest and evaluating bandwith and other criteria to determine the best possible system performance.

## 3.2 Root Locus Equations and the Starting and Terminating Points for Root Loci

We have already seen in Chapter 1 that the general expression for the system transfer function is, from Equation 1.3,

$$\frac{\theta_0}{\theta_i}(s) = \frac{G(s)}{1+G(s)H(s)}$$

Both $G(s)$ and $H(s)$ may be ratios of polynomials in s, and furthermore there must be a feedforward gain term in the feedforward transfer function $G(s)$. Thus $G(s)$ and $H(s)$ may be expanded to give

$$G(s) = K\left[\frac{p(s)}{q(s)}\right] \tag{3.1}$$

and

$$H(s) = \frac{v(s)}{w(s)} \tag{3.2}$$

as shown in Fig. 3.1. The numerical value of $K$ will depend upon whether the individual terms of $p(s)$, etc., are written in the form

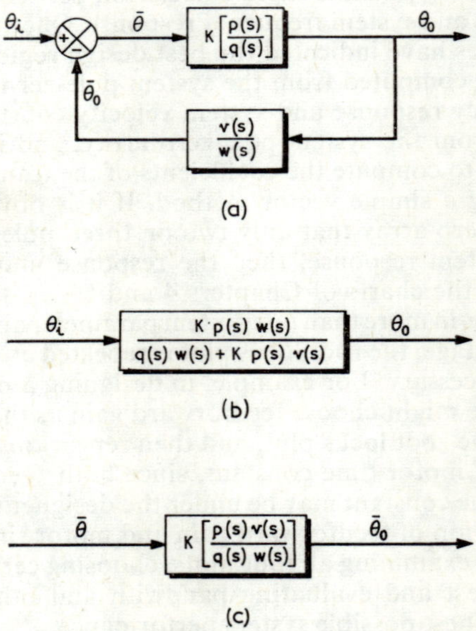


*Fig. 3.1. Block diagram of the general feedback control system in the form required for the root locus plot in terms of feedforward gain K: (a) general form of the system for the root locus method; (b) the system transfer function; (c) the open loop transfer function*

$1+\alpha s$ or in the form $s+1/\alpha$. Substituting Equations 3.1 and 3.2 into Equation 1.3 and simplifying, we obtain

$$\frac{\theta_0}{\theta_i}(s) = \frac{Kp(s)w(s)}{q(s)w(s)+Kp(s)v(s)} \tag{3.3}$$

Values of s which make the numerator of Equation 3.3 equal to zero are termed system zeros, whilst values of s which make the denominator of Equation 3.3 equal to zero are termed system poles. By inspection of Equation 3.3 we see immediately that the system zeros are defined by

$$Kp(\mathrm{s})w(\mathrm{s}) = 0 \tag{3.4}$$

i.e. the system zeros are the zeros of $G(\mathrm{s})$ and the poles of $H(\mathrm{s})$. Generally the roots of $p(\mathrm{s})$ and $w(\mathrm{s})$ are known from the individual component transfer functions, and for most systems we may immediately factorise the numerator of the system transfer function as follows:

$$\begin{aligned} 1+b_1\mathrm{s}+b_2\mathrm{s}^2+\ldots b_q\mathrm{s}^q &= \mu_1[Kp(\mathrm{s})w(\mathrm{s})] \\ &= \prod_{i=1} (1+T_i\mathrm{s})\times\prod^{i=q}\left[\frac{\mathrm{s}^2}{\omega_{ni}^2}+\left(\frac{2\zeta_i}{\omega_{ni}}\right)\mathrm{s}+1\right] \\ &= 0 \end{aligned} \tag{3.5}$$

where $\prod$ means the product of the individual zeros, and where $\mu_1$ is a linear scale factor which does not affect the system zeros in any way but which arises as a result of writing $b_0 = 1$.

The characteristic equation is given by the denominator of Equation 3.3, and we may also write the system poles as follows:

$$\begin{aligned} 1+a_1\mathrm{s}+a_2\mathrm{s}^2+\ldots a_n\mathrm{s}^n &= \mu_2[q(\mathrm{s})w(\mathrm{s})+Kp(\mathrm{s})v(\mathrm{s})] \\ &= \prod_{i=1} (1+T_i\mathrm{s})\times\prod^{i=n}\left[\frac{\mathrm{s}^2}{\omega_{ni}^2}+\left(\frac{2\zeta_i}{\omega_{ni}}\right)\mathrm{s}+1\right] \\ &= 0 \end{aligned} \tag{3.6}$$

where $\mu_2$ is a linear scale factor which does not affect the system poles in any way but which arises as a result of writing $a_0 = 1$. Unfortunately, the roots of Equation 3.6 are far from obvious, and the solution of the $n^{\text{th}}$ order polynomial by classical methods is tedious, to say the least. The solution must be repeated for every proposed value of gain likely to be of interest in a specific system. In contrast, for given component transfer functions represented by $p(\mathrm{s})$, $q(\mathrm{s})$, $w(\mathrm{s})$ and $v(\mathrm{s})$, the root locus method determines the system poles or roots of the characteristic equation of the system graphically, as $K$ is varied from zero to infinity. Thus the root locus is a design tool as well as a synthesis tool as the locus *assists* in determining the best value of feedforward gain.

For use with the graphical root locus technique, the system characteristic equation is rearranged to give

$$K\left[\frac{p(\mathrm{s})v(\mathrm{s})}{q(\mathrm{s})w(\mathrm{s})}\right] = -1 \tag{3.7}$$

where $K$ is the gain obtained when the open loop poles and the open loop zeros are written in the form $s+1/T$ and not in the form $1+Ts$.

As can be seen from Fig. 3.1, the left-hand side of Equation 3.7 is, of course, the open or broken loop relationship $\bar{\theta}_0/\bar{\theta}$ already met in Equation 1.37. In the root locus method, $-1$ is interpreted as $\pm$ 180°. From the open loop transfer function $\bar{\theta}_0/\bar{\theta}$ we can define the open loop poles as those values of s which make $\bar{\theta}_0/\bar{\theta}$ infinite, and open loop zeros as those values of s which make $\bar{\theta}_0/\bar{\theta}$ zero. Thus, from Equation 3.7 we obtain the equation for the open loop poles as

$$q(s)w(s) = 0 \tag{3.8}$$

and that for the open loop zeros as

$$p(s)v(s) = 0 \tag{3.9}$$

If we now revert to Equation 3.6, the starting and finishing points of the root locus are obvious. Thus, considering the equation

$$q(s)w(s)+Kp(s)v(s) = 0 \tag{3.10}$$

for $K = 0$, the roots of the characteristic equation are given by $q(s)w(s) = 0$, *i.e. the root locus starts at the open loop poles.* For $K = \infty$, $Kp(s)v(s)$ will be very much larger than $q(s)w(s)$, and the roots will therefore be given by $p(s)v(s) = 0$, *i.e the root locus terminates at the open loop zeros.*

To summarise, the system poles are to be obtained from a knowledge of open loop poles and zeros and, for a given value of feedforward gain, the system poles will then be known. Superimposing the system zeros will then yield the system pole-zero array, which

× OPEN LOOP POLES

⊙ OPEN LOOP ZEROS

• SYSTEM POLES

o SYSTEM ZEROS

*Fig. 3.2. Symbols for the representation of open loop and system poles and zeros*

can then be studied in a number of ways. A consistent set of symbols for open loop poles and zeros and system poles and zeros is therefore required, and a suitable set is shown in Fig. 3.2. Since the root locus technique is best taught by the use of specific examples, we shall briefly state the rules for constructing the root locus, and then proceed immediately to the construction of root loci for various systems.

## 3.3 Rules for Constructing Root Loci[27, 28]

The following rules are those most commonly used in drawing root loci. Although eleven rules are given, it is not always necessary to use all these rules, nor is it necessary to use them in the order given here.

### RULE 1. NUMBER OF BRANCHES OF THE ROOT LOCUS

There are the same number of branches of the root locus as there are open loop poles or open loop zeros.

### RULE 2. SYMMETRY OF THE ROOT LOCUS

For linear systems the root locus plot must be symmetrical about the real axis since complex roots must appear as pairs.

### RULE 3. STARTING AND FINISHING POINTS FOR THE ROOT LOCUS

Each branch of the root locus starts at an open loop pole (corresponding to $K = 0$) and terminates at an open loop zero (corresponding to $K = \infty$). If there are $n$ open loop poles and $m$ finite open loop zeros, then there are $n-m$ open loop zeros at infinity.

### RULE 4. EXISTENCE OF BRANCHES OF THE ROOT LOCUS ON THE REAL AXIS

A branch of the root locus exists at a point on the real axis provided that the total number of open loop poles and open loop zeros to the right of the point in question is an odd number. If there is an even number of poles plus zeros to the right of the point then the point is not on a branch of the locus.

### RULE 5. HIGH-GAIN ASYMPTOTES OF THE ROOT LOCUS

Since there are $n-m$ open loop zeros at infinity, it follows that there are $n-m$ high-gain asymptotes which the branches of the locus will approach as $K \to \infty$. The angles which these high-gain asymptotes make with the real axis (measured positively in the anticlockwise direction from the positive real axis) is given by $\pm k\pi/(n-m)$, where $k = 1, 3, 5$, etc.

### Rule 6. Intersection of the High-gain Asymptotes

The high-gain asymptotes intersect at a point $d$ on the real axis, given by

$$d = \frac{\sum \text{open loop poles} - \sum \text{open loop zeros}}{n-m}$$

### Rule 7. Intersection of the Root Locus with the Imaginary Axis

Intersections of the root locus with the imaginary axis correspond to conditions of limiting stability, and the intersection points can be found using the Routh Hurwitz criterion. Often, however, it will be found that the intersection points can be determined more quickly by using Rule 9 than by using the Routh Hurwitz criterion, in which case the value of gain determined for the crossing points can be fed back into the Routh Hurwitz array as a check on the root locus construction.

### Rule 8. Centre of Gravity of the Roots

The centre of gravity of the roots remains constant as $K$ varies from 0 to $\infty$. This position can be calculated by considering the case when $K = 0$, and the theorem may then be applied to locate the various roots associated with a particular value of $K$. The centre of gravity rule applies only if $n \geqslant m+2$.

### Rule 9. Graphical Determination of Points on the Root Locus

The above rules are used to avoid a random search for points on the root locus. As soon as a rough guide to the locus has been obtained, the further points on the locus may be determined by trial and error. A suitable trial point is chosen and vectors drawn from this trial point to all the open loop poles and zeros. A horizontal reference line is then taken, and angles from the reference line to the vectors are measured in the anti-clockwise sense. If the trial point is indeed on the root locus, then the following relationship will hold:

$$\sum \text{angles at loop poles} - \sum \text{angles at loop zeros} = 180^\circ + N\,360^\circ$$

where $N$ is a positive or negative integer.

### Rule 10. Point of Breakaway of the Root Locus from the Real Axis

If the locus leaves the real axis then the point of breakaway from the real axis can be obtained by considering a point a small distance from the real axis. The angles contributed by each open loop pole and zero are then expressed using small-angle approximations. The unknown quantity will then be the distance of the breakaway point from, say, the imaginary axis, and the solution will be found by an approximate method of calculation.

### Rule 11. Calculations of Gain for a Point on the Root Locus

Using the particular root locus definition of $K$ of Equation 3.7, the value of $K$ at any point on the locus is simply the product of the vectors from that point to the open loop poles divided by the product of the vectors from that point to the finite open loop zeros. If all the open loop zeros are at infinity, the product of the vectors to the open loop zeros is interpreted as unity.

The reader will see for himself that Rule 9 requires the addition and subtraction of angles in order to determine whether a trial point is on the locus or not. This procedure can be greatly speeded up by the use of an additive protractor. One such protractor available commercially is incorporated in the Spirule, a device developed by W. R. Evans, the father of the root locus technique. The Spirule also possesses facilities for multiplication and division of lengths, thus speeding up the computation of gain in accordance with Rule 11 for a given point on the root locus. The additive protractor facility is, however, the more important of the two, and allows the root loci of simpler systems to be drawn by a student in a time appropriate to examinations.

## 3.4 Root Locus for a Unity Numerator Second Order System in Terms of Feedforward Gain

EXAMPLE 3.4.1

Our first example of a root locus plot will be a simple second order system stabilised by rate feedback. Using the charts provided in Chapter 4, it would not in practice be necessary to draw the root locus for such a simple system in order to establish whether it is possible for such a system to meet a given specification, but this second order system does provide an opportunity to illustrate most of the constructional rules listed in Section 3.3. The system chosen is the simple missile roll autopilot shown in Fig. 3.3, which

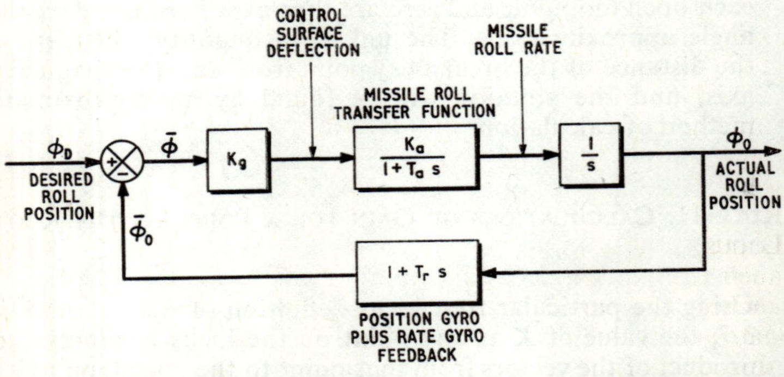


*Fig. 3.3. Block diagram of the second order missile roll autopilot loop used in Example 3.4.1*

has already been briefly mentioned in Chapters 1 and 2. $K_g$ and $T_r$ are parameters normally under the control system designer's jurisdiction, whilst $K_a$ and $T_a$ would normally be inherited from aerodynamic considerations. The open loop transfer function is, by inspection of Fig. 3.3,

$$\frac{\bar{\phi}_0}{\bar{\phi}}(s) = \frac{K\left(s+\dfrac{1}{T_r}\right)}{\left(s+\dfrac{1}{T_a}\right)s} = -1 \tag{3.11}$$

where $K = K_g K_a T_r / T_a$ is the value of gain which is computed from the root locus plot. From Equation 3.11 we see that the system has two open loop poles at $s = 0$ and $s = -1/T_a$, and an open loop zero at $s = -1/T_r$. Typical values of $K_a$ and $T_a$ for a particular flight case are 335 rad $sec^{-1}$ $rad^{-1}$ and 0·179 sec respectively and we shall assume that $T_r = \frac{1}{12}$ sec. Thus the open loop poles are at 0 and $-1/0{\cdot}179 = -5{\cdot}6$, and there is an open loop zero at $-12$. These open loop poles and open loop zeros are shown in Fig. 3.4.

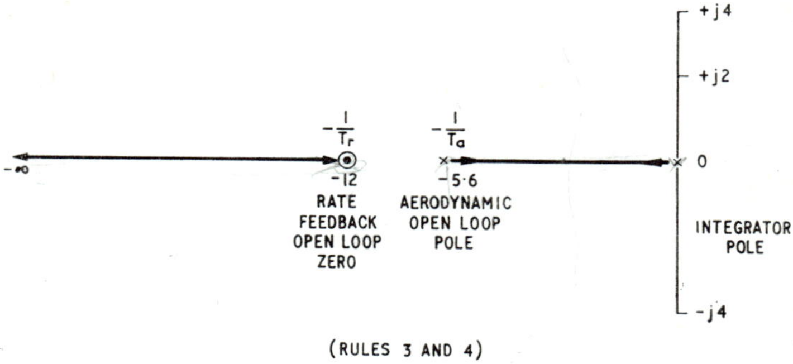


*Fig. 3.4. Root locus branches existing on the real axis for Example 3.4.1*

Using Rule 3 we know that the locus starts at 0 and $-5{\cdot}6$, and one branch terminates at $-12$. Rule 1 says that we must have two branches to the locus, thus indicating that one branch must terminate at a zero at infinity. From Rule 5, the high-gain asymptote to this zero at infinity is defined by the angle $\pm k\pi/(n-m)$; now $n-m = 2-1 = 1{\cdot}0$, hence the angle is $\pm\pi$, $\pm 3\pi$, etc., which is clearly along the negative real axis. Examining the real axis for the existence of branches of the locus according to Rule 4, we see that the locus can only exist between 0 and $-5{\cdot}6$, and between $-12$ and $-\infty$. Since we know that the two branches start at 0 and $-5{\cdot}6$ and terminate at $-\infty$ and $-12$, we can place arrows on the negative real axis near to these points indicating the arrival and departure of the branches of the locus. Fig. 3.4 summarises the locus construction at this stage.

Clearly the locus must break away from the negative real axis somewhere between 0 and $-5{\cdot}6$ to form complex roots. These complex roots must be symmetrical according to Rule 2, and will then describe a suitable contour to arrive at a point on the negative real axis between $-12$ and $-\infty$, whence the roots will again be negative real, one moving back towards $-12$, and the other moving out towards $-\infty$. To calculate the breakaway and return points we use Rule 10, noting that because there are two roots leaving the real axis, the breakaway (and return) locus must be perpendicular to the real axis at those points.

Fig. 3.5 shows the geometry for movement of the locus a small distance $h$ from the breakaway point which is at a distance $y$ from the imaginary axis. Note that the sign convention used for this

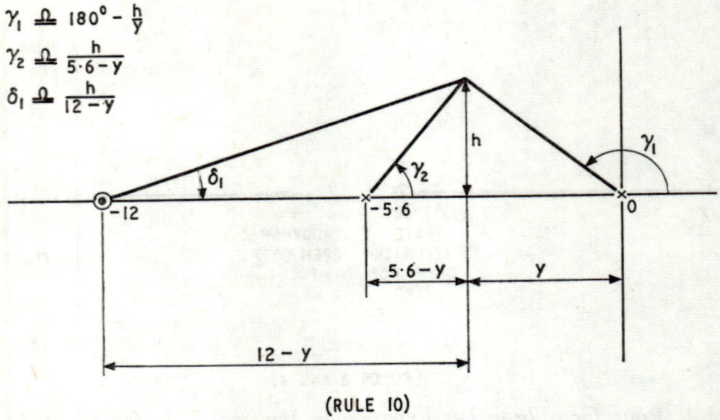


*Fig. 3.5. Vector diagram for the computation of the breakaway point on the real axis for Example 3.4.1*

computation, i.e. clockwise arrows for angles at zeros and anti-clockwise arrows for angles at poles, reduces the risk of making a mistake. The angle rule, Rule 9, states that

$$\gamma_1+\gamma_2-\delta_1 = 180°+N\,360° \tag{3.12}$$

Using small angle approximations, we may write

$$\left.\begin{aligned} \gamma_1 &= 180°-\frac{h}{y} \\ \gamma_2 &= \frac{h}{5{\cdot}6-y} \\ \delta_1 &= \frac{h}{12-y} \end{aligned}\right\} \tag{3.13}$$

Substituting Equation 3.13 into Equation 3.12, and eliminating $h$, which is common to all angles, we have

$$\frac{1}{y} = \frac{1}{5{\cdot}6-y}-\frac{1}{12-y} \tag{3.14}$$

Equation 3.14 may be rapidly solved by trial and error. The final solution, $y = 3{\cdot}25$, yields

$$\frac{1}{3{\cdot}25} \simeq \frac{1}{2{\cdot}35} - \frac{1}{8{\cdot}75}$$

$$0{\cdot}308 \simeq 0{\cdot}425 - 0{\cdot}114$$

and

$$0{\cdot}308 \simeq 0{\cdot}311$$

which is near enough, showing that the breakaway point is at $-3{\cdot}25$ from the imaginary axis.

The difference between the left-hand side and right-hand side which is acceptable in Equation 3.14 when solved by trial and error methods depends on the rate at which the terms are changing. In this example the trial values of $y = 3$ and $y = 4$ showed that the right-hand side was changing very rapidly, showing that a difference of 1% between the two sides represented an approximation to $y$ which was very close to the exact value.

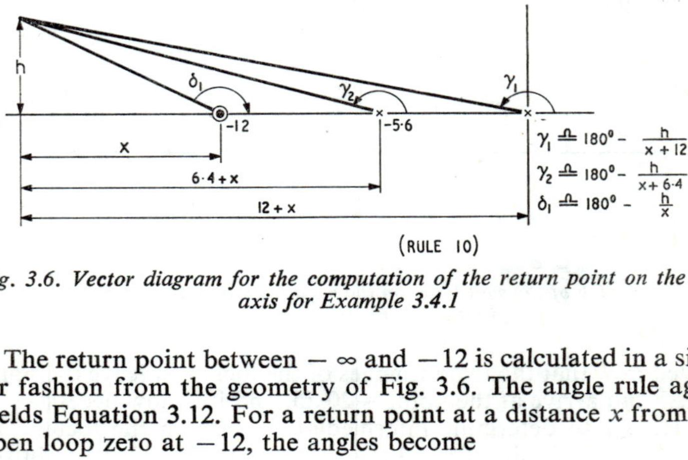


*Fig. 3.6. Vector diagram for the computation of the return point on the real axis for Example 3.4.1*

The return point between $-\infty$ and $-12$ is calculated in a similar fashion from the geometry of Fig. 3.6. The angle rule again yields Equation 3.12. For a return point at a distance $x$ from the open loop zero at $-12$, the angles become

$$\left.\begin{aligned} \gamma_1 &= 180° - \frac{h}{x+12} \\ \gamma_2 &= 180° - \frac{h}{x+6{\cdot}4} \\ \delta_1 &= 180° - \frac{h}{x} \end{aligned}\right\} \qquad (3.15)$$

Substituting Equation 3.15 into Equation 3.12, and eliminating $h$

which is again common, the angular requirement becomes

$$\frac{1}{x} = \frac{1}{x+12} + \frac{1}{x+6{\cdot}4} \tag{3.16}$$

Using trial and error to solve Equation 3.16, for $x = 8{\cdot}75$

$$\frac{1}{8{\cdot}75} \simeq \frac{1}{20{\cdot}75} + \frac{1}{15{\cdot}15}$$

$$0{\cdot}114 \simeq 0{\cdot}048 + 0{\cdot}66$$

and

$$0{\cdot}114 = 0{\cdot}114$$

which is as accurate an answer as can be obtained by the method. The return point is therefore at $-(12+8{\cdot}75) = -20{\cdot}75$. At the points $-20{\cdot}75$ and $-3{\cdot}25$ there will be a pair of equal roots, and, using the potential analogy, the locus will leave the breakaway and return points at angles of $\pm 90°$ to the real axis. Fig. 3.7 shows the root locus summarised this far.

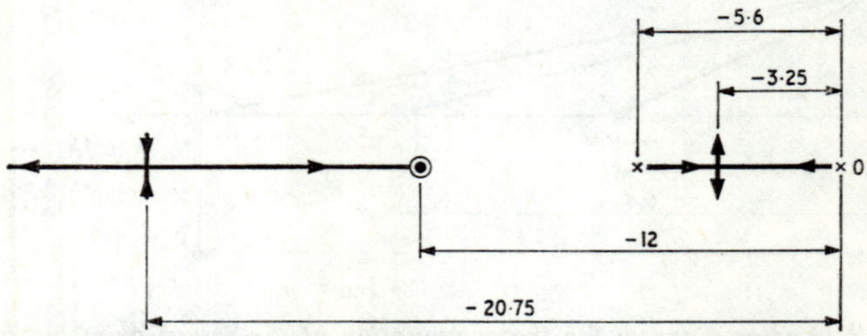


*Fig. 3.7. Sketch of the root locus on the negative real axis plus the breakaway and return points for Example 3.4.1*

We know that the two branches of the loci must leave the point $-3{\cdot}25$ and arrive at the point $-20{\cdot}75$, and it is now necessary to use Rule 9 to determine intermediate points on the locus. The best method of using Rule 9 is to choose a particular line in the root plane near to the known breakaway point, say the horizontal straight line passing through $+j2{\cdot}5$, and to determine the total angles for several trial points on this line. These trial points will start with total angles of, say, greater than 180°, and will end, as the trial point is moved along, with total angles of less than 180°. Interpolation to find the position of the root on the $+j2{\cdot}5$ line will then be straightforward. It should be noted that the starting point on the $+j2{\cdot}5$ line need not be taken at random, since we know that the locus is perpendicular at $(-3.25, j0)$ where it leaves the negative real axis. We would expect the locus to bend to the

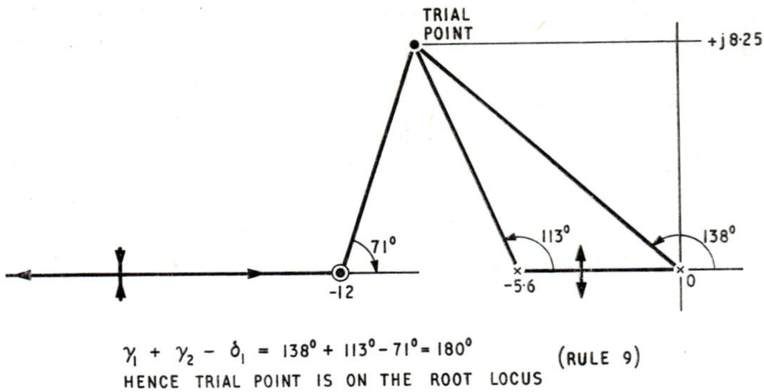


*Fig. 3.8. Vector diagram for a trial point on the root locus for Example 3.4.1*

left, and we should therefore start our trial points no further to the right than the point $-3{\cdot}25$. As a detailed example, the angles for the trial point at $-9{\cdot}25+j8{\cdot}25$ are shown in Fig. 3.8. The angles at the open loop poles and zeros, by direct measurement, are

$$\gamma_1 = 138^\circ$$

$$\gamma_2 = 113^\circ$$

and

$$\delta_1 = 71^\circ$$

The total angle at the trial point is thus $138^\circ+113^\circ-71^\circ = 180^\circ$, hence the trial point is indeed on the root locus. It is always advisable to sketch in the locus very roughly as the various points are determined, as this narrows down the search area for the next point, as shown in Fig. 3.9.

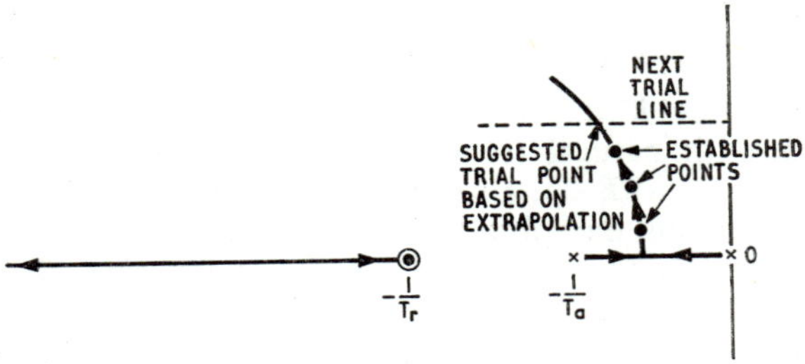


*Fig. 3.9. Sketching the root locus as points are located in order to determine the next trial point*

The line along which we search for points on the root locus should be varied to suit the particular part of the root plane which is under consideration. For example, the locus must obviously pass through the vertical line passing through the zero at $-12$, and it would therefore be advisable to search along this vertical line for a root.

We can, of course, also work back from the return point at $-20{\cdot}75$ if this is considered convenient. The final resultant locus is shown in Fig. 3.10, the branches forming a circle centred at

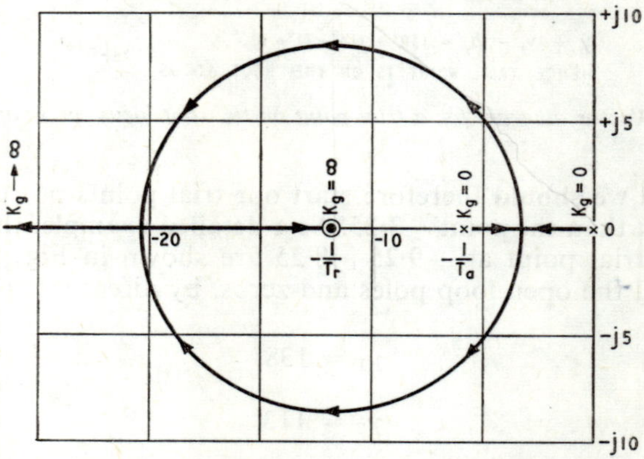


*Fig. 3.10. Complete root locus for Example 3.4.1 in terms of $K_g$*

$-12$, the lower half-plot being obtained by the symmetry condition of Rule 2. By inspecting the locus we can see that the system poles are situated in the left-hand plane for all positive values of $K$, and the system is thus always stable. We can also see that the poles can be categorised as follows:

*Region a.* For $K$ small, two negative real poles, one between 0 and $-3{\cdot}25$ and the other between $-3{\cdot}25$ and $-5{\cdot}6$.

*Region b.* For $K$ intermediate, there are complex conjugate poles, with natural frequencies varying between $3{\cdot}25$ and $20{\cdot}75$, and with damping ratios varying from $1{\cdot}0$ to $0{\cdot}68$. The minimum damping ratio corresponds to the line drawn from the origin tangential to the circular part of the root locus. This aspect will be examined further.

*Region c.* For $K$ large, there are again two negative real poles, one between $-12$ and $-20{\cdot}75$, and the other between $-20{\cdot}75$ and $-\infty$.

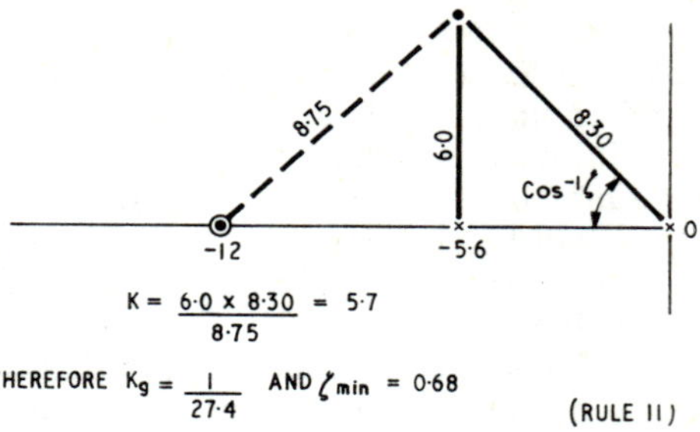


*Fig. 3.11. Vector diagram for the computation of the feedforward gain corresponding to the minimum possible damping ratio for Example 3.4.1*

We have still to calibrate the root locus in terms of feedforward gain. It will be remembered from Section 1.10, and Fig. 1.16, that the damping ratio for a pair of complex roots is defined as the cosine of the angle between the line joining the pole and the origin and the negative real axis. Thus we can see by inspection that the minimum value of damping for our system is defined by the line drawn from the origin tangential to the circle, which defines a damping angle of 47° and a damping ratio of 0·68, as shown in Fig. 3.11. The value of gain corresponding to this damping ratio may be determined using Rule 11 and the vector lengths shown in Fig. 3.11. Hence

$$K = \frac{\text{product of pole vectors}}{\text{product of zero vectors}} = \frac{6{\cdot}0 \times 8{\cdot}30}{8{\cdot}75} = 5{\cdot}70$$

To obtain $K_g$, we use the gain relationship defined by Equations 3.10 and 3.11, thus $K_g = KT_a/K_aT_r$. Substituting $K = 5{\cdot}7$, $T_a = 0{\cdot}179$, $T_r = 1/12$, and $K_a = 335$, we find $K_g = 1/27{\cdot}4$. The undamped natural frequency $\omega_r$ corresponding to this pole position is found by direct measurement to be 8·30 rad/sec.

At the risk of appearing pedestrian, we will now go back to the system transfer function in order that we can see exactly how much information has been obtained from the root locus plot. Referring to Equation 3.3, for the system under consideration $p(s) = w(s) = 1{\cdot}0$, hence from Equation 3.4 there are no system zeros, a process which we could have alternatively checked by obtaining the system transfer function

$$\frac{\phi_0}{\phi_D}(s) = \frac{1}{1+\left(T_r+\frac{1}{K_gK_a}\right)s+\left(\frac{T_a}{K_gK_a}\right)s^2} \tag{3.17}$$

from first principles. What the root locus plot tells us is that for

$T_r = 1/12$, $K_a = 335$ and $T_a = 0{\cdot}179$, a value of $K_g = 1/27{\cdot}4$ will result in a second order, unity numerator system with a damping ratio of 0·68 and an undamped natural frequency of 8·30 rad/sec, and this type of information will be repeated for any other point on the locus we may care to choose. For example, at the breakaway point there will be two equal roots giving a damping ratio of 1·0 and an undamped natural frequency of 3·25 rad/sec. The corresponding value of $K$ is $3{\cdot}25 \times 2{\cdot}35/8{\cdot}75 = 0{\cdot}870$, giving $K_g = 1/179$. All of this information could, in this case, have been obtained directly by computation from Equation 3.17; the only advantages of the root locus method for this simple system are:

a. It is immediately apparent that complex poles are possible.
b. It is immediately apparent that there is a minimum value of damping ratio that can be obtained from the system.
c. After a little contemplation it is apparent that when the poles are complex there is a trade-off balance between undamped natural frequency and damping ratio, and an optimum combination probably exists.

The latter point will be thoroughly covered later in this chapter, when ways of extracting further information from system pole-zero arrays will be discussed in detail, and also in Chapter 4, where rigorous design procedures for second order systems will be developed from first principles. If it is desired to find the effect on the root locus of changing the value of $T_r$, then the root locus must be redrawn for as many values of $T_r$ as are of interest. It is left to the reader to show for himself that if $T_r$ is made larger than $T_a$ the root locus cannot leave the negative real axis for any value of $K$, and the poles can therefore never be complex.

We shall now leave the simple second order system to study a slightly more complicated system, where the form of response changes radically, and which will thus justify a detailed correlation between the root locus plot and the system frequency response.

## 3.5 Root Locus for a Unity Numerator Third Order System in Terms of Feedforward Gain

EXAMPLE 3.5.1

If the servomechanism loop used to position the missile control surface of Example 3.4.1 is not infinitely fast in responding to the signal demanding the control surface deflection, the servo loop must be represented by an appropriate transfer function. Our first assumption will be that the servo loop transfer function is dominantly first order with a time constant $T_c$. The resulting third order system is shown in Fig. 3.12 and, by inspection, the open loop transfer function is

$$\frac{\bar{\phi}_0}{\bar{\phi}}(s) = \frac{K_g K_a(1+T_r s)}{(1+T_c s)(1+T_a s)s} \tag{3.18}$$

i.e. $p(s) = 1$, $q(s) = (1+T_a s)(1+T_c s)s$, $v(s) = 1+T_r s$, and $w(s) = 1{\cdot}0$. Rearranging Equation 3.18 in the form of Equation 3.7,

$$\frac{\bar{\phi}_0}{\bar{\phi}}(s) = \frac{K\left(s+\dfrac{1}{T_r}\right)}{\left(s+\dfrac{1}{T_a}\right)\left(s+\dfrac{1}{T_c}\right)s} \tag{3.19}$$

where $K$ is now $K_g K_a T_r/T_a T_c$, and is the value of gain which is computed from the root locus. We shall assume that $T_c = 0{\cdot}035$ $(= 1/28{\cdot}6)$ sec is considered to be a reasonable servo loop time constant, and take $T_a = 0{\cdot}179$, $T_r = 1/12$, and $K_a = 335$ as before. The open loop poles are at $-5{\cdot}6$ (aerodynamic), $-28{\cdot}6$ (servo) and 0 (integrator), and the open loop zero will be at $-12$.

There are three open loop poles and one finite open loop zero, hence there are three branches to the locus and there must be two open loop zeros at infinity. Using Rule 3, we know that the three branches start at 0, $-5{\cdot}6$ and $-28{\cdot}6$ respectively, and that one branch will end at $-12$, and the other two at infinity. Using Rule 4, we find that branches of the locus exist on the negative real axis between 0 and $-5{\cdot}6$, and between $-28{\cdot}6$ and $-12{\cdot}0$ only. The construction thus far is shown in Fig. 3.13.

There are two high-gain asymptotes, at angles given by $\pm k\pi/(n-m)$. Since $n-m = 2{\cdot}0$, then the angles are $\pm\pi/2$, $\pm 3\pi/2$, etc.; i.e. they are at 90° to the horizontal axis. The intersection point of the asymptote with the real axis is defined by the relationship

$$\begin{aligned} d &= \frac{\sum \text{open loop poles} - \sum \text{open loop zeros}}{n-m} \\ &= \frac{-(28{\cdot}6+5{\cdot}6+0-12)}{2} \\ &= -11{\cdot}1 \end{aligned}$$

The high-gain asympotes are shown in Fig. 3.14. It is now apparent from Fig. 3.13 and 3.14 that the branch between $-28.6$ and

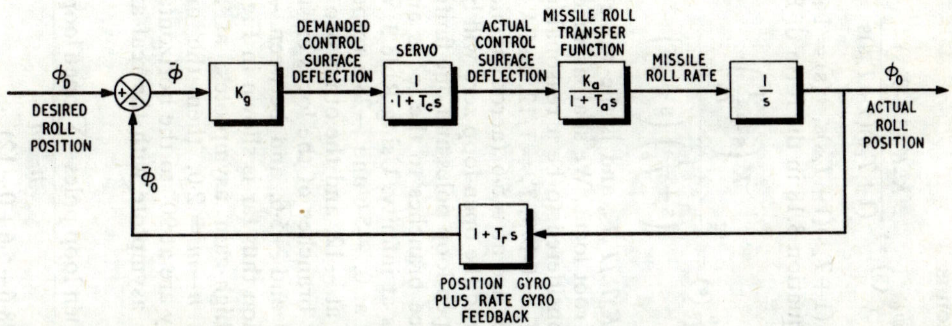


*Fig. 3.12. Block diagram of the third order missile roll autopilot loop used in Example 3.5.1*

$-12$ remains on the negative real axis, whilst the locus between 0 and $-5{\cdot}6$ must form two branches which must meet somewhere between these two values and then break away to form complex poles which will eventually proceed to infinity along the high-gain asymptotes. Clearly, the highest priority is now to calculate the

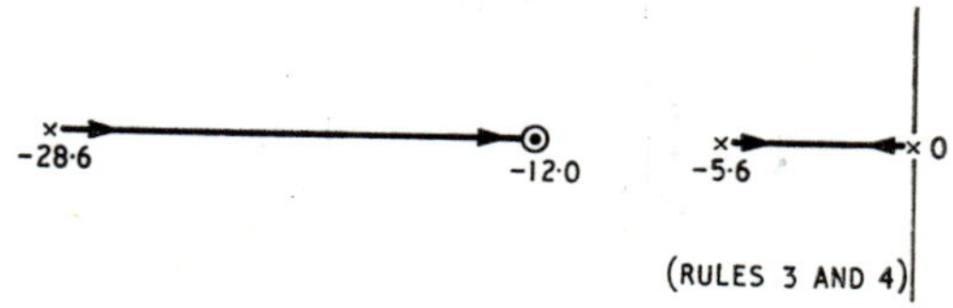


*Fig. 3.13. Root locus branches existing on the real axis for Example 3.5.1*

position of this breakaway point. The geometry required for computing this point is shown in Fig. 3.15. Using small-angle approximations, we may write

$$\left.\begin{aligned} \gamma_1 &= 180^\circ - \frac{h}{5{\cdot}6-y} \\ \gamma_2 &= \frac{h}{y} \\ \gamma_3 &= \frac{h}{23{\cdot}0+y} \\ \delta_1 &= \frac{h}{6{\cdot}4+y} \end{aligned}\right\} \tag{3.20}$$

Note that for convenience we have defined the distance $y$ from the open loop pole at $-5{\cdot}6$ on this occasion, whereas for Example 3.4.1 we defined the distance from the imaginary axis. The angular requirement, from Rule 9, states that

$$\gamma_1+\gamma_2+\gamma_3-\delta_1 = 180^\circ+N\ 360^\circ \tag{3.21}$$

Substituting Equation 3.20 into 3.21, we obtain the equation for $y$ as

$$\frac{1}{y} = \frac{1}{6{\cdot}4+y}+\frac{1}{5{\cdot}6-y}-\frac{1}{23{\cdot}0+y} \tag{3.22}$$

We would not expect the addition of the third open loop pole at $-28{\cdot}6$ to radically affect the position of the breakaway point, and so the first estimate for $y$ would be $5{\cdot}6-3{\cdot}25 = 2{\cdot}35$. This value of $y$ gives the left-hand side of Equation 3.22 as $0{\cdot}425$, and

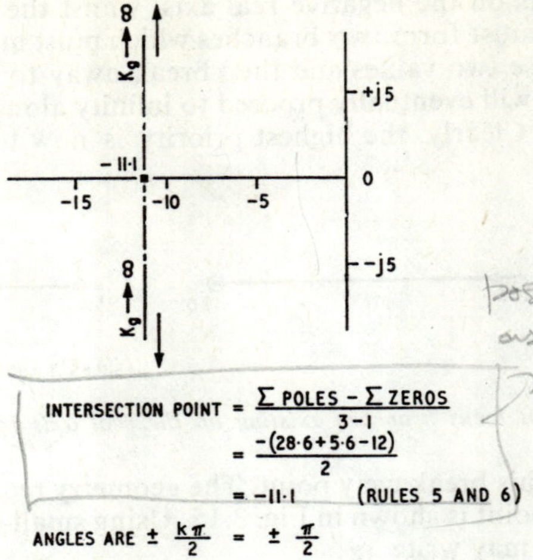


*Fig. 3.14. High-gain asymptotes for Example 3.5.1*

the right-hand side of Equation 3.22 as 0·383. Subsequent trial and error shows that $y$ is approximately 2·5, giving the breakaway point at a distance $-5{\cdot}6+2{\cdot}5 = -3{\cdot}1$ from the imaginary axis compared with a distance of $-3{\cdot}25$ for Example 3.4.1; Fig. 3.16 shows the root locus at this stage.

Further points on the locus are now obtained using Rule 9. Since the locus leaves the real axis at an angle of 90°, it is fairly straightforward to find points on the locus moving outward towards infinity particularly if the locus is sketched as points are found. For example, for the locus shown in Fig. 3.17, the first

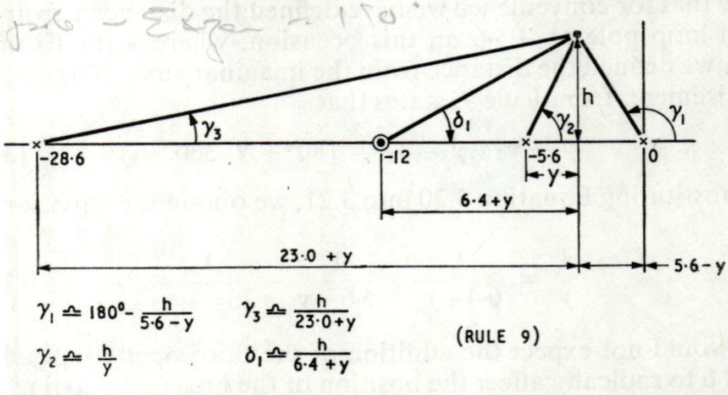


*Fig. 3.15. Vector diagram for the computation of the breakaway point on the real axis for Example 3.5.1*

Small angle approx
tan θ ≈

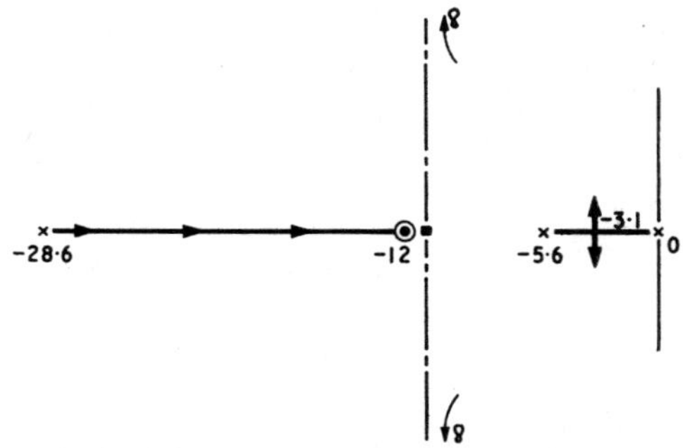


*Fig. 3.16. Sketch of the root locus on the negative real axis, the breakaway point and the high-gain asymptotes for Example 3.5.1*

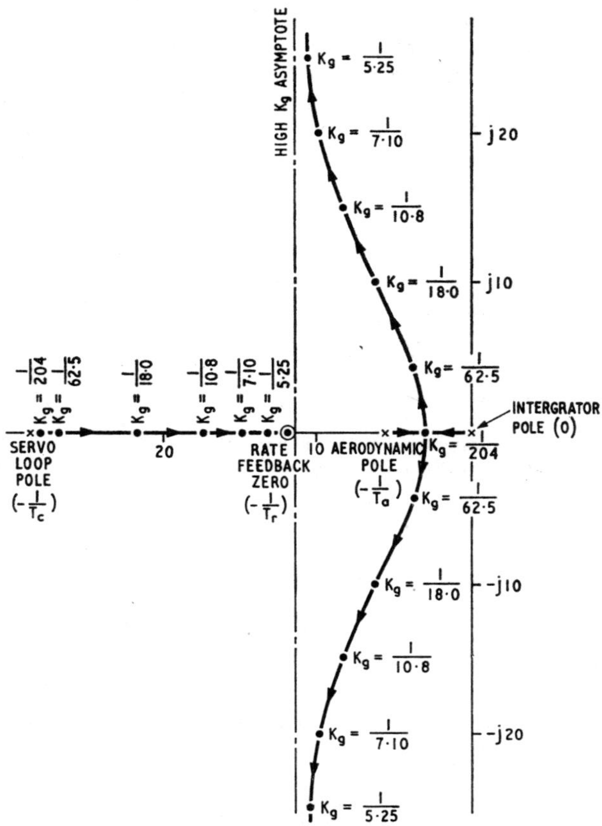


*Fig. 3.17. Root locus plot for the third order system of Example 3.5.1*

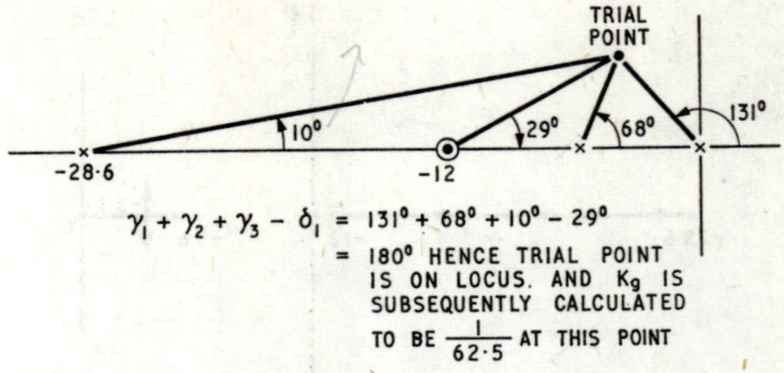


*Fig. 3.18. Vector diagram for the computation of the angles at the open loop poles and zeros for a trial point on the root locus of Example 3.5.1*

point marked on the locus corresponding to $K_g = 1/18{\cdot}0$ was the second trial point in that region, whilst the other points marked on the locus from there onward to infinity were, in fact, guessed correctly first time. The angle diagram for the complex pole corresponding to $K_g = 1/62{\cdot}5$ is shown in Fig. 3.18, and the angles, by direct measurement are, $\gamma_1 = 131°$, $\gamma_2 = 68°$, $\gamma_3 = 10°$, and $\delta_1 = 29°$, giving a total angle of 180°, showing that the point is indeed on the root locus.

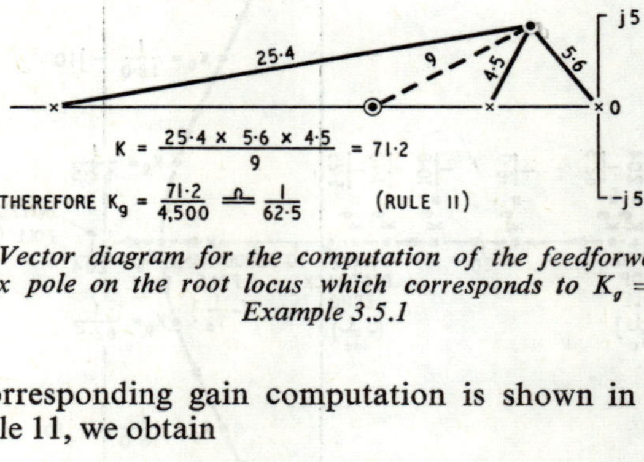


*Fig. 3.19. Vector diagram for the computation of the feedforward gain for the complex pole on the root locus which corresponds to $K_g = 1/62{\cdot}5$ for Example 3.5.1*

The corresponding gain computation is shown in Fig. 3.19. Using Rule 11, we obtain

$$K = \frac{\text{product of pole vectors}}{\text{product of zero vectors}} = \frac{25{\cdot}4 \times 5{\cdot}6 \times 4{\cdot}5}{9} = 71{\cdot}2$$

Now from Equations 3.18 and 3.19, $K_g = KT_aT_c/K_aT_r$, hence the value of $K_g$ is given by

$$K_g = \frac{71{\cdot}2 \times 0{\cdot}035 \times 0{\cdot}179 \times 12}{335} = \frac{1}{62{\cdot}5}$$

We now need to find the third pole corresponding to $K_g = 1/62{\cdot}5$. The simplest way to do this is to use Rule 11, since this is applicable, and determine the centre of gravity of the roots by considering the case for $K = 0$ as shown in Fig. 3.20. If $\bar{x}$ is the

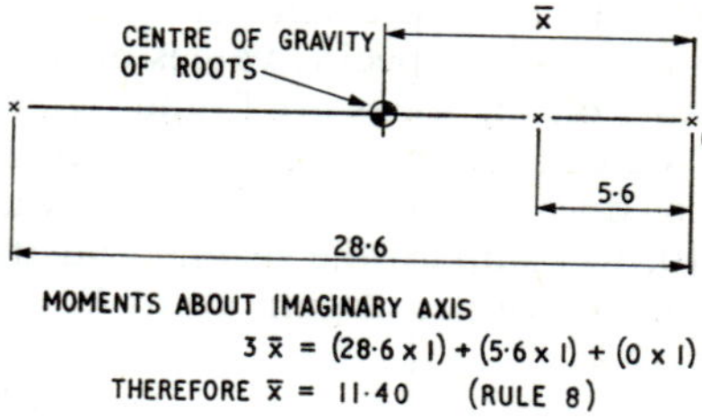


*Fig. 3.20. Computation of the position of the centre of gravity of the system poles for Example 3.5.1 by considering the case when $K = 0$*

distance of the centre of gravity from the imaginary axis, then taking moments about the imaginary axis gives us

$$3\bar{x} = (28{\cdot}6\times1)+(5{\cdot}6\times1)+(0\times1)$$

hence $\bar{x} = 11{\cdot}40$.

For $K_g = 1/62{\cdot}5$, two of the roots have moved to a distance of 3·9 from the imaginary axis as shown in Fig. 3.21. Taking moments

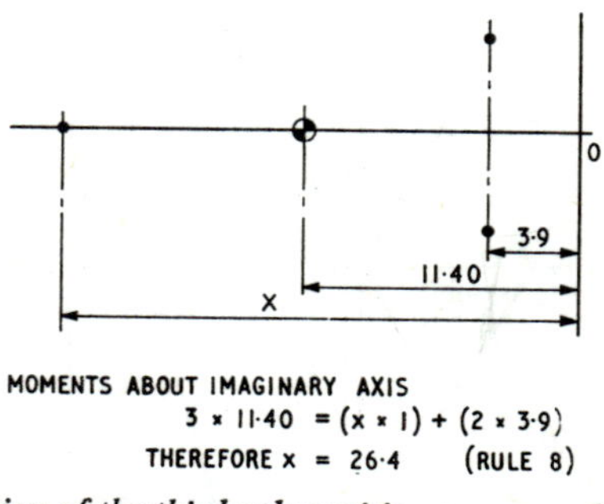


*Fig. 3.21. Computation of the third pole position corresponding to $K_g = 1/62{\cdot}5$ for Example 3.5.1*

about the imaginary axis to determine $x$, the position of the third pole from the imaginary axis, we have

$$3\times11{\cdot}40 = (1\times x)+(2\times3{\cdot}9)$$

hcnee $x = 26{\cdot}4$.

This process has been repeated for the wide range of values of $K_g$ shown on the locus in Fig. 3.17. The points marked on the locus were chosen for convenience, the intervals taken being such that a reasonable sample of responses might be obtained as $K_g$ was varied from zero to infinity. Clearly the values of $K_g$ evaluated for the points chosen on the locus are dictated by the geometry of the open loop poles and zeros. If specific values of $K_g$ are of interest, corresponding points on the locus are found by interpolation.

From Fig. 3.17 we can see that the system cannot be unstable, since the root locus does not cross the imaginary axis. Inspecting Figs. 3.17 and 3.10, to compare the second and third order root loci, we see that for low values of gain up to $K_g = 1/62{\cdot}5$ there is

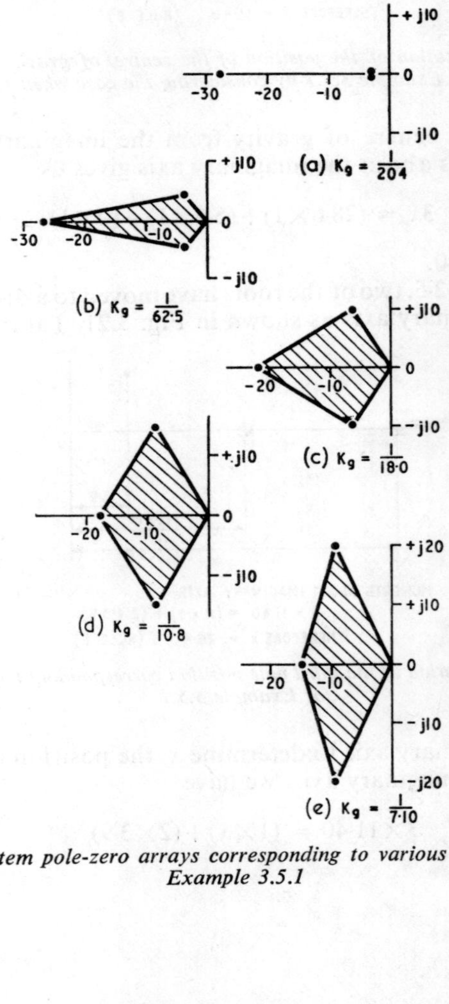


*Fig. 3.22. System pole-zero arrays corresponding to various values of $K_g$ for Example 3.5.1*

little difference between the root loci for the two systems. For $K_g > 1/18{\cdot}0$ the root loci are quite different, and for Example 3.5.1 there is no limit to the value of damping ratio obtainable for the complex poles. However, the presence of the third pole moving in towards the $-12$ point as the feedforward gain is increased will eventually lead to a situation where this pole is tending to dominate the response, as we shall see very shortly.

Referring to Equation 3.3, for the system under consideration $p(s) = w(s) = 1{\cdot}0$, hence from Equation 3.4 there are no system zeros. This can be checked from first principles, whence it will be found that the system transfer function in simplest form is

$$\frac{\phi_o}{\phi_D}(s) = \frac{1}{1+\left(T_r+\dfrac{1}{K_g K_a}\right)s+\left(\dfrac{T_a+T_c}{K_g K_a}\right)s^2+\left(\dfrac{T_a T_c}{K_g K_a}\right)s^3} \tag{3.23}$$

The system pole-zero array therefore consists of three poles only, and the array is shown in Fig. 3.22 for five different values of $K_g$. For $K_g \geqslant 1/204$, Equation 3.23 can be conveniently factorised into the form

$$\frac{\phi_o}{\phi_D}(s) = \frac{1}{(1+Ts)\left[1+\left(\dfrac{2\zeta}{\omega_n}\right)s+\dfrac{s^2}{\omega_n^2}\right]} \tag{3.24}$$

where $T$, $\zeta$, and $\omega_n$ can be read directly from the root locus plot. A summary of these characteristics is shown in Table 3.5.1.

Before constructing more complex root loci we shall consider how we may use the information we have obtained from the root locus plot of Example 3.5.1 to predict system performance.

**Table 3.5.1**

TRANSFER FUNCTION TERMS AS A FUNCTION OF $K_g$ FOR EXAMPLE 3.5.1

| $K_g$ | $1/T$ *(First order break frequency)* | $\omega_n$ *(Second order break frequency)* | *Damping angle* | *Damping ratio* |
|---|---|---|---|---|
| 1/204 | 28·0 | 3·1 | 0° | 1·0 |
| 1/62·5 | 26·4 | 5·6 | 49° | 0·66 |
| 1/18·0 | 21·5 | 11·6 | 57·5° | 0·54 |
| 1/10·8 | 17·6 | 17·1 | 60·5° | 0·49 |
| 1/7·10 | 14·8 | 22·25 | 63·5° | 0·45 |

## 3.6 Logarithmic Approximations to the System Frequency Response Given the System Pole-zero Array

We have already seen in Chapter 2 that the frequency response of linear transfer functions may be conveniently represented by plotting the amplitude ratio in decibels, and choosing a logarithmic frequency scale as shown in Fig. 3.23. The exact logarithmic plot can be determined using standard templates based

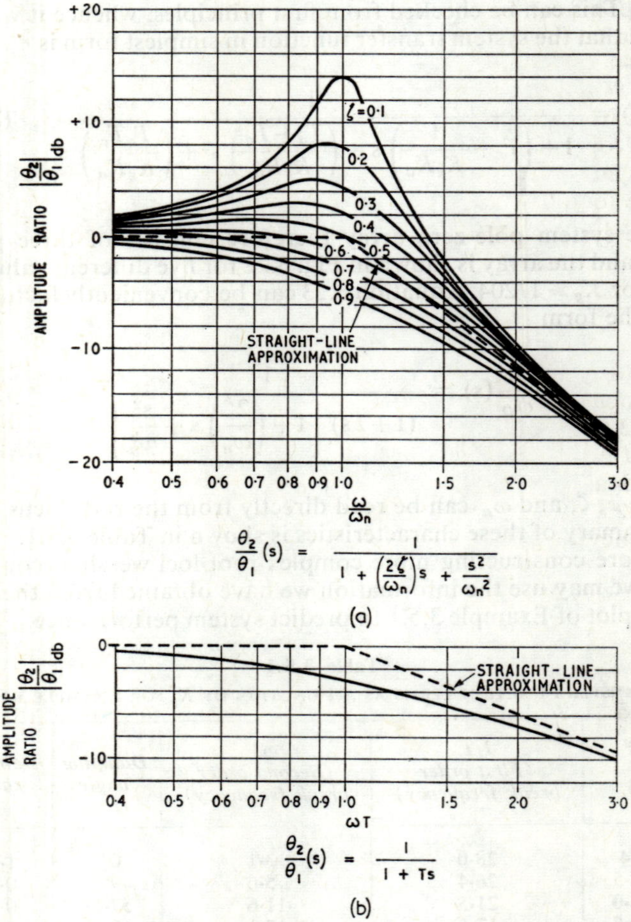


*Fig. 3.23. Logarithmic plot of the frequency response of first and second order linear transfer functions*

on Fig. 3.23, provided exact values of damping ratio are chosen for any complex poles which may be present in the system pole-zero array. At the initial synthesis stage it is often sufficient to plot the straight-line approximations to the logarithmic form of frequency response until such time as a promising design is found, whence more exact responses are evaluated.

In particular, a very crude approximation to a first order term $1/(1+Ts)$ is given by a line of slope 0 db from $\omega = 0$ to $\omega = 1/T$, and a slope of $-20$ db /decade from $\omega = 1/T$ to $\omega = \infty$. A similar approximation for a second order term $1/[1+(2\zeta/\omega_n)s+s^2/\omega_n^2]$ is given by a line of slope 0 db from $\omega = 0$ to $\omega = \omega_n$ and a slope of $-40$ db/decade from $\omega = \omega_n$ to $\omega = \infty$. The frequencies $\omega = 1/T$ and $\omega = \omega_n$ are known as the break frequencies or corner frequencies.

The logarithmic plotting of frequency responses has the considerable advantage that the response of transfer functions in series can be found by adding the individual responses. Thus the logarithmic form of Equation 3.24 is

$$20 \log_{10} \left| \frac{\phi_0}{\phi_D} \right|_{\omega} = -20 \log_{10} |1+Ts|_{j\omega} - 20 \log_{10} \left| 1 + \left(\frac{2\zeta}{\omega_n}\right) s + \frac{s^2}{\omega_n^2} \right|_{j\omega} \tag{3.25}$$

For a given value of $K_g$ we already know $T$ and $\omega_n$ form the root locus plot, and this information has already been given in Table 3.5.1. Thus the break frequencies $1/T$ and $\omega_n$ for the first and second order terms respectively are also known. Adding the two individual frequency responses together in accordance with Equation 3.25, we obtain the straight-line approximation to the system frequency responses for the five values of $K_g$ as shown in Fig. 3.24.

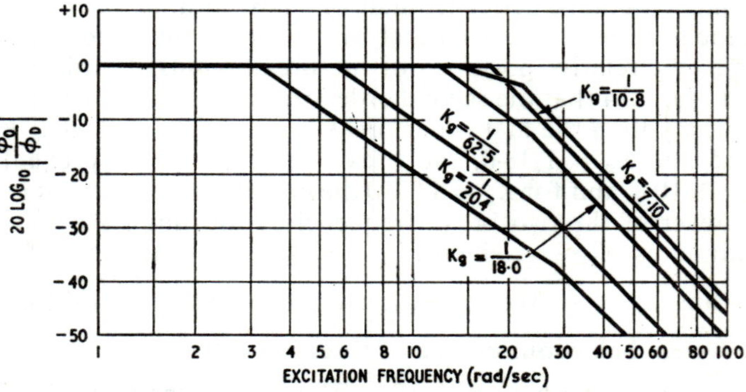


*Fig. 3.24. Straight-line approximation to the logarithmic plot of the frequency response of Example 3.5.1 for five values of $K_g$*

These straight-line approximations enable us to make a very crude estimate of system bandwidth by taking the frequency at which the gain is $-3$ db. This crude approximation to system bandwidth is shown in Fig. 3.25 as a function of $1/K_g$. It is immediately evident

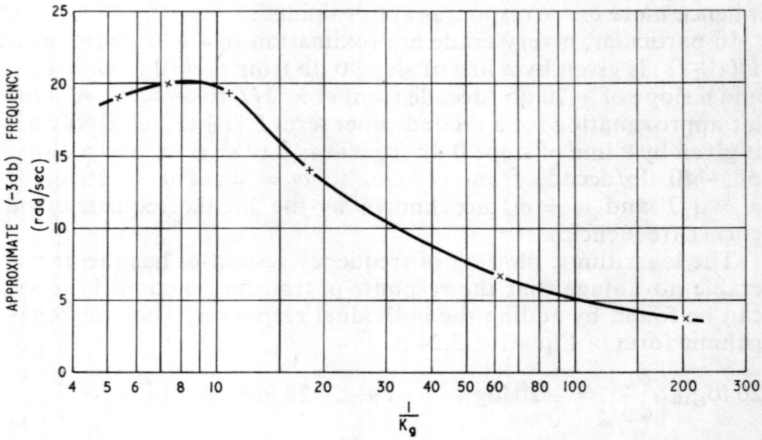


*Fig. 3.25. Approximate $-3$ db frequencies obtained from logarithmic approximation to the frequency response for Example 3.5.1*

that the maximum bandwidth possible with this system is somewhat less than 20 rad/sec, since by inspection we can see that one or more break points must occur at a frequency lower than 17 rad/sec. Thus we conclude that in particular the systems corresponding to $K_g = 1/18{\cdot}0$ and $K_g = 1/10{\cdot}8$ are worth further investigation in the frequency domain at this stage in order to estimate $M_p$ and to obtain a better estimate of bandwidth. Before determining the response more accurately, it is worthwhile directing attention to the relative positions of the break points. It will be seen from Fig. 3.24 that the second order break point occurs at a lower frequency than the first order break point for $K_g = 1/204$, $K_g = 1/62{\cdot}5$ and $K_g = 1/18{\cdot}0$, the break points are nearly equal for $K_g = 1/10{\cdot}8$, and for $K_g = 1/7{\cdot}10$ the first order break point is now the lower of the two break frequencies; and this state will now continue as $K_g \to \infty$, since Fig. 3.17 tells us that $\lim_{K_g \to \infty} 1/T = 12$ and $\lim_{K_g \to \infty} \omega_n = \infty$. We may therefore conclude by intuition that the system behaviour can be of three basic types.

a. Region where $K_g \ll 1/10{\cdot}8$. The first order factor is completely swamped by the second order factor, thus leaving a dominantly second order response. This response can be determined to a reasonable degree of accuracy by approximating the system to a second order transfer function.

b. Region where $K_g$ is of the order of $1/10{\cdot}8$. Provided $K_g$ is of the same order as $1/10{\cdot}8$, i.e. is not greater than about $5/10{\cdot}8$, say, and is not less than $1/5\times 10{\cdot}8$, both factors are important, and the response may appear quite different in form from that obtainable from either a first or second order transfer function. To attempt to reduce the order of the system transfer function by approximation may therefore result in misleading information.

c. Region where $K_g \gg 1/10{\cdot}8$. The second order factor is completely swamped by the first order factor, thus leaving a dominantly first order response. This response can then be determined to a reasonable degree of accuracy by approximating the system to a first order transfer function.

## 3.7 The Graphical Determination of System Frequency Response from the System Pole-zero Array

To obtain the system frequency response more accurately than was done in Fig. 3.24, it is possible to sketch in a more exact response for the individual first and second order transfer functions as shown in Fig. 3.23, and a crude estimate of $M_p$ can then be made. If the contributions from two or more transfer functions interact, as for example, when $K_g = 1/10{\cdot}8$, then the addition and subtraction involved becomes tedious, and unless great care is taken accuracy may be lost to such an extent that little has been gained in adding to the original straight-line approximations. Fortunately, it is possible to determine the steady state frequency response from the system pole-zero array using a simple graphical technique plus a slide-rule or the Spirule previously mentioned in Section 3.3. The graphical technique, which at this stage is offered without proof, is particularly easy to remember. The imaginary axis, to scale, is now the $j\omega$ axis, where $\omega$ is the excitation frequency. Vectors are now drawn from the frequency $\omega$ of interest to all the system poles and zeros. The amplitude ratio $|\theta_0/\theta_i|$, at frequency $\omega$, is then given by

$$\left|\frac{\theta_0}{\theta_i}\right|_{\omega} = \frac{\prod_{i=1}^{i=n} |s_{i.0}| \times \prod_{i=1}^{i=q} |z_{i.\omega}|}{\prod_{i=1}^{i=n} |s_{i.\omega}| \times \prod_{i=1}^{i=q} |z_{i.0}|} \tag{3.26}$$

where $|s_{i.\omega}|$ is the length of the vector from the $i^{th}$ pole to the point $\omega$ on the imaginary axis, $|s_{i.0}|$ is the length of the vector from the $i^{th}$ pole to the point $\omega$ when $\omega = 0$, $|z_{i.\omega}|$ is the length of the vector from the $i^{th}$ zero to the point $\omega$ on the imaginary axis and $|z_{i.0}|$ is the length of the vector from the $i^{th}$ zero to the point $\omega$ when $\omega = 0$. For a given system pole-zero array, $\prod_{i=1}^{i=n} |s_{i.0}|$ and $\prod_{i=1}^{b=1} |z_{i.0}|$ are constant, and we may therefore write

$$\left|\frac{\theta_o}{\theta_i}\right|_{\omega} = \mu \frac{\prod_{i=1}^{i=q} |z_{i.\omega}|}{\prod_{i=1}^{i=n} |s_{i.\omega}|} \tag{3.27}$$

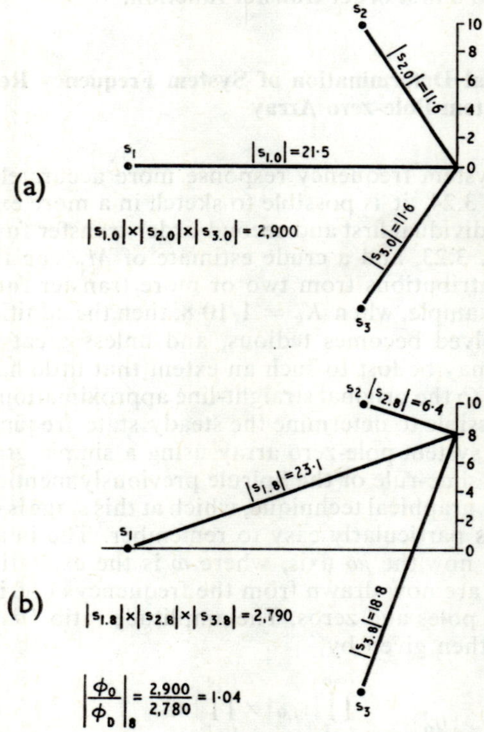


*Fig. 3.26. Vector diagram for the graphical computation of the steady state amplitude ratio for $\omega = 8$ rad/sec for Example 3.5.1, when $K_g = 1/18{\cdot}0$: (a) zero frequency vectors; (b) 8 rad/sec vectors*

where $\mu$ is the vector product at zero frequency. If there are no system zeros, the product of the zero vectors is taken as unity.

A sample computation for the system with $K_g = 1/18{\cdot}0$ is shown in Fig. 3.26, for a frequency of 8 rad/sec. The vector lengths are

$$|s_{1.0}| = 1/T = 21{\cdot}5 \qquad |s_{1.8}| = 23{\cdot}1$$
$$|s_{2.0}| = \omega_n = 11{\cdot}6 \qquad |s_{2.8}| = 6{\cdot}4$$
$$|s_{3.0}| = \omega_n = 11{\cdot}6 \qquad |s_{3.8}| = 18{\cdot}8$$

hence

$$\prod_{i=1}^{=3} |s_{i.0}| = 21{\cdot}5 \times 11{\cdot}6 \times 11{\cdot}6 = 2{,}900 = \mu$$

and

$$\prod_{i=1}^{=3} |s_{i.8}| = 23{\cdot}1 \times 6{\cdot}4 \times 18{\cdot}8 = 2{,}780$$

Thus from Equation 3.27, since there are no system zeros,

$$\left|\frac{\phi_o}{\phi_D}\right|_{\omega=8} = \frac{2{,}900}{2{,}780} = 1{\cdot}04$$

This sample point should convince the reader that this method of computing the steady state frequency response is much quicker than using Pythagoras' theorem algebraically and calculating individual moduli for the first and second order factors, particularly if a Spirule is available for multiplying the appropriate vector lengths.

Using this technique, the more exact frequency responses for the systems with $K_g = 1/18{\cdot}0$ and $K_g = 1/10{\cdot}8$ were constructed,

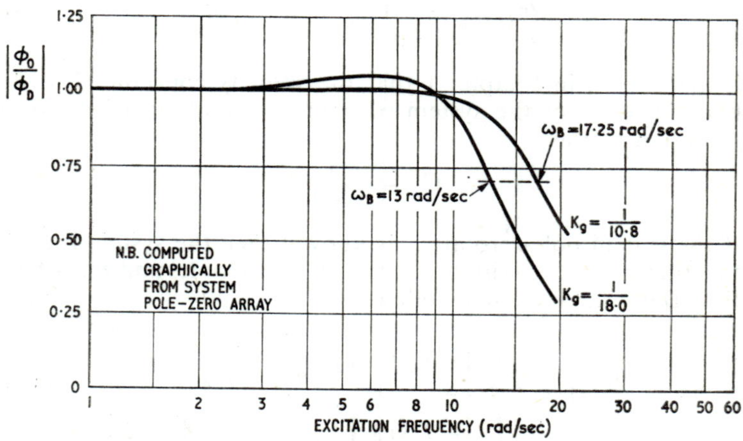


*Fig. 3.27. More exact frequency responses for two different values of $K_g$ for Example 3.5.1*

and it can be seen from Fig.3.27 that the first system has $M_p \simeq 1{\cdot}05$, $\omega_B \simeq 13$ rad/sec, whilst the second system has $M_p \simeq 1{\cdot}00$, $\omega_B \simeq 17{\cdot}5$ rad/sec. These two systems were chosen for further study on the evidence of Fig. 3.23, which certainly ruled out $K_g = 1/204$ and $K_g = 1/62{\cdot}5$ on the basis of too low a bandwidth to be of use in many applications. The system performance in the frequency domain obtainable with the system corresponding to $K_g = 1/10{\cdot}8$ may be acceptable as the bandwidth of 17·5 rad/sec is obtained with a very flat response to a frequency of 10 rad/sec. If a higher bandwidth is required, then the exercise would be repeated for different values of $T_r$, and if this failed to improve matters it would then be necessary to redesign the system to provide a system zero, or alternatively provide, for example, acceleration feedback in addition to the rate feedback term $T_r$. These aspects will be discussed more fully later, particularly in Chapters 4 and 5.

## 3.8 Determination of $C_v$ from the System Pole-zero Array

In Chapter 1 we discussed at some length the various ways in which system performance is specified, and concluded that in addition to stating a definite value of bandwidth, a specific requirement for a minimum value of velocity constant might be a common occurrence. We have just seen how bandwidth is obtained from the system pole-zero array, and we shall now proceed to determine $C_v$ as a function of $K_g$ for Example 3.5.1. For a general system with transfer function

$$\frac{\theta_0}{\theta_i}(\text{s}) = \frac{1+b_1\text{s}+b_2\text{s}^2+\ldots b_q\text{s}^q}{1+a_1\text{s}+a_2\text{s}^2+\ldots a_n\text{s}^n} \tag{3.28}$$

it was shown in Chapter 1 that the velocity constant $C_v$ is defined in terms of the system transfer function coefficients from Equation 1.28 as

$$C_v = \frac{1}{a_1-b_1} \tag{3.29}$$

If the system pole-zero array is known, then the system transfer function may be rewritten in terms of system poles and zeros from Equations 3.5 and 3.6 as follows

$$\frac{\theta_0}{\theta_i}(\text{s}) = \frac{\prod\limits_{z=1}(1+T_z\text{s})\times\prod\limits^{z=q}\left[\frac{\text{s}^2}{\omega_{nz}^2}+\left(\frac{2\zeta_z}{\omega_{nz}}\right)\text{s}+1\right]}{\prod\limits_{p=1}(1+T_p\text{s})\times\prod\limits^{p=n}\left[\frac{\text{s}^2}{\omega_{np}^2}+\left(\frac{2\zeta_p}{\omega_{np}}\right)\text{s}+1\right]} \tag{3.30}$$

Expanding Equation 3.30 it will be seen that the $b_1$ coefficient is given by the sum of the s terms of the numerator taken one at a time, and that the $a_1$ coefficient is given by the sum of the s terms of the denominator taken one at a time. Thus

$$\left.\begin{aligned} b_1 &= \sum_{z=1} T_z + \sum^{z=q} \frac{2\zeta_z}{\omega_{nz}} \\ a_1 &= \sum_{p=1} T_p + \sum^{p=n} \frac{2\zeta_p}{\omega_{np}} \end{aligned}\right\} \tag{3.31}$$

As will be shown subsequently, from Equation 3.31 it may be interpreted that $b_1$ is the sum of the reciprocals of the system zeros, whilst $a_1$ is the sum of the reciprocals of the system poles. Thus in terms of the system pole-zero array, we may write

$$-\frac{1}{C_v} = \sum_1^n \frac{1}{\text{system poles}} - \sum_1^q \frac{1}{\text{system zeros}} \tag{3.32}$$

If we apply Equation 3.32 to Example 3.5.1 we see that there are three system poles and no system zeros, and considering the case where two of the poles form a complex pair, we have

$$-\frac{1}{C_v} = \frac{1}{s_1} + \frac{1}{s_2} + \frac{1}{s_3}$$

Hence

$$\begin{aligned} \frac{1}{C_v} &= T + \frac{1}{\zeta\omega_n + j\omega_n\sqrt{1-\zeta^2}} + \frac{1}{\zeta\omega_n - j\omega_n\sqrt{1-\zeta^2}} \\ &= T + \frac{2\zeta\omega_n}{\zeta^2\omega_n^2 + \omega_n^2(1-\zeta^2)} \end{aligned}$$

and finally

$$\frac{1}{C_v} = T + \frac{2\zeta}{\omega_n} \tag{3.33}$$

Equation 3.33 shows that two complex poles result in a simple system pole reciprocal of $2\zeta/\omega_n$, thus confirming the interpretation of Equations 3.31 and 3.32 as the sum of the reciprocals of the system zeros and system poles respectively. $C_v$ can also be plotted directly from the system transfer function, since in this case, for a given value of $K_g$, the $a_1$ transfer function coefficient is known from Equation 3.23 as

$$a_1 = T_r + \frac{1}{K_g K_a} \tag{3.34}$$

whence

$$\frac{1}{C_v} = T_r + \frac{1}{K_g K_a} \tag{3.35}$$

$C_v$ has been calculated from both Equation 3.35 and from Equation 3.33, in the latter case using the values of $T$, $\zeta$, and $\omega_n$ obtained from the root locus plot of Fig. 3.17, and tabulated in Table 3.5.1. Fig. 3.28 shows $C_v$ as a function of $1/K_g$. The close agreement between the two methods of calculation shows that the root

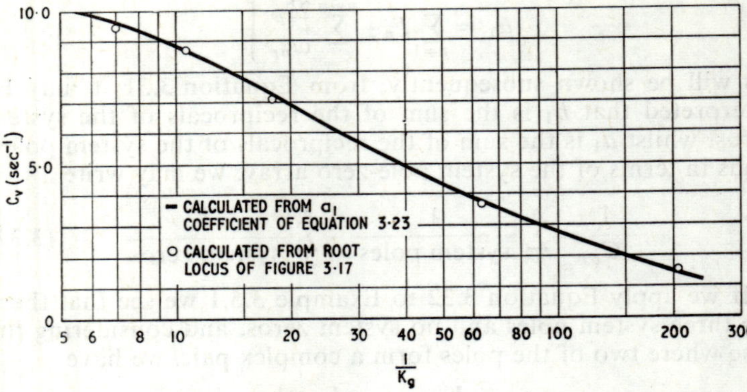


*Fig. 3.28. $C_v$ as a function of $1/K_g$ for Example 3.5.1*

locus plot of Fig. 3.17 is correct, and the simple calculation of Equation 3.35 in this instance has provided a very rapid method of checking the constructional accuracy. As a design aid, Fig. 3.28 indicates the value of $K_g$ needed to satisfy a given $C_v$ requirement for the system under consideration, and whether the system, as it stands, is capable of fulfilling this requirement. The maximum value of $C_v$ possible with this system occurs when $\omega_n = \infty$, leaving $C_v$ as $1/T_r = 12{\cdot}0$; but this is, of course, an impracticable system. $C_v$ can be increased either by decreasing $T_r$ or by introducing a system zero. This latter problem we shall consider next.

## 3.9 System Pole-zero Array for a Three-pole One-zero System in Terms of Feedforward Gain

EXAMPLE 3.9.1

The addition of a system zero can completely alter system response; in general, bandwidth will increase, $M_p$ will increase, and $C_v$ will increase. Fig. 3.29 shows a third order missile roll autopilot which differs from Fig. 3.12 in that the rate feedback

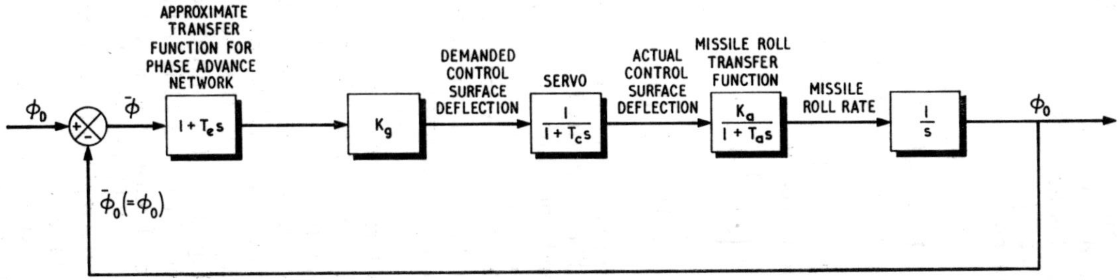


$$\frac{\phi_0}{\phi_D}(s) = \frac{1 + b_1 s}{1 + a_1 s + a_2 s^2 + a_3 s^3}$$

WHERE $b_1 = T_e$, $a_1 = T_e + \frac{1}{K_g K_a}$, $a_2 = \frac{T_a + T_c}{K_g K_a}$, $a_3 = \frac{T_a T_c}{K_g K_a}$

*Fig. 3.29. Block diagram of the third order missile roll autopilot with a system zero used in Example 3.9.1*

term has been removed, and a phase advance network of approximate transfer function $1+T_e s$ has been inserted in the feedforward path. We have assumed that the lag term in the network is fast compared with the speed of response of the autopilot, and that the attenuation due to this network has been compensated by amplification to produce $G_c(s) \simeq 1+T_e s$. The open loop transfer function for this system is

$$\frac{\bar{\phi}_0}{\bar{\phi}}(s) = \frac{K_g K_a(1+T_e s)}{(1+T_c s)(1+T_a s)s} \tag{3.36}$$

i.e. $p(s) = 1+T_e s$, $q(s) = (1+T_c s)(1+T_a s)s$, $v(s) = 1{\cdot}0$, and $w(s) = 1{\cdot}0$.

Rearranging Equation 3.36 into the form of Equation 3.7, the broken loop transfer function becomes

$$\frac{\bar{\phi}_0}{\bar{\phi}}(s) = \frac{K\left(s+\dfrac{1}{T_e}\right)}{\left(s+\dfrac{1}{T_a}\right)\left(s+\dfrac{1}{T_c}\right)s} = -1 \tag{3.37}$$

where $K = K_g K_a T_e/T_a T_c$, and is the value of gain which is computed from the root locus. The similarity between Equations 3.37 and 3.19 should be noted and underlined. For a given flight case for $K_a$ and $T_a$, and a given servo loop time constant, the root loci will be identical provided that $T_r = T_e$. For this example we shall indeed assume that $T_e = 1/12$, which is the same value as used for $T_r$ in Example 3.5.1, and will investigate the effect on performance of using the phase advance network as compared with the use of the rate feedback term. Since the root locus is identical with that shown in Fig. 3.17, it will not be repeated here. The system zero is given by $p(s)w(s) = 0$, i.e. $1+T_e s = 0$, hence $z_1 = -1/T_e$. For the five values of feedforward gain considered in detail in Example 3.5.1, i.e. $K_g = 1/204$, $K_g = 1/62{\cdot}5$, $K_g = 1/18{\cdot}0$, $K_g = 1/10{\cdot}8$ and $K_g = 1/7{\cdot}10$, the new system pole-zero arrays are shown in Fig. 3.30. The only difference between these arrays and those shown in Fig. 3.22 is the addition of the zero at $-1/T_e = -12$ in each case. The new values of $C_v$ are easily obtained from Equation 3.32 and are plotted against $1/K_g$ in Fig. 3.31, showing that in the region of $K_g$ of particular interest it is easily possible to double the value of $C_v$ by adding the system zero at $-12$. It should be noted at this stage by comparison with Figs. 1.18 and 1.19, that the use of the phase advance network approximated by $1+T_e s$ in the feedforward path is an example of series compensation, whilst the use of the rate feedback term $T_r s$ is an example of feedback compensation.

The straight-line approximations to the logarithmic plot of frequency response of Example 3.9.1 are shown in Fig. 3.32. It can be seen that, provided the lowest break point of the system is greater than about 12, the bandwidth is increased, as indicated by the approximate $-3$ db frequencies plotted in Fig. 3.33. For $K_g = 1/10{\cdot}8$, the $-3$ db frequency is increased from about 20 to 25 rad/sec. Rounding off approximations indicates that the true bandwidth will be somewhat different from these values of 20 and 25 rad/sec. To show this effect more accurately, the frequency response for the system with $K_g = 1/18{\cdot}0$ has been constructed from the system pole-zero array, and is shown in Fig. 3.34. The true bandwidth for the system with a zero at $-12$ is about

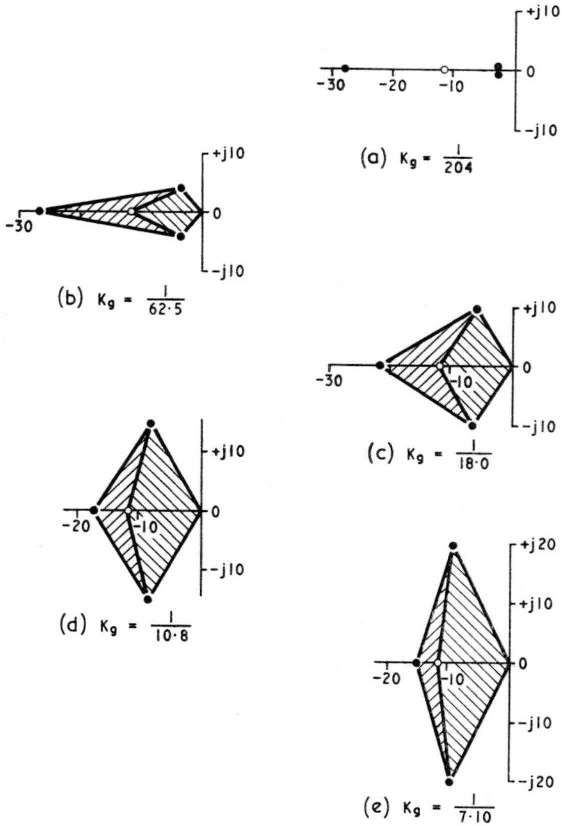


*Fig. 3.30. System pole-zero arrays corresponding to various values of $K_g$ for Example 3.9.1: (a) $K_g = 1/204$; (b) $K_g = 1/62{\cdot}5$; (c) $K_g = 1/18{\cdot}0$; (d) $K_g = 1/10{\cdot}8$; (e) $K_g = 1/7{\cdot}10$*

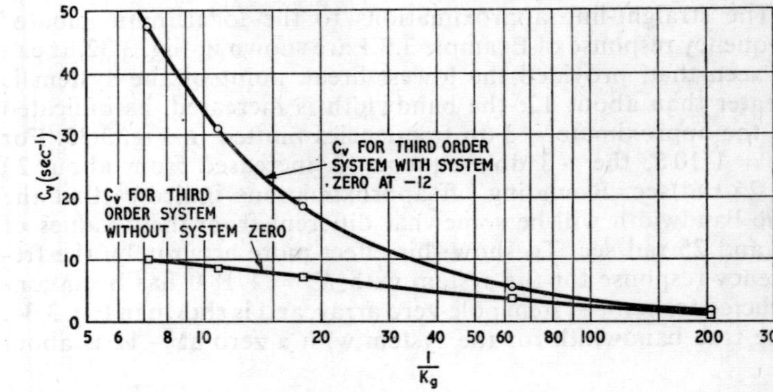


*Fig. 3.31. $C_v$ as a function of $K_g$ for Example 3.9.1, showing the effect on $C_v$ of a system zero at −12*

17·25 rad/sec compared with the −3 db frequency of 15·5 rad/sec. This increase is due to the effect of the zero at −12 which is contributing an amplification at the bandwidth frequency which is ignored in the straight-line approximations of Fig. 3.32. Without the system zero present, the system bandwidth is 13 rad/sec and the bandwidth increase obtained by adding the system zero at −12 is at the expense of increasing $M_p$ from approximately 1·05 to approximately 1·26, but we would also have the additional advantage of increasing $C_v$ by a factor greater than two. Adding a system zero at −12 is not by any means the only way in which the system performance could be adjusted for a given value of $K_g$. It is reasonable to assume that the missile parameters $K_a$ and $T_a$

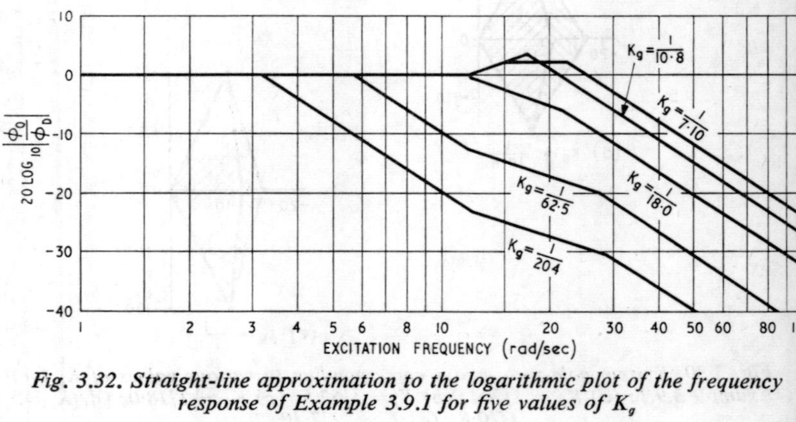


*Fig. 3.32. Straight-line approximation to the logarithmic plot of the frequency response of Example 3.9.1 for five values of $K_g$*

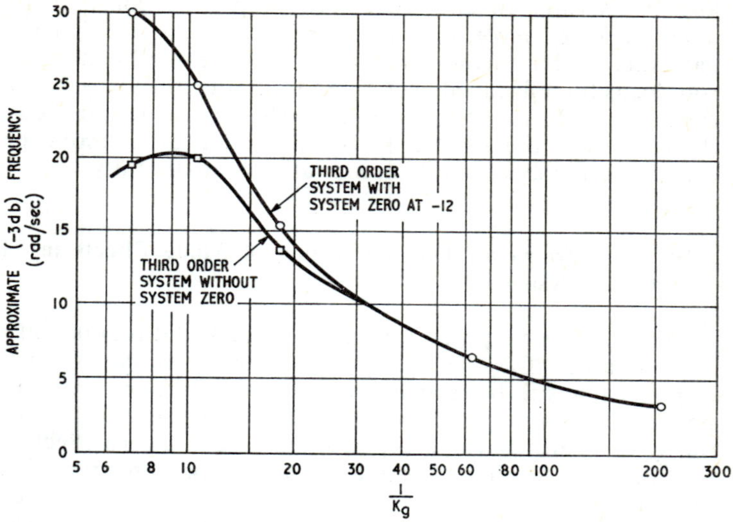


*Fig. 3.33. Approximate −3 db frequencies as a function of $K_g$ for Example 3.9.1, showing the effect of a system zero at −12*

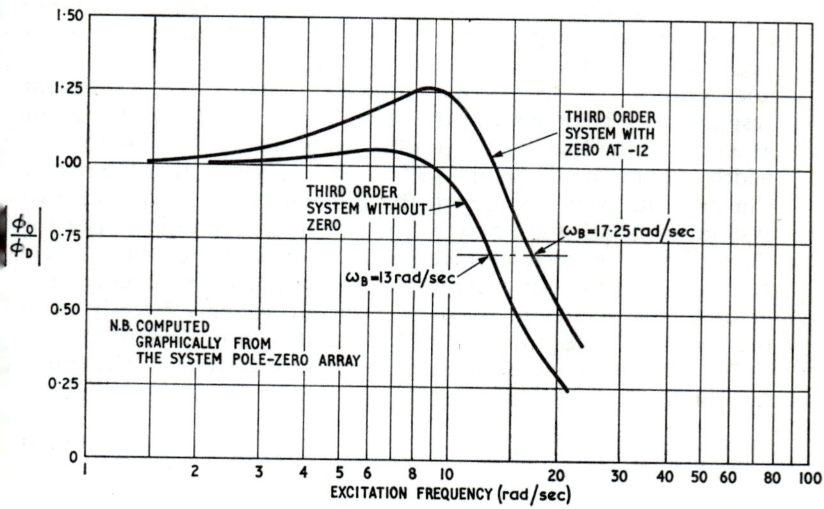


*Fig. 3.34. More exact frequency response for $K_g = 1/18{\cdot}0$ for Example 3.9.1, showing the effect of a system zero at −12*

are fixed (or nearly so) at the design stage, but considerable adjustment is available for the rate feedback term $T_r$, and it may also be possible to alter the servo loop time constant $T_c$ by altering, for example, the piston cross-sectional area. To investigate this aspect further, in Section 3.11 we shall construct the root locus for a fixed value of $K_g$ and calibrate the locus for variations in $T_c$, whilst in Section 3.12 we shall construct the locus for the case when the servo loop is represented by a second order transfer function with a damping ratio less than unity.

## 3.10 Some Frequency Domain Dominant Mode Effects in Third Order Systems

From the system pole-zero array of Fig. 3.30 it is apparent that for low values of $K_g$ the third pole is some way to the left, and the response tends to be relatively unaffected by the presence of this pole, whereas at high values of $K_g$ the third pole is very close to the system zero. In the latter case the pole and zero effects tend to cancel out, leaving a dominantly unity numerator-type second order response, whilst the former case raises the interesting question of how far to the left must the third pole be before the response is unaffected by the pole.

The approximation to the true system response, which is obtained by considering dominant mode effects only, is partly a function of the system pole-zero array including the remoteness of neglected poles and zeros, and is partly a function of the performance criterion under consideration. We shall discuss transient response in a later section; in the meantime we shall confine our attention to the effect of the position of a third pole on the bandwidth of a nominally second order system in order that the reader can quantitatively appreciate the difficulties involved. The two systems we shall study may be looked upon as limiting cases for the second order system, since we shall consider the unity numerator and zero velocity error systems respectively. Recognition of dominant modes in third order linear systems directly from the system transfer function will be dealt with in Chapter 5.

The general second order unity numerator system has the transfer function

$$\frac{\theta_0}{\theta_i}(s) = \frac{1}{1+\left(\dfrac{2\zeta}{\omega_n}\right)s+\dfrac{s^2}{\omega_n^2}} \tag{3.38}$$

and this system will have a bandwidth $\omega_B$. If a third pole is introduced at a distance $P$ times further away from the origin than the real part of the complex poles (the complex pole positions being unchanged), then an additional factor $1+s/P\zeta\omega_n$ will be introduced into the system transfer function, which now becomes

$$\frac{\theta_0}{\theta_i}(s) = \frac{1}{\left(1+\frac{s}{P\zeta\omega_n}\right)\left[1+\left(\frac{2\zeta}{\omega_n}\right)s+\frac{s^2}{\omega_n^2}\right]} \tag{3.39}$$

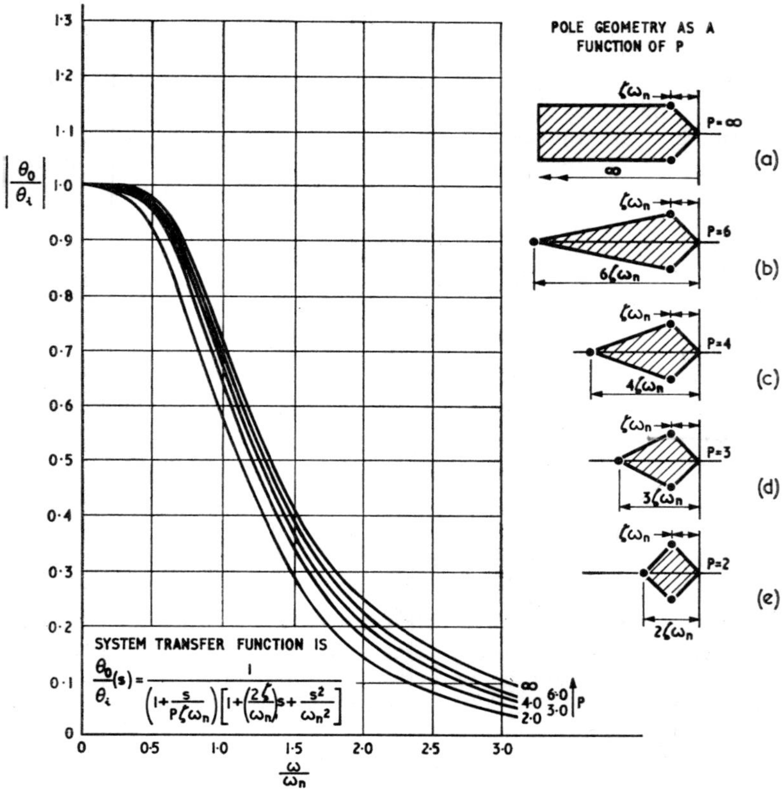


*Fig. 3.35. Effect of the third pole position on the frequency response of a nominally second order zero displacement error system with a damping ratio of 0·7*

$P$ is thus a unique measure of the third pole position. For a damping ratio of 0·7, the frequency responses for various values of $P$ from 2 to 6 have been plotted in Fig. 3.35, and the approximate percentage reduction in bandwidth as a function of $P$ is

**Table 3.10.1**

EFFECT OF THIRD POLE POSITION ON BANDWIDTH FREQUENCY OF A UNITY NUMERATOR SECOND ORDER SYSTEM WITH DAMPING RATIO 0·7

| $P$ | 2 | 3 | 4 | 6 | ∞ |
|---|---|---|---|---|---|
| *Percentage reduction in* $\omega_B$ | 17 | 9 | 6 | 3 | 0 |

shown in Table 3.10.1. This shows that above $P = 4$ the reduction in bandwidth can be neglected at the design stage for this particular system pole-zero array.

The second order zero velocity lag system, as shown in Chapter 1, requires that $b_1 = a_1$, and the system transfer function is

$$\frac{\theta_o}{\theta_i}(s) = \frac{1+\left(\frac{2\zeta}{\omega_n}\right)s}{1+\left(\frac{2\zeta}{\omega_n}\right)s+\frac{s^2}{\omega_n^2}} \tag{3.40}$$

On introduction of the third pole at a distance $P$ times further away than the complex poles, the system transfer function becomes

$$\frac{\theta_o}{\theta_i}(s) = \frac{1+\left(\frac{2\zeta}{\omega_n}\right)s}{\left(1+\frac{s}{P\zeta\omega_n}\right)\left[1+\left(\frac{2\zeta}{\omega_n}\right)s+\frac{s^2}{\omega_n^2}\right]} \tag{3.41}$$

For a damping ratio of 0·7, the frequency responses for various values of $P$ from 2 to 6 have been plotted in Fig. 3.36, and the approximate percentage reduction in bandwidth as a function of $P$ is shown in Table 3.10.2. This shows that even with $P = 6$

**Table 3.10.2**

EFFECT OF THIRD POLE POSITION ON BANDWIDTH FREQUENCY OF A ZERO VELOCITY ERROR SECOND ORDER SYSTEM WITH DAMPING RATIO 0·7

| $P$ | 2 | 3 | 4 | 6 | ∞ |
|---|---|---|---|---|---|
| *Percentage reduction in* $\omega_B$ | 31 | 21 | 15 | 8 | 0 |

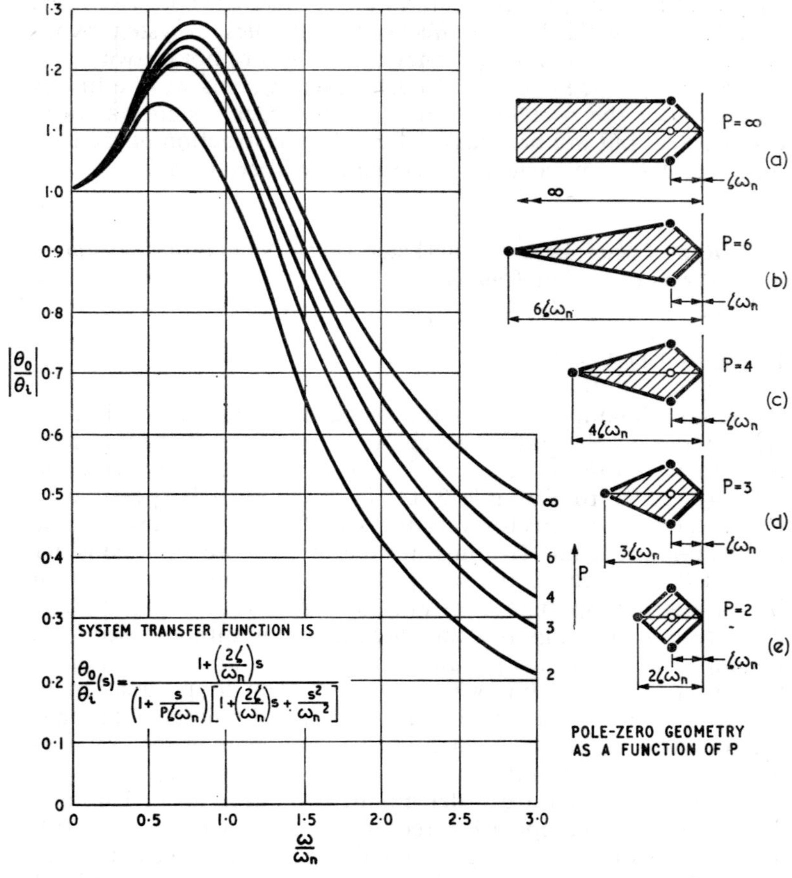


*Fig. 3.36. Effect of the third pole position on the frequency response of a nominally second order zero velocity error system with a damping ratio of 0·7*

there is still an 8% reduction in bandwidth. Thus for a system with an $M_p$ greater than 1·0, whether due primarily to a zero operating on a highly damped pair of complex poles or whether due primarily to a lightly damped pair of complex poles, the effect of the third pole is much more significant than for a deadbeat system. Conversely, neglected poles must be further to the left for systems with significant $M_p$ than for deadbeat systems. The reader should carefully note that none of the bandwidth reductions recorded in Table 3.10.1 and Table 3.10.2 are observable from straight-line approximations to the logarithmic plot of frequency response, and some intuition is therefore required in the correct interpretation of the effect of far-off poles on system bandwidth if the more exact frequency responses are not drawn.

Dominant mode effects in fourth order systems will be briefly studied after the root locus for a fourth order system has been constructed, but our next task will be the construction of the root locus for a system variable other than feedforward gain.

## 3.11 Root Locus for a Third Order Unity Numerator System in Terms of a Time Constant

EXAMPLE 3.11.1

By suitable manipulation at the block diagram stage it is possible to redraw the system in such a way that the root locus may be obtained in terms of a variable other than feedforward gain. With reference to Example 3.5.1, it would then be possible to examine the root locus for variations in $T_c$ or $T_r$ for a given value of feedforward gain. To illustrate the method we shall consider $K_g$ and $T_r$ to be fixed, and study the root locus plot for variation in the servo loop time constant $T_c$ from zero to infinity. The procedure is to manipulate the block diagram until a closed loop system is formed which has the same system transfer function as the original system, but which has a broken loop equation in which $T_c$ appears as an explicit multiplier. It should be noted that the resultant broken loop equation has no meaning for any purpose other than obtaining the root locus in terms of $T_c$.

We shall use the third order missile roll autopilot of Example 3.5.1. The first manipulation required is shown in Fig. 3.37(a), and consists of the lumping together of all the linear elements in the feedforward path except the block containing the variable of interest. Since any first order lag with transfer function $1/(1+T_c s)$ can be reduced to a simple feedback loop with feedforward path of unity and feedback path $T_c s$, the servo loop can be opened up as shown in Fig. 3.37(b). $T_c$ can then be removed from the feedforward path by feeding back the $\phi_0$ signal and shaping it with the transfer function $(1+T_a s)s/K_g K_a$ as shown in Fig. 3.37(c), the signal entering the $T_c s$ block being the same as for Fig. 3.37(b). The major loop containing $G(s) = K_g K_a/(1+T_a s)s$ and $H(s) = 1+T_r s$ is then reduced to form a block as shown in Fig. 3.37(d), and on inspection it is found that this block has a transfer function corresponding to the nominal second order system which arises if $T_c = 0$. It is convenient to define this nominal second order system as

$$\left[\frac{\phi_0}{\phi_D}(s)\right]_{T_c=0} = \frac{\omega_n^2}{s^2+2\zeta\omega_n s+\omega_n^2} \tag{3.42}$$

where

$$\omega_n^2 = \frac{K_g K_a}{T_a}$$

and

$$\frac{2\zeta}{\omega_n} = T_r + \frac{1}{K_g K_a}$$

The broken loop equation yielding an explicit multiplier in $T_c$ is, from Fig. 3.37(e),

$$\frac{\overline{\phi}_0}{\overline{\phi}}(s) = \frac{\omega_n^2}{s^2+2\zeta\omega_n s+\omega_n^2} \times \frac{T_c s^2(1+T_a s)}{K_g K_a} \tag{3.43}$$

which is the equation to be solved from the root locus plot.

Before plotting the locus it is very important to check that the block diagram shown in Fig. 3.37(e) satisfies the condition of invariance for the system transfer function. For the final block diagram, the feedforward and feedback transfer functions are

$$G(s) = \frac{K_g K_a}{T_a s^2+(1+K_g K_a T_r)s+K_g K_a} \tag{3.44}$$

and

$$H(s) = \frac{T_c s^2(1+T_a s)}{K_g K_a} \tag{3.45}$$

Since

$$\frac{\phi_0}{\phi_D}(s) = \frac{G(s)}{1+G(s)H(s)}$$

we have

$$\frac{\phi_0}{\phi_D}(s) = \frac{\dfrac{K_g K_a}{T_a s^2+(1+K_g K_a T_r)s+K_g K_a}}{1+\dfrac{T_c s^2(1+T_a s)}{T_a s^2+(1+K_g K_a T_r)\,s+K_g K_a}}$$

which, on cross-multiplication and simplification, becomes

$$\frac{\phi_0}{\phi_D}(s) = \frac{1}{1+\left(T_r+\dfrac{1}{K_g K_a}\right)s+\left(\dfrac{T_a+T_c}{K_g K_a}\right)s^2+\left(\dfrac{T_a T_c}{K_g K_a}\right)s^3} \tag{3.46}$$

Equation 3.46 agrees exactly with Equation 3.23, thus confirming the validity of the block diagram manipulation of Fig. 3.37. Returning now to the special broken loop form of Equation 3.43, the root locus requires that we solve the equation in the form

$$\frac{\bar{\phi}_0}{\bar{\phi}} = \frac{Ks^2\left(s+\dfrac{1}{T_a}\right)}{\omega_n^2+2\zeta\omega_n s+s^2} \tag{3.47}$$

and since $\omega_n^2 = K_g K_a/T_a$ it is found by comparison of Equations 3.47 with Equation 3.43 that in this case $K = T_c$, which is indeed fortuitous! From Equation 3.47 we see that there is an open loop zero at $-1/T_a$, and a double open loop zero at 0. There are complex poles at $-\zeta\omega_n \pm j\omega_n\sqrt{1-\zeta^2}$ and the third pole must start at minus infinity, since $T_c = 0$ corresponds to the simple second order system with two finite system poles. If $K_a = 335$, $T_a = 1/5{\cdot}6$ and $K_g = 1/18{\cdot}0$, then

$$\omega_n = \sqrt{\frac{335\times 5{\cdot}6}{18}} = 10{\cdot}2 \text{ rad/sec}$$

and if $T_r = 1/12$, then

$$\zeta = \frac{\omega_n}{2}\left(T_r+\frac{1}{K_g K_a}\right) = \frac{10{\cdot}2}{2}(0{\cdot}084+0{\cdot}054) = 0{\cdot}72$$

The complex poles are thus at a radius $\omega_n = 10{\cdot}2$ and at an angle $\pm\cos^{-1} 0{\cdot}72 = \pm 44°$ to the negative real axis, and the zeros are two at 0 and the third at $-1/T_a = -5{\cdot}6$.

Using the rule for the existence of branches of the locus on the negative real axis, we see that there can only be a branch on the

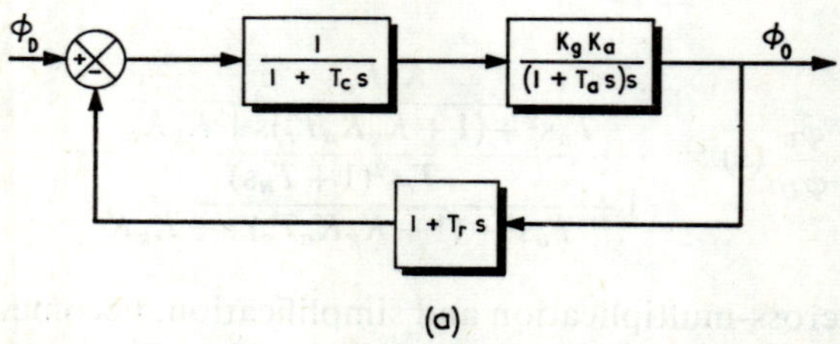


*Fig. 3.37. Block diagram manipulation of a third order missile autopilot to obtain the root locus in terms of $T_c$ to form Example 3.11.1: (a) lumping together all feedforward elements except the servo loop; (b) opening the servo loop; (c) removal of $T_c$ from the feedforward path; (d) reduction to single loop form; (e) final block diagram in terms of the nominal second order system*

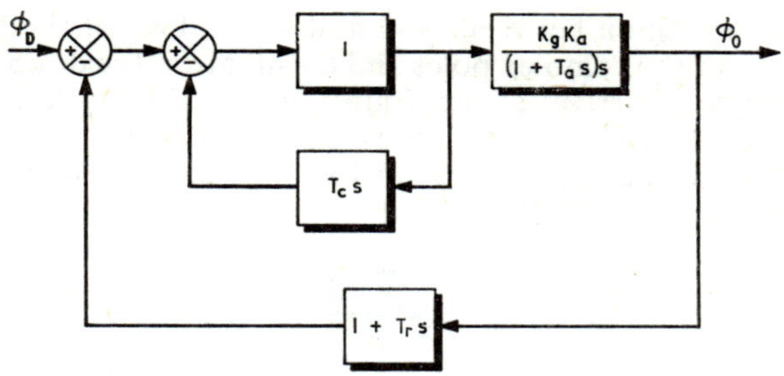

φD
+
−
1
KgKa / (1 + Ta s)s
φ0
Tc s
1 + Tr s


(b)

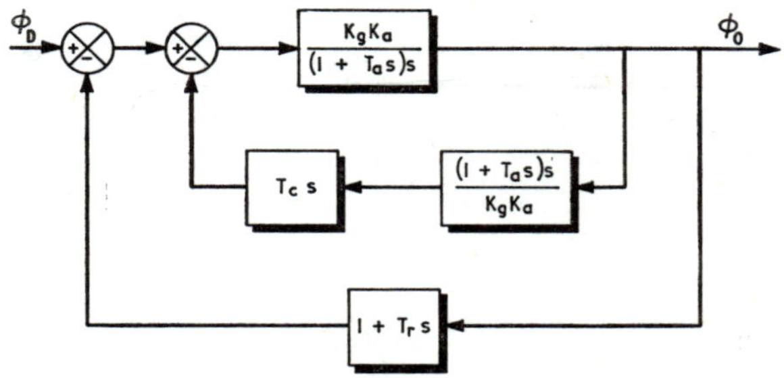

φD
+
−
KgKa / (1 + Ta s)s
φ0
Tc s
(1 + Ta s)s / KgKa
1 + Tr s


(c)

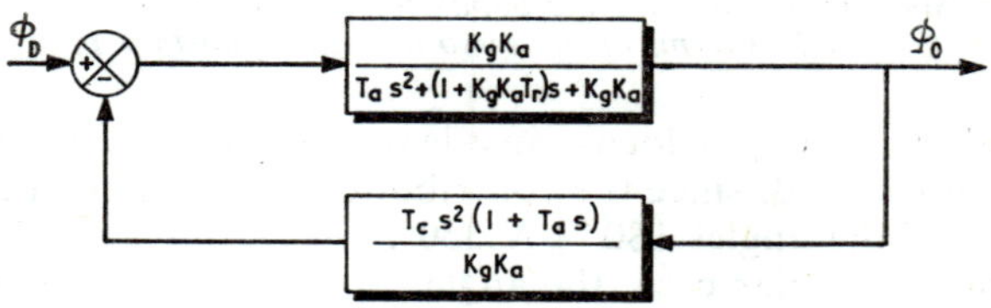

φD
+
−
KgKa / Ta s2 + (1 + KgKaTr)s + KgKa
φ0
Tc s2 (1 + Ta s) / KgKa


(d)

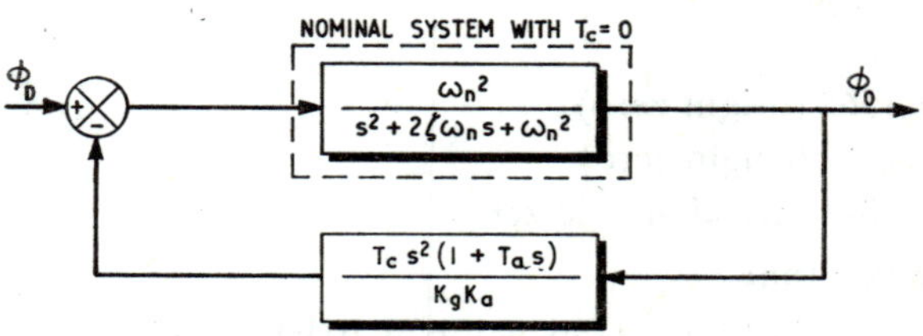

NOMINAL SYSTEM WITH Tc = 0
φD
+
−
ωn2 / s2 + 2ζωn s + ωn2
φ0
Tc s2 (1 + Ta s) / KgKa


(e)

axis for the region between $-\infty$ and $-5{\cdot}6$. Two of the branches must start at the complex poles and terminate at the double zero at 0. The angle of arrival at the double zero will be $\pm 90°$, since the angles for double poles or zeros on the real axis must be perpendicular to the axis, as seen in Example 3.4.1. Computation of

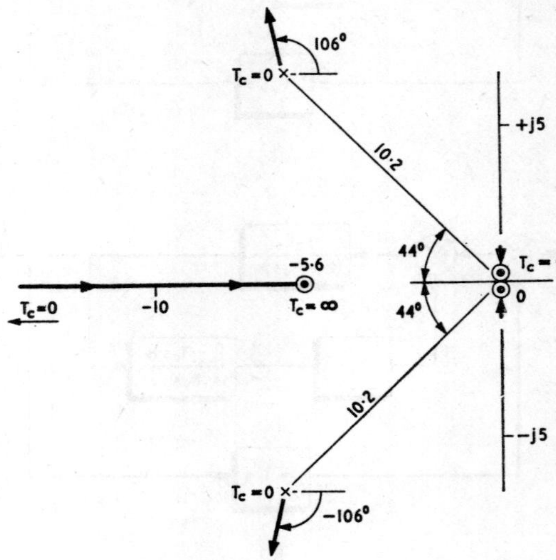


*Fig. 3.38. Starting and terminating points for the root locus plot of Example 3.11.1 in terms of the servo loop time constant $T_e$*

the angle at which the locus must leave a complex open loop pole is straightforward, since the contribution at that pole must make the sum of the angles $180° \pm N\,360°$.

In this particular case, the angles at the positive complex pole are given by:

$\delta_1$ = unknown
$\delta_2 = 90°$
$\gamma_1 = 136°$ (origin zero)
$\gamma_2 = 136°$ (origin zero)
$\gamma_3 = 104°$ (aerodynamic zero)

Thus, given that

$$\delta_1 + 90° - (136° + 136° + 104°) = \pm 180°$$

we obtain $\delta_1 = +106°$. Conversely, the angle at the negative complex pole is $-106°$. The locus thus far is shown in Fig. 3.38

and the complete locus is shown in Fig. 3.39. It can be observed that the locus does cut the imaginary axis and therefore the system can be made unstable, provided that the servo loop time constant is made large enough.

Since the open loop equation does not have at least two more finite poles than finite zeros, the centre of gravity theorem cannot be used to determine the third root corresponding to a given value of $T_c$, and it would be necessary in practice to evaluate as many points as are of interest on all branches of the loci and then interpolate where required until a state is reached where particular system pole-zero arrays can be constructed. By adding the system zeros, the effect of $T_c$ on the system pole-zero array can then be determined visually as was done for the case of $K_g$ variations in Figs. 3.22 and 3.30.

For our present purpose it is sufficient to check the value of $T_c$ necessary for a particular system pole configuration, given $K_g$, $K_a$, $T_a$ and $T_r$. From Example 3.4.1 and Table 3.5.1 we know that if $K_g = 1/18{\cdot}0$, $T_c = 0{\cdot}035$ will result for our standard values of

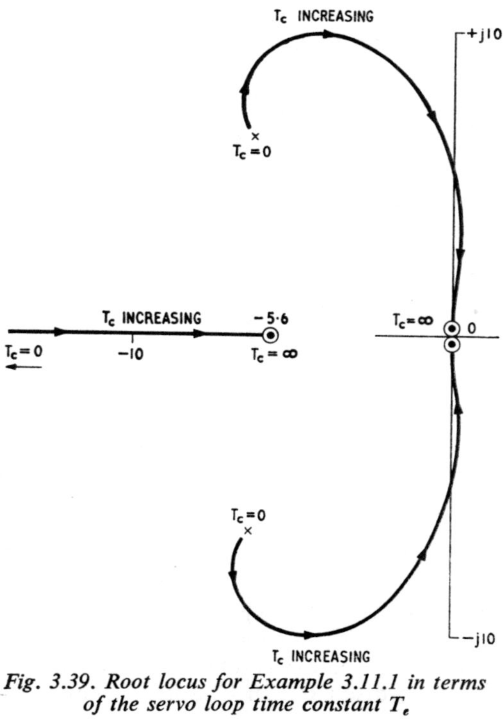


*Fig. 3.39. Root locus for Example 3.11.1 in terms of the servo loop time constant $T_c$*

radius 11·6 and a damping angle of 57·5°. Thus we have two checks on the root locus of Fig. 3.39, since in the first place these poles should actually lie on the locus, and secondly the value of $T_c$ at these points should be 0·035. The computation of $T_c$ for the positive complex pole is shown in Fig. 3.40, and the computation for the pole at −21·5 is shown in Fig. 3.41. In both cases the value of $T_c$ agrees with the predetermined value of 0·035; this is particularly heartening, as both philosophy and accuracy are confirmed simultaneously.

For the type of system considered so far the designer has control over $K_g$ and $T_r$, whilst there may be several discretely separated possible values of $T_c$ corresponding to different designs of servo

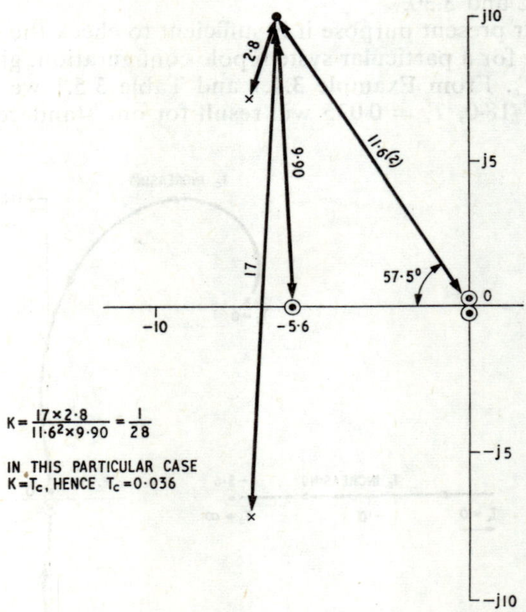


*Fig. 3.40. Vector diagram for the computation of $T_c$ for Example 3.11.1 for complex poles with a damping ratio of 0·54*

loop. Unless the specialised techniques of Chapter 5 are used, it may be necessary to repeat the root locus plot in terms of $K_g$ for several values of $T_r$ and several values of $T_c$ before the best design is obtained. Generally, the most economic design will be the design which corresponds to pole-zero arrays minimising the ITAE performance index as discussed in Chapter 1. This is not to say that any other design is unacceptable; for most specifications there will be many systems with satisfactory characteristics, and the choice between competitive systems will often be made on the grounds of economics or convenience.

The root locus plot of Fig. 3.39, in addition to providing information on the best value for the servo loop time constant $T_c$, if such a best value of $T_c$ exists, can also be used to estimate the effect

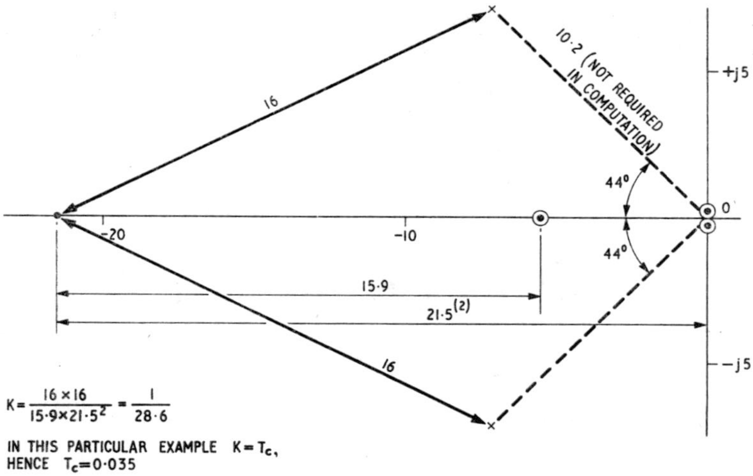


*Fig. 3.41. Vector diagram for the computation of $T_c$ for Example 3.11.1 for a pole at −21·5*

of errors in setting $T_c$ on the system poles. This phenomenon is termed 'drift', and it is important to choose a value of $T_c$ which will need to be considerably in error for system instability to occur, i.e. resultant system poles should be well away from the imaginary axis. Note that errors in $T_c$ can occur as a result of statistically by variable effects in manufacturing processes as well as of a physical malsetting of, say, a potentiometer.

In Chapter 1 it was stated that one of the outstanding advantages of the use of feedback was that changes in load characteristics, etc., would be attenuated in effect on the system transfer function. Since the root locus determines the system poles as one system parameter is varied, the construction may be used to determine the sensitivity of the system poles to changes in one or more parameters, such as a load characteristic. Ideally this sensitivity should be zero, but as this is impossible, it should be designed to be small within the system bandwidth.

## 3.12 Root Locus for a Fourth Order System in Terms of Feedforward Gain

EXAMPLE 3.12.1

So far it has been assumed that the servo loop considered in the various examples has been designed such that the relationship between actual and desired control surface position is a first order lag. There is no reason why the servo loop should not be designed

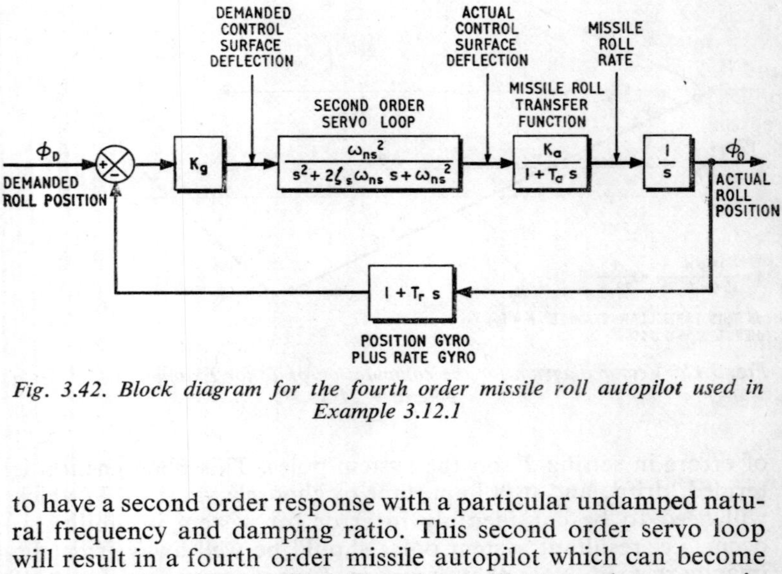


*Fig. 3.42. Block diagram for the fourth order missile roll autopilot used in Example 3.12.1*

to have a second order response with a particular undamped natural frequency and damping ratio. This second order servo loop will result in a fourth order missile autopilot which can become unstable when $K_g$ is varied. In order to illustrate the manner in which the autopilot can become unstable as $K_g$ is increased, we shall assume that the servo loop has a relatively low damping ratio of 0·5, and a relatively low undamped natural frequency of 30 rad/sec. The block diagram for this fourth order system is shown in Fig. 3.42, and we shall assume that the system is compensated by rate feedback rather than by the use of a phase advance network. The open loop transfer function for this example is

$$\frac{\bar{\phi}_0}{\bar{\phi}}(s) = \frac{K_g K_a \omega_{ns}^2 (1+T_r s)}{(1+T_a s)(s^2+2\zeta_s\omega_{ns}s+\omega_{ns}^2)s} \tag{3.48}$$

Rewriting the equation in root locus form, we have

$$\frac{\phi_0}{\phi}(s) = \frac{K\left(s+\frac{1}{T_r}\right)}{s\left(s+\frac{1}{T_a}\right)(s^2+2\zeta_s\omega_{ns}s+\omega_n s^2)} \tag{3.49}$$

where

$$K = \frac{K_g K_a \omega_{ns}^2 T_r}{T_a}$$

There will be an open loop zero at $-1/T_r = -12$ as before, and there will be open loop poles at $-1/T_a = -5{\cdot}6$ and at 0, and a pair of complex open loop poles at a radius of 30 and at angles $\pm\cos^{-1}0{\cdot}5 = \pm 60°$ to the negative real axis. Branches of the locus exist on the real axis between $-\infty$ and $-12$ and between $-5{\cdot}6$ and 0, since in these regions there is an odd number of open loop poles plus the number of open loop zeros to the right of these regions.

In this example, owing to the position in the s plane of the complex open loop poles, their effect on the breakaway points on the real axis will be small, leaving only the open loop poles at 0 and $-5{\cdot}6$, and the open loop zero at $-12$; the breakaway points are therefore the same as for Example 3.4.1, and the vector computations will be those shown in Figs. 3.5 and 3.6. The breakaway point is thus $-3{\cdot}25$ and the return point is $-20{\cdot}75$. For the general effect of complex open loop poles and zeros on breakaway points on the real axis, the reader is referred to Appendix I for a detailed analysis.

From Equation 3.49, it is apparent that there are four open loop poles but only one finite open loop zero. There are therefore three open loop zeros at infinity, and three high-gain asymptotes. The intersection point of the asymptotes is given by

$$d = \frac{\sum \text{open loop poles} - \sum \text{finite open loop zeros}}{n-m}$$

hence

$$d = -\left(\frac{2\zeta\omega_n + 0 + 1/T_a - 1/T_r}{3}\right)$$

$$= -\left(\frac{2\times 15 + 0 + 5{\cdot}6 - 12}{3}\right)$$

$$= -7{\cdot}86$$

The angles of the asympotes to the real axis are given by $\pm k\pi/(n-m) = \pm\pi/3, \pm\pi$, etc.; i.e. $\pm 60°$ and $180°$ respectively.

It is possible to compute the angles at which the branches of the locus leave the complex open loop poles by considering the angles contributed by the other open loop poles and open loop zero as shown in Fig. 3.43. The governing equation is

$$\gamma_1+\gamma_2+\gamma_3+\gamma_4-\delta_1 = 180^\circ \pm N\,360^\circ \tag{3.50}$$

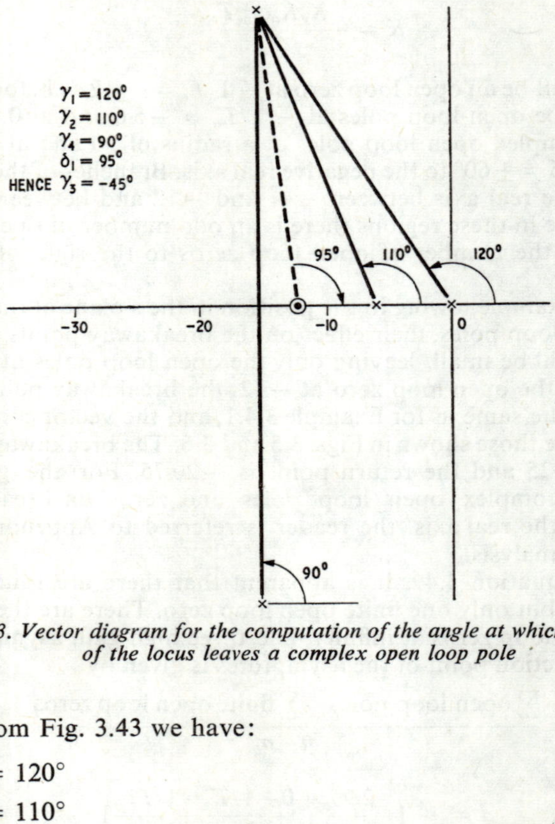


*Fig. 3.43. Vector diagram for the computation of the angle at which the branch of the locus leaves a complex open loop pole*

and from Fig. 3.43 we have:

$\gamma_1 = 120^\circ$

$\gamma_2 = 110^\circ$

$\gamma_3 =$ unknown at the upper complex pole

$\gamma_4 = 90^\circ$

$\delta_1 = 95^\circ$

Hence $\gamma_3 = -45^\circ$, and by symmetry the locus will leave the conjugate pole at an angle of $+45^\circ$. The root locus thus far is shown in Fig. 3.44, and the complete root locus constructed by using the graphical method for the determination of intermediate points is shown in Fig. 3.45. It can be seen that if the gain is made high enough the system will become unstable, owing to the servo poles moving into the right-hand plane. The integrator pole and the

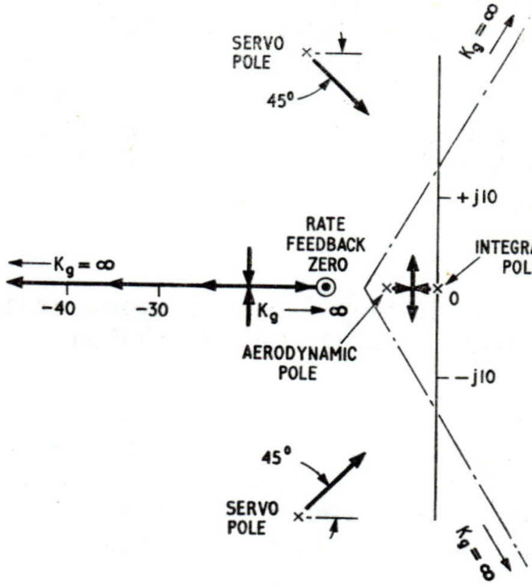


*Fig. 3.44. Starting and terminating points for the root locus plot of Example 3.12.1 in terms of feedforward gain*

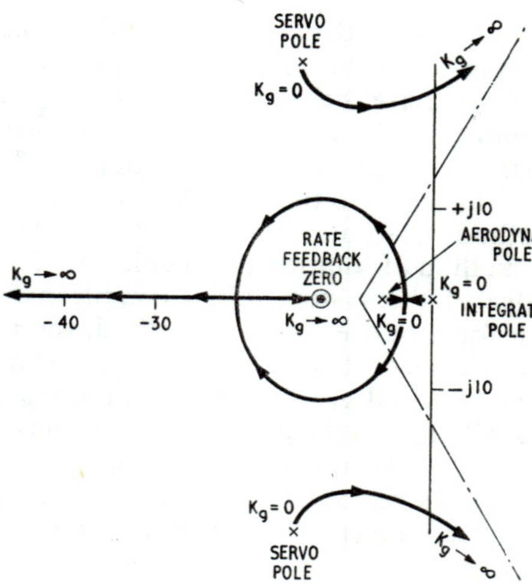


*Fig. 3.45. Complete root locus plot for Example 3.12.1 in terms of feedforward gain*

aerodynamic pole move together, meet, and diverge to form a complex pole pair which traces out the ellipse before meeting again on the negative real axis, one pole then moving into the open loop zero and the other moving out to $-\infty$. This is a particularly interesting example, since it is possible to have a system with two second order modes if desired, by judicious choice of feed-forward gain. There is a danger, however, that if the servo open loop poles are in practice more lightly damped than anticipated at the design stage, the system may become unstable because of the system poles resulting from the servo poles drifting into the right-hand plane.

## 3.13 Computation of the Step Function Response From the System Pole-zero Array Using the Residue Method

Provided the system pole-zero array is known, the transient response to a step function input can be determined in a straight-forward manner. Provided multiple system poles do not exist (in which case the reader is referred to Evans[27] or Lawden[29]), the step function response of a linear system with poles $s_1$, $s_2$, $s_3$, etc., may be written

$$\theta_0 = \theta_i(1 + A_1 e^{s_1 t} + A_2 e^{s_2 t} + A_3 e^{s_3 t} + \cdots) \tag{3.51}$$

where the poles may be real or complex and have been obtained from the solution of the system characteristic equation. The coefficients $A_1, A_2, A_3$, etc., are the coefficients of the partial fraction expansion technique used in the Laplace transform method of solution, and are a function of the system poles and zeros. These particular coefficients are also the arbitrary constants occurring in the classical method of solution of differential equations, and in this alternative method of solution the coefficients are determined from a knowledge of the system initial conditions. Using either the Laplace transform or the classical method, the procedure for obtaining the coefficients is tedious. However, the partial-fraction expansion can be obtained graphically from the system pole-zero array, and thus all the coefficients in the response determined using protractor and scale rule only.

The proof of the evaluation is based on the Laplace transform technique; we may summarise the results of interest as follows.[30]

$$A_1 = \frac{|s_1|\,|s_2|\,|s_3|\ldots|s_n|}{|z_1|\,|z_2|\ldots|z_m|} \times \frac{(s_1 - z_1)(s_1 - z_2)\ldots(s_1 - z_m)}{s_1(s_1 - s_2)(s_1 - s_3)\ldots(s_1 - s_n)}$$

and

$$A_r = \frac{|s_1|\,|s_2|\,|s_3|\ldots|s_n|}{|z_1|\,|z_2|\ldots|z_m|} \times$$

$$\frac{(s_r - z_1)(s_r - z_2)\ldots(s_r - z_m)}{s_r(s_r - s_1)\ldots(s_r - s_{r-1})(s_r - s_{r+1})\ldots(s_r - s_n)} \tag{3.52}$$

These vectors are expressible simply as a length and an angle. For example, $s_1 - s_2$ is the length of the line joining the poles s and $s_2$, together with the angle to the horizontal made by the vector at $s_2$, measured positive anticlockwise.

For real poles, the angular contributions will be a positive or negative integer product of 180°, leaving the coefficient as plus or minus times a numeric as the case may be. A pair of complex poles will result in a pair of coefficients of equal magnitude, the angular contributions being equal in magnitude but of opposite sign. The two poles may be combined to give the normal cosine form involving $\zeta$ and $\omega_n$ by means of the following rule:

$$A_r e^{s_r t} + A_{r+1} e^{s_{r+1} t} = 2Re^{-\zeta\omega_n t} \cos(\omega_n \sqrt{1-\zeta^2}\,t + v) \tag{3.53}$$

where $A_r = A_{r+1} = R$, $v =$ the angular contribution of $A_r$, and $s_r =$ the complex pole in the top half of the root plane.

EXAMPLE 3.13.1

Evaluate the transient response of the third order linear system with transfer function

$$\frac{\theta_0}{\theta_i}(s) = \frac{1}{1 + 2s + 1{\cdot}5s^2 + s^3}$$

using the graphical residue method to determine the arbitrary constants.

Using the methods of Chapters 5 and 6, the system transfer function is factorised to give

$$\frac{\theta_0}{\theta_i}(s) = \frac{1}{1 + 2s + 1{\cdot}5s^2 + s^3} = \frac{1}{(s + 0{\cdot}70)(s^2 + 0{\cdot}79s + 1{\cdot}44)} \tag{3.54}$$

There are no system zeros, but the system poles are given by Equation 3.54 as

$$s_1 = -0{\cdot}70$$

$$\left.\begin{aligned} s_2 &= -0{\cdot}40 + j1{\cdot}13 \\ s_3 &= -0{\cdot}40 - j1{\cdot}13 \end{aligned}\right\} \begin{aligned} &\text{in polar form, } \omega_n = 1{\cdot}20,\ \zeta = 0{\cdot}33, \\ &\omega_n\sqrt{1-\zeta^2} = 1{\cdot}13,\ \zeta\omega_n = 0{\cdot}40 \end{aligned}$$

Because there are complex roots, the transient response will take the form

$$\theta_0(t) = \theta_i[1 + A_1 e^{s_1 t} + 2Re^{-\zeta\omega_n t}\cos(\omega_n\sqrt{1-\zeta^2}t + \nu)]$$

The vector diagram for the computation of $A_1$ is shown in Fig. 3.46(a), and therefore, measuring directly from the diagram,

$$A_1 = \frac{|s_1|\,|s_2|\,|s_3|}{s_1(s_1-s_2)(s_1-s_3)}$$

$$= \frac{0{\cdot}70 \times 1{\cdot}20^2}{(0{\cdot}70\underline{/180^\circ})\,(1{\cdot}18\underline{/-105^\circ})\,(1{\cdot}18\underline{/105^\circ})}$$

$$= -1{\cdot}035$$

Similarly, from the vector diagram for the computation of $A_2$ shown in Fig. 3.46(b),

$$A_2 = \frac{|s_1|\,|s_2|\,|s_3|}{s_2(s_2-s_1)(s_2-s_3)}$$

$$= \frac{0{\cdot}70 \times 1{\cdot}20^2}{(1{\cdot}20\underline{/110^\circ})(1{\cdot}18\underline{/75^\circ})(2{\cdot}26\underline{/90^\circ})}$$

$$= 0{\cdot}312\underline{/+85^\circ}$$

From symmetry conditions we may write $A_3 = 0{\cdot}312\underline{/-85^\circ}$. The corresponding vector diagram is, however, shown in Fig. 3.46(c) for completeness. Combining $s_2$ and $s_3$, according to Equation 3.53, we have

$$A_2 e^{s_2 t} + A_3 e^{s_3 t} = 0{\cdot}624 e^{-\zeta\omega_n t}\cos(\omega_n\sqrt{1-\zeta^2}t + \nu)$$

Substituting the real and complex roots into the time solution, the complete solution becomes

$$\theta_0(t) = \theta_i\,[1 - 1{\cdot}035 e^{-0{\cdot}70t} + 0{\cdot}624 e^{-0{\cdot}40t}\cos(1{\cdot}13t + 85^\circ)] \qquad (3.55)$$

Step | Exponential decay of opposite sign to step | Decaying oscillation

The transient response defined by Equation 3.55 has been evaluated in 1 sec steps and is shown in Fig. 3.46(d). At $t = 0$, $\theta_0 = 0$, which is a useful check on the validity of our solution. Maximum percentage overshoot is 6% and time to first zero is 3·4 sec, and we may take the ±5% settling time as 7 sec. For a second order system, the times at which the response cuts the zero error line

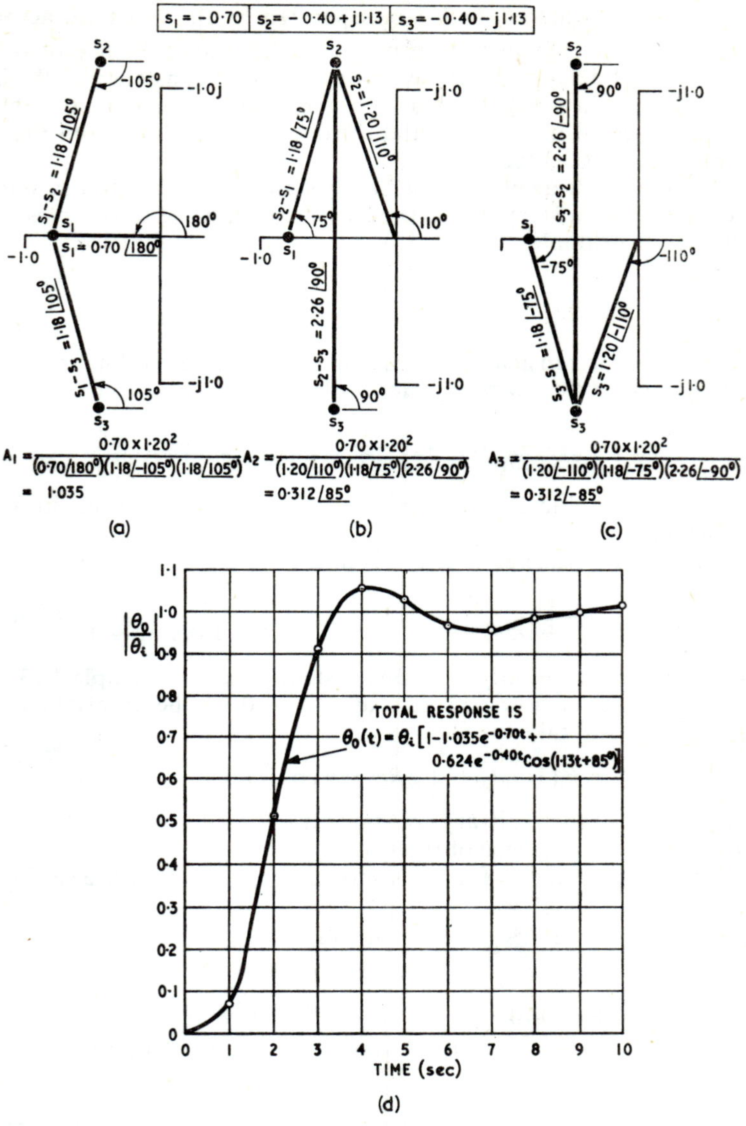


*Fig. 3.46. Transient response of the third order linear type (0,3) system with transfer function $\theta_0/\theta_i\,(s)=1/(1+2s+1{\cdot}5s^2+s^3)$ showing vector construction for the determination of $A_1$, $A_2$ and $A_3$: (a) vector diagram for the computation of $A_1$; (b) vector diagram for the computation of $A_2$; (c) vector diagram for the computation of $A_3$; (d) plot of the transient response*

(i.e. times at which $\theta_0 = \theta_i$) are separated by the constant factor $2\pi/\omega_n\sqrt{1-\zeta^2}$, but in our third order system the presence of the exponential delay has completely distorted the constancy of these times. Thus first and second zero error times are separated by about 2 sec, whilst second and third zero error times are separated by about 3·3 sec.

We shall now determine the highly significant effect of a system zero on the transient response of a system with the same poles as the above example.

EXAMPLE 3.13.2

Evaluate the transient response of the third order linear zero velocity lag system with transfer function

$$\frac{\theta_0}{\theta_i}(s) = \frac{1+2s}{1+2s+1{\cdot}5s^2+s^3}$$

using the graphical residue method to determine the arbitrary constants.

The system transfer function is now

$$\frac{\theta_0}{\theta_i}(s) = \frac{1+2s}{1+2s+1{\cdot}5s^2+s^3} = \frac{1+2s}{(s+0{\cdot}70)(s^2+0{\cdot}79s+1{\cdot}44)} \qquad (3.56)$$

This system has exactly the same system poles as Example 3.13.1 but there will be a system zero at $z_1 = -0{\cdot}5$. The transient response will still take the form

$$\theta_0(t) = \theta_i[1+A_1e^{s_1t}+2Re^{-\zeta\omega_nt}\cos(\omega_n\sqrt{1-\zeta^2}t+\nu)]$$

as before, but the arbitrary constants $A_1$, $R$, and $\nu$ will be different because of the presence of the system zero.

The vector diagram for $A_1$ is shown in Fig. 3.47(a). Measuring directly,

$$A_1 = \frac{|s_1|\ |s_2|\ |s_3|}{|z_1|} \times \frac{s_1-z_1}{s_1(s_1-s_2)(s_1-s_3)}$$

$$= \frac{1{\cdot}20^2 \times 0{\cdot}70}{0{\cdot}50} \times \frac{0{\cdot}20\underline{/180^\circ}}{(0{\cdot}70\underline{/180^\circ})(1{\cdot}18\underline{/-105^\circ})(1{\cdot}18\underline{/105^\circ})}$$

$$= +0{\cdot}413$$

*Fig. 3.47. Transient response of the third order linear zero velocity lag system with transfer function $\theta_0/\theta i(s) = (1+2s)/(1+2s+1{\cdot}5s^2+s^3)$, showing vector construction for determination of $A_1$ and $A_2$: (a) vector diagram for the computation of $A_1$; (b) vector diagram for the computation of $A_2$; (c) plot of the transient response*

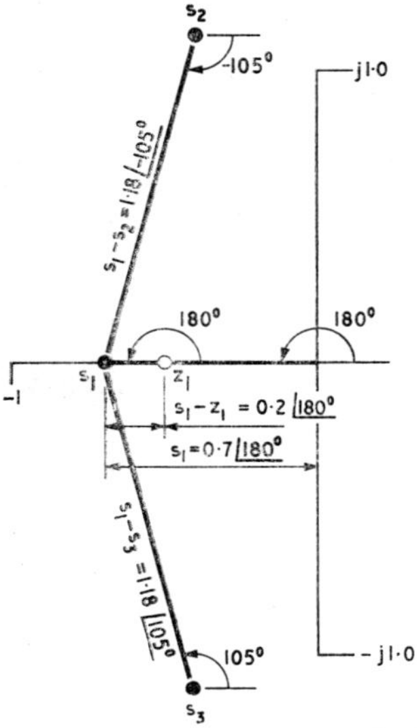


| |
|---|
| $s_1 = -0{\cdot}70$ |
| $s_2 = -0{\cdot}40 + j1{\cdot}13$ |
| $s_3 = -0{\cdot}40 - j1{\cdot}13$ |
| $z_1 = -0{\cdot}50$ |

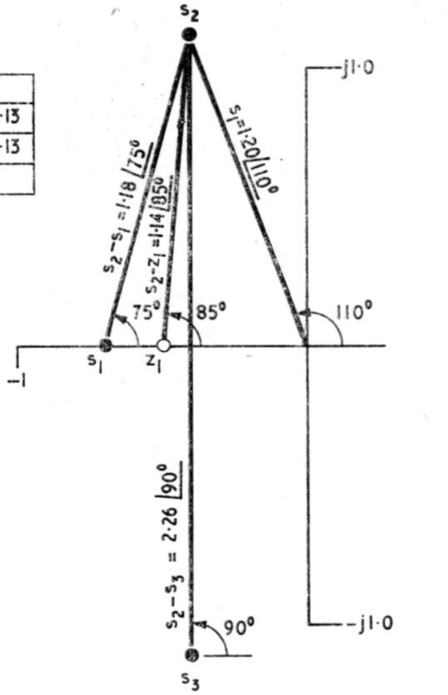


$$A_1 = \frac{1{\cdot}20^2 \times 0{\cdot}70}{0{\cdot}50} \times \frac{0{\cdot}20 \angle 180^\circ}{(0{\cdot}70 \angle 180^\circ)(1{\cdot}18 \angle -105^\circ)(1{\cdot}18 \angle 105^\circ)} = +0{\cdot}413$$

(a)

$$A_2 = \frac{1{\cdot}20^2 \times 0{\cdot}70}{0{\cdot}50} \times \frac{1{\cdot}14 \angle 85^\circ}{(2{\cdot}26 \angle 90^\circ)(1{\cdot}18 \angle 75^\circ)(1{\cdot}20 \angle 110^\circ)} = -0{\cdot}716 \angle -10^\circ$$

(b)

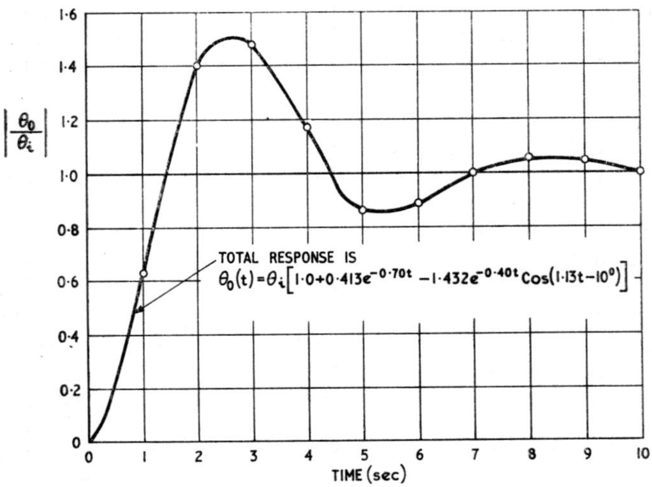


(c)

Similarly, from Fig. 3.47(b),

$$A_2 = \frac{|s_1|\,|s_2|\,|s_3|}{|z_1|} \times \frac{s_2 - z_1}{s_2(s_2-s_1)(s_2-s_3)}$$

$$= \frac{1{\cdot}20^2 \times 0{\cdot}70}{0{\cdot}50} \times \frac{1{\cdot}14\underline{/85^\circ}}{(2{\cdot}26\underline{/90^\circ})(1{\cdot}18\underline{/75^\circ})(1{\cdot}20\underline{/110^\circ})}$$

$$= 0{\cdot}716\underline{/-190^\circ}$$

$$= -0{\cdot}716\underline{/-10^\circ}$$

The vector diagram for $A_3$ is omitted since, from symmetry, $A_3 = -0{\cdot}716\underline{/+10^\circ}$, giving the oscillatory mode

$$A_2 e^{s_2 t} + A_3 e^{s_3 t} = 1{\cdot}432 e^{-\zeta\omega_n t} \cos(\omega_n \sqrt{1-\zeta^2}\,t - 10^\circ)$$

The complete transient response is therefore

$$\theta_0(t) = \theta_i\,[1 + 0{\cdot}413 e^{-0{\cdot}70t} - 1{\cdot}432 e^{-0{\cdot}40t} \cos(1{\cdot}13t - 10^\circ)] \quad (3.57)$$

Step — Exponential decay of same sign as step — Decaying oscillation

Note how the presence of the zero has considerably affected the arbitrary constants, reversing the signs of both magnitudes and altering the phase angle by more than $\pi/2$. The corresponding transient response is shown in Fig. 3.47(c) plotted in 1 sec intervals. Maximum percentage overshoot is now $52\frac{1}{2}\%$, and time to first zero is reduced to 1·4 sec, but the $\pm 5\%$ settling time is increased to about 9 sec.

These changes in performance criteria are characteristic of the effect of adding a system zero. In Example 3.13.1 all three poles contributed very significantly to the transient response. As evidence of this, the maximum percentage overshoot of the second order system with $\zeta = 0{\cdot}33$ is 37%, whereas the presence of the third pole has attenuated this to 6%. When the system zero is added, all three poles still contribute to the transient response, but the damping effect of the real pole is reduced owing to the presence of the zero; the zero will also liven up the response of the complex poles, because the zero is relatively near the origin. Adding the zero has, in this case, made $C_v = \infty$ since $a_1 = b_1$, but we have achieved a major increase in $C_v$ at the expense of a very considerable increase in maximum percentage overshoot and settling time.

In Chapters 4 and 5 we shall develop techniques for the design of type (1,2) systems and certain third order systems to meet a maximum percentage overshoot criterion, which will eliminate the need for the reader to actually evaluate the transient responses of these systems in detail. Chapter 6 will present the general principles of s plane compensation; where the dominant mode philosophy is used we shall endeavour to drive $n-2$ poles and $m-1$ zeros to infinity or to form dipoles, leaving the dominant mode to be evaluated directly from Chapter 4. Further additional data reported by Elgred and Stephens[31] can also be used as a general guide to the transient responses of type (0,1) to type (2,3) systems inclusive.

### 3.14 Dominant Mode Effects in Fourth Order Systems

In order to further simplify the understanding of high order system pole-zero arrays it is useful at this juncture to examine the effect of the position of secondary poles on the performance of a fourth order linear system. The primary poles are defined as the poles of the system transfer function

$$\frac{\theta_0}{\theta_i}(s) = \frac{\omega_n^2}{\omega_n^2 + 2\zeta_1\omega_n s + s^2} \tag{3.58}$$

and the fourth order transfer function, with secondary poles at $\zeta_2$, $\omega_{n2}$, is defined by

$$\frac{\theta_0}{\theta_i}(s) = \frac{\omega_{n2}^2\omega_n^2}{(\omega_n^2 + 2\zeta_1\omega_n s + s^2)(\omega_{n2}^2 + 2\zeta_2\omega_{n2} s + s^2)} \tag{3.59}$$

It is convenient to define $\omega_{n2} = \lambda\omega_n$, and Equation 3.59 can then be rewritten as

$$\frac{\theta_0}{\theta_i}(s) = \frac{\lambda^2\omega_n^4}{(\omega_n^2 + 2\zeta_1\omega_n s + s^2)(\lambda^2\omega_n^2 + 2\zeta_2\lambda\omega_n s + s^2)} \tag{3.60}$$

The nominal system with poles $\zeta_1$ and $\omega_n$ will have a nominal bandwidth $\omega_B$, and we have already seen in Section 3.10 that the addition of a third pole on the negative real axis would result in a reduction in bandwidth. With the addition of a pair of secondary complex poles to the nominal system, it is possible to increase or decrease the bandwidth from the nominal value, depending on the damping ratio of the secondary poles. This effect has been investigated by Cooper,[32] and a typical Cooper chart is shown in Fig. 3.48, which is for a nominal system with dominant poles of damping ratio 0·5. The contours show the allowable positions for the secondary complex poles for a fixed percentage variation from the nominal system bandwidth. It is immediately apparent from Fig. 3.48 that, for an allowable change in bandwidth of $\pm 2\%$,

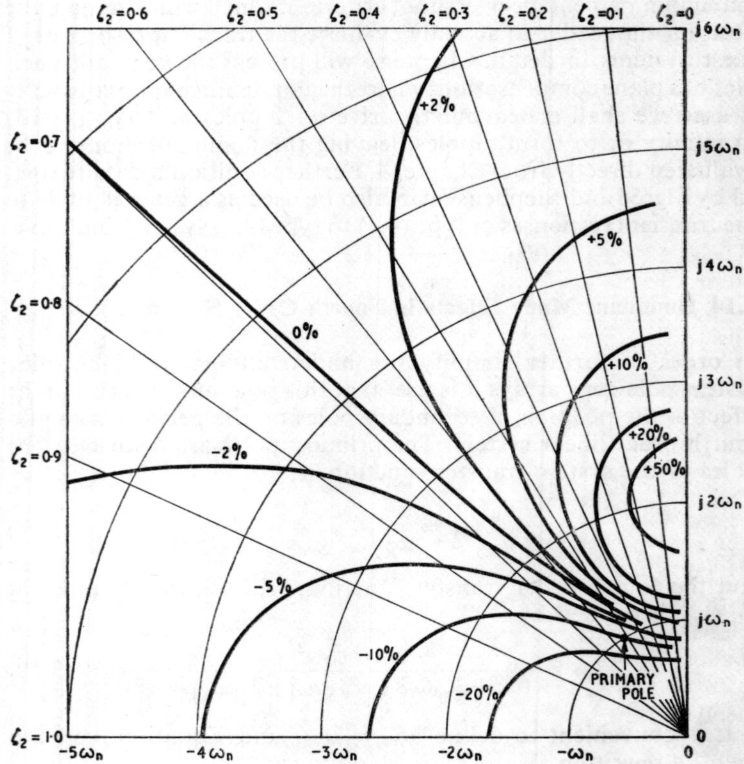


$$\frac{\theta_o}{\theta_i}(s) = \frac{1}{\left[1+\left(\frac{2\zeta_1}{\omega_n}\right)s+\frac{s^2}{\omega_n{}^2}\right]\left[1+\left(\frac{2\zeta_2}{\lambda\omega_n}\right)s+\frac{s^2}{\lambda^2\omega_n{}^2}\right]}, \quad \lambda = \frac{\omega_{n2}}{\omega_n}$$

*Fig. 3.48. Contours showing allowable positions of secondary complex closed loop poles of the fourth order system for a given percentage variation in the system bandwidth*

the secondary poles can be positioned anywhere between the $+2\%$ contour and the $-2\%$ contour.

Thus, when located in this region, the effect of secondary poles on system bandwidth is certainly negligible at the design stage. In the thesis by Cooper similar contours have been constructed for other values of $\zeta_1$, and they exhibit somewhat similar patterns to those shown in Fig. 3.48 for $\zeta_2 = 0{\cdot}5$.

From our knowledge of the behaviour of third order systems with two dominant poles and a dominant zero, as discussed in eSction 3.10, we should expect, for a similar change in bandwidth, to find that the secondary poles must be further to the left than for a system without a dominant zero. Nevertheless, we should expect to find vast regions in the s plane where secondary poles would not affect the system bandwidth significantly.

Although there is no variation in bandwidth as the secondary poles are moved along a constant bandwidth contour, there may well be a considerable variation in some other parameter of interest, such as percentage overshoot or velocity constant.

The effect of secondary poles on the transient response of the system has been investigated by Mulligan[33] for two particular values of $\zeta_1$, and contours have been constructed in the root plane using a potential analogue method of solution. The work of Cooper does allow a simpler approach to the problem of isolating regions in which the secondary poles may be situated without significantly affecting the transient response characteristics. A fourth order system with two pairs of complex poles will have a transient response of the form

$$\theta_0 = \theta_i[1 - A_1 e^{-\zeta_1 \omega_{n1} t} \cos(\omega_{n1}\sqrt{1-\zeta_1^2}\,t + \nu_1) - A_2 e^{-\zeta_2 \omega_{n2} t} \times \cos(\omega_{n2}\sqrt{1-\zeta_2^2}\,t + \nu_2)] \tag{3.61}$$

For this system the effect of the secondary poles on the transient response is therefore measured by the relative significance of the last factor in Equation 3.61. If we are particularly interested in percentage overshoot, then the significance of the last factor will depend on the relative magnitude of $A_1$ and $A_2$, the relative magnitude of rates of decay $\zeta_1\omega_{n1}$ and $\zeta_2\omega_{n2}$, and the respective phase angles $\nu_1$ and $\nu_2$. Using the vector method outlined in Section 3.13 for the computation of $A_1$ and $A_2$, Cooper has shown that the ratio of $A_1$ to $A_2$ is given by

$$\frac{A_1}{A_2} = \frac{\lambda^2\sqrt{1-\zeta_2^2}}{\sqrt{1-\zeta_1^2}} \tag{3.62}$$

where $\lambda = \omega_{n2}/\omega_n$. Thus if $\lambda = 3$, and $\zeta_1 = \zeta_2$, then $A_1 = 9A_2$, and since the rate of decay of the second factor is much faster than for the first factor, and also since it may be shown that under these circumstances the phase angle $\nu$ is little affected by the presence of the secondary poles, we may conclude that the maximum overshoot will occur at substantially the same time as for the nominal system; the second factor effect will have mostly disappeared by this time, and the percentage overshoot will therefore be substantially the same as for the nominal system. Clearly, from Equation 3.62, the value of $\lambda$ necessary before the secondary poles have little effect on percentage overshoot is also a function of the damping ratios $\zeta_1$ and $\zeta_2$. If $\zeta_1 > \zeta_2$, then $\lambda$ can become less than three, and if $\zeta_2 > \zeta_1$, $\lambda$ may need to be increased to maintain parity on percentage overshoot.

Sample transient responses for $\zeta_1 = \zeta_2 = 0{\cdot}5$ are shown in Fig. 3.49, and it can be seen that for $\lambda > 3{\cdot}0$ there is little further

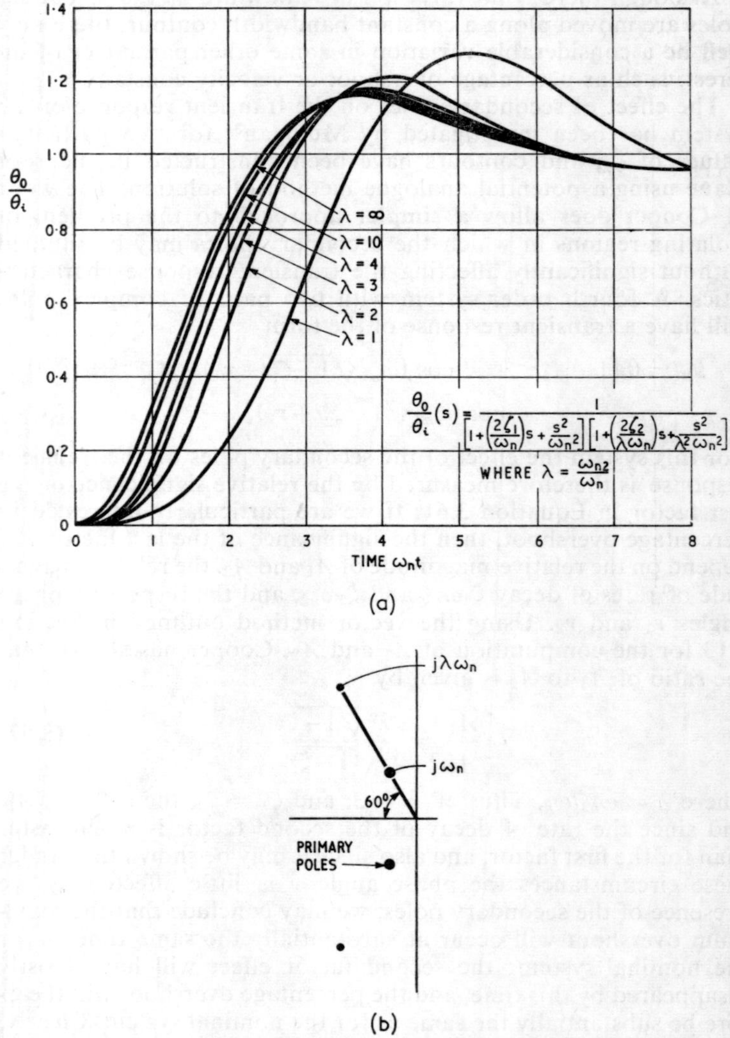


*Fig. 3.49. Effect of secondary complex poles on the fourth order system transient response for $\zeta_1 = \zeta_2 = 0{\cdot}5$: (a) transient response; (b) pole-zero array*

effect on the percentage overshoot, which has settled down remarkably quickly to a value near that obtained when $\lambda = \infty$, which is, of course the response of the nominal second order system. Fig. 3.50 shows the effect on the transient response of varying $\zeta_2$ with $\lambda$ fixed ($= 3$), and shows, as expected, that the higher $\zeta_2$, the lower the percentage overshoot, although the difference in percentage overshoot even between the cases corresponding to $\zeta_2 = 0{\cdot}3$ and $\zeta_2 = 0{\cdot}7$ is relatively small by practical engineering standards.

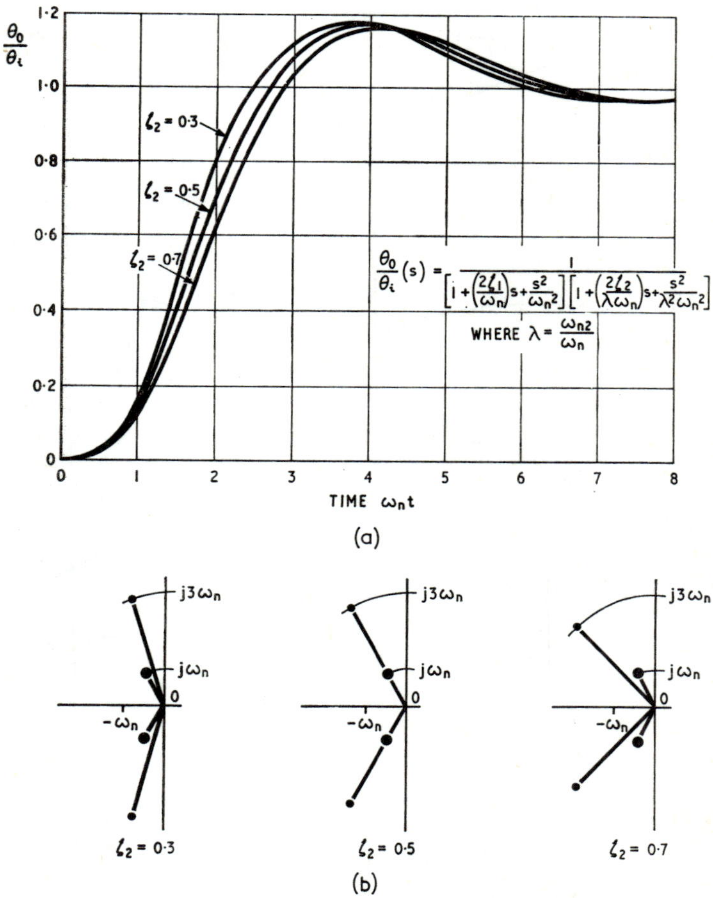


*Fig. 3.50. Effect of secondary complex poles on the fourth order system transient response for $\zeta_1 = 0{\cdot}5$ and $\lambda = 3$: (a) transient response; (b) pole-zero array*

An important point to note is that the transients shown in Fig. 3.49, for $\zeta_1 = \zeta_2 = 0{\cdot}5$, as $\lambda$ is varied from 2 to $\infty$, correspond to systems whose bandwidth is little different (within 2%) of the nominal system resulting from the dominant pole pair alone, and yet the rise time of the system, however this is defined, is affected considerably by $\lambda$. Thus it appears that fourth order systems with a dominant pole pair can easily be designed such that the percentage overshoot and bandwidth are little affected by the presence of the secondary poles, but at the same time these systems will have a vastly different behaviour in the early stages of

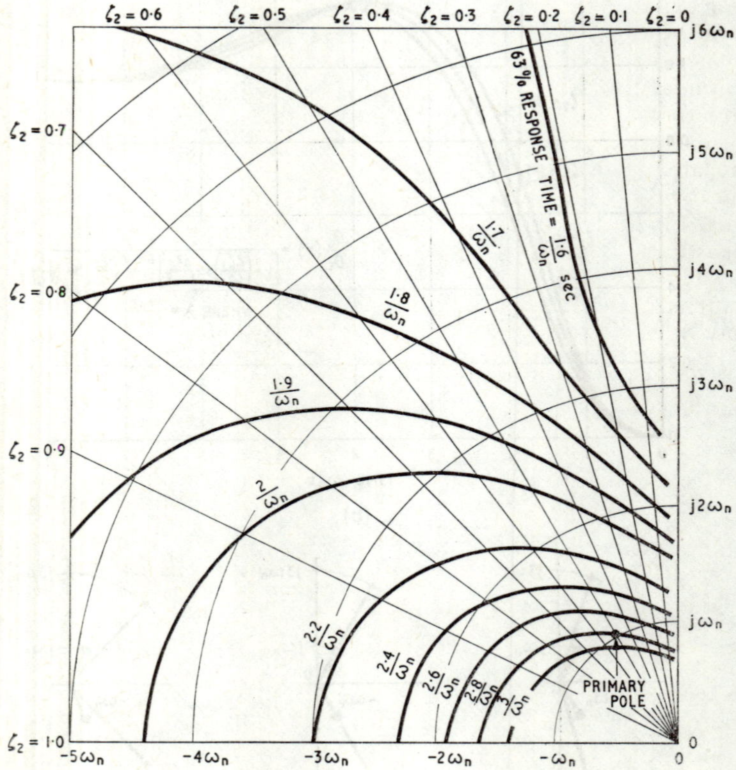


$$\frac{\theta_o}{\theta_i}(s) = \frac{1}{\left[1+\left(\frac{2\zeta_1}{\omega_n}\right)s + \frac{s^2}{\omega_n^2}\right]\left[1+\left(\frac{2\zeta_2}{\lambda\omega_n}\right)s + \frac{s^2}{\omega_n^2}\right]}, \quad \lambda = \frac{\omega_{n2}}{\omega_n}$$

*Fig. 3.51. Contours showing allowable positions of secondary complex closed loop poles of the fourth order system for constant 63% response times*

the transient response compared with the nominal second order system. This is clearly shown in Fig. 3.51, in which contours of allowable positions of the secondary complex poles for a given 63% response time are shown in the system pole plane. For the system corresponding to $\zeta_1 = \zeta_2 = 0{\cdot}5$, $\lambda = 6$, there will be a 63% response time, which is over 10% larger than the nominal value, thus showing that if this response time is an important criterion for a given application, secondary poles this distance away still have a significant effect and should not be neglected. On the contrary, since, by definition, the secondary poles will, in general, be further from the origin (measured along the negative real axis) than the dominant poles, the system settling time will soon be unaffected by the presence of the secondary poles as $\lambda$ is increased, and $\lambda = 3{\cdot}0$ will often be sufficient to allow the secondary pole effect to be neglected at the design stage. Summarising, therefore, we may say that secondary complex poles affect the early transient response in the region $t \simeq 0$, and do not affect the late transient response in the region $t \gg 0$. Quite clearly, then, our definition of 'far to the left' depends not only on the disturbance function but also on the performance criterion used to measure the response to the disturbance function.

The effect of the secondary pole positions on the velocity constant is often of importance at the design stage. Equation 3.60 may be expanded and written in unity numerator form as

$$\frac{\theta_0}{\theta_i}(s) = \frac{1}{\left[1+\frac{s}{\omega_n}\left(2\zeta_1+\frac{2\zeta_2}{\lambda}\right)+\frac{s^2}{\omega_n^2}\left(1+\frac{1}{\lambda^2}+\frac{4\zeta_1\zeta_2}{\lambda}\right)+\frac{s^3}{\omega_n^3}\left(\frac{2\zeta_1}{\lambda^2}+\frac{2\zeta_2}{\lambda}\right)+\frac{s^4}{\lambda^2\omega_n^4}\right]} \quad (3.63)$$

Now, from Section 1.16, $C_v = 1/(a_1-b_1)$, and $b_1 = 0$, hence

$$C_v = \frac{\omega_n}{2\zeta_1+\frac{2\zeta_2}{\lambda}} \quad (3.64)$$

where $C_v = \omega_n/2\zeta_1$ for the nominal system. For a given reduction in $C_v$ caused by the presence of the secondary poles, the locus in the system pole plane will be $\zeta_2/\lambda =$ constant, and thus for a given $C_v$ specification there will be a definite trade-off between the damping ratio and undamped natural frequency of the secondary poles, a small reduction in $C_v$ from the nominal $C_v$ necessitating high $\lambda$ or low $\zeta_2$. This is clearly shown in Fig. 3.52, which is again for a system with $\zeta_1 = 0{\cdot}5$; for example if $C_v = 0{\cdot}9\omega_n$, we

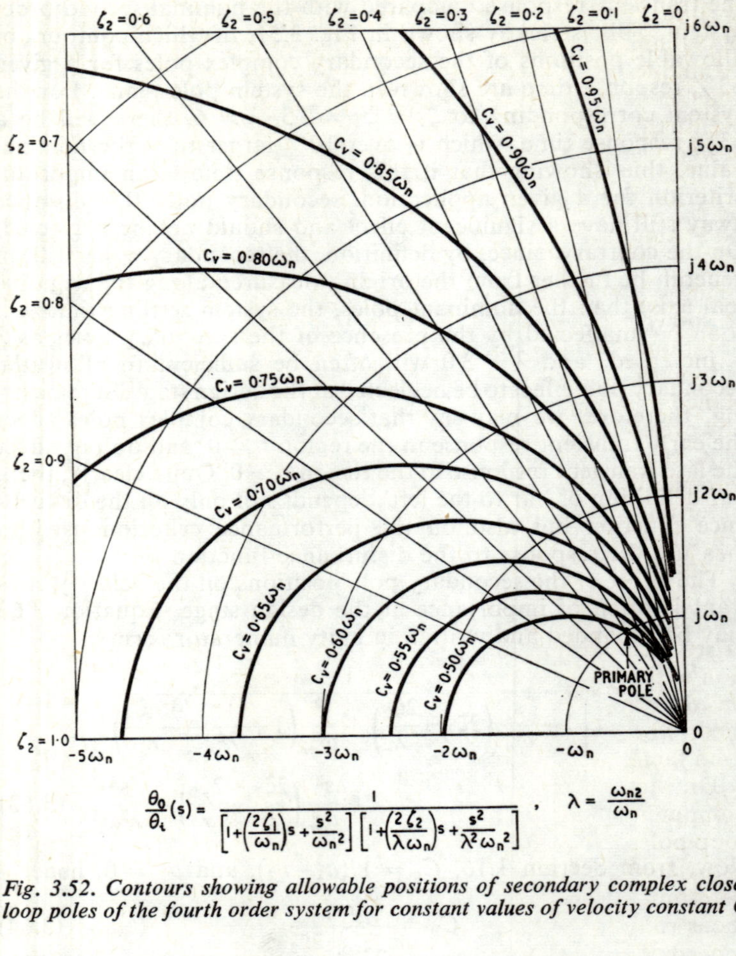


$$\frac{\theta_o}{\theta_i}(s) = \frac{1}{\left[1+\left(\frac{2\zeta_1}{\omega_n}\right)s+\frac{s^2}{\omega_n^2}\right]\left[1+\left(\frac{2\zeta_2}{\lambda\omega_n}\right)s+\frac{s^2}{\lambda^2\omega_n^2}\right]}, \quad \lambda = \frac{\omega_{n2}}{\omega_n}$$

*Fig. 3.52. Contours showing allowable positions of secondary complex closed loop poles of the fourth order system for constant values of velocity constant $C_v$*

have such choices as $\zeta_2 = 0{\cdot}2$, $\lambda = 3{\cdot}6$, and $\zeta_2 = 0{\cdot}3$, $\lambda = 5{\cdot}4$. The effect of secondary poles and zeros upon $C_v$ is so significant that we must make due allowance when positioning the dominant poles and zeros. $C_v$ can be easily reduced by 25% as a result of the secondary poles. We can either put a safety margin on the $C_v$ required from the dominant poles and zeros, or we can combine s plane design techniques with broken loop techniques, such as Bode plots, where $C_v$ can be kept constant whilst various compensation transfer functions are adjusted to meet the other specifications.

It is hoped that the examples quoted in this section and in Sections 3.10 and 3.13 will have given the reader some feel for the elimination of far-off poles and zeros so that rapid synthesis techniques may be used to position correctly the dominant poles and zeros, whatever number of dominant poles and zeros are eventually determined to be necessary for a viable and economic design.

### 3.15 Automatic Computation of the Root Locus

Using a digital computer for the computation of the root locus is sometimes beneficial if a reliable polynomial factorisation programme is available, and if the computer is provided with automatic plotting facilities so that the results are available at a glance. The computation process is remote from the designer, however, and it is sometimes apparent, even halfway through the plot, that modifications to the system are very necessary. This information will not normally be available from the digital computer until all the computations are finished and the results plotted either automatically or by hand. If a process can be found whereby the designer can inspect the results as they are determined, then modifications to the design can be made immediately without wasting computing time. A further disadvantage which arises from using the digital computer for root locus determination is the large amount of time which is often necessarily spent on algebraic manipulation before the problem can be programmed, whereas the root locus works directly from simple open loop factors.

Dr. Merle M. Morgan[34] has designed a potential plane analogue computer which directly plots the system poles for given open loop poles and zeros, thus overcoming many of the disadvantages associated with the use of the digital computer for root locus determination. Using the special computer and three of the root locus rules, it is possible to construct the root locus very rapidly indeed. The conventional root locus is drawn in the s plane; the potential plane analogue computer does not plot the root locus in the s plane, but in the log s plane. This plane transformation is necessary because the computer is based on the potential analogue method which involves logarithmic relationships. The transformation is reasonably easy to understand, as radial lines of constant damping angle in the s plane become horizontal straight lines in the log s plane, and the horizontal quantity used is log s itself, as shown in Fig. 3.53. Lines of constant real quantities and lines of constant imaginary quantities in the s plane

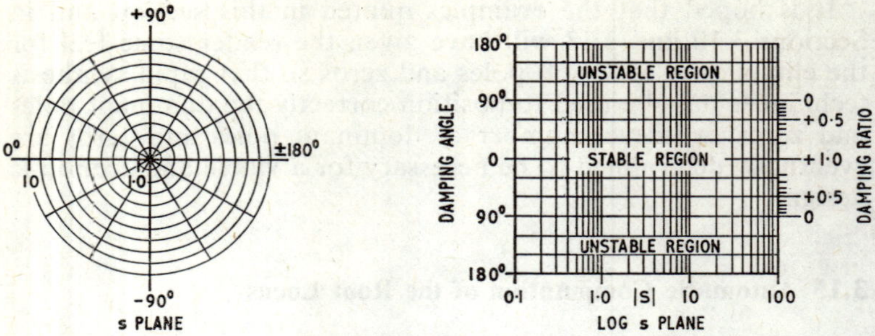


*Fig. 3.53.* s *plane to log* s *plane transformation for lines of constant damping angle and lines of constant radius*

form orthogonal parabolae as in Fig. 3.54. Both these coordinate systems are available as standard graph paper for use with the potential plane analogue computer. In using the computer the open loop poles and zeros are marked on whichever coordinate system is preferred, and the parts of the locus lying on the real axis are marked on in accordance with Rule 4. Once the probes representing the open loop poles and zeros have been correctly set, a stylus is moved over the graph paper, and a meter reading is observed.

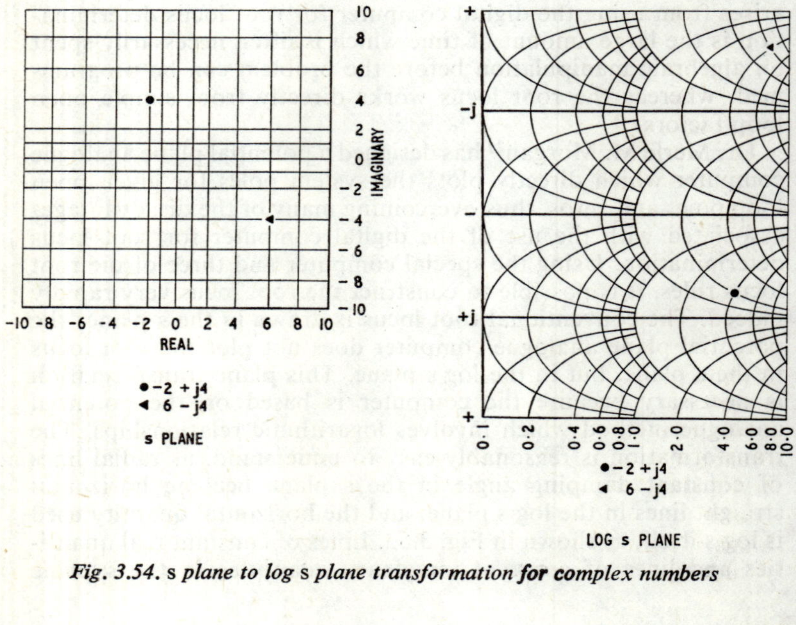


*Fig. 3.54.* s *plane to log* s *plane transformation for complex numbers*

If the meter reading is 180°, then the point immediately beneath the stylus is on the locus, and a mark is made automatically on the paper. Thus the log s plane is searched for roots with the special computer in exactly the same way as the s plane is searched for roots with the Spirule, except that the operation is many times quicker. An unstable root corresponds to a damping angle greater than 90°, as shown in Fig. 3.53, and thus roots in the right-hand half of the s plane correspond to loci entering the region 90° → 180° and the region −90° → −180° in the log s

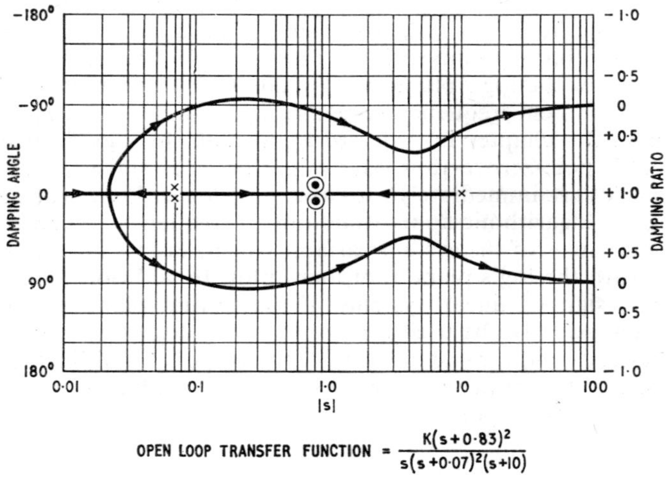


*Fig. 3.55. Root locus plot in the log* s *plane for a system with four open loop poles and two open loop zeros (after Morgan)*

plane. As a simple example, the root locus in the log s plane for the system with the open loop equation

$$\frac{\bar{\theta}_0}{\bar{\theta}}(s) = \frac{K(s+0{\cdot}83)^2}{s(s+0{\cdot}07)^2(s+10)} \tag{3.65}$$

is shown in Fig. 3.55. This example is due to Morgan, and shows a very real advantage of the log s plane over the s plane, in so far as there is a constant resolution per decade in the log s plane which thus portrays such vastly different open loop poles as −0·07 and −10 with the same degree of clarity. Once the locus is drawn it is a simple matter to use the computer to calibrate the locus in terms of feedforward gain, and thus establish desired values of feedforward gain for satisfactory system performance.

'ESIAC' is particularly useful for systems where more than one variable is under study at the synthesis stage, since the second and subsequent variables, if directly available in the open loop form, can be adjusted by moving the setting probes. It is also possible to translate the origin of the log s plane in order to study troublesome regions in detail. For example, in structural resonance problems in missiles, there may be high-frequency open loop poles and zeros which are very lightly damped, and the locus as the system pole migrates from the open loop pole to the open loop zero may cross the imaginary axis, which anyway is very near because of the low damping ratio. Using origin translation in the s plane it is possible to study these poles much more thoroughly than by conventional root locus techniques.

The use of the log s plane means that the logarithmic approximation to system frequency response, which we used extensively earlier in the chapter in order to appreciate the significance of particular pole-zero arrays, is straightforward, since the break frequencies are immediately available on a logarithmic scale, and a series of logarithmic approximations to frequency response for different values of the system variable can be drawn immediately underneath the log s plane root locus plot. In addition to its use in the design of linear systems, the potential plane analogue computer may also be used for systems with dead-time, sampled data systems, and for systems with certain types of non-linearities. The method recommended for the solution of problems involving dead-time, or transportation lag, is particularly elegant.

## 3.16 Derivation of the Open Loop Frequency Response from the Open Loop Poles and Zeros

The system poles and zeros uniquely determine system performance and, provided that the resultant system pole-zero arrays can be interpreted readily at the design stage, there is little justification for the use of open loop design methods for many linear systems. However, the use of an open loop approach in addition to a study of system pole-zero arrays can be justified under particular circumstances, such as a need to examine threshold values in non-linear systems, the stability of non-linear systems, and the initial testing of systems where many assumptions have been made at the design stage. These assumptions, particularly with regard to load characteristics, may have been made in all good faith, but their cumulative effect may be to show that a system which is unstable in practice is stable in theory. It is therefore wise under such circumstances to test the prototype system under open loop conditions to avoid damage that might be caused when the system operates under closed loop, and hence possibly unstable, conditions. If the Bode approximation to the logarithmic frequency plot is not considered accurate enough in particular circumstances, the more exact theoretical open loop frequency response can be obtained graphically from the open loop pole-zero array in a manner similar to that used for computing system frequency response in Section 3.7.

In general, the open loop equation for a linear system may be written

$$\frac{\bar{\theta}_0}{\bar{\theta}}(\mathrm{s}) = K\frac{p(\mathrm{s})v(\mathrm{s})}{q(\mathrm{s})w(\mathrm{s})} \tag{3.66}$$

For a position control system there will be at least one pole of Equation 3.66 at $\mathrm{s} = 0$. If the open loop poles and zeros defined by Equation 3.66 are drawn, exactly as for the construction of the root locus, then vectors drawn from the open loop poles and zeros to the appropriate point on the $j\omega$ axis will give the magnitude of the amplitude ratio at frequency $\omega$, and the angular contribution can also be measured directly. The results may be summarised as follows: the attenuation due to a pole for excitation frequency $\omega$ is equal to the vector length from that pole to the origin divided by the vector length from that pole to the $j\omega$ point under consideration and, conversely, the amplification due to a zero at frequency $\omega$ is equal to the vector length from that zero to the $j\omega$ point divided by the vector length from that zero to the origin. This statement applies to all finite poles and zeros. Thus in vector terminology, the amplification due to a zero is $|\mathrm{z}_{i.\omega}|/|\mathrm{z}_{i.0}|$, and the attenuation due to a pole is $|\mathrm{s}_{i.0}|/|\mathrm{s}_{i.\omega}|$. Since there will be $g$ open loop poles at the origin, we may write the amplitude ratio at frequency $\omega$ as

$$\left|\frac{\bar{\theta}_0}{\bar{\theta}}\right|_{\omega} = \frac{K'}{\omega^g} \times \frac{\prod_{i=n-g}^{i=n} |\mathrm{s}_{i.0}|}{\prod_{i=1}^{i=m} |\mathrm{z}_{i.0}|} \times \frac{\prod_{i=1}^{i=m} |\mathrm{z}_{i.\omega}|}{\prod_{i=n-g}^{i=n} |\mathrm{s}_{i.\omega}|} \tag{3.67}$$

where $K'$ is the feedforward gain when the poles and zeros are written in the form $1+T\mathrm{s}$ and not in the form $\mathrm{s}+1/T$. If we wish to use the value of $K$ defined by the root locus, this is readily obtained, since

$$K = K' \times \frac{\prod_{i=n-g}^{i=n} |\mathrm{s}_{i.0}|}{\prod_{i=1}^{i=m} |\mathrm{z}_{i.0}|}$$

Equation 3.67 can then be written as

$$\left|\frac{\bar{\theta}_0}{\bar{\theta}}\right|_\omega = \frac{K}{\omega^g} \times \frac{\prod_{i=1}^{i=m} |z_{i.\omega}|}{\prod_{i=n-g}^{i=n} |s_{i.\omega}|} \tag{3.68}$$

The phase lag of the open loop response can be readily calculated by reading off the angles at the open loop poles and zeros for a given value of $\omega$. The phase lag of $\bar{\theta}_0$ behind $\bar{\theta}$ is thus merely the algebraic sum of the angles given by

$$(\text{phase lag})_\omega = (90°\times g) + \sum_{i=n-g}^{i=n} \gamma_i - \sum_{i=1}^{i=m} \delta_i \tag{3.69}$$

If the reader should consider these formulae somewhat complicated, he can rest assured that it is very much quicker to actually construct the response than to describe how it is constructed, and a simple example should help prove the point. Let us consider the now familiar system studied in Example 3.5.1, in which there are three open loop poles, one of which is at the origin, and an open loop zero due to the rate feedback term $T_r$. If $K_a = 335$, $T_a = 1/5{\cdot}6$, $T_c = 1/28{\cdot}8$ and $T_r = 1/12$, then for $K_g = 1/18{\cdot}0$, which is a case investigated by the root locus method, we can write, from Equation 3.68,

$$\left|\frac{\theta_0}{\theta}\right|_\omega = \frac{1}{\omega}\left(\frac{K_g K_a T_r}{T_a T_c}\right) \times \frac{|z_{1.\omega}|}{|s_{2.\omega}|\,|s_{3.\omega}|} \tag{3.70}$$

where $z_1 = -1/T_r = -12$, $s_2 = -1/T_a = -5{\cdot}6$, $s_3 = -1/T_c = 28{\cdot}8$, and $K_g K_a T_r/T_a T_c$ is the value of $K$ for the root locus form. This amplitude ratio is evaluated in Fig. 3.56 for a frequency of 10 rad/sec. For the case considered (Table 3.5.1, third row)

$$\frac{K_g K_a T_r}{T_a T_c} = \frac{335\times 5{\cdot}6\times 28{\cdot}8}{18\times 12} = 251$$

From Fig. 3.56 the respective vectors are $|z_{1.10}| = 15{\cdot}6$, $|s_{2.10}| = 11{\cdot}4$, $|s_{3.10}| = 30{\cdot}5$, and the product required is

$$\frac{15{\cdot}6}{11{\cdot}4\times 30{\cdot}5} = \frac{1}{22{\cdot}2}$$

Since $\omega = 10$ rad/sec, the amplitude ratio is

$$\left|\frac{\bar{\theta}_0}{\bar{\theta}}\right| = \frac{251}{10\times 22{\cdot}2} = 1{\cdot}14$$

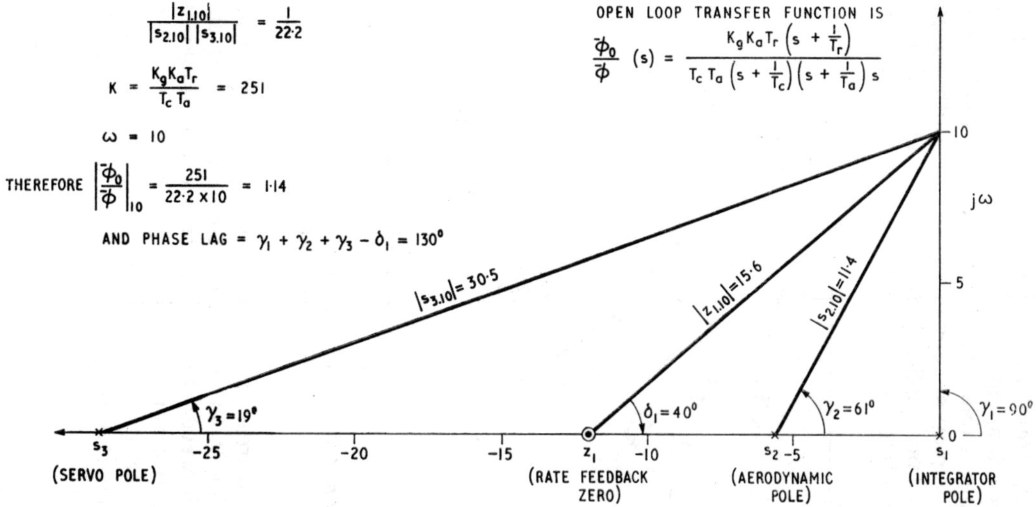


*Fig. 3.56. Vector diagram for the computation of the open loop amplitude ratio and the phase lag for Example 3.5.1 from the open loop pole-zero array (excitation frequency = 10 rad/sec)*

and the corresponding phase lag is $90°+61°+19°-40° = 130°$. The corresponding Nyquist plot is shown in Fig. 3.57, and shows a phase margin at unity gain of 40°; the plot can be constructed in a few minutes from the open loop pole-zero array. For this system the open loop measure of stability of phase margin is 48°, whilst from Table 3.5.1 it can be seen that the damping ratio of the complex system poles is 0·54. Both measures are acceptable for many applications; limits of 40° and 0·50 respectively are often placed on these measures of stability, since these

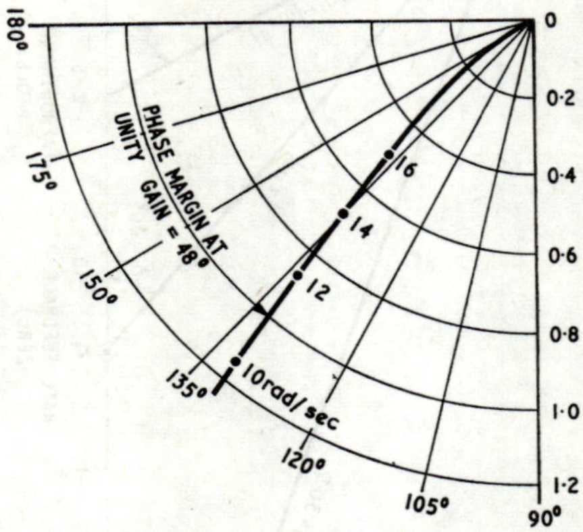


$$\frac{\bar{\phi}_0}{\bar{\phi}}(s) = \frac{251(s+12)}{s(s+5{\cdot}6)(s+28{\cdot}8)}$$

*Fig. 3.57. Nyquist plot for Example 3.5.1 for $K_g = 1/18{\cdot}0$ computed graphically from the open loop pole-zero array*

limits will allow reasonable maladjustment component values without resulting in any unstable system. The reader should be in no doubt, however, that whatever the occasional merits of using open loop criteria, particularly with non-linear systems, in general the use of system criteria is more fundamental, more meaningful, and less ambiguous since it is *system performance* which is required.

## 3.17 Conclusions

In this chapter we have seen how the root locus factorises the system characteristic equation for all values of a system variable, the variable normally taken being feedforward gain. When the system zeros are added to the system poles obtained for a particular value of that system variable, the system pole-zero array is formed, and system performance and system transfer function are both uniquely defined. The system poles and system zeros nearest to the origin tend to dominate the system response and an approximate assessment of system performance can often be made by neglecting the effect of secondary pole position on performance. Several systems and performance criteria have been examined, thus providing the reader with some background information on the assessment of pole-zero geometry. Synthesis using the root locus method consists essentially of constructing the root locus for one or more system variables, isolating the more useful values of those variables by some rapid assessment techniques such as the logarithmic approximation to the frequency response, and then examining these promising system pole-zero arrays in more detail. For this purpose more exact system frequency responses, and the system velocity constants, can be easily determined from the system pole-zero array, and the partial-fraction expansion coefficients for the transient response are also readily obtained, if required. Unless additional specialised techniques, such as those of Chapter 5 are used, the design process outlined above is often essentially iterative in nature, and successful previous designs will assist in reducing the number of iterative steps required before the specification is completely satisfied. In Chapter 6 we shall study several general methods for achieving satisfactory system design utilising the s plane almost exclusively.

Conventional analogue and digital computers can be of great assistance at the design stage, but even then the root locus is very useful for roughly fixing certain parameters to avoid using excessive computer time in finding an acceptable and physically realisable system pole-zero array. The potential plane analogue computer is a special analogue computer which plots the root locus in the log s plane very rapidly indeed, and has the added advantage of constant resolution of roots per decade, thus allowing the effect of both large and small open loop poles and zeros to be studied equally well.

The root locus method utilises information about the system which is readily available in simple form, since the open loop poles and zeros are normally known in factored form. Open loop design methods also emanate from a knowledge of these open loop poles and zeros, and a graphical construction for the open loop response has been shown, thus allowing the reader to examine many aspects of feedback control system behaviour using very simple tools and techniques.

CHAPTER 4

# The General Second Order Linear System

## 4.1 Introductory Remarks

The analysis and synthesis of second order linear control systems is of great importance to the control engineer. Many basic problems in control engineering can be understood by studying second order systems, and the lessons learned subsequently applied to higher order systems. As we have already seen in Chapter 3, it is often possible to design high order systems such that the response of the system is dominated by a second order mode, and it is nearly always possible to estimate roughly the type and magnitude of one or more of the compensation elements required for high order systems by second or third order approximations. Because of the simplicity of second order linear systems it is possible to present performance characteristics as simple non-dimensionalised curves, thus making the determination of the appropriate system transfer function a simple process, normally requiring a few arithmetical calculations sometimes supplemented by a subsidiary graph. We shall interpret the general second order system to possess one zero and two poles. If we specify the performance of the system using three independent performance characteristics such as bandwidth, percentage overshoot, and velocity constant, a unique transfer function corresponding to a specific zero and specific pole positions will be specified. If only two of these quantities are specified, there will be many system transfer functions which will yield the required performance. Conversely, if the zero is undesirable, a limitation is placed on the system transfer function, and hence a similar limitation is placed on system performance. A necessary condition for the control system designer to have complete jurisdiction over the positions of the poles and zero, is that there should be three infinitely adjustable parameters in the system. It is convenient to think of such a system as possessing three degrees of design freedom. Such a system is automatically and uniquely specified by three independent performance characteristics of the type discussed above.

Once the system transfer function has been determined from the performance specification, an actual design in terms of individual components can then be made and, as we shall see later in the chapter, it will be possible to design several systems which yield the same system transfer function. The choice between alternative systems is then made by comparing such characteristics as cost, weight, noise rejection, and load disturbance rejection. It is possible that the design which requires the highest feedforward gain necessitates the next size of motor, proving unnecessarily cumbersome or expensive. Such choices can only be made in the light of experience. It is important to remember that performance is only one aspect of a control system, even if the most critical aspect, and the best design will be that which achieves the best balance between conflicting requirements. A control system with too high a power consumption may well be as ineffective as a control system which is too slow.

## 4.2 Second Order Transfer Functions

The general second order type (1, 2) system with two poles and one zero can be written, in polynomial form,[35] as

$$\frac{\theta_0}{\theta_i}(s) = \frac{1+b_1 s}{1+a_1 s+a_2 s^2} \tag{4.1}$$

and alternatively, in non-dimensional form,

$$\frac{\theta_0}{\theta_i}(s) = \frac{1+\alpha\left(\dfrac{2\zeta}{\omega_n}\right)s}{1+\left(\dfrac{2\zeta}{\omega_n}\right)s+\dfrac{s^2}{\omega_n^2}} \tag{4.2}$$

where $\zeta$ is termed the damping ratio, $\omega_n$ is the undamped natural frequency, and $\alpha$ is a design variable which has special significance in so far as $\alpha = 0$ in Equation 4.2 will result in a zero displacement error system, whilst $\alpha = 1$ will result in a zero velocity error system. Furthermore, $1-\alpha$ is a measure of the steady state lag of the system. For servomechanisms, $\alpha$ and $\zeta$ are generally 1·0 or less. The choice of Equation 4.1 as the basic second order system is essential in order to meet a specification written in terms of three independent performance criteria.[36]

Given the components and hence a basic configuration, the analysis problem is to determine the coefficients of Equation 4.1 or 4.2 and thus to determine the performance characteristics of interest. The design problem is, given the desired performance characteristics, firstly to determine the coefficients of Equation 4.1 or 4.2, and then to design a system yielding this desired transfer function.

The system performance is completely defined by the transfer function, and it follows that the performance is expressible in terms of the transfer function coefficients. For second order systems it is often convenient to present the performance in terms of $\zeta$, $\alpha$, and $\omega_n$, but the conversion to the polynomial coefficients $a_1$, $a_2$, and $b_1$ is straightforward. The direct use of the polynomial coefficients is sometimes preferable since certain system performance characteristics, notably steady state velocity lag, are obtainable directly from these coefficients. Equation 4.1 could, of course, be rewritten to include a second order numerator term, $b_2s^2$, resulting in a system with two poles and two zeros and offering the designer further possibilities for shaping the system response. This system is, however, relatively rarely met in practice either in true or approximate form and the analysis and synthesis methods to be presented in this chapter will concentrate on the one-zero two-pole system. If the reader should need to synthesise the two-zero two-pole system, the more general methods of Chapter 3 can be used to achieve a satisfactory design.

In this chapter, as in Chapters 3 and 5 to a lesser extent, we shall often use compensation transfer functions of the form $G_c(s) = 1+Ts$ in the feedforward path in order to shape the system transfer function into the desired form. The use of such a transfer function in the feedforward path requires some explanation. If an instrument such as a tachogenerator is available then, as shown in Equation 2.57, a signal proportional to rate is available. Such a tachogenerator fixed to the output shaft will measure output rate $T_r s\theta_0$ and, fixed to the input shaft, will measure input rate $T_i s\theta_i$. Provided the tachogenerator lag is negligible, the use of compensation terms such as $1+T_r s$ and $1+T_i s$ is exact. If there are identical tachogenerators on both shafts, and the output rate signal is subtracted from the input rate signal, a true error rate signal $T_e s\theta$ is generated, hence under these circumstances a compensation transfer function $G_c(s) = 1+T_e s$ is available. Compensation transfer functions $1+T_i s$ and $1+T_e s$ derived in this way will result in a genuine one-zero two-pole system [type (1,2)] which exactly fits the standard transfer functions of Equations 4.1 and 4.2. Now, it is often necessary to generate rate signals using passive networks, rather than by introducing one or more special instruments such as rate gyroscopes or tachogenerators. As we have already seen in Chapter 2, practical networks only approximate to the desired rate signal, the approximation being good over a particular frequency range. Equations 2.43 and 2.45 are for typical practical compensation networks generating such an approximation to a proportional plus rate transfer function. A typical transfer for such a network in terms of numbers would be $\frac{1}{15}(1+Ts)/[1+(T/15)s]$. It is reasonable to assume that we would include extra signal amplification to compensate for the attenuation, and therefore we may write the network plus amplification transfer function as $(1+Ts)/[1+(T/15)s]$. Clearly if the desired speed of response of the system is much less than $15/T$ we may neglect the lag at this stage and write the compensation transfer function as $G_c(s) = 1+Ts$. The introduction of the true compensation network has transformed an uncompensated (0,2) system into a (1,3) compensated system which we are approximating by a (1,2) system simply by neglecting the third term of the characteristic equation. This is justified where the dominant poles are unaffected by the presence of the additional lag term in the feedforward path; the effect of such a lag term has already been thoroughly investigated in Example 3.11.1. The reader should carefully note the difference in approach between this very simple technique of neglecting the third coefficient of the characteristic equation, as distinct from the much more accurate s plane approximation used in Chapter 6. However, for many simple linear servos with passive network compensation, the former approximation is good enough for the rapid design techniques described later.

## 4.3 Derivation of the System Transfer Function for Some Common Second Order Servomechanisms

### Example 4.3.1

The basic second order servomechanism is shown in Fig. 4.1(a). An inertia load of $J$ slugs ft$^2$ is position-controlled by a motor which generates K lb ft torque per radian excitation signal, the signal being amplified between the comparator and the motor in order to provide the power level to operate the motor. The gain $K$ is thus the product of the signal amplifier gain and the motor gain;

the multiplication being justified by virtue of the linearity of the system. A viscous resistive torque of $f$ lb ft rad$^{-1}$ sec is assumed to be present at the load. Should gearing exist between the motor

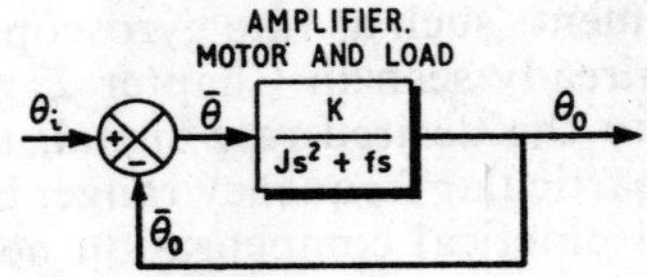


$$\frac{\theta_o}{\theta_i}(s) = \frac{1}{1 + a_1 s + a_2 s^2}$$

$$= \frac{1}{1 + \left(\frac{2\zeta}{\omega_n}\right)s + \frac{s^2}{\omega_n^2}}$$

$$= \frac{1}{1 + \left(\frac{f}{K}\right)s + \left(\frac{J}{K}\right)s^2}$$

(a)

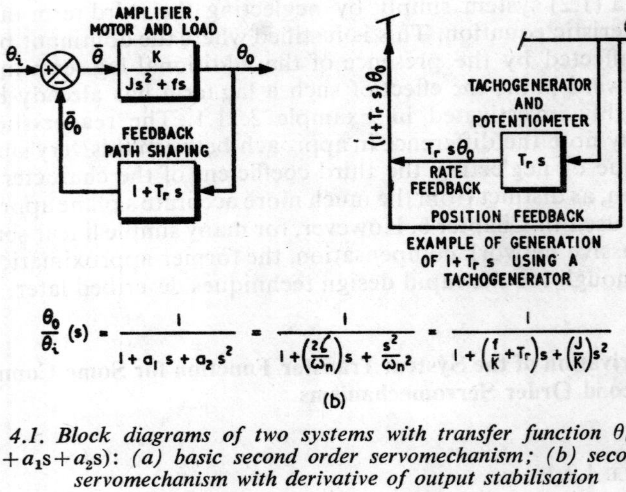


$$\frac{\theta_o}{\theta_i}(s) = \frac{1}{1 + a_1 s + a_2 s^2} = \frac{1}{1 + \left(\frac{2\zeta}{\omega_n}\right)s + \frac{s^2}{\omega_n^2}} = \frac{1}{1 + \left(\frac{f}{K} + T_r\right)s + \left(\frac{J}{K}\right)s^2}$$

(b)

*Fig. 4.1. Block diagrams of two systems with transfer function $\theta_0/\theta_i(s) = 1/(1+a_1 s+a_2 s)$: (a) basic second order servomechanism; (b) second order servomechanism with derivative of output stabilisation*

shaft and the load, the motor torque, load inertia and viscous resistive torque must be referred to a common point, such as the load shaft, using normal dynamic equivalence formulae, and we shall assume this procedure has been carried out. $J$ is reckoned to include the motor moment of inertia if this is significant. Torque equilibrium between motor and load dictates that

$$K\bar{\theta} = J\frac{d^2\theta_0}{dt^2}+f\frac{d\theta_0}{dt} \tag{4.3}$$

The excitation signal $\bar{\theta}$ in Equation 4.3 is, in fact, the error of the system, since the signal being fed back to the comparator is the system output, and input minus output is error. Substituting $s = d/dt$ in Equation 4.3, the feedforward transfer function is

$$G(s) = \frac{K}{Js^2+fs} \tag{4.4}$$

Since the feedback transfer function is unity, the system transfer function is

$$\frac{\theta_0}{\theta_i}(s) = \frac{G(s)}{1+G(s)} = \frac{K}{K+fs+Js^2}$$

and, putting in standard form,

$$\left.\begin{aligned}\frac{\theta_0}{\theta_i}(s) &= \frac{1}{1+\left(\frac{f}{K}\right)s+\left(\frac{J}{K}\right)s^2}\\ &= \frac{1}{1+\left(\frac{2\zeta}{\omega_n}\right)s+\frac{s^2}{\omega_n^2}}\\ &= \frac{1}{1+a_1s+a_2s^2}\end{aligned}\right\} \tag{4.5}$$

yielding the transfer function coefficients

$$\left.\begin{aligned}\frac{J}{K} &= a_2 = \frac{1}{\omega_n^2}\\ \frac{f}{K} &= a_1 = \frac{2\zeta}{\omega_n}\end{aligned}\right\} \tag{4.6}$$

If the viscous damping coefficient $f$ is variable, then, since the feedforward gain $K$ is normally adjustable, the designer has two independent variables under his jurisdiction, and can therefore theoretically design a system to meet two independent performance characteristics. However, since $b_1 = 0$ it will not be possible to produce a system with zero steady state velocity lag. Furthermore, the adjustment of $f$ to provide a reasonable value of damping ratio (say 0·5 to 0·7) will be found to be excessively power-consuming, as all the torque used to overcome the viscous damping term is dissipating useful energy from the motor. This system is therefore rarely found in practice, the use of 'synthetic damping' being preferable in order to produce a reasonable response without excessive waste of energy. An increase in viscous damping also increases the steady state velocity lag since $a_1$ is increased.

EXAMPLE 4.3.2

A system with 'synthetic damping' is shown in Fig. 4.1(b). A signal proportional to the velocity of the load is fed back to the comparator and subtracted in an identical fashion to the load position feedback signal. $T_r$ arises from the gain of the tachogenerator used to measure the speed of the output shaft, multiplied by the potentiometer setting which is used to scale the velocity feedback signal to the required level. The feedback of velocity as a compensator term is also known as 'velocity feedback', and intuitively can be thought of as a process in which the future output position of the load is being crudely predicted and this information fed back to adjust the motor excitation signal.

From Fig. 4.1(b), the feedforward transfer function is

$$G(s) = \frac{K}{Js^2+fs} \tag{4.7}$$

and the feedback transfer function is

$$H(s) = 1+T_r s \tag{4.8}$$

Thus the system transfer function is

$$\frac{\theta_0}{\theta_i}(s) = \frac{G(s)}{1+G(s)H(s)} = \frac{K}{Js^2+fs+K(1+T_r)s}$$

which, in standard form, is

$$\left.\begin{aligned} \frac{\theta_0}{\theta_i}(s) &= \frac{1}{1+a_1 s+a_2 s^2} \\ &= \frac{1}{1+\left(\frac{2\zeta}{\omega_n}\right)s+\frac{s^2}{\omega_n^2}} \\ &= \frac{1}{1+\left(T_r+\frac{f}{K}\right)s+\left(\frac{J}{K}\right)s^2} \end{aligned}\right\} \tag{4.9}$$

yielding the transfer function coefficients

$$\left.\begin{aligned} \frac{J}{K} &= a_2 = \frac{1}{\omega_n^2} \\ T_r+\frac{f}{K} &= a_1 = \frac{2\zeta}{\omega_n} \end{aligned}\right\} \tag{4.10}$$

Thus for the same value of feedforward gain as in the previous example, the introduction of the velocity feedback term has increased the damping ratio of the system and has thus reduced the tendency of the system to oscillate when subjected to a step stimulus. The damping ratio of the complex poles is also a measure of the absolute stability of the system. Thus for a constant $K$ the introduction of negative velocity feedback increases the absolute stability of the system. Since $a_1$ has been increased by this method, we have increased the steady state velocity lag, as $b_1$ is zero. This basic design is therefore still unacceptable for systems where zero velocity lag is required. For $f$ constant the designer has $K$ and $T_r$ as system variables under his jurisdiction, and can only guarantee to meet two independent performance requirements simultaneously.

### Example 4.3.3

We have seen from our discussion of Examples 4.3.1 and 4.3.2 that it will often be necessary to introduce additional shaping networks in order to meet specifications which include three independent performance characteristics. In particular, we wish to incorporate a shaping network which will introduce a $b_1$ term into the system transfer function. This can be achieved by the introduction of a phase advance network of the form $1+T_e s$ into the feedforward path. Fig. 4.2(a) shows a system where the phase advance term is in the feedforward path inside the loop and is in fact operating on the system error signal, thus generating a true derivative of error term. The feedforward transfer function is

$$G(s) = \frac{K(1+T_e s)}{Js^2+fs} \tag{4.11}$$

and since the feedback transfer function is unity, the system transfer function is

$$\frac{\theta_0}{\theta_i}(s) = \frac{G(s)}{1+G(s)} = \frac{K(1+T_e s)}{Js^2+fs+K(1+T_e s)}$$

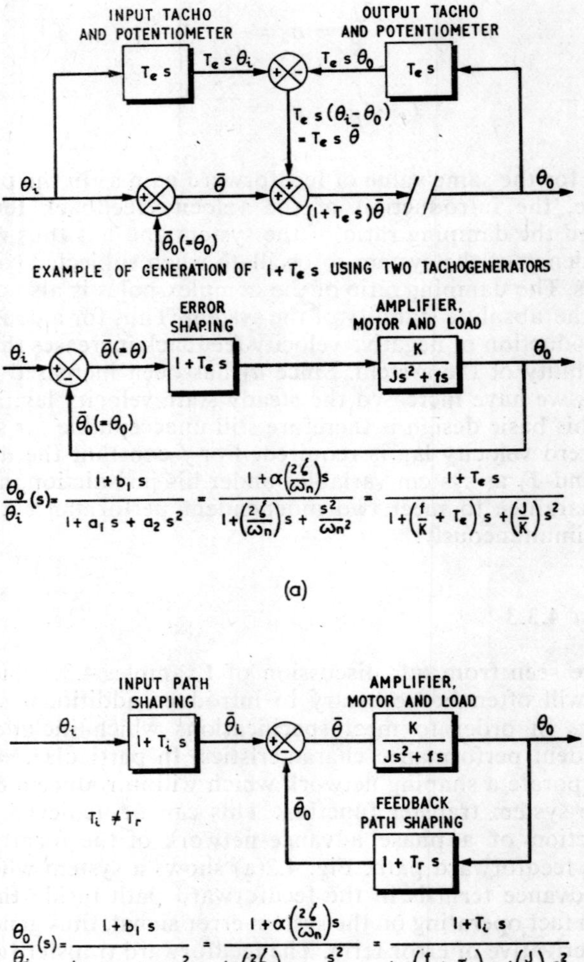


*Fig. 4.2. Block diagrams of two systems with transfer function $\theta_0/\theta_i(s) = (1+b_1s)/(1+a_1s+a_2s^2)$: (a) second order servomechanism with derivative of error stabilisation; (b) second order servomechanism with derivative of output stabilisation and derivative of input response shaping*

which, in standard form, is

$$\left.\begin{aligned}\frac{\theta_o}{\theta_i}(\mathrm{s}) &= \frac{1+T_e\mathrm{s}}{1+\left(T_e+\dfrac{f}{K}\right)\mathrm{s}+\left(\dfrac{J}{K}\right)\mathrm{s}^2}\\ &= \frac{1+b_1\mathrm{s}}{1+a_1\mathrm{s}+a_2\mathrm{s}^2}\\ &= \frac{1+\alpha\left(\dfrac{2\zeta}{\omega_n}\right)\mathrm{s}}{1+\left(\dfrac{2\zeta}{\omega_n}\right)\mathrm{s}+\dfrac{\mathrm{s}^2}{\omega_n^2}}\end{aligned}\right\} \tag{4.12}$$

yielding the transfer function coefficients

$$\left.\begin{aligned}\frac{J}{K} &= a_2 = \frac{1}{\omega_n^2}\\ \frac{f}{K}+T_e &= a_1 = \frac{2\zeta}{\omega_n}\\ T_e &= b_1 = \alpha\left(\frac{2\zeta}{\omega_n}\right)\\ \alpha &= \frac{b_1}{a_1} = \frac{T_e}{\dfrac{f}{K}+T_e}\end{aligned}\right\} \tag{4.13}$$

Quite clearly, by inspecting the relationships defined by Equation 4.13, we can manipulate the two system variables $K$ and $T_e$ to obtain a suitable performance, but it should be noted that we do not have complete design freedom. As we shall show shortly, if the specification has three independent requirements it may be interpreted as specific values of $\alpha$, $\zeta$, and $\omega_n$. If we use $\omega_n$ to fix $K$ knowing $J$, then we can see that there will be definite combinations of $\alpha$ and $\zeta$ which will result from a value of $T_e$, and it is unlikely that the combination of $\alpha$ and $\zeta$ required by the specification will be realised. A compromise is therefore necessary, but it would be better under these circumstances to introduce a further compensating network in order to give the extra degree of freedom.

A special case exists for the system of Fig. 4.2(a) when $f = 0$. Under these conditions $b_1 = a_1$ and $\alpha = 1{\cdot}0$, resulting in a zero steady state velocity error system.

EXAMPLE 4.3.4

The system shown in Fig. 4.2(b) does allow the designer the necessary three degrees of design freedom, since we are incorporating rate feedback damping as well as a feedforward derivative term.

Note that in this example the feedforward derivative term occurs outside the loop and is therefore providing a signal to the comparator which is proportional to the input signal plus the derivative of the input signal.

From Fig. 4.2(b) the pre-loop transfer function $G_p(s)$ is given by the expression

$$\frac{\bar{\theta}_i}{\theta_i}(s) = G_p(s) = 1+T_i s \tag{4.14}$$

The loop feedforward transfer function is

$$\frac{\theta_0}{\theta}(s) = G(s) = \frac{K}{Js^2+fs} \tag{4.15}$$

and the feedback path transfer function is

$$\frac{\bar{\theta}_0}{\theta_0}(s) = H(s) = 1+T_r s \tag{4.16}$$

Closing the loop, the transfer function relating shaped input to output is

$$\frac{\theta_0}{\bar{\theta}_i}(s) = \frac{G(s)}{1+G(s)H(s)} \tag{4.17}$$

and the system transfer function is then

$$\frac{\theta_0}{\theta_i}(s) = \frac{\theta_0}{\bar{\theta}_i}(s) \times \frac{\bar{\theta}_i}{\theta_i}(s) = \frac{G_p(s)G(s)}{1+G(s)H(s)} \tag{4.18}$$

Hence for the particular system shown

$$\frac{\theta_0}{\theta_i}(s) = \frac{(1+T_i s)K}{Js^2+fs+K(1+T_r s)}$$

which, in standard form, becomes

$$\left.\begin{aligned}\frac{\theta_o}{\theta_i}(s) &= \frac{1+T_i s}{1+\left(\dfrac{f}{K}+T_r\right)s+\left(\dfrac{J}{K}\right)s^2} \\ &= \frac{1+b_1 s}{1+a_1 s+a_2 s^2} \\ &= \frac{1+\alpha\left(\dfrac{2\zeta}{\omega_n}\right)s}{1+\left(\dfrac{2\zeta}{\omega_n}\right)s+\dfrac{s^2}{\omega_n^2}}\end{aligned}\right\} \tag{4.19}$$

yielding the transfer function coefficients

$$\left.\begin{aligned}\frac{J}{K} &= a_2 = \frac{1}{\omega_n^2} \\ \frac{f}{K}+T_r &= a_1 = \frac{2\zeta}{\omega_n} \\ T_i &= b_1 = \alpha\left(\frac{2\zeta}{\omega_n}\right) \\ \alpha &= \frac{b_1}{a_1} = \frac{T_i}{f/K+T_r}\end{aligned}\right\} \tag{4.20}$$

Thus by correctly adjusting $K$, $T_r$, and $T_i$, we can design the system to meet three independent performance specifications.

EXAMPLE 4.3.5

So far in this chapter we have investigated four different systems which yield second order linear transfer functions. Methods used to stabilise and shape the transfer function in these examples have included both series and feedback compensation. Finally we shall consider the case where the compensation takes the form of subsidiary feedback to give a minor feedback loop in the main feedforward path of the system, thus necessitating the use of block diagram reduction techniques in determining the system transfer function. Fig. 4.3(a) shows a servomechanism where output velocity is measured but is not fed back to the main comparator; instead it is fed back negatively to the amplifier input, thus

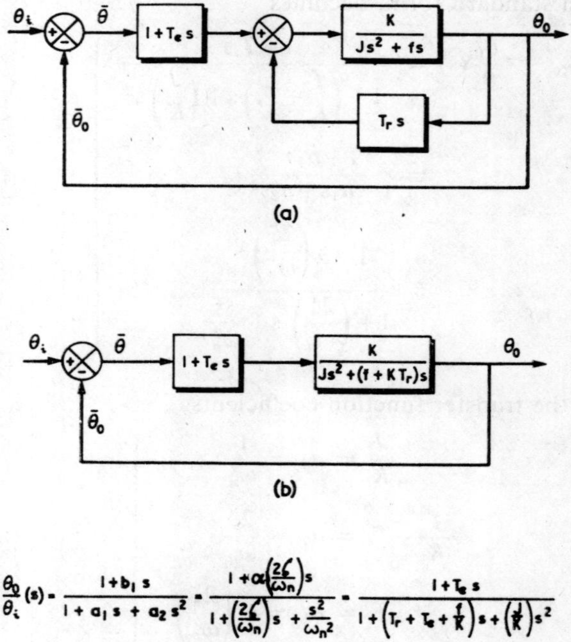


*Fig. 4.3. A minor loop compensated system with the transfer function $\theta_0/\theta_i(s) = (1+b_1s)/(1+a_1s+a_2s^2)$: (a) block diagram of servomechanism with minor loop compensation; (b) simplified equivalent block diagram*

forming a minor loop around the amplifier, motor and load. For the minor loop, the feedforward transfer function is given by

$$G_3(s) = \frac{K}{Js^2+fs} \tag{4.21}$$

and the minor loop feedback transfer function is

$$H_3(s) = T_r s \tag{4.22}$$

giving the minor loop transfer function

$$G_2(s) = \frac{G_3(s)}{1+H_3(s)G_3(s)} = \frac{K}{Js^2+(f+KT_r)s} \tag{4.23}$$

The complete system transfer function is given by

$$\frac{\theta_0}{\theta_i}(s) = \frac{G_1(s)G_2(s)}{1+G_1(s)G_2(s)} \tag{4.24}$$

where $G_1(s) = 1+T_e s$. On substitution, we obtain

$$\left.\begin{aligned}\frac{\theta_0}{\theta_i}(s) &= \frac{1+T_e s}{1+\left(T_r+T_e+\dfrac{f}{K}\right)s+\left(\dfrac{J}{K}\right)s^2}\\ &= \frac{1+b_1 s}{1+a_1 s+a_2 s^2}\\ &= \frac{1+\alpha\left(\dfrac{2\zeta}{\omega_n}\right)s}{1+\left(\dfrac{2\zeta}{\omega_n}\right)s+\dfrac{s^2}{\omega_n^2}}\end{aligned}\right\} \tag{4.25}$$

yielding the transfer function coefficients

$$\left.\begin{aligned}T_e = b_1 &= \alpha\left(\frac{2\zeta}{\omega_n}\right)\\ T_r+T_e+\frac{f}{K} = a_1 &= \frac{2\zeta}{\omega_n}\\ \frac{J}{K} = a_2 &= \frac{1}{\omega_n^2}\\ \alpha = \frac{b_1}{a_1} &= \frac{T_e}{T_r+T_e+f/K}\end{aligned}\right\} \tag{4.26}$$

Correct adjustment of $T_r$, $T_e$ and $K$ can thus lead to a system which, in general, is capable of meeting three independent performance requirements. The reader should note, however, that the case for a zero velocity error system ($a_1 = b_1$) can only exist if $T_r = f = 0$, and we have thus imposed a constraint on the design of the system. The reader is left to verify for himself that this constraint could be removed by providing derivative of input in the feedforward path instead of derivative of error.

We have discussed several of the many ways in which the general second order linear transfer function as defined by Equation 4.1 or 4.2 can be obtained, using a simple position servomechanism as a model for our analysis. In practice there will probably be a preferred form of system based on experience and convenience for a given application. For regulating systems the command signal is normally zero, and under these circumstances phase advance would be placed inside the loop in the feedforward path, whilst for fire control systems the input is zero for compensatory tracking and it is often not practicable to build in output rate damping, leaving derivative of error shaping as the obvious solution. In many servomechanisms the ease with which tachogenerators may be geared to the servomotor to provide a feedback signal proportional to output velocity makes output rate feedback an attractive proposition for providing sufficient damping in the $a_1$ term. Having shown that there may be several solutions of merit for a particular problem, we shall now proceed to the performance analysis problem: i.e. given the transfer function, to determine the performance characteristics of interest.

## 4.4 Analysis: Pole-zero Configurations for the General Second Order System

The system transfer function defined by Equation 4.2 as

$$\frac{\theta_0}{\theta_i}(s) = \frac{1+\alpha\left(\frac{2\zeta}{\omega_n}\right)s}{1+\left(\frac{2\zeta}{\omega_n}\right)s+\frac{s^2}{\omega_n^2}}$$

has one zero at

$$z_1 = -\frac{\omega_n}{2\zeta\alpha} \tag{4.27}$$

and two poles at

$$\left.\begin{aligned} s_1 &= -\zeta\omega_n + j\omega_n\sqrt{1-\zeta^2} \\ s_2 &= -\zeta\omega_n - j\omega_n\sqrt{1-\zeta^2} \end{aligned}\right\} \tag{4.28}$$

provided that $\zeta < 1{\cdot}0$ which is usual, at least for servomechanisms. As we have seen, the complex pole positions may be alternatively defined in polar coordinates by a radius $\omega_n$ from the origin, and an angle of $\pm\cos^{-1}\zeta$ measured from the negative real axis. Fig. 4.4 shows the pole-zero locations as a function of $\alpha$ and $\zeta$ for the general second order system of Equation 4.1 or 4.2. The areas bounded by the pole-zero array have been shaded for clarity. In only one of those shown is the zero nearer the origin than are the poles, and this is for $\zeta = 1{\cdot}0$, $\alpha = 1{\cdot}0$. For $\zeta = 1{\cdot}0$, $\alpha = 0{\cdot}5$, both poles and the zero are coincident. One pole is exactly nulled by the presence of the zero, leaving the system with one negative real pole. The effect of the zero on the system is best determined by comparing the relative distances of the zero and the poles from the origin, as measured along the negative real axis. This ratio is, in terms of $\alpha$ and $\zeta$,

$$\frac{z_1}{R_{s_1}} = \frac{\omega_n/2\zeta\alpha}{\zeta\omega_n} = \frac{1}{2\zeta^2\alpha} \tag{4.29}$$

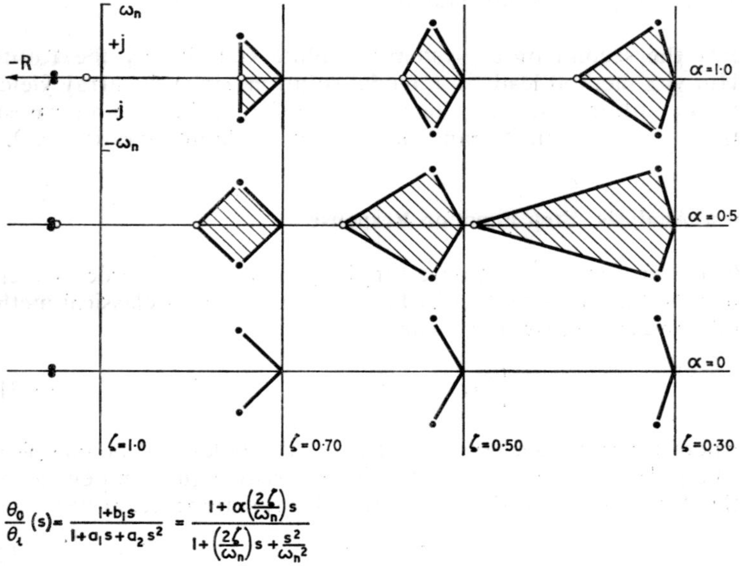


*Fig. 4.4. Pole-zero locations for second order control systems*

where $R_{s_1}$ is the real part of the complex poles. Equation 4.29 is expressible directly in terms of transfer function coefficients since $\alpha = b_1/a_1$ and $\omega_n$ can be eliminated between $a_1$ and $a_2$ to express $\zeta$ in terms of these two coefficients.

The result is

$$\frac{z_1}{R_{s_1}} = \frac{2a_2}{a_1 b_1} \tag{4.30}$$

Thus if it is required for the zero to be ten times further from the origin than the real part of the complex poles, the coefficients are defined by $5b_1 a_1 = a_2$. Shortly we shall relate performance criteria, such as percentage overshoot, bandwidth, and velocity constant, directly to the system transfer function. Figs. 4.10, 4.11, 4.12, 4.13 and 4.15 show performance criteria as functions of $\alpha$, $\zeta$ and $\omega_n$.

If it is required to determine the criteria not in terms of the system transfer function but in terms of pole-zero geometry, as in Truxal[37] or Horowitz,[38] then Equation 4.30 can be used to define a new variable $\lambda = z_1 R_{s_1}$ in place of $\alpha$. The aforementioned figures can then be redrawn as functions of $\lambda$, $\zeta$ and $\omega_n$.

## 4.5 Analysis: Absolute Stability

For the second order system, absolute stability by the Routh Hurwitz method leads to a trivial solution, since the array yields the condition that the system is stable if $a_1$ and $a_2$ are both positive. In practice this means that in terms of damping ratio, $\zeta > 0$.

## 4.6 Analysis: Step Function Response

The step function response for the general second order system may be determined either by Laplace transform or classical methods. In general, we may write

$$\frac{\theta_0}{\theta_i}(t) = 1 + A_1 e^{s_1 t} + A_2 e^{s_2 t} \tag{4.31}$$

where for servomechanisms, at least, the poles will be complex. The poles are the roots of the characteristic equation defined by the denominator of the system transfer function; i.e. either

$$a_2 s^2 + a_1 s + 1 = 0 \tag{4.32}$$

or

$$\frac{s^2}{\omega_n^2} + \left(\frac{2\zeta}{\omega_n}\right) s + 1 = 0 \tag{4.33}$$

Applying the rule for the solution of a quadratic to Equation 4.33, we obtain the 'roots' or poles as

$$s = \frac{-\dfrac{2\zeta}{\omega_n} \pm \sqrt{\dfrac{4\zeta^2}{\omega_n^2} - \dfrac{4}{\omega_n^2}}}{2/\omega_n^2} = -\zeta\omega_n \pm \omega_n \sqrt{\zeta^2 - 1}$$

where $\zeta = 1{\cdot}0$ represents the critical case. For servomechanisms, $\zeta < 1{\cdot}0$; hence the roots are normally written as

$$s = -\zeta\omega_n \mp j\omega_n\sqrt{1-\zeta^2} \tag{4.34}$$

From symmetry considerations we may combine $A_1$ and $A_2$ and rewrite Equation 4.31 in terms of Equation 4.34:

$$\frac{\theta_0}{\theta_i}(t) = 1 + Ae^{-\zeta\omega_n t}\left(e^{+j\omega_n\sqrt{1-\zeta^2}\,t} + e^{-j\omega_n\sqrt{1-\zeta^2}\,t}\right)$$

Remembering that $2\cos\phi = e^{j\phi} + e^{-j\phi}$ and two arbitrary constants are required in the solution of a second order differential equation, the complete solution can be written as

$$\frac{\theta_0}{\theta_i}(t) = 1+Ae^{-\zeta\omega_n t}\cos\left(\omega_n\sqrt{1-\zeta^2}t+\nu\right) \tag{4.35}$$

$A$ and $\nu$ are determined from a knowledge of the initial conditions, and the zeros contribute to these constants. If $b_1 = \alpha = 0$, and the system is thus a simple second order zero displacement

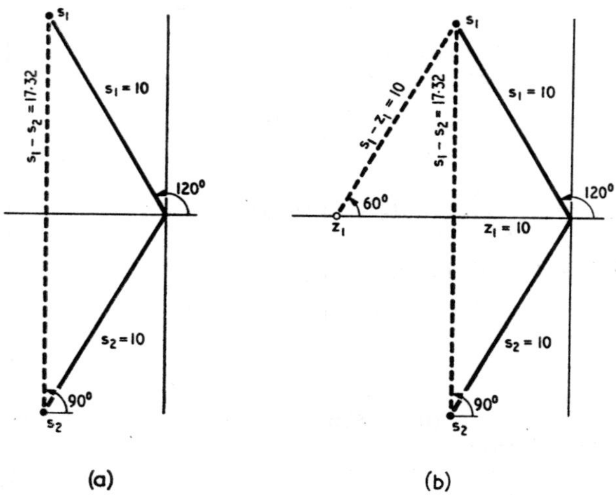


*Fig. 4.5. The pole-zero array: (a) for $\alpha = 0$, $\zeta = 0{\cdot}5$ and $\omega_n = 10$; (b) for $\alpha = 1{\cdot}0$, $\zeta = 0{\cdot}5$ and $\omega_n = 10$*

error system, the solution of Equation 4.35 in terms of $\zeta$ and $\omega_n$ is particularly straightforward; hence

$$\frac{\theta_0}{\theta_i}(t) = 1-\frac{e^{-\zeta\omega_n t}}{\sqrt{1-\zeta^2}}\cos\left(\omega_n\sqrt{1-\zeta^2}\,t-\tan^{-1}\frac{\zeta}{\sqrt{1-\zeta^2}}\right) \tag{4.36}$$

If $\alpha$ is non-zero then Equation 4.36 becomes more complicated algebraically, although arithmetically it is still straightforward, and the arbitrary constants may be obtained with ease from the pole-zero array as described in Chapter 3. As examples of the computation of the constants of the solution directly from the pole-zero array, consider the case where $\zeta$ is 0·5. Fig. 4.5 shows the pole-zero arrays for $\alpha = 0$ and $\alpha = 1{\cdot}0$.

EXAMPLE 4.6.1

The pole-zero array for $\alpha = 0$, $\zeta = 0{\cdot}5$ and $\omega_n = 10$ is shown in Fig. 4.5(a). Reading directly from the figure,

$$A_1 = \frac{|s_1||s_2|}{s_1(s_1 - s_2)}$$

$$= \frac{10^2}{(10\underline{/120^\circ})(17{\cdot}32\underline{/90^\circ})}$$

$$= -\frac{0{\cdot}58}{\underline{/30^\circ}}$$

$$= -0{\cdot}58\underline{/-30^\circ}$$

Similarly,

$$A_2 = -0{\cdot}58\underline{/+30^\circ}$$

Combining $A_1$ and $A_2$ vectorially, we have,

$$A = 2(-0{\cdot}58) = -1{\cdot}16$$

and

$$\nu = -30^\circ$$

Now $\omega_n\sqrt{1-\zeta^2} = 10 \times 0{\cdot}866 = 8{\cdot}66$ rad/sec and $\zeta\omega_n = 5{\cdot}0$. Hence the solution is

$$\frac{\theta_o}{\theta_i}(s) = 1 - 1{\cdot}16e^{-5{\cdot}0t}\cos(8{\cdot}66t - 30^\circ) \qquad (4.37)$$

EXAMPLE 4.6.2

Fig. 4.5(b) shows the pole-zero array for $\alpha = 1{\cdot}0$, $\zeta = 0{\cdot}5$ and $\omega_n = 10$. From the Figure,

$$A_1 = \frac{|s_1||s_2|}{|z_1|} \times \frac{s_1 - z_1}{s_1(s_1 - s_2)}$$

$$= \frac{10^2}{10} \times \frac{10\underline{/60^\circ}}{10(17{\cdot}32\underline{/210^\circ})}$$

$$= \frac{0{\cdot}58}{\underline{/150^\circ}}$$

$$= -0{\cdot}58\underline{/+30^\circ}$$

Similarly,

$$A_2 = -0{\cdot}58\underline{/-30^\circ}$$

Combining vectorially, we have

$$A = 2(-0{\cdot}58) = -1{\cdot}16$$

and

$$\nu = +30^\circ$$

Again $\omega_n\sqrt{1-\zeta^2} = 8{\cdot}66$ and $\zeta\omega_n = 5{\cdot}0$; hence the complete solution is

$$\frac{\theta_0}{\theta_i}(\mathrm{s}) = 1-1{\cdot}16e^{-5{\cdot}0t}\cos(8{\cdot}66t+30^\circ) \qquad (4.38)$$

These two examples have shown the ease with which the solution coefficients may be calculated. Fig. 4.6 shows the transient response of the general second order system as a function of $\alpha$ and

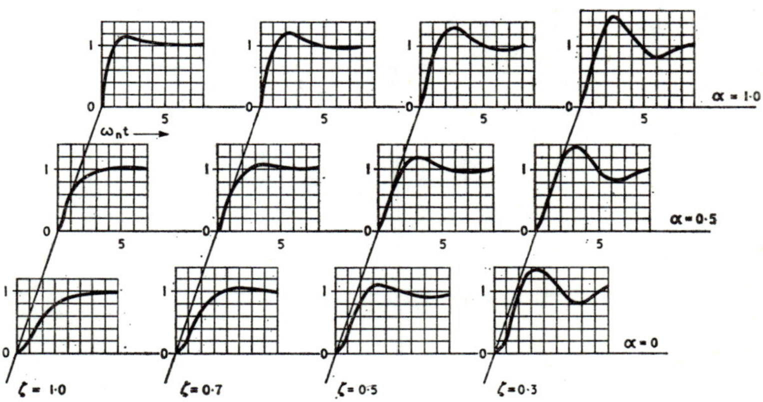


$$\frac{\theta_0}{\theta_i}(s) = \frac{1+b_1 s}{1+a_1 s + a_2 s^2} = \frac{1+\alpha\left(\frac{2\zeta}{\omega_n}\right)s}{1+\left(\frac{2\zeta}{\omega_n}\right)s + \frac{s^2}{\omega_n^2}}$$

*Fig. 4.6. Step function response of second order control systems*

$\zeta$ plotted against normalised time $\omega_n t$. In the system where $\alpha = 0$, $\zeta = 1{\cdot}0$, the transient response is never greater than the magnitude of the input step, giving rise to the concept of $\zeta = 1{\cdot}0$ as the critical damping ratio. Note that for $\alpha > 0$, an overshoot does occur. In general we conclude, by studying Fig. 4.6, that the effect of adding a zero (adding a finite $b_1$) to the system is to increase the maximum overshoot and to decrease the time taken, for example, to reduce first the error to zero. Thus we can use the presence of the zero to liven the transient response in addition to, or as an alternative to, increasing the undamped natural frequency $\omega_n$.

## 4.7 Analysis: Frequency Response

Substituting $j\omega$ for s in Equation 4.1 we obtain the steady state frequency response as

$$\left|\frac{\theta_0}{\theta_i}(\omega)\right| = \sqrt{\frac{1+b_1^2\omega^2}{(1-a_2\omega^2)^2+a_1^2\omega^2}} \tag{4.39}$$

and

$$\psi = \tan^{-1}\left(\frac{a_1\omega}{1-a_2\omega^2}\right)-\tan^{-1}(b_1\omega) \tag{4.40}$$

In standard form,

$$\left|\frac{\theta_0}{\theta_i}(\omega)\right| = \sqrt{\frac{1+\left(\frac{\alpha 2\zeta}{\omega_n}\right)^2\omega^2}{\left(1-\frac{\omega^2}{\omega_n^2}\right)^2+\left(\frac{2\zeta}{\omega_n}\right)^2\omega^2}} \tag{4.41}$$

and

$$\psi = \tan^{-1}\left(\frac{2\zeta\omega/\omega_n}{1-\omega^2/\omega_n^2}\right)-\tan^{-1}(2\alpha\zeta\omega/\omega_n) \tag{4.42}$$

where $\psi$ has been defined such that phase lag is positive in accordance with Fig. 1.24.

Fig. 4.7 shows the frequency response as a function of $\zeta$ and $\alpha$ in terms of the normalised frequency $\omega/\omega_n$. The peak amplitude ratio decreases with $\zeta$ and increases with $\alpha$. It may be easily shown that for $\alpha = 0$, $\zeta = 1/\sqrt{2}$, $M_p$ is never greater than unity. To increase the system bandwidth it is apparent from Fig. 4.7 that this can be achieved by increasing $\alpha$ or $\omega_n$ or by decreasing $\zeta$. In the latter case we should be altering the stability margin of the system however this is defined, since we are reducing the damping ratio of the oscillatory mode. Note carefully that $M_p$ can effectively approach infinity if $\alpha$ is made high enough relative to $\zeta$, thus showing that a response which for all practical purposes is unstable (i.e. $M_p$ is much greater than desirable) can result from a system with absolute stability (i.e. $\zeta>0$).

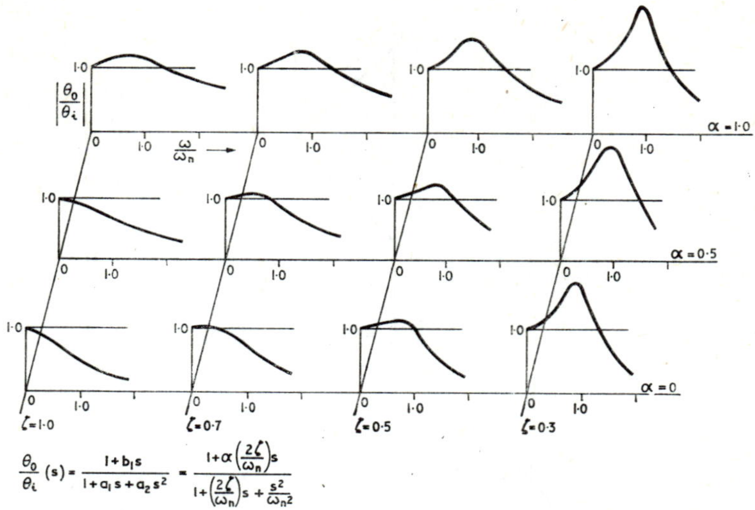


*Fig. 4.7. Frequency responses of second order control systems*

## 4.8 ITAE Optimum Second Order Forms

The second order unity numerator control system, when optimised applying the ITAE criterion to a unit step function response, yields the transfer function

$$\frac{\theta_o}{\theta_i}(s) = \frac{1}{1+\left(\dfrac{1{\cdot}4}{\omega_n}\right)s+\dfrac{s^2}{\omega_n^2}} \tag{4.43}$$

i.e. a damping ratio of 0·70. The pole positions, step function response and frequency response are shown in Fig. 4.8. The damping angle is 46°, the maximum percentage overshoot is 4·6% and the bandwidth is 1·01 $\omega_n$. It can be seen by inspection that the response is certainly acceptable for many applications, and the use of the ITAE criterion leads to a simple choice of component characteristics once the required bandwidth has been determined from input data or assumed input data.

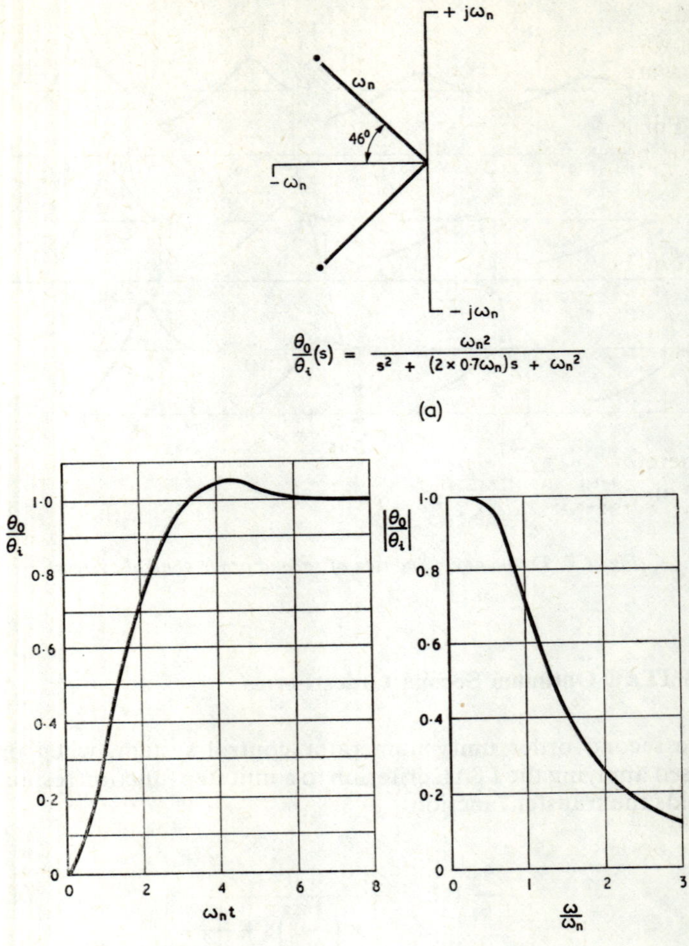


*Fig. 4.8. ITAE optimum zero displacement error second order system; (a) pole locations; (b) step function response; (c) frequency response*

EXAMPLE 4.8.1

A zero displacement error servomechanism is to be designed according to the transfer function minimising the ITAE criterion. The moment of inertia referred to the output shaft is 2·5 slugs ft$^2$ and the viscous damping coefficient at the load shaft is 2·5 lb ft

$\text{rad}^{-1}$ sec. A bandwidth of 15 rad/sec is required, and the system shown in Fig. 4.1(b) is acceptable. Determine the required feed-forward gain $K$ and output derivative time constant $T_r$. Determine also the steady state velocity lag for an input of 0·1 rad/sec.

For an ITAE optimum system $\omega_B = 15$ rad/sec and $\omega_B/\omega_n = 1{\cdot}01$, hence

$$\omega_n = \frac{15}{1{\cdot}01} = 14{\cdot}85 \text{ rad/sec}$$

From Equation 4.10, from the $a_2$ coefficient, $K/J = \omega_n^2$, hence

$$K = \omega_n^2 J = 14{\cdot}85 \times 14{\cdot}85 \times 2{\cdot}5 = 551 \text{ lb ft/rad}$$

Also, from Equation 4.10, from the $a_1$ coefficient we have

$$\frac{f}{K} + T_r = \frac{2\zeta}{\omega_n}$$

Therefore

$$T_r = \frac{2\zeta}{\omega_n} - \frac{f}{K} = \frac{1{\cdot}40}{14{\cdot}85} - \frac{2{\cdot}5}{551} = 0{\cdot}0898 \text{ sec}$$

The velocity constant $C_v$ is given by

$$C_v = \frac{1}{a_1} = \frac{\omega_n}{2\zeta} = \frac{14{\cdot}85}{1{\cdot}40} = 10{\cdot}6 \text{ sec}^{-1}$$

hence

$$\theta_{ss} = \frac{\Omega}{C_v} = \frac{0{\cdot}10}{10{\cdot}6} = 0{\cdot}0094 \text{ rad}$$

If the ITAE criterion is now applied to a zero velocity lag system such that the ITAE is minimised in following a ramp input, the optimum transfer function is

$$\frac{\theta_0}{\theta_i}(\text{s}) = \frac{1 + \left(\dfrac{1{\cdot}7}{\omega_n}\right)\text{s}}{1 + \left(\dfrac{1{\cdot}7}{\omega_n}\right)\text{s} + \dfrac{\text{s}^2}{\omega_n^2}} \tag{4.44}$$

i.e., a damping ratio of 0·85. The pole-zero array, step function response and frequency response are shown in Fig. 4.9. Bandwidth is $2{\cdot}25\omega_n$, maximum precentage overshoot is 16·7% and the damping angle is 32°. $C_v$ is, of course, infinite from the definition applied to the zero velocity lag. The system chosen by the ITAE criterion is suitable for most applications, in particular the 16·7% maximum overshoot being about the tolerated figure for instrument-type servomechanisms.

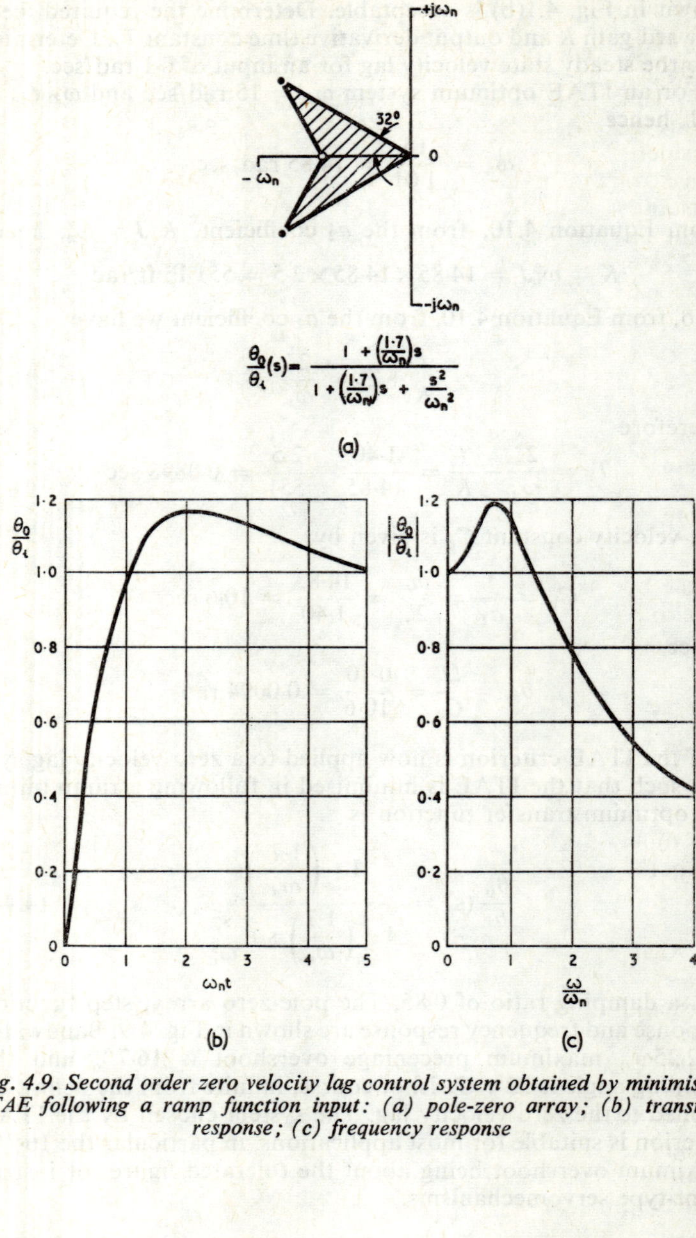


*Fig. 4.9. Second order zero velocity lag control system obtained by minimising ITAE following a ramp function input: (a) pole-zero array; (b) transient response; (c) frequency response*

EXAMPLE 4.8.2

It is required to design a zero velocity lag servomechanism based on the minimisation of the ITAE criterion in following ramp function. The load is 2·5 slugs ft$^2$ inertia, and the viscous damping torque is 2·5 lb ft rad$^{-1}$ sec. If the system shown in Fig. 4.2(b) is considered suitable, determine the necessary values of feedforward gain $K$, rate feedback time constant $T_r$, and input derivative time constant $T_i$ for a bandwidth of 15 rad/sec.

The bandwidth of the optimum system is 2·25 $\omega_n$, hence $\omega_n = 15/2{\cdot}25 = 6{\cdot}67$ rad/sec. From the $a_2$ coefficient $K = \omega_n^2 J = 6{\cdot}67 \times 6{\cdot}67 \times 2{\cdot}5 = 111{\cdot}2$ lb ft/rad; from Equation 4.20 for the $a_1$ coefficient, we have

$$\frac{1{\cdot}40}{\omega_n} = \frac{f}{K} + T_r = \frac{2{\cdot}5}{111{\cdot}2} + T_r = 0{\cdot}022 + T_r$$

Now, we can substitute for $\omega_n = 6{\cdot}67$ rad/sec, in order to obtain the desired bandwidth

$$\frac{1{\cdot}40}{\omega_n} = \frac{1{\cdot}40}{6{\cdot}67} = 0{\cdot}210$$

Hence by subtraction, $T_r = 0{\cdot}210 - 0{\cdot}022 = 0{\cdot}188$ sec. We now have to compute $T_i$, which can be obtained from the $b_1$ coefficient of Equation 4.20 as

$$T_i = \frac{1{\cdot}40}{\omega_n} = \frac{1{\cdot}40}{6{\cdot}67} = 0{\cdot}21 \text{ sec}$$

The velocity lag is, of course, zero in this case. The reader should note that the feedforward gains, $K$, required for Examples 4.8.1 and 4.8.2 are vastly different for systems with identical bandwidth. This difference is due solely to the effect of the finite zero in speeding up the system performance.

## 4.9 Analysis: Common Performance Characteristics

We shall now investigate some of the more common performance characteristics of the general second order system. Fig. 4.10 shows the percentage overshoot in following a step function evaluated in terms of $\alpha$ and $\zeta$. For $\alpha = 0$ the analytical solution for the percentage overshoot is straightforward. The solution for $\alpha \neq 0$ is not straightforward algebraically; if a precise value for overshoot is required the reader is recommended to evaluate numerically the coefficients in the step function response as was shown in Section

4.6, and evaluate the peak by differentiation or otherwise. However, the emphasis in this chapter is on the use of graphical techniques, and interpolation of Fig. 4.10 is quite suitable for engineering design purposes.

Bandwidth is readily evaluated by substituting $j\omega$ in the system transfer function and equating the square of the amplitude

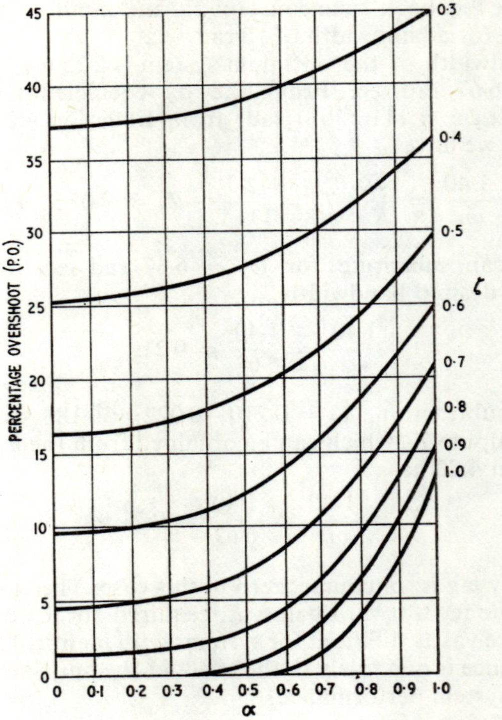


$$\frac{\theta_0}{\theta_i}(s) = \frac{1 + b_1 s}{1 + a_1 s + a_2 s^2} = \frac{1 + \alpha\left(\frac{2\zeta}{\omega_n}\right)s}{1 + \left(\frac{2\zeta}{\omega_n}\right)s + \frac{s^2}{\omega_n^2}}$$

*Fig. 4.10. Percentage overshoot for second order control systems*

ratio to $\frac{1}{2}$, according to the definition in Chapter 1. Thus in polynomial form, $\omega_B$ is the positive root defined by

$$\frac{1+b_1^2\omega_B^2}{(1-a_2\omega_B^2)^2+a_1^2\omega^2} = \frac{1}{2} \tag{4.45}$$

In normalised form, the bandwidth equation is

$$\frac{1+4\alpha^2\zeta^2\left(\dfrac{\omega_B}{\omega_n}\right)^2}{\left[1-\left(\dfrac{\omega_B}{\omega_n}\right)^2\right]^2+4\zeta^2\left(\dfrac{\omega_B}{\omega_n}\right)^2}=\frac{1}{2} \tag{4.46}$$

This yields a quartic in $\omega_B/\omega_n$ and, fortuitously, this reduces to a quadratic in $(\omega_B/\omega_n)^2$, the solution of which is shown in Fig. 4.11. Bandwidth is always increased by increasing $b_1/a_1$ or $\alpha$, but the

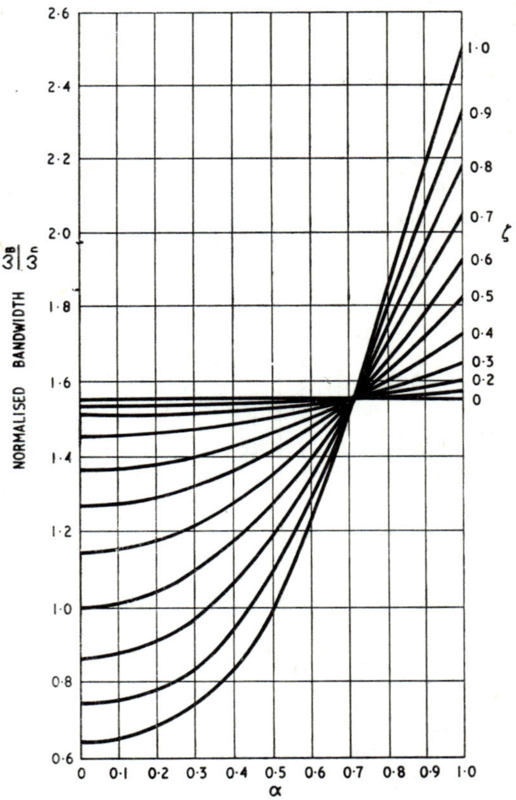


$$\frac{\theta_o}{\theta_i}(s) = \frac{1+b_1 s}{1+a_1 s+a_2 s^2} = \frac{1+\alpha\left(\dfrac{2\zeta}{\omega_n}\right)s}{1+\left(\dfrac{2\zeta}{\omega_n}\right)s+\dfrac{s^2}{\omega_n^2}}$$

*Fig. 4.11. Normalised bandwidth for second order control systems*

effect of $\zeta$ on bandwidth depends on the value of $\alpha$. Thus for $\alpha = 0$, the bandwidth is increased by decreasing $\zeta$, whilst for $\alpha = 1{\cdot}0$ the bandwidth is decreased by decreasing $\zeta$. The transition value for $\alpha$ in this respect is $1/\sqrt{2}$.

The velocity constant $C_v$ has already been defined in Chapter 1 as

$$\theta_{ss} = \frac{\Omega}{C_v} \tag{4.47}$$

where $\theta_{ss}$ is the steady state velocity lag in following a ramp function $\theta_i = \Omega t$; steady state velocity lag means the residual steady state misalignment between output and input when following a ramp function input. $C_v$ was determined in Chapter 1 as a function of the transfer function coefficients $a_1$ and $b_1$, but we shall repeat the derivation here as it is convenient for many readers to keep this Chapter 4 as self-contained as possible.

Let us consider an input $\theta_i = \Omega t$ applied to the general second order system defined by Equation 4.1. Since $\theta = \theta_i - \theta_0$, the transfer function relating error to input is

$$\frac{\theta}{\theta_i}(\mathrm{s}) = \frac{\theta_i - \theta_0}{\theta_i}(\mathrm{s}) = 1 - \frac{\theta_0}{\theta_i}(\mathrm{s}) = \frac{(a_1 - b_1)\mathrm{s}}{1 + a_1\mathrm{s} + a_2\mathrm{s}^2} \tag{4.48}$$

Hence

$$\theta(\mathrm{s}) = \left[\frac{(a_1 - b_1)\mathrm{s}}{1 + a_1\mathrm{s} + a_2\mathrm{s}^2}\right]\theta_i(\mathrm{s})$$

We shall evaluate the steady state error, and $C_v$, in two different ways. Firstly, by using the Laplace transform final value theorem,

$$\lim_{t \to \infty} \theta(t) = \lim_{\mathrm{s} \to 0} \theta_i(\mathrm{s})\mathrm{s}\left[\frac{(a_1 - b_1)\mathrm{s}}{1 + a_1\mathrm{s} + a_2\mathrm{s}^2}\right] \tag{4.49}$$

For a ramp function input, $\theta_i(\mathrm{s}) = \Omega/\mathrm{s}^2$; hence, substituting and evaluating the limit,

$$\theta_{ss} = \lim_{t \to \infty} \theta(t) = \Omega(a_1 - b_1) \tag{4.50}$$

Using the classical theory of differential equations as the second method for evaluating the steady state velocity lag, we write the differential equation as

$$(1 + a_1\mathrm{s} + a_2\mathrm{s}^2)\theta(\mathrm{s}) = (a_1 - b_1)\mathrm{s}\theta_i(\mathrm{s})$$

Assume $\theta_{ss} = A$, a constant. Then

$$\frac{\mathrm{d}\theta_{ss}}{\mathrm{d}t} = 0$$

and

$$\frac{d\theta_{ss}}{dt^2} = 0$$

Now $s\theta_i(s) = d\theta_i/dt = \Omega$ for $\theta_i = \Omega t$; hence

$$A = (a_1 - b_1)\Omega$$

and, as before,

$$\theta_{ss} = \Omega(a_1 - b_1)$$

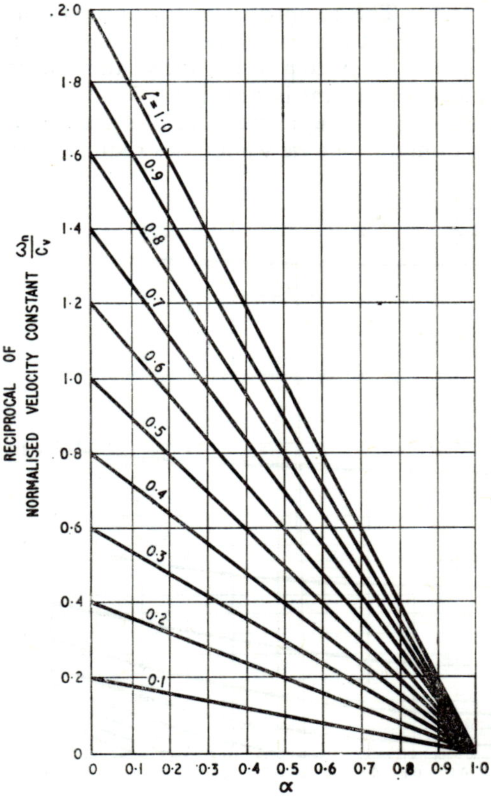


$$\frac{\theta_0}{\theta_i}(s) = \frac{1 + b_1 s}{1 + a_1 s + a_2 s^2} = \frac{1 + \alpha\left(\frac{2\zeta}{\omega_n}\right)s}{1 + \left(\frac{2\zeta}{\omega_n}\right)s + \frac{s^2}{\omega_n^2}}$$

*Fig. 4.12. Normalised velocity constant for second order control systems*

The velocity constant $C_v$ is obtained by equating Equations 4.47 and 4.50, thus

$$C_v = \frac{1}{a_1 - b_1} \tag{4.51}$$

whilst from the normalised Equation 4.2 we have

$$C_v = \frac{1}{\dfrac{2\zeta}{\omega_n} - \alpha\left(\dfrac{2\zeta}{\omega_n}\right)} = \frac{\omega_n}{2\zeta(1-\alpha)} \tag{4.52}$$

The reciprocal of the normalised velocity constant $(\omega_n/C_v)$ is plotted in Fig. 4.12. The reciprocal is linear in both $\zeta$ and $\alpha$, $C_v = \infty$ corresponding to zero velocity lag, and is ensured by making $b_1 = a_1$. For a system with unity numerator transfer function, the velocity constant is simply $\omega_n/2\zeta$.

System bandwidth is a unique measure of system performance in the frequency domain, but we sometimes require that the peak

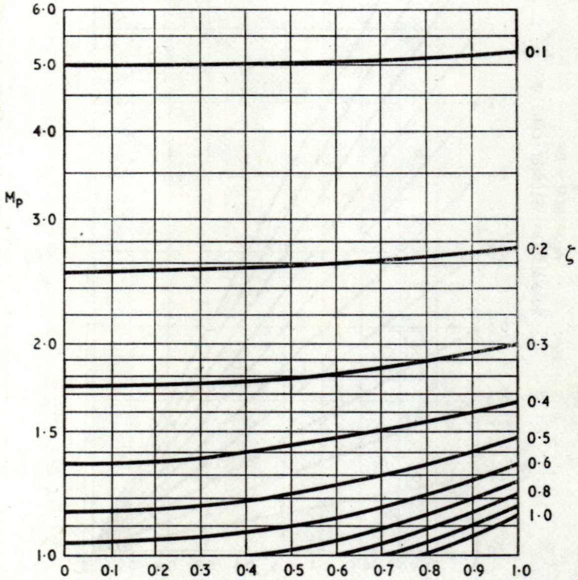


$$\frac{\theta_o}{\theta_i}(s) = \frac{1 + b_1 s}{1 + a_1 s + a_2 s^2} = \frac{1 + \alpha\left(\dfrac{2\zeta}{\omega_n}\right)s}{1 + \left(\dfrac{2\zeta}{\omega_n}\right)s + \dfrac{s^2}{\omega_n^2}}$$

*Fig. 4.13. Peak amplitude ratio $M_p$ for second order control systems*

amplitude ratio $M_p$ of a system lies within specified limits. $M_p$ may be determined analytically by evaluating the amplitude ratio as a function of frequency, as in Equation 4.39, and using the differential calculus to find the frequency $\omega_p$ at which the amplitude ratio is a maximum. $M_p$ is then found by substituting $\omega_p$ in Equation 4.39. The results are shown in Fig. 4.13. It can be seen that $M_p$ is increased by increasing $\alpha$ or decreasing $\zeta$. A typical maximum $M_p$ often specified in practice is about 1·40.

## 4.10 Synthesis: The Second Order Unity Numerator System

In Section 4.9 we evaluated several important characteristics of the general second order system. For synthesis purposes we shall assume that the control system designer receives a performance specification based on three independent performance criteria. Three criteria should certainly be enough to allow a reasonable design study to be made, whilst only two criteria supplied to the designer may represent a certain amount of slackness on the part of the user, since it is often possible to design several systems with widely differing behaviour yet which satisfy the two criteria chosen. The three performance criteria most often specified are percentage overshoot, bandwidth and velocity constant, and we shall base our synthesis technique on these specifications. The reader can devise similar techniques, based for example on the use of $M_p$, or even on the phase lag at bandwidth, as alternative specifications.

As we have already seen, the absence of the numerator term $b_1$ can handicap the control system designer when attempting to meet a stringent specification. The introduction of the numerator term can, for certain projects, mean a basic change in design philosophy, particularly where noise is present, or where the additional components are an embarrassment, and it is important to evaluate the limitations imposed by the absence of $b_1$. When synthesising systems to meet the three main specifications quoted above, it is convenient to combine the bandwidth and the velocity constant into a new variable $\omega_B/C_v$ and then solve for the two design parameters $\omega_B/C_v$ and percentage overshoot. Fig. 4.14 shows $\omega_B/C_v$ plotted against percentage overshoot for the unity numerator system, and the limitations imposed by the absence of $b_1$ are now obvious. For example, if the maximum percentage overshoot that can be tolerated is 10%, $\omega_B/C_v$ is more or less fixed at about 1·40. However, Chapter 6 will discuss how an integrating dipole can be added to boost $C_v$ without greatly affecting $\omega_B$ or the percentage overshoot.

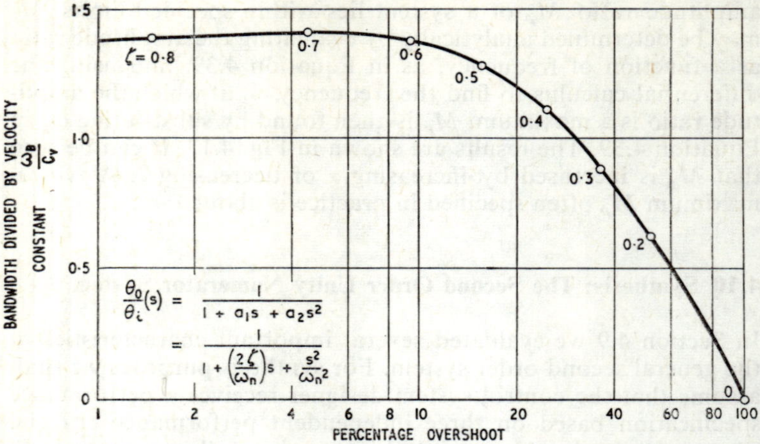


*Fig. 4.14. Some design limitations imposed by the second order unity numerator control system*

## 4.11 Synthesis: General Design of Second Order Control Systems

The performance characteristics shown and discussed in Section 4.9, together with the specialised techniques to be introduced in this section, in general render the use of the root locus method redundant in the design of second order systems to meet the three main specifications of bandwidth, velocity constant and percentage overshoot. Fig. 4.15 shows the ratio $\omega_B/C_v$ plotted for the general second order system. With the percentage overshoot curves of Fig. 4.10, and, at most one subsidiary graph, the determination of $a_1$, $a_2$ and $b_1$, or alternatively $\zeta$, $\alpha$ and $\omega_n$, is quite straightforward. To illustrate the technique we shall consider a typical specification, and having determined the system transfer function from the standard curves, we shall determine several possible designs of servomechanism which will yield the required transfer function.

EXAMPLE 4 .11.1

Determine the transfer function of the general second order control system meeting the specification of 15% overshoot, 15 rad/sec bandwidth and velocity constant 30 sec$^{-1}$. Hence evaluate the system frequency response using the pole-zero method.

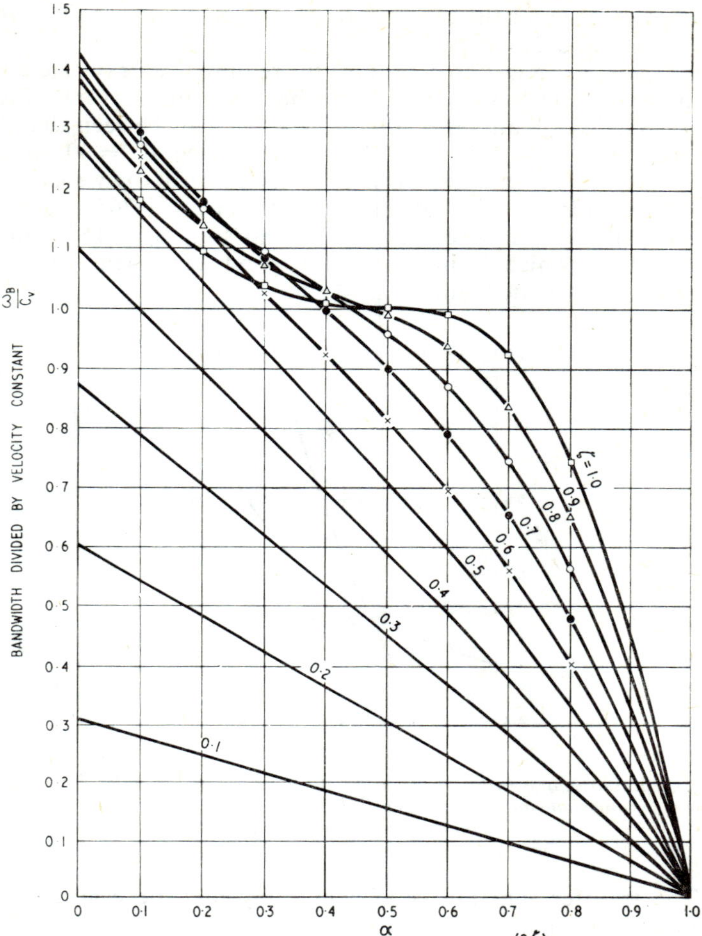


$$\frac{\theta_o}{\theta_i}(s) = \frac{1 + b_1 s}{1 + a_1 s + a_2 s^2} = \frac{1 + \alpha\left(\frac{2\zeta}{\omega_n}\right)s}{1 + \left(\frac{2\zeta}{\omega_n}\right)s + \frac{s^2}{\omega_n^2}}$$

*Fig. 4.15. Bandwidth divided by velocity constant for second order control systems*

Since $C_v = 30$ and $\omega_B = 15$, $\omega_B/C_v = 0{\cdot}50$. From Fig. 4.15, the combinations of $\alpha$ and $\zeta$ corresponding to $\omega_B/C_v = 0{\cdot}5$ can be determined, and similarly from Fig. 4.10 the combinations of $\alpha$ and $\zeta$ corresponding to 15% overshoot can be determined. These two loci are plotted in the $\alpha - \zeta$ plane as shown in Fig. 4.16.

The intersection of the two loci represents the design point, with $\alpha = 0{\cdot}765$, and $\zeta = 0{\cdot}65$.

It is now necessary to determine the required value of undamped natural frequency $\omega_n$, which can be obtained in several different ways. To obtain $\omega_n$ from Fig. 4.12, it can be seen that for $\zeta = 0{\cdot}65$ and $\alpha = 0{\cdot}765$, $\omega_n/C_v = 0{\cdot}305$. Since $C_v = 30$, $\omega_n = 30 \times 0{\cdot}305 = 9{\cdot}15$ rad/sec. To check this value of $\omega_n$ we can use Fig. 4.11, which shows that for $\zeta = 0{\cdot}65$ and $\alpha = 0{\cdot}765$, $\omega_B/\omega_n = 1{\cdot}635$. Since $\omega_B$ is specified as 15 rad/sec, $\omega_n = 15/1{\cdot}635 = 9{\cdot}16$ rad/sec which is in reasonably good agreement with the previously calculated value.

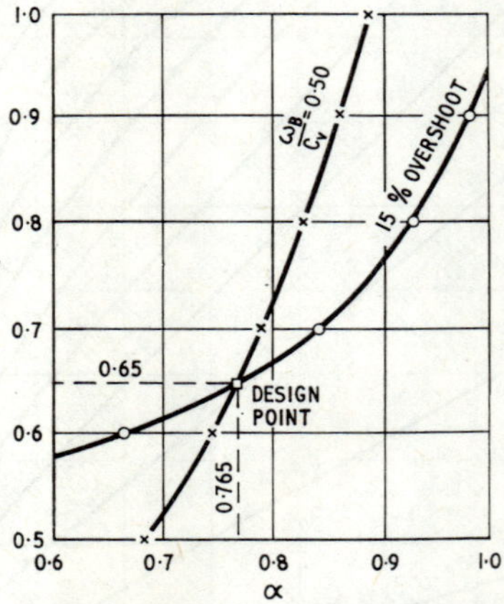


*Fig. 4.16. Subsidiary graph required to determine the design point for a second order servomechanism to meet specified bandwidth, percentage overshoot and velocity constant requirements*

The transfer function coefficients may now be evaluated by comparison, since

$$\frac{\theta_0}{\theta_i}(s) = \frac{1+b_1 s}{1+a_1 s+a_2 s^2} = \frac{1+\alpha\left(\dfrac{2\zeta}{\omega_n}\right)s}{1+\left(\dfrac{2\zeta}{\omega_n}\right)s+\dfrac{s^2}{\omega_n^2}}$$

Hence

$$b_1 = \alpha\left(\frac{2\zeta}{\omega_n}\right) = \frac{0{\cdot}765 \times 2 \times 0{\cdot}65}{9{\cdot}15} = 0{\cdot}109$$

$$a_1 = \frac{2\zeta}{\omega_n} = \frac{2 \times 0{\cdot}65}{9{\cdot}15} = 0{\cdot}142$$

and

$$a_2 = \frac{1}{\omega_n^2} = \frac{1}{9{\cdot}15^2} = 0{\cdot}012$$

Hence the desired system transfer function is

$$\frac{\theta_0}{\theta_i}(s) = \frac{1+0{\cdot}109s}{1+0{\cdot}142s+0{\cdot}012s^2} \qquad (4.53)$$

The poles are located at

$$\begin{aligned} s_1, s_2 &= -(0{\cdot}65\times 9{\cdot}15)\pm j9{\cdot}15\sqrt{1-0{\cdot}65^2} \\ &= -5{\cdot}95\pm j9{\cdot}15\times 0{\cdot}76 \\ &= -5{\cdot}95\pm j6{\cdot}95 \end{aligned}$$

and the zero is located at

$$z_1 = -\frac{1}{0{\cdot}109} = -9{\cdot}17$$

The pole-zero array is shown in Fig. 4.17(a), and the computation table determining the frequency response from the pole-zero array based on the method of Chapter 3 is shown in Fig. 4.17(b). Fig. 4.17(c) shows the frequency response in detail, and it can be seen that the bandwidth is approximately 15 rad/sec, thus checking our original calculations. Having evaluated the transfer function from the specification, it is well worth computing the frequency response in order to check the transfer function and to obtain a greater insight into system behaviour.

Having obtained the requisite transfer function from the performance specification, we shall now investigate several ways in which the transfer function defined by Equation 4.53 may be obtained when the system under consideration is a tracking radar with a moment of inertia of 2·5 slugs ft$^2$ and viscous damping coefficient of 2·5 lb ft rad$^{-1}$ sec.

EXAMPLE 4.11.2

Evaluate the component values required to achieve the transfer function of Equation 4.53 when the servomechanism is basically stabilised by output rate feedback, and derivative of input shaping is used to give the desired performance. Thus the problem is, given $J = 2{\cdot}5$ slugs ft$^2$ and $f = 2{\cdot}5$ lb ft rad$^{-1}$ sec, to determine the required values of feedforward gain $K$, input shaping time constant $T_i$, and rate feedback time constant $T_r$.

The system is shown in Fig. 4.2(b) and the transfer function is known from Equation 4.19. Therefore

$$\frac{\theta_0}{\theta_i} = \frac{1+T_i s}{1+\left(\dfrac{f}{K}+T_r\right)s+\left(\dfrac{J}{K}\right)s^2} = \frac{1+0{\cdot}109s}{1+0{\cdot}142s+0{\cdot}012s^2}$$

Hence

$$T_i = 0{\cdot}109 \text{ sec}$$

$$K = \frac{J}{0{\cdot}012} = \frac{2{\cdot}5}{0{\cdot}012} = 208 \text{ lb ft/rad}$$

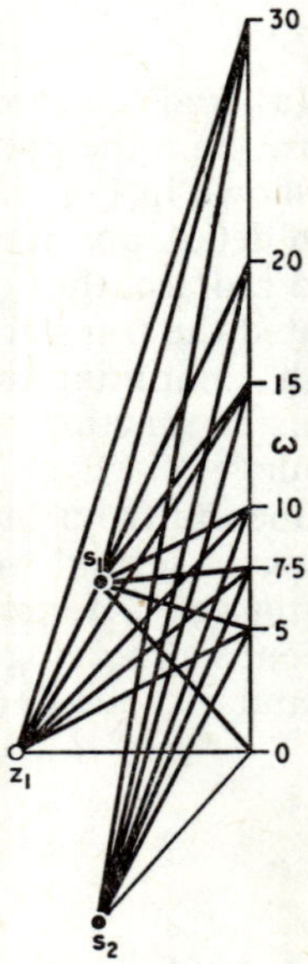


$$\frac{\theta_0}{\theta_i}(s) = \frac{1+0{\cdot}109s}{1+0{\cdot}142s+0{\cdot}012s^2}$$

(a)

| $\omega$ | $\lvert z_\omega\rvert$ | $\lvert s_{1\omega}\rvert$ | $\lvert s_{2\omega}\rvert$ | $\frac{\lvert z_{1\omega}\rvert}{\lvert s_{1\omega}\rvert\lvert s_{2\omega}\rvert}$ | $\frac{\theta_0}{\theta_i}$ |
|---|---|---|---|---|---|
| 0 | 9·17 | 9·15 | 9·15 | $\frac{1}{9{\cdot}13}$ | 1·00 |
| 5 | 11·0 | 6·4 | 13·40 | $\frac{1}{7{\cdot}80}$ | 1·17 |
| 7·5 | 12·0 | 6·0 | 15·50 | $\frac{1}{7{\cdot}75}$ | 1·18 |
| 10 | 13·7 | 6·8 | 18·00 | $\frac{1}{8{\cdot}94}$ | 1·04 |
| 15 | 17·6 | 10·1 | 22·80 | $\frac{1}{13{\cdot}10}$ | 0·70 |
| 20 | 22·1 | 14·4 | 27·60 | $\frac{1}{17{\cdot}9}$ | 0·51 |
| 30 | 31·4 | 23·9 | 37·40 | $\frac{1}{28{\cdot}4}$ | 0·32 |

(b)

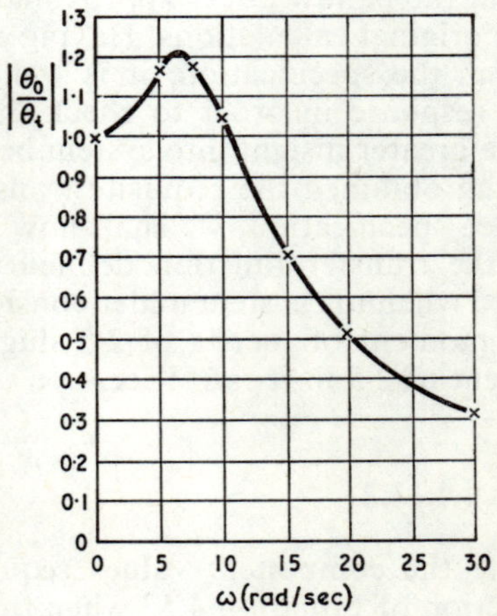


(c)

*Fig. 4.17. Computation of the frequency response from the pole-zero array for the second order transfer function with 15% overshoot, 15 rad/sec bandwidth and 30 sec$^{-1}$ velocity constant: (a) pole-zero construction; (b) table showing computation; (c) frequency response*

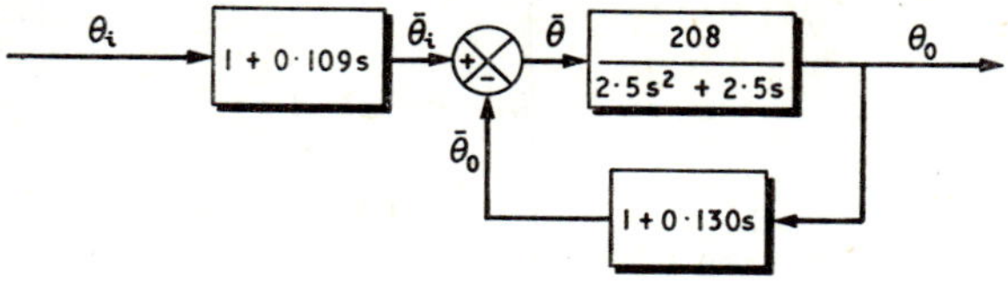


(a)

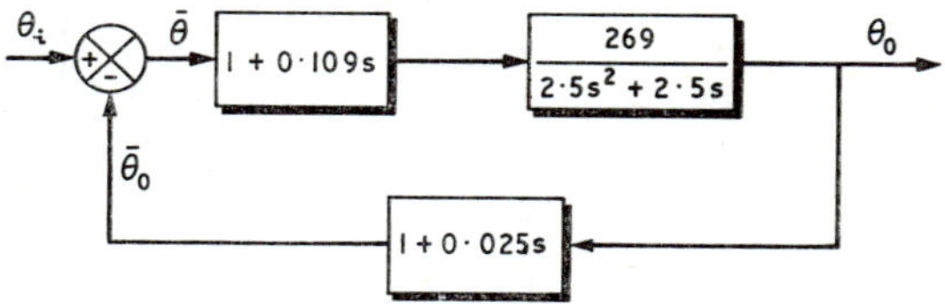


(b)

*Fig. 4.18. Two systems satisfying a specification of 15% overshoot, 15 rad/sec bandwidth and a velocty constant of 30 sec$^{-1}$: (a) servo with derivative of input and derivative of output signal shaping; (b) servo with derivative of output shaping plus loop feedforward derivative signal shaping*

and

$$T_r = 0{\cdot}142 - \frac{f}{K} = 0{\cdot}142 - \frac{2{\cdot}5}{208} = 0{\cdot}130 \text{ sec}$$

The block diagram is shown in Fig. 4.18(a).

EXAMPLE 4.11.3

Since $f$ is finite it is not possible to obtain a zero velocity lag system with a second order system without derivative of input shaping, but since our system specification allows a finite lag we could consider a system with derivative shaping inside the loop plus rate feedback damping to stabilise the system. This system was not discussed in Section 4.3, and so we shall derive the transfer function from first principles. Note that as the signal fed back to the comparator is output displacement plus output rate, the derivative network in the feedforward path will not differentiate the true error signal, and therefore we should not refer to this system as possessing error derivative signal shaping.

Using the normal terminology the feedforward transfer function is thus

$$G(s) = \frac{K(1+T_e s)}{Js^2+fs}$$

and the feedback transfer function is

$$H(s) = 1+T_r s$$

Hence

$$\left.\begin{aligned} \frac{\theta_0}{\theta_i}(s) &= \frac{G(s)}{1+G(s)H(s)} \\ &= \frac{1+T_e s}{1+\left(\frac{f}{K}+T_e+T_r\right)s+\left(T_r T_e+\frac{J}{K}\right)s^2} \\ &= \frac{1+0{\cdot}109s}{1+0{\cdot}142s+0{\cdot}012s^2} \end{aligned}\right\} \quad (4.54)$$

Hence

$$T_e = 0{\cdot}109 = b_1$$

$$T_r T_e+\frac{2{\cdot}5}{K} = 0{\cdot}012 = a_2$$

and

$$\frac{2{\cdot}5}{K}+T_e+T_r = 0{\cdot}142 = a_1$$

Substituting for $T_e$ in $a_1$ and $a_2$, we obtain the simultaneous equations

$$0{\cdot}109T_r+\frac{2{\cdot}5}{K} = 0{\cdot}012$$

$$T_r+\frac{2{\cdot}5}{K} = 0{\cdot}142-0{\cdot}109 = 0{\cdot}033$$

By subtraction we obtain

$$-0{\cdot}891T_r = -0{\cdot}021$$

Hence

$$T_r = \frac{0{\cdot}021}{0{\cdot}891} = 0{\cdot}025 \text{ sec}$$

and

$$\frac{2{\cdot}5}{K} = 0{\cdot}012-(0{\cdot}109\times 0{\cdot}025) = 0{\cdot}012-0{\cdot}0027$$

Hence

$$K = \frac{2{\cdot}5}{0{\cdot}0093} = 269 \text{ lb ft/rad}$$

The block diagram of the system is shown in Fig. 4.18(b). Note that from Equation 4.54, the $T_r$ and $T_e$ terms couple together to affect the undamped natural frequency of the system.

EXAMPLE 4.11.4

If it is desired to avoid the coupling of the $T_e$ and $T_r$ terms as was shown to occur in Example 4.11.3, we can use the output rate feedback signal to form a minor feedback loop around the amplifier, motor and load as shown in Fig. 4.3. The output rate signal is then not further differentiated by passing through the $1+T_e$s network. Reducing the minor loop to a single block, and then solving for the major loop, we obtain the system transfer function given in Equation 4.25 as

$$\frac{\theta_0}{\theta_i}(\mathrm{s}) = \frac{1+T_e\,\mathrm{s}}{1+\left(T_e+T_r+\dfrac{f}{K}\right)\mathrm{s}+\left(\dfrac{K}{J}\right)\mathrm{s}^2} = \frac{1+0{\cdot}109\mathrm{s}}{1+0{\cdot}142\mathrm{s}+0{\cdot}012\mathrm{s}^2}$$

Therefore

$$T_e = 0{\cdot}109 \text{ sec}$$

$$K = \frac{2{\cdot}5}{0{\cdot}012} = 208 \text{ lb ft/rad}$$

and

$$\begin{aligned} T_r &= 0{\cdot}142-0{\cdot}109-\frac{2{\cdot}5}{208} \\ &= 0{\cdot}142-0{\cdot}109-0{\cdot}012 \\ &= 0{\cdot}142-0{\cdot}121 \\ &= 0{\cdot}021 \text{ sec} \end{aligned}$$

The resultant system is shown in Fig. 4.19. It will be noted that the system is in many ways similar to the system of Fig. 4.18(a), since the expressions for the system transfer function in terms of the individual components are identical. However, there are major differences in the open loop responses for these two systems, as we shall see in the next section.

In this section we have seen the ease with which the system transfer function may be obtained from a particular set of desired

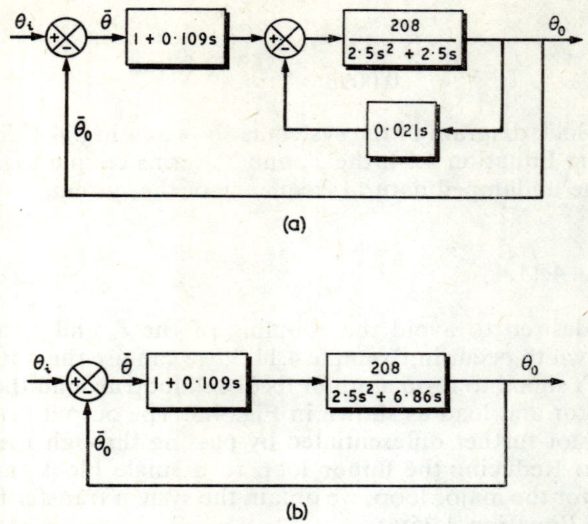


*Fig. 4.19. Minor loop compensated system satisfying a specification of 15% overshoot, 15 rad/sec bandwidth and 30 sec$^{-1}$ velocity constant: (a) block diagram of a system with minor loop compensation; (b) simplified equivalent block diagram*

performance characteristics, and have investigated three different systems yielding the transfer function corresponding to 15% overshoot, 15 rad/sec bandwidth, and 30 sec$^{-1}$ velocity constant. The reader may like to try and devise other second order systems which also have this desired transfer function.

## 4.12 Analysis: Open Loop Response of the Second Order System

In feedforward and feedback transfer function terms, the system transfer function may be written, as shown in Chapter 1, as

$$\frac{\theta_0}{\theta_i}(s) = \frac{G(s)}{1+G(s)H(s)}$$

and the system characteristic equation, on which absolute stability depends, is given by the denominator term. Thus the equation solved by classical or root locus methods is

$$1+G(s)H(s) = 0 \tag{4.55}$$

'Open' or 'broken' loop methods are based on the steady state response to sinusoidal excitation, and the characteristic equation is rewritten as

$$G(j\omega)H(j\omega) = -1 \tag{4.56}$$

Now, by inspection of any of the block diagrams of feedback control systems in this book, it will be apparent that $G(s)H(s) = \bar{\theta}_0/\bar{\theta}$, and hence Equation 4.56 is the equation of motion obtained by physically breaking the loop at the main comparator, injecting a sinusoidal input $\bar{\theta}$, and calculating or measuring the amplitude and phase characteristics of the feedback signal $\bar{\theta}_0$. Thus we may rewrite Equation 4.56 in terms of $\bar{\theta}_0/\bar{\theta}$ such that the open or broken loop stability equation is

$$\frac{\bar{\theta}_0}{\bar{\theta}}(j\omega) = -1 \tag{4.57}$$

As we have already seen in Chapter 1, the open loop steady state response may be expressed as an amplitude ratio and a phase lag, and thus $\bar{\theta}_0/\bar{\theta}(j\omega)$ may be plotted in polar form, and a physical argument used to show that for a simple system continuous oscillations will exist in the system proper if the locus of $\bar{\theta}_0/\bar{\theta}(j\omega)$ passes through the $(-1, j0)$, or Nyquist, point. Similarly, it may be argued, again for a simple system, that the amount by which the locus passes within the $(-1, j0)$ point may be taken as an indication of system stability.

In Section 4.11 we investigated three second order systems which fulfilled the requirements of 15% overshoot, 15 rad/sec bandwidth and 30 $\text{sec}^{-1}$ velocity constant. These three systems possess identical transfer functions and clearly behave in exactly the same way in following the input signal $\theta_i$. We shall now show that these three systems have completely different open loop characteristics, thus reminding the reader of the care to be taken when assessing system suitability by open loop methods. It is helpful whenever an open loop method is used to sketch the block diagram of the system on the Nyquist, Bode, or Nichols diagram, showing quite clearly where the loop has been broken. To make certain that the open loop in any particular circumstance is correct, it is only necessary to check that the response measured or calculated is, in fact, the solution of the characteristic equation as written in Equation 4.57.

Returning to our three systems of Section 4.11, an inspection of Figs. 4.18(a), 4.18(b) and 4.19 allows the three respective broken loop transfer functions to be written down as shown in Fig. 4.20. Parts (a), (b) and (c) of Fig. 4.20 correspond to the systems of Figs. 4.18(a), 4.18(b) and 4.19 respectively. For the system of Fig. 4.18(a), substituting $j\omega$ for s, the open loop locus is defined by

$$\left.\begin{aligned}\left|\frac{\bar{\theta}_0}{\bar{\theta}}(j\omega)\right| &= \frac{208}{2{\cdot}5\omega}\sqrt{\frac{1+0{\cdot}130^2\omega^2}{1+\omega^2}}\\ \underline{/\bar{\theta}}-\underline{/\bar{\theta}_0} &= -90^\circ-\tan^{-1}\omega+\tan^{-1}0{\cdot}130\omega\end{aligned}\right\} \quad (4.58)$$

This response is plotted in Fig. 4.21(a), and it can be seen that the phase margin at unity gain is 64°. Since the numerator term

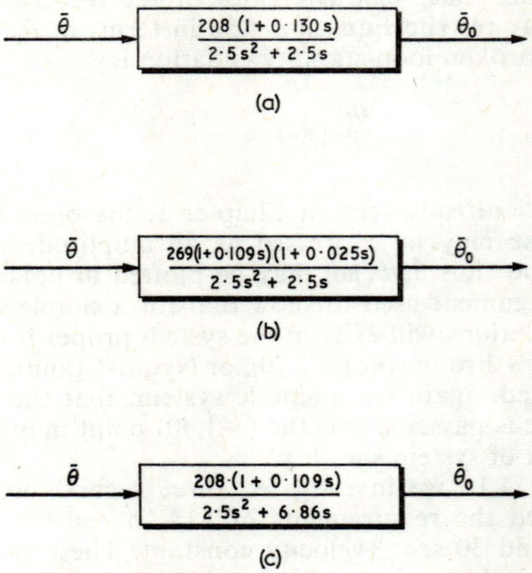


*Fig. 4.20. Broken or open loop transfer function for the three alternative servos satisfying a specification of 15% overshoot, 15 rad/sec bandwidth and 30 sec⁻¹ velocity constant: (a) servo with derivative of input and derivative of output signal shaping; (b) servo with derivative of output shaping plus loop feedforward derivative signal shaping; (c) servo with derivative of error shaping and minor loop compensation*

for $\bar{\theta}_0/\bar{\theta}$ is of first order and the denominator term is of second order, as $\omega \to \infty$ the phase lag approaches $180^\circ - 90^\circ = 90^\circ$ as shown.

The system of Fig. 4.18(b) has an open loop locus defined by

$$\left.\begin{aligned}\left|\frac{\bar{\theta}_0}{\bar{\theta}}(j\omega)\right| &= \frac{269}{2{\cdot}5\omega}\sqrt{\frac{(1+0{\cdot}109^2\omega^2)(1+0{\cdot}025^2\omega^2)}{(1+\omega^2)}}\\ \underline{/\bar{\theta}}-\underline{/\bar{\theta}_0} &= -90^\circ-\tan^{-1}\omega+\tan^{-1}0{\cdot}109\omega+\tan^{-1}0{\cdot}025\omega\end{aligned}\right\} \quad (4.59)$$

which is plotted in Fig. 4.21(b). Phase margin at unity gain is 82°, and since both numerator and denominator of $\bar{\theta}_0/\bar{\theta}$ are second order, the phase lag as $\omega \to \infty$ approaches $180° - 180° = 0°$ as shown.

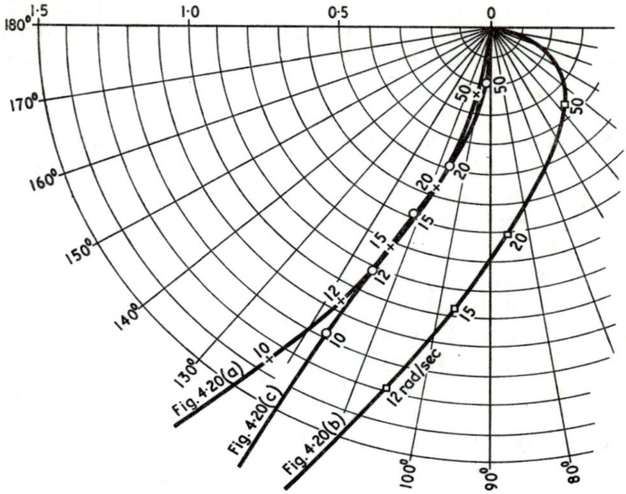


Fig. 4·20(a) $$\frac{\bar{\theta}_0}{\bar{\theta}}(s) = \frac{208(1+0{\cdot}130s)}{2{\cdot}5s^2 + 2{\cdot}5s}$$

Fig. 4·20(b) $$\frac{\bar{\theta}_0}{\bar{\theta}}(s) = \frac{269(1+0{\cdot}109s)(1+0{\cdot}025s)}{2{\cdot}5s^2 + 2{\cdot}5s}$$

Fig. 4·20(c) $$\frac{\bar{\theta}_0}{\bar{\theta}}(s) = \frac{208(1+0{\cdot}109s)}{2{\cdot}5s^2 + 6{\cdot}86s}$$

*Fig. 4.21. Broken or open loop response for the three alternative servos satisfying a specification of 15% overshoot, 15 rad/sec bandwidth and 30 sec$^{-1}$ velocity constant*

Finally, the system of Fig. 4.19 has an open loop locus defined by

$$\left.\begin{aligned} \left|\frac{\bar{\theta}_0}{\bar{\theta}}(j\omega)\right| &= \frac{208}{2{\cdot}5\omega}\sqrt{\frac{1+0{\cdot}109^2\omega^2}{1+2{\cdot}74^2\omega^2}} \\ \underline{/\bar{\theta}} - \underline{/\bar{\theta}_0} &= -90° - \tan^{-1} 2{\cdot}74\omega + \tan^{-1} 0{\cdot}109\omega \end{aligned}\right\} \quad (4.60)$$

which is plotted in Fig. 4.21(c). Phase margin at unity gain is 65°, and the infinity frequency phase lag is 90°, as for the system of Fig. 4.18(a).

Fig. 4.20 is most interesting in portraying the open loop responses of systems with the same system transfer function. The behaviour of these systems in rejecting noise or disturbance signals is, of course, different, and where serious noise or disturbance problems are anticipated, it would be worthwhile evaluating the noise transmittance transfer function $\theta_0/N(s)$ and the disturbance transmittance transfer function $\theta_0/U(s)$ before making a final decision on the better system.

Since the systems under investigation in this chapter are second order, the phase lag cannot be greater than 180° even at infinite frequency, and thus the locus cannot cross the 180° phase lag line. The gain margin for all these systems is thus always unity.

When the feedback transfer function is unity, correlation between open loop locus and system harmonic response is straightforward when plotted in Nyquist or Nichols form, provided certain graphical aids are available. Should the feedback transfer function be frequency-dependent, the inverse Nyquist plot can be manipulated to yield the system harmonic response. It should be made quite clear, however, that the charts of system performance in terms of $\alpha$, $\zeta$ and $\omega_n$ provided in this chapter allow the designer to design his system without recourse to open loop methods; we have in this instance only looked at open loop methods in order to assist in recognising the second order system in all its forms.

## 4.13 Analysis: Sensitivity of System Performance to Changes in Component Values

The control engineer is acutely interested in the effect of changes in component values on the performance of his system. Whilst it is to be hoped that quality control during manufacture will ensure reasonable consistency in component characteristics, the reader is certain to be aware that a balance must be achieved between cost and consistency, and that as a result the control engineers must be prepared to stipulate, and accept, reasonable tolerances on individual components. The interpretation of 'reasonable tolerances' may alter by a factor of decades depending on the application of the control system.

Wes hall use the system of Fig. 4.18(a) as a mathematical model for investigating the effects of component inaccuracies and setting inaccuracies. We shall further assume, for the purpose of illustration, that the inaccuracies occur in the feedforward gain $K$, and in the input derivative time constant $T_i$. In practice, the inherent viscous damping $f$ would certainly be very suspect in value, and might vary considerably from servo to servo. The effects of component variations on parameters and characteristics have been evaluated by substituting the new value of component in the system transfer function. For higher order systems, the root locus method is useful in reducing the labour involved in estimating changes in parameters, particularly when the inaccuracy occurs in the feedforward gain.[39]

Fig. 4.22 shows the effect of component inaccuracies on the transfer function coefficients. For this particular system, $K$ only

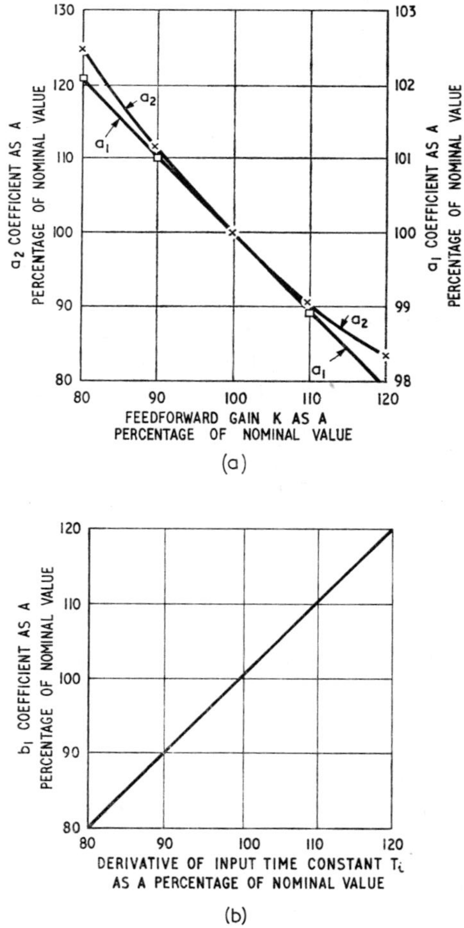


*Fig. 4.22. Effect of component inaccuracies on the transfer function coefficients for the system of Fig. 4.18(a): (a) effect of inaccuracies in $K$; (b) effect of inaccuracies in $T_i$*

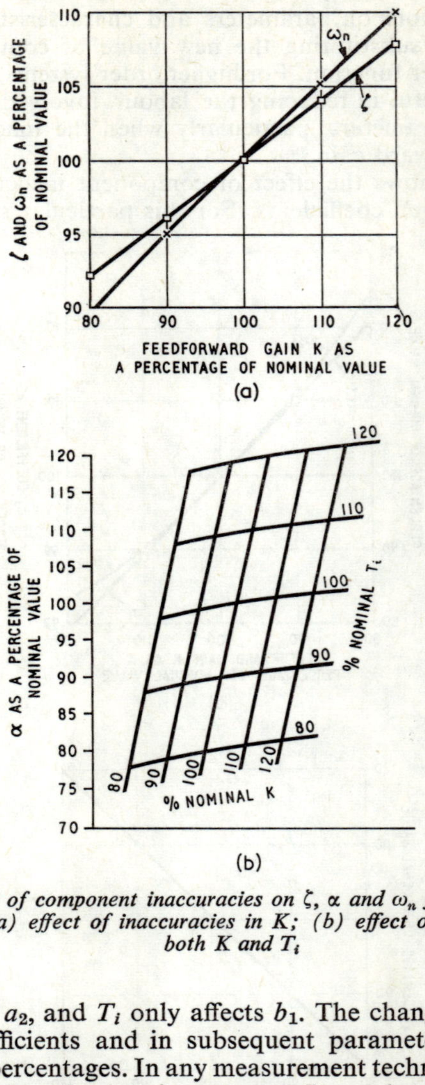


*Fig. 4.23. Effect of component inaccuracies on ζ, α and $\omega_n$ for the system of Fig. 4.18(a): (a) effect of inaccuracies in K; (b) effect of inaccuracies in both K and $T_i$*

affects $a_1$ and $a_2$, and $T_i$ only affects $b_1$. The changes in transfer function coefficients and in subsequent parameters have been expressed as percentages. In any measurement technique the absolute value is also of prime importance. It can be seen from Fig. 4.22 that the percentage change in $b_1$ is identical to the percentage change in $T_i$; the effect of changes in $K$ is more marked on $a_2$ than on $a_1$ by a factor of about 10.

The effect on the normalised coefficients $\alpha$, $\zeta$ and $\omega_n$ is shown in Fig. 4.23. $\zeta$ and $\omega_n$ vary only with $K$, and the percentage change in both these quantities is about one half the corresponding percenage change in $K$. Both $K$ and $T_i$ affect $\alpha$, the effect of $T_i$ being far more significant, as we would expect. Changes in $T_i$ result in a corresponding percentage change in $\alpha$, whilst percentage changes in $K$ result in less than 1/10 of this percentage change in $\alpha$.

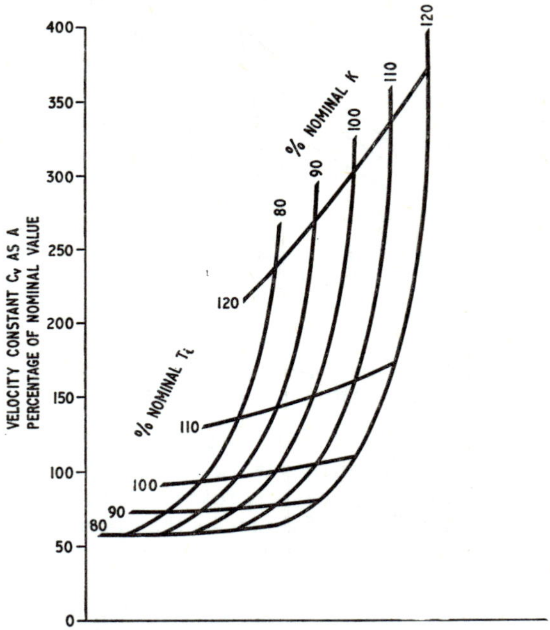


*Fig. 4.24. Effect of component inaccuracies on velocity constant for the system of Fig. 4.18(a)*

Velocity constant changes are shown in Fig. 4.24; the effect of $T_i$ is always considerable, whilst the effect of changes in $K$ depends to some extent on the actual value of $T_i$. For example, when $T_i$ is low, changes in $K$ have little effect on $C_v$, whilst when $T_i$ is high, changes in $K$ have a marked effect on $C_v$. The general amplification of component errors revealed in the $C_v$ plot does indicate that the measurement of velocity lag is often a desirable feature of test equipment.

Bandwidth variations are shown in Fig. 4.25. Bandwidth is quite sensitive to changes in $K$ and $T_i$ in something like the same proportion. A change in feedforward gain $K$ is immediately apparent from the open loop response provided that all other components are at their nominal values. Should the designer now place upper

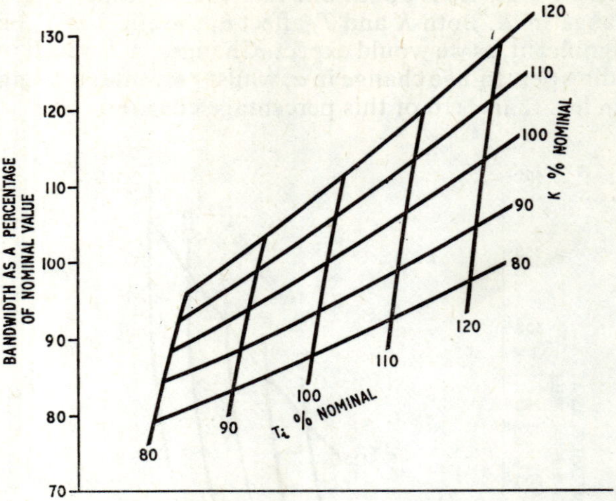


*Fig. 4.25. Effect of component inaccuracies on bandwidth for the system of Fig. 4.18(a)*

and lower limits on, for example, system bandwidth, figures such as Fig. 4.25 allow the designer to work back and evaluate tolerances on the individual components. Cross-coupling effects between $K$ and $T_i$ as shown in Fig. 4.25 do mean that a certain amount of component selection may be justified as an alternative to imposing excessively tight limits on $T_i$ and $K$. In practice, $K$ is normally variable to some extent, and the problem during test may be setting $K$ to give a satisfactory result for the particular value of $T_i$ for the servo in question.

## 4.14 Analysis: Control of a Variable Load

Before leaving second order systems we shall briefly examine the variation in system performance caused by variation in the load transfer function. We shall consider the effect of speed and altitude on the transient response of the roll autopilot analysed in Chapter 2. The system transfer function is

$$\frac{\phi_0}{\phi_D}(s) = \frac{1}{\left(\dfrac{T_a}{K_g K_a}\right)s^2 + \left(T_r + \dfrac{1}{K_g K_a}\right)s + 1} \tag{4.61}$$

where $T_a$ = aerodynamic time constant
$K_a$ = aerodynamic gain

$T_r$ = rate feedback time constant
$K_g$ = feedforward gain adjustment = $\xi_D/\bar{\phi}$
$\xi_0$ = actual control surface deflection
$\xi_D$ = demanded control surface deflection

The actual control surface deflection is achieved by means of an electro-hydraulic actuator incorporated in a minor feedback loop. We shall assume that this minor loop is very fast compared to the autopilot loop, thus we may write $\xi_D = \xi_0$ and arrive at Equation 4.61. As shown in Chapter 2, the missile transfer function in roll is $K_a/(1+T_a s)$, and this is the load transfer function which in this case relates missile roll rate to actual control surface deflection.

Typical values for $K_a$ and $T_a$ as functions of speed and altitude for a surface-to-air missile are shown in Table 4.14.1.[7] In order to

**Table 4.14.1**

VARIATION OF MISSILE ROLL PARAMETERS WITH FLIGHT CONDITION

| *Mach no.* | *Height* | $K_a$ rad sec$^{-1}$ rad$^{-1}$ | $T_a$ sec |
|---|---|---|---|
| 1·0 | Sea level | 458 | 0·210 |
| 1·4 | 20,000 ft | 307 | 0·217 |
| 1·4 | 40,000 ft | 266 | 0·416 |

obtain a relatively high bandwidth, the autopilot feedforward gain and rate feedback coefficients have been selected as $K_g = 2{\cdot}29$ and $T_r = 0{\cdot}027$ sec respectively. Appropriate autopilot block diagrams for the three flight cases are shown in Fig. 4.26. The corresponding transfer functions for the autopilot as a function of flight condition are obtained by substitution in Equation 4.61 to yield,

(a) Mach 1·0, sea level

$$\frac{\phi_0}{\phi_D}(s) = \frac{1}{1+0{\cdot}0261s+0{\cdot}000229s^2} \tag{4.62}$$

(b) Mach 1·4, 20,000 ft

$$\frac{\phi_0}{\phi_D}(s) = \frac{1}{1+0{\cdot}0266s+0{\cdot}000353s^2} \tag{4.63}$$

(c) Mach 1·4, 40,000 ft

$$\frac{\phi_0}{\phi_D}(s) = \frac{1}{1+0{\cdot}0269s+0{\cdot}000782s^2} \tag{4.64}$$

The transient responses are shown in Fig. 4.27; the 40,000 ft case is the slowest, and corresponds to $\zeta = 0{\cdot}475$, $\omega_n = 35{\cdot}76$ rad/sec. At 20,000 ft, $\zeta$ increases to 0·71 and $\omega_n$ increases to 53·2 rad/sec, whilst at sea level $\zeta$ has increased to 0·860 and $\omega_n$ to 66 rad/sec In the latter case there is only a very slight overshoot of about $\frac{1}{2}$%.

The variation in performance shown in Fig. 4.27 is just acceptable for many applications. Because a feedback control system

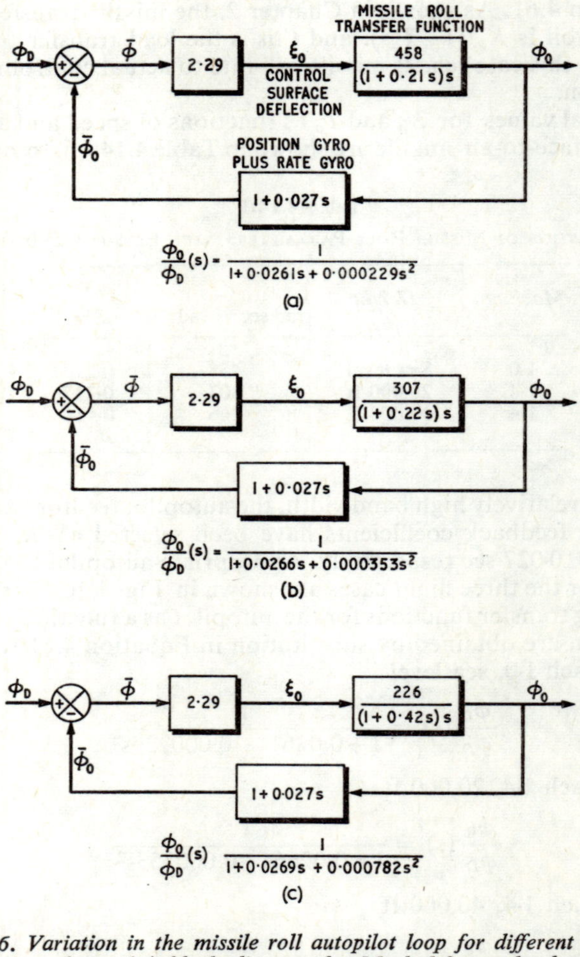


*Fig. 4.26. Variation in the missile roll autopilot loop for different points on the flight envelope: (a) block diagram for Mach 1·0, sea level; (b) block diagram for Mach 1·4, 20,000 ft; (c) block diagram for Mach 1·4, 40,000 ft*

is used, the variation in performance is considerably less than would result from the use of an open loop method of control since, as was discussed in Chapter 1, the presence of feedback can attenuate the effect of changes in the load transfer function on system performance. The values of $K_g$ and $T_r$ used in this particular example were chosen on an *ad hoc* basis, and less variation in performance might be expected if a more sophisticated study using root locus techniques was carried out. For further information on

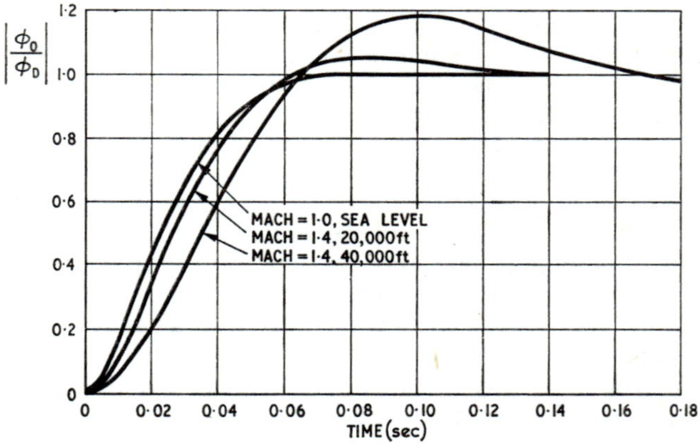


*Fig. 4.27. Effect of flight conditions on the response of the missile roll autopilot loop shown in Fig. 4.26*

this type of problem the reader is referred to two papers by Horowitz.[40, 41] In practice, if the variation in performance is still embarrassing even when sophisticated linear design techniques are used, a passive adaptive compensation device might be incorporated in the autopilot loop. Satisfactory values of $K_g$ and $T_r$ would be determined for each point on the flight envelope in terms of altitude and speed, using flight test and wind tunnel data. A device would then be designed to measure altitude and speed and actuate a potentiometer setting for $K_g$ and/or $T_r$ in order to achieve the predetermined value. One common device uses the static pressure of the surrounding air to deflect a bellows, which, through a suitable linkage, moves the potentiometer arm and thus adjusts the feedforward gain to a suitable value. This particular device compensates for altitude changes only, and is hence termed an altitude gain unit.

## 4.15 Concluding Remarks

The material presented and discussed in this chapter provides a substantial foundation on which a general knowledge of control engineering can be built. Indeed, the particular graphs given are sufficient to design a wide variety of control systems, and designers of certain simple types of control systems will have found all the science of control engineering they are likely to require.

There are, of course, limitations to the application of the methods of this chapter. Firstly, we have only considered the response of the system to command signals, and there will be occasions when calculations on system response to noise signals and load disturbances would be justified. We can, however, reasonably assume that system bandwidth would be chosen as the best compromise between command signal transmission and noise signal rejection, and that if load disturbances are significant these are reduced by making the gain before the disturbance point as large as consistent with the other performance requirements. Secondly, in deriving the individual component transfer functions certain assumptions are made in neglecting higher order terms, and it is possible that several neglected terms may combine to give a significant third order term which may possibly affect stability, but which is more likely to change the form of system response. Thirdly, the graphical analysis and synthesis methods included in this chapter are provided basically for linear system design. In turn, the linear system requires the torque or force acting on the load to be proportional to the shaped error signal, as we have seen in several examples. In practice many such proportional systems tend to be non-linear for very small inputs due to threshold effects such as dead zone, and also tend to be non-linear for very large inputs due to saturation phenomena. However, the assumption of linearity at the design stage is often justifiable as the design labour requirements are thereby greatly reduced. For certain highly non-linear systems the graphs may be used in conjunction with small perturbation theory or describing function theory to yield the system performance as a function of input amplitude. The extension of this approach to such systems will be fully covered in the chapter on non-linear control systems.

CHAPTER 5

# Transfer Function Techniques Applied to Third Order Linear Systems

## 5.1 Introduction

In Chapter 4 it was observed that second order linear systems could be designed rapidly from a given set of specifications involving bandwidth, velocity constant and either percentage overshoot or maximum amplitude ratio, provided that three independent adjustable parameters were available in the system. Not only is this approach extremely useful for genuine second order systems, but also for higher order systems which can be approximated by two dominant poles and one dominant zero. Since it has been found that the vast majority of systems may be approximated by at the most three poles and one zero,[42] there is a strong incentive to extend the methods of Chapter 4 to genuine third order systems, since these methods would then allow us to determine the major design features of a system required to meet a given specification. Of course, in considering third order systems we are introducing a further variable which complicates not only system synthesis but also the presentation of results. However, using the normalisation technique of Chapter 1, together with the resulting normalised coefficient plane, it is possible to present graphically all the important responses and performance criteria for third order unity numerator and third order zero velocity error systems. It is thus possible to avoid the construction of the root locus for third order systems except for the special case where there are three dominant poles plus a dominant zero, and the zero is not positioned so as to make $C_v = \infty$. An outstanding advantage of the normalised coefficient plane is that system synthesis in terms of two design parameters is particularly easy, the two parameters being varied simultaneously so that an optimum combination of the parameters can be chosen. Thus use of the coefficient plane can not only avoid the construction of many root loci at the design stage, but also presents the exact performance criteria as a set of design charts wherein the effect of the third pole is not neglected. For example, in designing the missile roll autopilot used to illustrate the root locus method in Chapter 3, both $K_g$ and $T_r$ can be varied simultaneously until a suitable combination of bandwidth and, say, $M_p$ is found.[43]

The powerful exact synthesis techniques presented in this chapter are restricted to the type (0,3) system and the zero velocity lag system, which is a special case of the type (1,3) system. Fortunately, these two systems represent two fairly broad limits of design practice, and therefore between them cover a wide range of possible designs. For these two systems a complete set of design charts relating performance criteria to normalised coefficients is given. *In the interests of accuracy of resolution, the design charts are retricted to the region of the coefficient plane wherein all three poles make a significant contribution to the performance criteria of interest.* Naturally, the boundaries chosen to isolate this region are somewhat elastic, and in fact are very much a function of any system zeros present. For the two systems under detailed study, it is found that the boundaries are quite different for the type (0,3) system compared with the zero velocity lag system.

The coefficient plane concept is not restricted to the above completely charted systems. We can also use the coefficient plane to determine exactly the roots of those systems wherein all three poles are significant but which also possess one or more independent system zeros. (The third order zero velocity lag system has the system zero positioned automatically by the system poles.) A method for the determination of poles in this region of the coefficient plane is a worthwhile design aid, since the iterative process for numerical computation of roots to be described in Chapter 6 does not converge so rapidly where roots are relatively close together. Using this particular chart to determine the system poles, we add the system zeros to form the system pole-zero array and then use the techniques of Chapter 3 to select the most promising pole-zero arrays for further studies. This method of determination of system poles is particularly useful when two or more design variables are in need of simultaneous adjustment, whereas the root locus is more suited to those third order systems wherein we are left at a particular stage of the design with only one adjustable design variable.

Finally, we shall use the coefficient plane to determine approximately the system poles in the region where all three poles are real, and in regions where there are complex poles and a real pole but in which the real pole, or the complex poles, have an insignificant effect on system performance. We can thus estimate performance criteria directly from the approximations to first and second order systems. In the latter case, for systems approximating in the coefficient plane to the general type (1,2) system, the complete techniques of Chapter 4 become available for analysis and synthesis. Note that even if certain poles are neglected, the *exact* value of $C_v$ is known because the transfer function coefficients are exactly known.

## 5.2 The Normalisation Technique Applied to Third Order Systems

The normalisation technique and the coefficient plane were used by Graham and Lathrop in their classic paper on the synthesis of optimum forms of system response, and the approach has since been considerably extended by Towill.[43–45] The first step is to normalise the system characteristic equation, which, in standard form, is

$$1+a_1 s+a_2 s^2+a_3 s^3 = 0 \tag{5.1}$$

Introducing the normalised Laplace operator S, and the normalisation frequency $\omega_0$ defined by

$$\omega_0 S = s \tag{5.2}$$

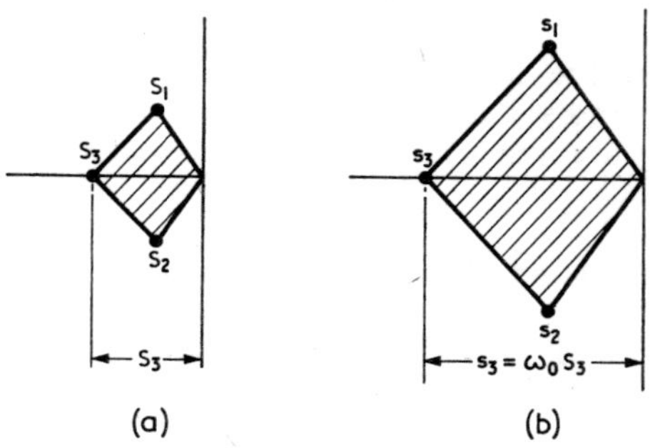


*Fig. 5.1. True* s *root plane and normalised* S *root plane for the third order characteristic equation, the planes being geometrically similar: (a) the* S *plane; (b) the* s *plane*

where $\omega_0 = (1/a_3)^{1/3}$, the system characteristic equation may be rewritten in normalised form as

$$1+cS+bS^2+S^3 = 0 \tag{5.3}$$

where $c = a_1\omega_0$ and $b = a_2\omega_0^2$.

The roots of Equation 5.3 are geometrically similar to those of Equation 5.1, as shown in Fig. 5.1; the scale factor relating the S plane to the s plane is the normalisation frequency $\omega_0$. Not only does this normalisation technique allow all geometrically similar systems to be represented by one single system pole-zero array (with different normalisation frequencies), but it has been found that the normalisation technique applied to any order of system results in a normalised characteristic equation with very easily managed coefficients which minimises storage problems arising in many computational methods. As a simple example, consider the system characteristic equation

$$1+0{\cdot}4s+0{\cdot}12s^2+0{\cdot}008s^3 = 0 \tag{5.4}$$

The normalisation frequency is given by

$$\omega_0 = \left(\frac{1}{a_3}\right)^{1/3} = \left(\frac{1{,}000}{8}\right)^{1/3} = 5$$

Hence $c = a_1\omega_0 = 0{\cdot}4\times 5 = 2{\cdot}0$, and $b = a_2\omega_0^2 = 0{\cdot}12\times 5^2 = 3{\cdot}0$. Thus the normalised characteristic equation is

$$1+2S+3S^2+S^3 = 0 \tag{5.5}$$

Note that the normalisation frequency $\omega_0$ is a pure number, the concept of $\omega_0$ as a frequency is one of convenience rather than mathematical rigour. A comparison of Equations 5.5 and 5.4 should leave the reader in no doubt that the magnitude of the coefficients of the normalised equation leads to much easier manipulation and solution of the characteristic equation whether the true roots are much larger, or much smaller, than unity.

### 5.3 Relative Stability of Unity Feedback Third Order Systems

The Routh Hurwitz stability criterion applied to the normalised characteristic equation yields

$$\begin{array}{l|ll} S^3 & 1 & c \\ S^2 & b & 1 \\ S^1 & (bc-1)/b & \\ S^0 & 1 & \end{array}$$

Thus, for absolute stability,

$$bc > 1$$

When plotted in the $bc$ plane, or coefficient plane $bc = 1{\cdot}0$, the absolute stability line is the hyperbola passing through the point $b = 1$, $c = 1$. If *discussion in this section only is restricted to unity feedback systems, or to non-unity feedback systems manipulated to give a unity feedback path*, the Routh Hurwitz criterion may be extended to provide unambiguous gain margin contours in the coefficient plane.[44] As shown in Fig. 5.2, if the system has a gain

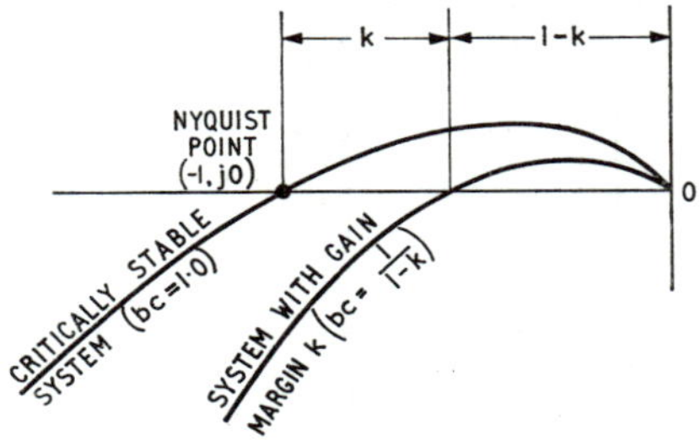


*Fig. 5.2. Gain margin diagram for the third order unity feedback path system*

margin $k$, then the system gain is $1-k$ times the critical gain defined by the Routh Hurwitz criterion. Since the critical system has already been defined by the relationship $bc = 1{\cdot}0$, it follows that for a given gain margin $k$, the corresponding contour in the coefficient plane is

$$bc = \frac{1}{1-k}$$

which is the hyperbola passing through the point $b = 1/\sqrt{1-k}$, $c = 1/\sqrt{1-k}$. A family of gain margin contours is shown in Fig. 5.3. The contour $bc = 1{\cdot}5$ defines a system with a gain margin of $0{\cdot}33$, and although the shape of the open loop response curve is also important in defining a suitable system, this particular gain margin is normally the absolute minimum met in practice; indeed a satisfactory design will often require a much higher value of gain margin. Thus the $bc = 1{\cdot}5$ contour may be looked upon as a boundary not to be crossed during the design process, and the contour will therefore become an integral feature of the coefficient plane.

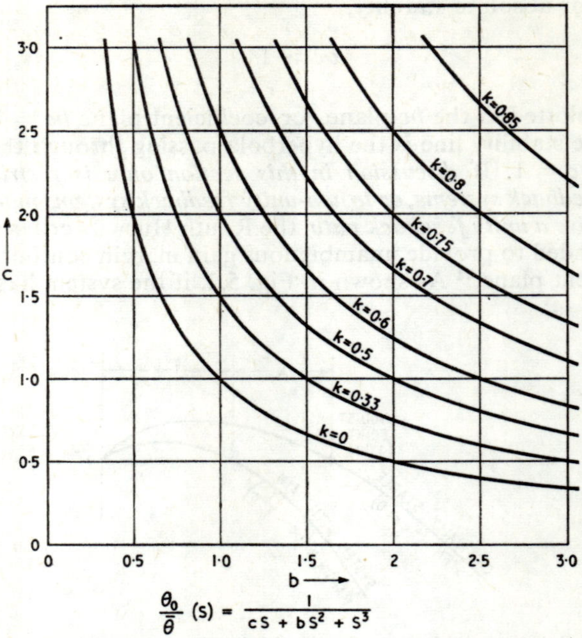


*Fig. 5.3. Gain margin contours in the coefficient plane for the third order system with unity feedback path*

## 5.4 Qualitative Assessment of Transient and Frequency Responses for Third Order Linear Zero Displacement Error and Zero Velocity Error Systems

The general third order linear system with the transfer function

$$\frac{\theta_0}{\theta_i}(s) = \frac{1+b_1 s}{1+a_1 s+a_2 s^2+a_3 s^3} \tag{5.6}$$

can be normalised to give

$$\frac{\theta_0}{\theta_i}(S) = \frac{1+\alpha c S}{1+cS+bS^2+S^3} \tag{5.7}$$

where $\alpha := b_1/\omega_0 c$ and, using Equation 1.28,

$$C_v = \frac{1}{a_1-b_1} = \frac{\omega_0}{c(1-\alpha)} \tag{5.8}$$

If we define the coefficient plane as the plane with variables $b$ and $c$, we can represent any response or performance criterion required directly in this plane and thus immediately recognise the effect of the normalised coefficients on the response or criterion of interest. We can do this because, *pro tem.*, we have removed the effect of time and frequency by using the normalisation technique. Furthermore, we can divide performance criteria into two classes, those which are independent of the normalisation frequency, and those which are linear functions of the normalisation frequency. In the former category are maximum percentage overshoot and peak amplitude ratio, whilst in the latter category are velocity constant, bandwidth and response time. Thus we shall calibrate the coefficient plane in terms of $C_v/\omega_0$, $\omega_B/\omega_0$, $\omega_0 t_{63}$, etc., for the frequency-dependent criteria, but shall use absolute values for percentage overshoot and $M_p$.

The coefficient plane will handle performance criteria as functions of $b$ and $c$ provided $\alpha$ is constant. In this chapter we shall present design charts for by far the two most common cases, the type (0,3) third order zero displacement error system and the special type (1, 3) system with zero velocity lag. For the first system $\alpha = 0$, for the second system $\alpha = 1{\cdot}0$. Third order systems with intermediate values of $\alpha$ must be handled using the techniques of Chapter 6, the application of the coefficient plane in these circumstances being restricted to solution of the characteristic equation and to providing a clear indication directly from the transfer function coefficients as to suitable approximations on particular mode dominance.

The results presented in this chapter have been obtained directly from the digital computer, but could quite easily have been obtained much more laboriously using the hand techniques of Chapter 3. Probably the outstanding advantage of the coefficient plane, apart from the usual clarity of the subsequent design process, is the ability to portray performance criteria directly, even when the third pole is highly significant, as was found to be the case in Example 3.13.1. To obtain some understanding of the effect of the normalised coefficients on system response, the transient and frequency responses of the $\alpha = 0$ and $\alpha = 1$ systems are shown pictorially in Figs. 5.4 and 5.5 respectively. Zero displacement system responses are shown in full lines, and zero velocity lag system responses are shown in dotted lines. By comparing the two sets of responses, the considerable effect of the presence of the system zero at $-1/c$ for the zero velocity lag system is clearly seen. At a particular point in the coefficient plane it is observed that the effect of the zero is to increase normalised bandwidth and decrease 63% response time at the expense of increasing $M_p$ and

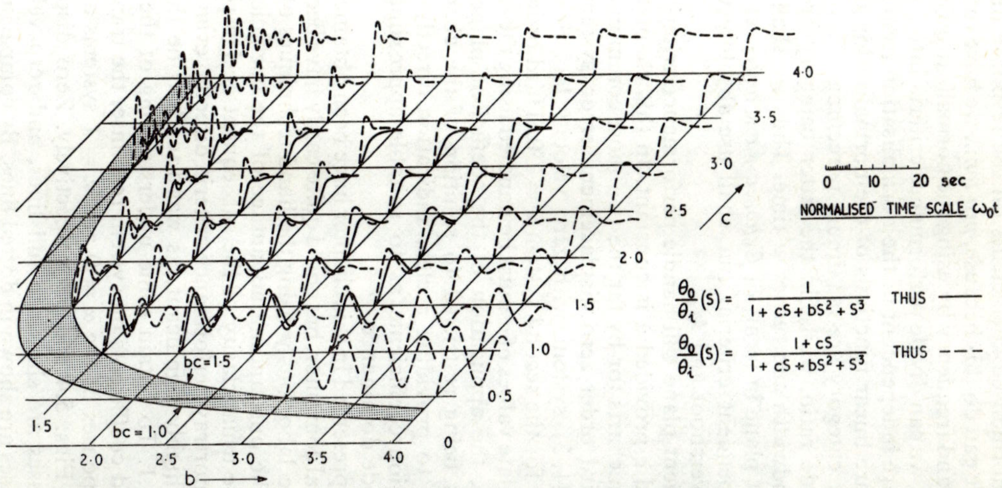


*Fig. 5.4. Transient responses of the third order unity numerator and the third order zero velocity error systems shown pictorially in the coefficient plane*

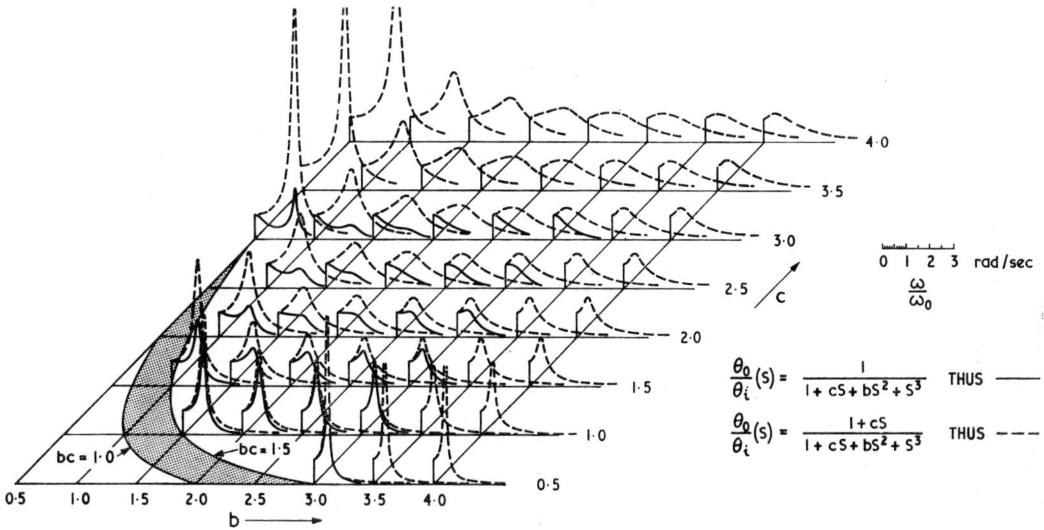


*Fig. 5.5 Frequency responses of the third order unity numerator and the third order zero velocity error system shown pictorially in the coefficient plane*

maximum percentage overshoot. Note, for example, at $b = c = 3{\cdot}0$, how the addition of a system zero has livened up an obviously deadbeat system so that the percentage overshoot is now 25%. Such a zero can also easily double the normalised bandwidth and halve the normalised 63% response time, but the reader should always remember that the true bandwidths and the true response times are both very much a function of the normalising frequency $\omega_0$.

## 5.5 Fundamental Properties of the Third Order Unity Numerator System

For this part of our work on third order systems the choice of boundaries for the coefficient plane has been made on a somewhat arbitrary basis, but the justification for the chosen values is as follows. A unity numerator system with $b = c = 3{\cdot}0$ would have three roots at $-1$, and would therefore be an absolutely deadbeat system. Furthermore, for the unity numerator system, the optimum systems defined by the common performance indices such as ITAE lie well within the region in the coefficient plane bounded by the ordinates $b = 3{\cdot}0$, $c = 3{\cdot}0$, and thus unity numerator systems with three significant poles will lie within this region. This does not mean that unity numerator systems must lie within this region for satisfactory performance, but outside this region it will be found that often one or two poles can be neglected and the system designed on the assumption that the response is dominated by a first or second order mode. It can be seen at a glance from Fig. 5.4 that the unity numerator system transient response can take several forms, depending on the position of the system in the coefficient plane. For example, for $b = 0{\cdot}5$, $c = 3{\cdot}0$, the system response has an exponential term on which is superimposed a lightly damped high-frequency second term, resulting in a type of response which is quite unlike any of the second order responses discussed in Chapter 4. In particular, for this system the maximum percentage overshoot does not occur during the first overshoot. This type of response is sometimes met in hydraulic servomechanisms incorporated in a minor loop such as is used for positioning an aerodynamic control surface,[46, 47] and is acceptable under these circumstances because these high-frequency oscillations are adequately filtered by the major loop and are not reflected in the vehicle motion.

At $b = c = 2{\cdot}5$, there is a very small overshoot of about 1%, whilst for $b = 3{\cdot}0$, $c = 1{\cdot}5$, the system response has a maximum percentage overshoot of 28%, and a type of response similar to a second order system with a damping ratio of approximately 0·36. Provided three independent design parameters are available, Fig. 5.4 may be used to choose the coefficient corresponding to a desired form of response, and $\omega_0$ can then be adjusted to give the required time scale. We have already mentioned that a hydraulic servo might be designed with $b = 0{\cdot}5$, $c = 3{\cdot}0$, but other

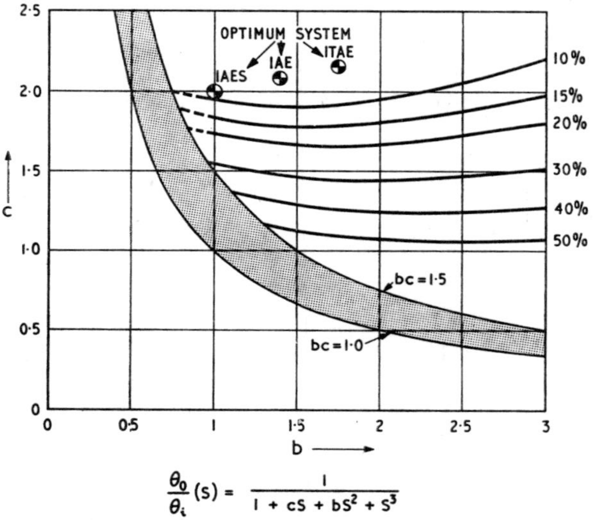


*Fig. 5.6. Maximum percentage overshoot contours in the coefficient plane for the third order unity numerator system*

equally suitable values of $b$ and $c$ for the same application which would have a higher gain margin can be found by progressing along the line $b = 1{\cdot}0$ within the region between $c = 2{\cdot}0$ and $c = 2{\cdot}5$.

If the ITAE performance index is applied to the normalised unity numerator system in following a step function input, the optimum system is

$$\frac{\theta_0}{\theta_i}(S) = \frac{1}{1+2{\cdot}15S+1{\cdot}75S^2+S^3} \tag{5.9}$$

and the normalised 63% response time, for this system $\omega_0 t$, is 2·42 sec. For systems which are major loops in themselves, the type of response defined by Equation 5.9 is satisfactory, and if three independent design parameters are available for adjustment, component values can be determined by solving three simple algebraic equations. If only two independent parameters are adjustable, then a compromise is necessary and this will be discussed fully in Section 5.7. Maximum percentage overshoot contours are shown in Fig. 5.6. Care should be taken in the interpretation of these curves in the region where the maximum percentage overshoot does not occur at the first overshoot as shown by the dotted lines in the figure. Provided the type of response is specified, however, there will be no ambiguity in the results.

There are four possible types of frequency response possible with the unity numerator third order system, and these are shown together with the appropriate boundaries in the coefficient plane in Fig. 5.7. The responses may be categorised as follows.[44, 48]

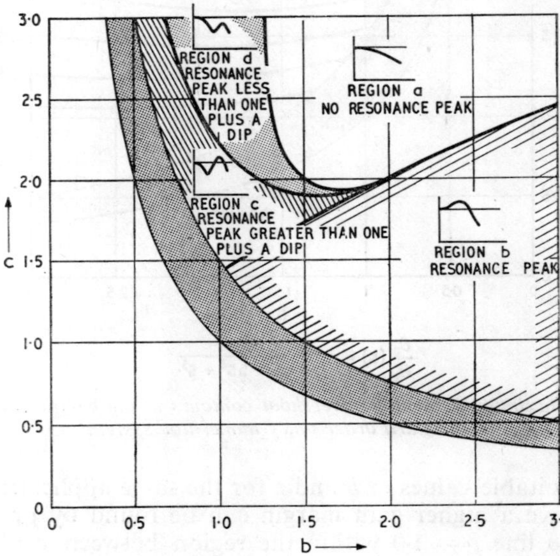


$$\frac{\theta_o}{\theta_i}(s) = \frac{1}{1 + cS + bS^2 + S^3}$$

*Fig. 5.7. Division of the coefficient plane into four regions corresponding to the bas c types of frequency responses for the third order unity numerator system*

*Region a.* No resonance peak ($M_p = 1{\cdot}0$ at zero frequency).
*Region b.* Resonance peak greater than one.
*Region c.* Resonance peak greater than one, plus a dip below one.
*Region d.* Resonance peak less than one, plus a dip.

Generally, Region b is preferred, but this is by no means obligatory, in fact the ITAE optimum system lies in Region a, and the

IAES optimum system lies in Region c. Due to the existence of Region d, a discontinuity in bandwidth occurs in the region of $M_p = 0{\cdot}707$, since for this particular form of response there are three frequencies with amplitude ratio $= 0{\cdot}707$ provided $M_p > 0{\cdot}707$, and hence the highest of these frequencies, by definition is $\omega_B$, whilst when $M_p < 0{\cdot}707$, the bandwidth drops to the lowest of these three frequencies. This discontinuity in bandwidth is clearly shown in Fig. 5.8. Since the $M_p$ contours are almost orthogonal to these bandwidth contours for large regions

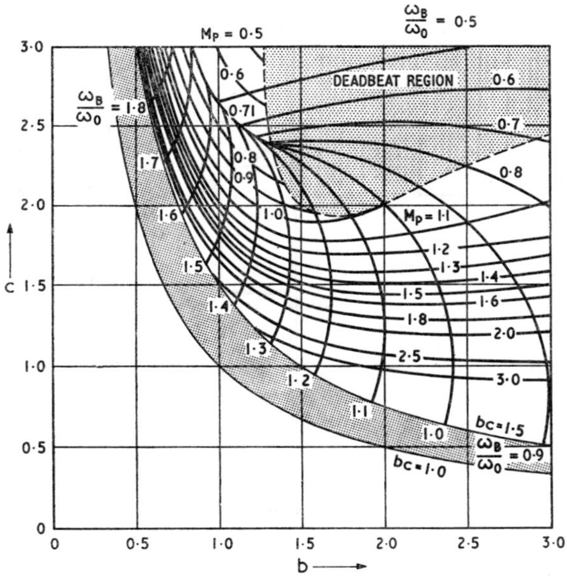


$$\frac{\theta_o}{\theta_i}(s) = \frac{1}{1 + cs + bs^2 + s^3}$$

*Fig. 5.8. Peak amplitude ratio and normalised bandwidth contours in the coefficient plane for the third order unity numerator system*

in the coefficient plane, it is convenient to superimpose both sets of contours on the same graph, and thus the $M_p$ contours are also shown in Fig. 5.8. These two sets of contours will subsequently be used in Section 5.8 for the synthesis of a third order system with only two adjustable design parameters. The ITAE optimum system, as previously explained, will have a peak amplitude ratio of unity at zero frequency, and the system bandwidth will be $1{\cdot}04\omega_0$.

## 5.6 Design of Third Order Unity Numerator Autopilot Using the ITAE Criterion and the Desired System Bandwidth

Provided that three independent adjustable design parameters are incorporated in the proposed design, it is a very simple matter to fix these parameters to give the ITAE optimum response form and a desired system bandwidth, i.e. under these conditions there is no synthesis problem, since every parameter is uniquely defined. The missile roll autopilot shown in Fig. 5.9 has three such adjustable parameters in the feedforward gain $K_g$, the rate feedback time constant $T_r$, and the acceleration feedback constant $T_\alpha$. Missile roll rate will normally be measured by a rate gyroscope, and missile roll acceleration will normally be measured by an

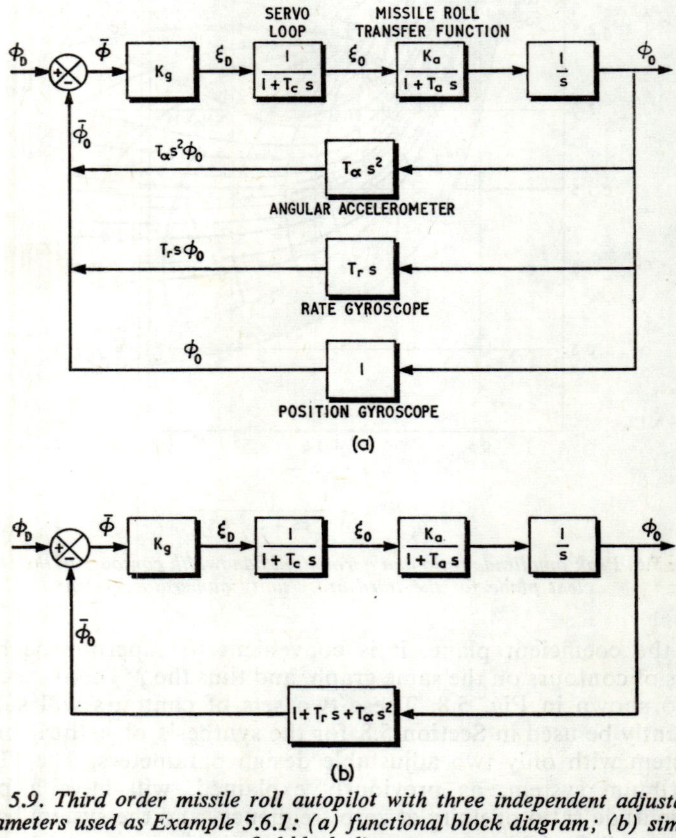


*Fig. 5.9. Third order missile roll autopilot with three independent adjustable parameters used as Example 5.6.1: (a) functional block diagram; (b) simplified block diagram*

angular accelerometer. The constants $T_r$ and $T_\alpha$ will be obtained from the instrumentation signals amplified or attenuated as necessary. For the system shown in Fig. 5.9, we may write

$$\frac{\phi_0}{\bar{\phi}}(s) = G(s) = \frac{K_g K_a}{(1+T_c s)(1+T_a s)s} \tag{5.10}$$

and

$$\frac{\bar{\phi}_0}{\phi_0} = H(s) = 1+T_r s+T_\alpha s^2 \tag{5.11}$$

Using the general expression for the system transfer function,

$$\frac{\phi_0}{\phi_D}(s) = \frac{G(s)}{1+G(s)H(s)}$$

$$= \frac{1}{1+\left(T_r+\frac{1}{K_g K_a}\right)s+\left(T_\alpha+\frac{T_a+T_c}{K_g K_a}\right)s^2+\left(\frac{T_a T_c}{K_g K_a}\right)s^3} \tag{5.12}$$

Comparing with the standard form of Equation 5.3, we have

$$\left.\begin{aligned} \omega_0 &= \left(\frac{1}{a_3}\right)^{1/3} = \left(\frac{K_g K_a}{T_a T_c}\right)^{1/3} \\ b &= a_2\omega_0^2 = \left(T_\alpha+\frac{T_a+T_c}{K_g K_a}\right)\omega_0^2 \\ c &= a_1\omega_0 = \left(T_r+\frac{1}{K_g K_a}\right)\omega_0 \end{aligned}\right\} \tag{5.13}$$

EXAMPLE 5.6.1

It is desired that the autopilot shown in Fig. 5.9 should have the ITAE optimum form and a bandwidth of 30 rad/sec. For the flight case corresponding to $T_a = 0{\cdot}179$ sec, $K_a = 335$ rad sec$^{-1}$ rad$^{-1}$ and $T_c = 0{\cdot}035$ sec, determine the appropriate values of $K_g$, $T_\alpha$ and $T_r$, and the true transfer function coefficients.

The procedure, using Equation 5·13, is as follows. For the ITAE optimum system, $\omega_B/\omega_0 = 1{\cdot}04$, hence $\omega_0 = 30/1{\cdot}04 = 28{\cdot}8$ rad/sec. To determine $K_g$, the $a_3$ coefficient is used; thus

$$\omega_0^3 = 28{\cdot}8^3 = \frac{K_g K_a}{T_a T_c} = \frac{335\times K_g}{0{\cdot}035\times 0{\cdot}179}$$

Hence

$$K_g = \frac{1}{2{\cdot}23}$$

$T_\alpha$ is determined from the $a_2$ equation

$$b = 1{\cdot}75 = a_2\omega_0^2$$

Therefore

$$T_\alpha + \frac{T_a+T_c}{K_g K_a} = \frac{b}{\omega_0^2} = \frac{1{\cdot}75}{28{\cdot}8^2} = 0{\cdot}00211$$

Hence

$$T_\alpha = 0{\cdot}00211 - \frac{0{\cdot}035+0{\cdot}179}{335\times 1/2{\cdot}23}$$

$$= 0{\cdot}00211 - 0{\cdot}00143$$

$$= 0{\cdot}00068 \text{ rad rad}^2 \text{ sec}^{-2}$$

$T_r$ is determined from the $a_1$ equation

$$c = 2{\cdot}15 = a_1\omega_0$$

Therefore

$$T_r + \frac{1}{K_g K_a} = \frac{2{\cdot}15}{28{\cdot}8} = 0{\cdot}0746$$

Hence

$$T_r = 0{\cdot}0746 - \frac{1}{335\times 1/2{\cdot}23} = 0{\cdot}0679 \text{ rad sec rad}^{-1}$$

The true transfer function coefficients are

$$a_1 = \frac{c}{\omega_0} = 0{\cdot}0746$$

$$a_2 = \frac{b}{\omega_0^2} = 0{\cdot}00211$$

and

$$a_3 = \frac{1}{\omega_0^3} = 0{\cdot}000042$$

The system transfer function is thus

$$\frac{\phi_0}{\phi_D}(\text{s}) = \frac{1}{1+0{\cdot}0746\text{s}+0{\cdot}00211\text{s}^2+0{\cdot}000042\text{s}^3} \tag{5.14}$$

## 5.7 Using the Coefficient Plane for the Analysis of Third Order Unity Numerator Systems

It will be remembered that in Chapter 3 a third order missile autopilot with only two adjustable parameters, $K_g$ and $T_r$, was studied using the root locus method. The coefficient plane may be used for both analysis and synthesis of such a system. We shall firstly consider the analysis problem.

EXAMPLE 5.7.1

Using the coefficient plane, determine $M_p$ and $\omega_B$ for a third order missile roll autopilot with the following parameter values: $K_g = 1/18{\cdot}0$, $K_a = 335$, $T_a = 0{\cdot}179$, $T_c = 0{\cdot}035$, $T_r = 0{\cdot}0835$.

The procedure is to determine $\omega_0$, $b$ and $c$ in that order, and to locate the point $b,c$ in the coefficient plane. $M_p$ and $\omega_B/\omega_0$ are next read off Fig. 5.8, and $\omega_B$ is then determined by simple multiplication. The system block diagram is shown in Fig. 3.12, and the system transfer function is

$$\frac{\phi_0}{\phi_D}(s) = \frac{1}{1+\left(T_r+\dfrac{1}{K_gK_a}\right)s+\left(\dfrac{T_c+T_c}{K_gK_a}\right)s^2+\left(\dfrac{T_aT_c}{K_gK_a}\right)s^3} \tag{5.15}$$

Now

$$\omega_0 = \left(\frac{K_gK_a}{T_aT_c}\right)^{1/3} = \left(\frac{335}{18{\cdot}0\times 0{\cdot}179\times 0{\cdot}035}\right)^{1/3} = 14{\cdot}36$$

$$b = \omega_0^2\left(\frac{T_a+T_c}{K_gK_a}\right) = \frac{14{\cdot}36^2\times 18\times 0{\cdot}214}{335} = 2{\cdot}37$$

and

$$c = \omega_0\left(T_r+\frac{1}{K_gK_a}\right) = 14{\cdot}36\left(0{\cdot}0835+\frac{18}{335}\right) = 1{\cdot}97$$

By interpolation in Fig. 5.8, it is seen that $M_p \simeq 1{\cdot}05$ and $\omega_B/\omega_0$ $0{\cdot}920$, hence $\omega_B = 0{\cdot}920\times 14{\cdot}36 = 13{\cdot}20$ rad/sec. These results are in good agreement with the values obtained from the frequency response shown in Fig. 3.27, which was computed from the system pole-zero array. For $b = 2{\cdot}37$, $c = 1{\cdot}97$, it can be seen from Fig. 5.4 that the maximum percentage overshoot ocurs at the first overshoot, and from Fig. 5.6 this maximum percentage overshoot is found to be about 11%.

## 5.8 Using the Coefficient Plane for the Synthesis of Third Order Unity Numerator Systems[43]

The coefficient plane allows the synthesis of third order unity numerator and third order zero velocity error systems to be undertaken much more rapidly than by using the root locus method, and with considerably more accuracy. For systems with only two design parameters it will not be possible to specify both the exact form of response and the true bandwidth, and it is necessary to investigate the trade-off between bandwidth and some other criterion. Once the normalised coefficients $b$ and $c$ have been determined, and $\omega_0$ is known, the system performance is adequately described using the coefficient plane only. As our first example of synthesis using the coefficient plane, we shall consider the design of the third order autopilot when only two design variables are available for adjustment. The system block diagram is shown in Fig. 3.12, and the system transfer function is given by Equation 5.15.

EXAMPLE 5.8.1

For the flight case corresponding to $K_a = 335$ rad sec$^{-1}$ rad$^{-1}$, $T_a = 0{\cdot}179$ sec, and $T_c = 0{\cdot}035$, determine, using the coefficient plane, the best values of $T_r$ and $K_g$. (We shall interpret the best values of $T_r$ and $K_g$ to be the values which result in a high system bandwidth coupled with a reasonable value of $M_p$ or maximum percentage overshoot, whichever is of more interest.)

The procedure is the same as for Example 5.7.1, except that the computations are repeated for as many values of $K_g$ and $T_r$ as are of interest. From Equation 5.15 we may write

$$\omega_0 = \left(\frac{K_g K_a}{T_a T_c}\right)^{1/3}$$

$$b = \omega_0^2\left(\frac{T_a+T_c}{K_g K_a}\right)$$

and

$$c = \omega_0\left(T_r+\frac{1}{K_g K_a}\right)$$

whence $\omega_B/\omega_0$ and $M_p$ (assuming $M_p$ is the other criterion of interest) can be determined from Fig. 5.8. Using these results it is then possible to plot the graph relating $M_p$ and $\omega_B$ for the different values of $K_g$ and $T_r$.

It would be very reasonable to assume that we are particularly interested in the type of frequency response corresponding to Region b of Fig. 5.7, i.e. the form of one significant peak and with no dip in the response prior to the peak. Thus during the synthesis stage we can ignore those combinations of $b$ and $c$ which fall significantly outside the Region b of Fig. 5.7. A reasonable matrix for $T_r$ and $K_g$ would be $T_r$ varying from 0·05 to 0·08 in steps of 0·01, and $K_g$ equal to 0·04, 0·10, 0·20 and 0·40. Some idea of likely values for $T_r$ and $K_g$ can be obtained by taking a sample point in the region of interest and computing $K_g$ and $T_r$ from the values of $b$ and $c$ at the sample point.

The complete computation is tabulated in Table 5.8.1; values of *b* and *c* were marked on the coefficient plane as they were calculated, to indicate those combinations of $T_r$ and $K_g$ regarded as unsuitable owing to the form of the resultant frequency response. These points are then immediately evident because the *bc* point falls outside Region b of Fig. 5.7. In fact, in Table 5.8.1 it was found convenient to retain two points in the coefficient plane which are just outside Region b, as their retention aided subsequent interpolation. Fig. 5.10 shows $M_p$ and $\omega_B$ plotted as functions of

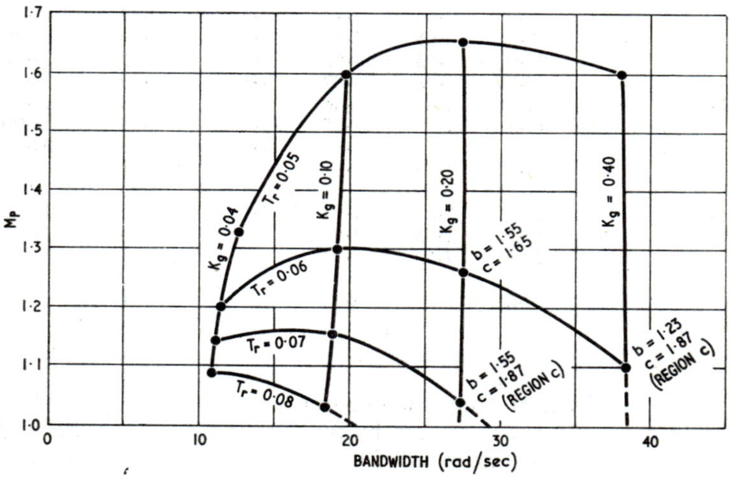


*Fig. 5.10. Peak amplitude ratio and bandwidth as functions of $K_g$ and $T_r$ for the third order unity numerator autopilot with only two adjustable parameters used as Example 5.8.1*

$K_g$ and $T_r$; the trade-off between these two performance criteria is immediately obvious, since we can make the bandwidth quite high by increasing $K_g$, but this will result in a high $M_p$. Also, as reference back to the coefficient plane for the appropriate values of *b* and *c* shows, this $M_p$ together with the high bandwidth will result from the normally unacceptable form of frequency response corresponding to Region c. By inspecting Fig. 5·10, we may conclude that $K_g = 0{\cdot}20$, $T_r = 0.06$ would be reasonable design values, as the bandwidth is then 27·5 rad/sec, and $M_p = 1{\cdot}26$. This system corresponds to normalised coefficients $b = 1{\cdot}55$, $c = 1{\cdot}65$, and is thus in the desired Region c and is, in fact, fairly near the boundary between Region b and Region c. Whilst this system is nearly the best that may be obtained with only two variable design parameters, bearing in mind reasonable restrictions on

**Table 5.8.1**

DETERMINATION OF $M_p$ AND $\omega_B$ AS FUNCTIONS OF $K_g$ AND $T_r$ FOR EXAMPLE 5.7.1 USING THE COEFFICIENT PLANE TO ESTIMATE $M_p$ AND $\omega_B/\omega_0$

| $T_r$ | $K_g$ | $\omega_0$ | $b$ | $c$ | $M_p$ | $\omega_B/\omega_0$ | $\omega_B$ |
|---|---|---|---|---|---|---|---|
| 0·05 | 0·04 | 12·9 | 2·65 | 1·61 | 1·32 | 0·92 | 11·83 |
| | 0·10 | 17·5 | 1·95 | 1·40 | 1·60 | 1·12 | 19·60 |
| | 0·20 | 22·0 | 1·55 | 1·43 | 1·65 | 1·24 | 27·4 |
| | 0·40 | 27·8 | 1·23 | 1·60 | 1·60 | 1·37 | 38·0 |
| 0·06 | 0·04 | 12·9 | 2·65 | 1·73 | 1·20 | 0·89 | 11·43 |
| | 0·10 | 17·5 | 1·95 | 1·57 | 1·30 | 1·10 | 19·20 |
| | 0·20 | 22·0 | 1·55 | 1·65 | 1·26 | 1·25 | 27·50 |
| | 0·40 | 27·8 | 1·23 | 1·87 | 1·10 | 1·39 | 38·6 |
| 0·07 | 0·04 | 12·9 | 2·65 | 1·86 | 1·14 | 0·88 | 11·32 |
| | 0·10 | 17·5 | 1·95 | 1·75 | 1·15 | 1·08 | 18·90 |
| | 0·20 | 22·0 | 1·55 | 1·87 | 1·03 | 1·23 | 27·30 |
| | 0·40 | 27·8 | 1·23 | 2·15 | | Unsuitable | |
| 0·08 | 0·04 | 12·9 | 2·65 | 1·99 | 1·09 | 0.85 | 10.95 |
| | 0·10 | 17·5 | 1·95 | 1·92 | 1·03 | 1.04 | 18.30 |
| | 0·20 | 22·0 | 1·55 | 2·09 | | Unsuitable | |
| | 0·40 | 27·8 | 1·23 | 2·43 | | Unsuitable | |

$M_p$ and the form of frequency response, it would be possible to obtain a slightly better system by plotting the boundary between Regions b and c in the bandwidth–$M_p$ plane of Fig. 5.10. However, this will not normally be necessary, as in Section 5.9 we shall design a suitable system merely by inspecting the coefficient plane. This design by inspection procedure need not necessarily make Fig. 5.10 redundant. Quite apart from illustrating performance criteria in relation to system parameters, it would be invaluable if a further criterion, such as $C_v$, were specified.

## 5.9 Design of a Unity Numerator Third Order System by Inspection of the Coefficient Plane

If at the design stage we are prepared to state, for example, $M_p$, and also say that we require the highest bandwidth obtainable from a system with a given type of frequency response, then in general a particular point in the normalised coefficient plane will be automatically defined. For a third order unity numerator system the provision of only two design parameters may normally be interpreted as meaning that we can fix two of the normalised parameters $b$, $c$ and $\omega_0$, and have to accept whatever value results for the third parameter. Thus if a particular type of response is specified, the appropriate part of the $bc$ plane is examined to determine the combination of $b$ and $c$ which will give the highest value of normalised bandwidth $\omega_B/\omega_0$. Provided $\omega_0$ is changing slowly near this point, or is increasing in the same direction in the $bc$ plane as $\omega_B/\omega_0$ itself, the true bandwidth $\omega_B$ will also be maximised at the same point in the coefficient plane.

EXAMPLE 5.9.1

It is required to determine the values of $T_r$ and $K_g$ for the third order unity numerator autopilot with only two design variables such that the frequency response corresponds to Region b. $M_p$ is not to exceed 1·20, and the bandwidth is to be near maximum for the flight case corresponding to $K_a = 335$ rad sec$^{-1}$ rad$^{-1}$, $T_a = 0{\cdot}179$ sec, $T_c = 0{\cdot}035$ sec, $T_r = 0{\cdot}0835$ rad sec rad$^{-1}$.

Fig. 5.11 shows the boundary between frequency responses corresponding to Regions b and c superimposed on the $\omega_B/\omega_0$ and $M_p$ contours. If we traverse the $M_p = 1{\cdot}20$ line in the region of the boundary, then as we move nearer the boundary we are also increasing the normalised bandwidth $\omega_B/\omega_0$, and hence the likely maximum true bandwidth will be given by the intersection of the $M_p = 1{\cdot}20$ contour with the boundary of interest. The coordinates of this point are $b = 1{\cdot}45$, $c = 1{\cdot}71$. Equation 5.15 is the system transfer function for this particular autopilot, and we may write the normalised coefficients as

$$b = \omega_0^2 a_2 = \omega_0^2\left(\frac{T_a+T_c}{K_g K_a}\right) = 1{\cdot}45 \qquad (5.16)$$

$$c = \omega_0 a_1 = \omega_0\left(T_r + \frac{1}{K_g K_a}\right) = 1{\cdot}71 \qquad (5.17)$$

and

$$\omega_0 = \left(\frac{1}{a_3}\right)^{1/3} = \left(\frac{K_g K_a}{T_a T_c}\right)^{1/3} = \text{unknown} \qquad (5.18)$$

$\omega_0$ may be eliminated between Equations 5.18 and 5.16, leaving an equation with only $K_g$ unknown. Thus we may write

$$K_g = \frac{(T_a+T_c)^3}{K_a b^3 (T_a T_c)^2}$$

For convenience $T_a T_c = 1/160$, and hence

$$K_g = \frac{0{\cdot}214^3 \times 160^2}{335 \times 1{\cdot}45^3} = \frac{1}{4{\cdot}10}$$

Using Equation 5.18, $\omega_0$ is determined directly, since

$$\omega_0 = \left(\frac{K_g K_a}{T_a T_c}\right)^{1/3} = \left(\frac{335\times 160}{4{\cdot}10}\right)^{1/3} = 23{\cdot}6$$

Substituting this value of $\omega_0$ into Equation 5.17 we obtain

$$T_r = \frac{c}{\omega_0} - \frac{1}{K_g K_a} = \frac{1{\cdot}71}{23{\cdot}6} - \frac{4{\cdot}10}{335} = 0{\cdot}0603$$

The normalised bandwidth $\omega_B/\omega_0$ for the point $b = 1{\cdot}45$, $c = 1{\cdot}71$, from Fig. 5.11, is 1·28, giving the true bandwidth

$$\omega_B = \omega_0\left(\frac{\omega_B}{\omega_0}\right) = 23{\cdot}6\times 1{\cdot}28 \text{ rad/sec}$$

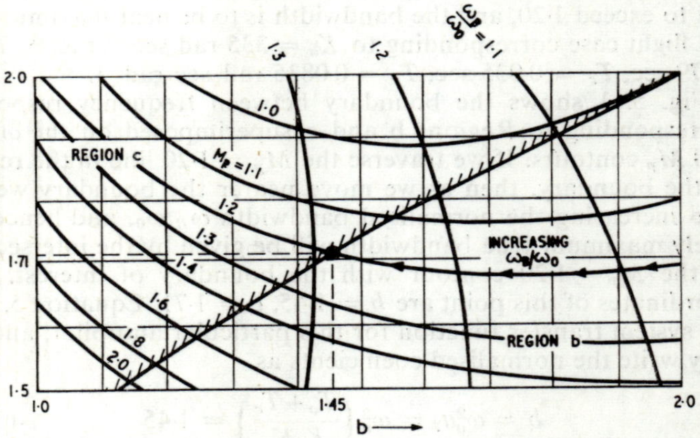


*Fig. 5.11. Peak amplitude ratio and normalised bandwidth contours in the area around the boundary separating frequency responses corresponding to Regions b and c of Fig. 5.7*

If there is any doubt as to the behaviour of $\omega_0$ in the region of the design point, then this simple computation should be repeated for a point on the $M_p$ curve a little way inside the boundary. In this case the doubt is removed immediately, since a small movement to the right along the $M_p = 1{\cdot}2$ contour will achieve a major increase in $b$, and only a small change in $c$. From the $K_g$ equation it follows that an increase in $b$ reduces $K_g$ and hence reduces the normalisation frequency $\omega_0$ and, from Fig. 5.11, it is also apparent that this movement to the right along the $M_p = 1{\cdot}20$ line reduces the normalised bandwidth $\omega_B/\omega_0$. The overall effect of movement to the right away from the boundary is therefore to reduce the true bandwidth, showing that the system defined by

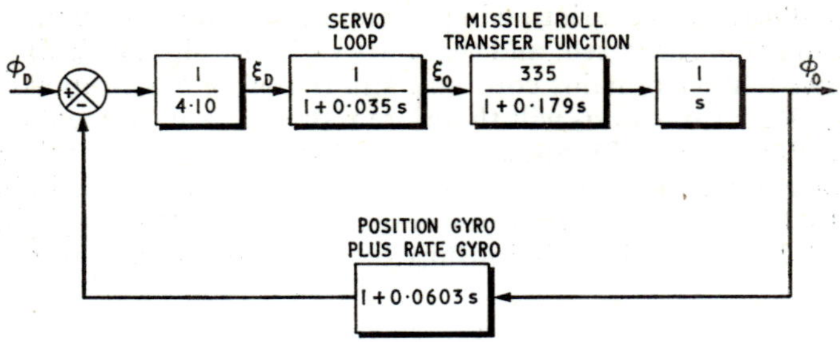


$$\frac{\phi_0}{\phi_D}(s) = \frac{1}{1+0{\cdot}0725s+0{\cdot}0026s^2+0{\cdot}000076s^3}$$

*Fig. 5.12. Block diagram of 'best' autopilot based on $M_p = 1{\cdot}20$ and a frequency response corresponding to Region b of Fig. 5.7*

Equations 5.16, 5.17 and 5.18 is the fastest system with an $M_p = 1{\cdot}20$ and the desired type of frequency response which can be obtained with only two design variables. Fig. 5.12 shows the final block diagram for the resultant missile autopilot.

## 5.10 Variation of Percentage Overshoot of a Third Order Unity Numerator System Caused by Variation of Transfer Function Coefficients

The use of $M_p$ as a performance criterion in the last three sections was a matter of convenience, and the specification in each case could well have been written in terms of percentage overshoot instead, particularly as the regions of interest coincided with the type of transient response where the maximum percentage overshoot occurs at the first peak. Since maximum percentage overshoot is such a common performance specification, it is of interest to use the coefficient plane to determine the variation in percentage overshoot with changes in the system transfer function, particularly since practical measurements on the system or the system analogue using a transfer function computer evaluates the transfer function coefficients directly. For the system designed in Section 5.9 and shown in Fig. 5.12, the nominal transfer function is

$$\frac{\phi_0}{\phi_D}(s) = \frac{1}{1+0{\cdot}0725s+0{\cdot}0026s^2+0{\cdot}000076s^3} \tag{5.19}$$

and the percentage overshoot for this system, from Fig. 5.6, is 17·6%.

To illustrate the variation in percentage overshoot, let us assume that $a_1$ and $a_3$ are fixed, and that $a_2$ is variable. Since $a_3$ is fixed, it follows that $\omega_0$ remains constant, and therefore $b$ is proportional to $a_2$. Variations in $a_2$ will therefore represent a movement along the ordinate $c = 1{\cdot}71$, and from Fig. 5.6 it can be seen that the percentage overshoot curves are very flat in this region, thus indicating that percentage overshoot is rather insensitive to changes in $a_2$. Fig. 5.13 shows the relationship between percentage overshoot

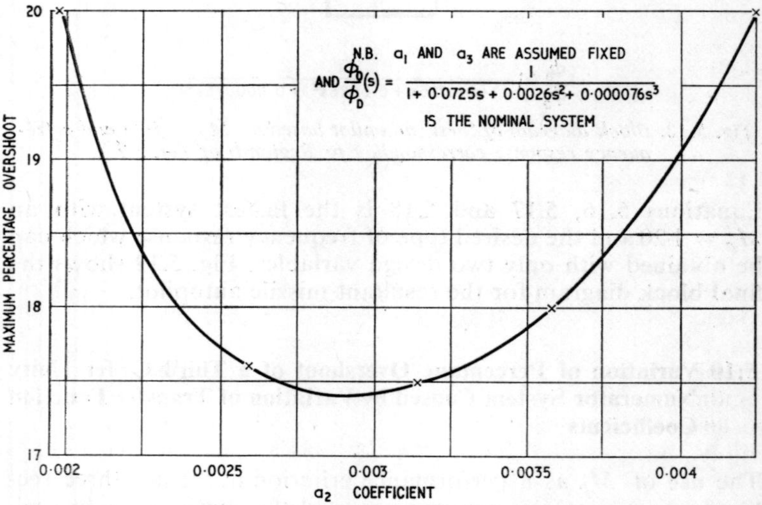


*Fig. 5.13. Effect of variation of the $a_2$ transfer function coefficient on maximum percentage overshoot for Example 5.9.1*

and $a_2$ in the region where the percentage overshoot is less than 20%. Two things should be noted from Fig. 5.13, firstly that the percentage overshoot criterion is so insensitive to changes in $a_2$ that a 50% increase in $a_2$ from the nominal value only increases the percentage overshoot from 17·6% to 18·75% and, secondly, the curve is not symmetrical about the nominal coefficient, and a decrease in $a_2$ will have more effect than a corresponding increase in $a_2$.

If $a_2$ and $a_3$ are kept constant and $a_1$ varied, then the locus in the coefficient plane will be along the line $b = 1{\cdot}45$, since $b$ is proportional to $a_2$ and $c$ is proportional to $a_1$. It can be seen from Fig. 5.6 that the percentage overshoot is very sensitive to changes in $c$ and hence changes in $a_1$, a decrease in $a_1$ corresponding to an increase in percentage overshoot. Thus the $a_1$ coefficient in a third order system can be regarded as a direct damping term, irrespective of the relative significance of the various poles. Variation in $a_3$ will result in changes in both $b$ and $c$, since the normalisation frequency $\omega_0$ will change, and the locus in the $bc$ plane will then be a general curve. Should variations in two coefficients, say $a_1$ and $a_3$, occur simultaneously, then the locus in the $bc$ plane will form a mesh or carpet corresponding to the matrix of values of these coefficients. Variation of all three coefficients simultaneously will result in a set of such meshes, all of which can be determined using simple arithmetic. Once the locus has been superimposed on Fig. 5.6, the effect of the variation in transfer function coefficients on the maximum percentage overshoot can be plotted as shown in Fig. 5.13.

## 5.11 Variation of Bandwidth of Third Order Unity Numerator System Caused by Variation of Transfer Function Coefficients

As a further example of the use of the coefficient plane, let us consider the effect of variation in transfer function coefficients on system bandwidth. The reader has probably already noted from Fig. 5.8 that the system bandwidth contours are approximately orthogonal to the percentage overshoot curves of Fig. 5.6 in the region of the design point for Example 5.8.1, and thus although variation in percentage overshoot as $a_2$ is adjusted is small, there will be considerable variation in bandwidth. This is shown in Fig. 5·14, and it can be seen that a 50% increase in $a_2$ will result in a decrease in bandwidth from 30·3 rad/sec to 23·9 rad/sec.

Variations in $a_1$, provided $a_2$ and $a_3$ are kept constant, will not be so significant, as $b$ will be constant, and in the region of the design point $b = 1{\cdot}45$, $c = 1{\cdot}71$, the $\omega_B/\omega_0$ contours of Fig. 5.8 are running almost vertically. For other regions in the $bc$ plane the effect of changes in $a_2$ on system bandwidth may be much more significant than the effect near this particular design point. The significance of changes in $a_2$ (and hence $b$) on system bandwidth is particularly spectacular in the deadbeat region around $b = 2$, $c = 2{\cdot}5$.

Once again we may summarise the study of the effect of transfer function variation by saying that variations in $a_1$ alone will result in a locus along the line $c =$ constant, variations in $a_2$ alone will result in a locus along the line $b =$ constant, whilst variations in $a_2$ and $a_3$ simultaneously will form a mesh in the $bc$ plane. Variation of $a_3$ alone will change $b$ and $c$ and the locus will then be a general curve, whilst variation of $a_3$ and $a_1$ or $a_2$ together will form a mesh in the $bc$ plane. Variation of all three coefficients simultaneously

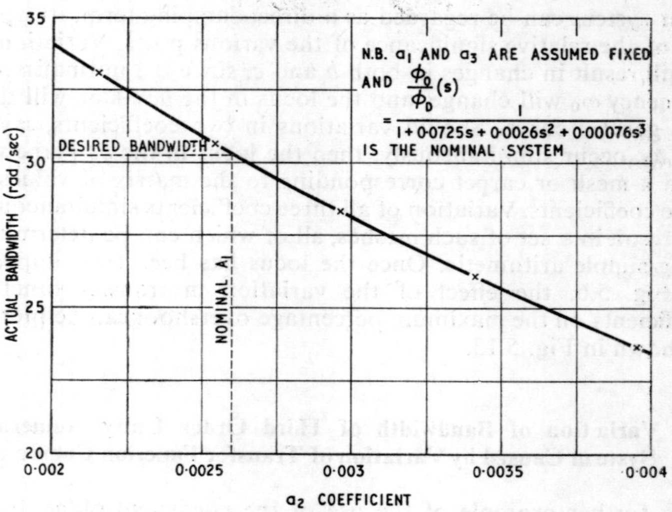


*Fig. 5.14. Effect of variation of the $a_2$ transfer function coefficient on system bandwidth for Example 5.9.1*

will result in a series of meshes in the *bc* plane, all of which can be determined using simple arithmetic, as was the case in determining the effect of changes in transfer function coefficients on the maximum percentage overshoot. If required, superimposition of the locus or mesh on Fig. 5.8 will allow a series of relationships between system bandwidth and transfer function coefficients to be plotted similar to that shown in Fig. 5.14.

## 5.12 Fundamental Properties of the Third Order Zero Velocity Error System[49]

The normalised transfer function for the third order zero velocity lag system is

$$\frac{\theta_o}{\theta_i}(s) = \frac{1+cS}{1+cS+bS^2+S^3} \tag{5.20}$$

and the system pole-zero array for a given point in the coefficient plane will possess the same poles as the unity numerator system with the same coefficients, but in addition will have a system zero at $-1/c$. Due to the zero at $-1/c$ the response at any point in the coefficient plane will be radically different from the response of the unity numerator system, and this is clearly shown in the transient response of Fig. 5.4, and in the frequency responses of Fig. 5.5. This radical difference due to the presence of the system zero is entirely in accordance with expectations based on Chapters 3 and 4, but the magnitude of the change is of particular interest. For example, for the point $b = 2$, $c = 1{\cdot}5$ the more important criteria are shown in Table 5.12.1. Thus the bandwidth

**Table 5.12.1**

COMPARISON OF PERFORMANCE CRITERIA FOR THE THIRD ORDER UNITY NUMERATOR AND THIRD ORDER ZERO VELOCITY ERROR SYSTEMS AT THE POINT $b = 2{\cdot}0$, $c = 1{\cdot}5$ IN THE COEFFICIENT PLANE

| *System* | *P.O.* | $M_p$ | $\omega_B/\omega_0$ | $C_v$ |
|---|---|---|---|---|
| Unity numerator | 27% | 1·40 | 1·09 | $\omega_0/1{\cdot}5$ |
| Zero velocity error | 53% | 2·10 | 1·45 | $\infty$ |

is increased by over 30% but the percentage overshoot is doubled as a consequence. It is immediately evident from Figs. 5.4 and 5.5 that the useful part of the coefficient plane has moved further away from the stability boundary $bc = 1{\cdot}0$, and this is to be expected in order that the maximum percentage overshoot should be kept down to a reasonable figure.

There are two possible interpretations of the ITAE optimum system for systems with zero velocity error.[21] Firstly, we may apply the ITAE performance index to the response of the system in following a step function input, and it is then found that the system which minimises the integral is

$$\frac{\theta_0}{\theta_i}(s) = \frac{1+3{\cdot}0S}{1+3{\cdot}0S+1{\cdot}75S^2+S^3} \tag{5.21}$$

In fact, the value of $c$ which minimises the integral is infinite, but any value of $c$ equal to or greater than 3 is considered acceptable. The maximum percentage overshoot is about 39%. However, it is plausible to argue that since the requirement of a zero velocity error system is that there should be a zero steady state lag following a ramp function input, it follows that the transient behaviour during the early stages of the ramp function input is also of interest. If the ITAE performance index is applied to this transient behaviour, it is found that the integral is minimised by the system with transfer function

$$\frac{\theta_0}{\theta_i}(S) = \frac{1+2{\cdot}94S}{1+2{\cdot}94S+1{\cdot}30S^2+S^3} \tag{5.22}$$

and this system has a maximum percentage overshoot of about 50%, which is rather on the high side. Equations 5.21 and 5.22 assist in indicating the region of interest in the coefficient plane, and to allow for the shift in emphasis from points such as $b = 1{\cdot}75$, $c = 2{\cdot}15$, which were met in the unity numerator system studies, the coefficient plane has been extended to the ordinates $b = 4{\cdot}0$, $c = 4{\cdot}0$. It must be repeated, however, that perfectly acceptable systems may lie outside this region of the coefficient plane, but generally they will be systems with a response dominated by the system zero and one or two poles.

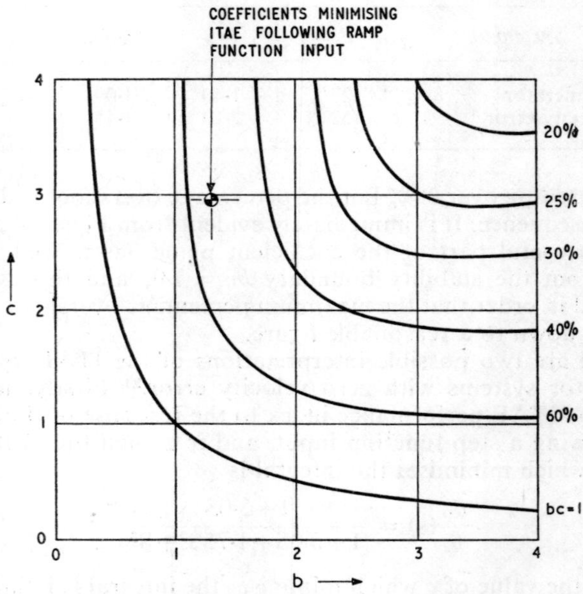


$$\frac{\theta_o}{\theta_i}(s) = \frac{1 + cs}{1 + cs + bs^2 + s^3}$$

*Fig. 5.15. Constant maximum percentage overshoot contours for the third order zero velocity lag system*

Fig. 5.15 shows the maximum percentage overshoot contours plotted in the coefficient plane: there is no ambiguity in the transient response for the region in the coefficient plane bounded by the ordinates $b = 4$, $c = 4$, since the maximum percentage overshoot occurs at the first peak. The 25% contour may be crudely approximated by the hyperbola $bc = 9$ in the region of interest, and this contour may be looked upon as design contour for many applications. We must generally accept larger percentage overshoots for zero velocity error systems than were allowed in the more general examples in Chapter 4 but, if component values allow it, the percentage overshoot can be much further reduced by moving the design point into the region around $b = c = 6.0$ where the three system poles are very much all negative real, as we shall see later when discussing system poles in relation to the position of the design point in the coefficient plane.

Only one type of frequency response is encountered in the part of the coefficient plane bounded by the ordinates $b = 4$, $c = 4$, and this is the type of response with $M_p$ greater than unity, and without a dip below unity before the peak amplitude ratio is reached. Hence there will be no discontinuity in bandwidth in this part of the coefficient plane. Owing to the geometric properties of the normalised bandwidth and $M_p$ contours it is again convenient to superimpose the two sets of contours as shown in Fig. 5.16. It is interesting to note that the bandwidth contours are almost straight for large parts of the region, with a slope of approximately 45°. Comparison of the normalised bandwidth

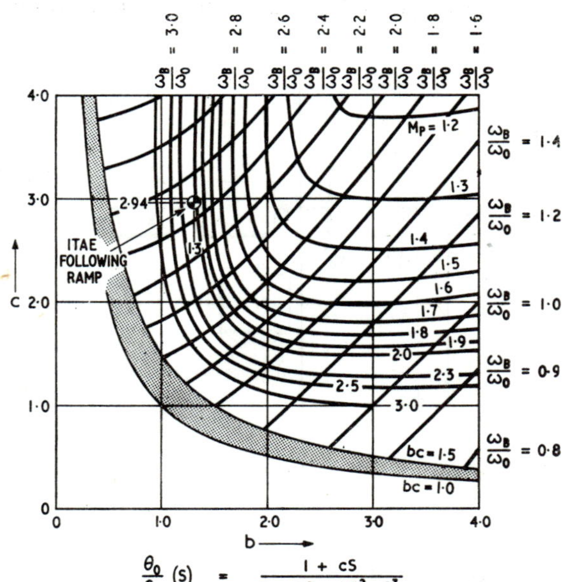


$$\frac{\theta_0}{\theta_i}(s) = \frac{1 + cs}{1 + cs + bs^2 + s^3}$$

*Fig. 5.16. Normalised bandwidth and constant peak amplitude ratio contours for the third order zero velocity error system*

contours with the $M_p$ or percentage overshoot contours shows that the normalised bandwidth may be kept constant whilst $M_p$ or the percentage overshoot is reduced by moving along a normalised bandwidth contour into the interior of the coefficient plane, thus confirming that if the component values allow the design point to move into this region beyond, say $b = 4{\cdot}0$, $c = 4{\cdot}0$ to $b = 6{\cdot}0$, $c = 6{\cdot}0$, then a reasonably low value of $M_p$ or the percentage overshoot can be obtained in conjunction with a reasonable bandwidth. However, as we are moving a long way from the ITAE optimum design point, whichever way this is defined, it may be argued that we are not making the best use of the system poles, but each case should be treated on individual merit.

## 5.13 Analysis of a Third Order Zero Velocity Error System Using the Coefficient Plane

The appropriate coefficient plane contours can be used for the analysis and synthesis of third order zero velocity error systems in exactly the same way as the respective contours were used for the third order unity numerator system. In Chapter 4 several of the examples were based on the design of a servomechanism

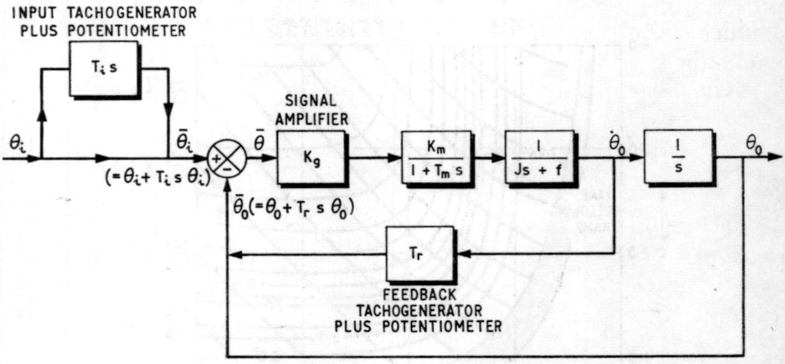


SYSTEM TRANSFER FUNCTION $\quad \dfrac{\theta_o}{\theta_i}(s) = \dfrac{1 + T_i s}{1 + \left(T_r + \dfrac{f}{K_g K_m}\right)s + \left(\dfrac{f\,T_m + J}{K_g K_m}\right)s^2 + \left(\dfrac{T_m J}{K_g K_m}\right)s^3}$

AND FOR ZERO VELOCITY LAG, $\quad T_i = T_r + \dfrac{f}{K_g K_m}$

*Fig. 5.17. Block diagram of the third order linear servomechanism with derivative of input signal shaping which can be adjusted to give zero velocity lag*

which would control a load with moment of inertia 2·5 slugs ft². By assuming that the servomotor was infinitely fast in responding to any demand made upon it, the resultant servomechanism was of second order. If the servomotor does possess a finite lag $T_m$, then the block diagram becomes as shown in Fig. 5.17. For the purpose of Example 5.13.1 we shall assume that compensation is obtained by using the rate feedback and derivative of input signal shaping. Note carefully that three adjustable parameters $K_g$, $T_i$ and $T_r$, are under the designer's jurisdiction, but these three parameters do not provide three independent quantities $b$, $c$ and $\omega_0$, since $T_i$ is adjusted to balance $T_r + f/K_gK_m$, thus making $b_1 = a_1$ which is the requirement for a zero velocity error system, leaving no adjustment for $b$ which is then determined by the fixed quantities $f$, $T_m$, $J$ and $K_m$, together with the normalisation frequency $\omega_0$. This could be overcome by introducing a fourth variable in the form of acceleration feedback, which would then give direct control over $b$.

EXAMPLE 5.13.1

In the servomechanism shown in Fig. 5.17, $J = 2{\cdot}5$ slugs ft², $f = 25$ lb ft rad⁻¹ sec, $K_gK_m = 250$ lb ft/rad, $T_m = 0{\cdot}1$ sec, and $T_r = 0{\cdot}13$ sec. Assuming $C_v = \infty$, calculate $T_i$, $b$, $c$ and $\omega_0$, and hence using the coefficient plane determine system bandwidth, $M_p$, and maximum percentage overshoot.

The system transfer function is

$$\frac{\theta_0}{\theta_i}(\mathrm{s}) = \frac{G_p(\mathrm{s})G(\mathrm{s})}{1+G(\mathrm{s})H(\mathrm{s})}$$

where

$$G_p(\mathrm{s}) = 1+T_i(\mathrm{s})$$

$$G(\mathrm{s}) = \frac{K_gK_m}{(1+T_m\mathrm{s})(J\mathrm{s}+f)\mathrm{s}}$$

and

$$H(\mathrm{s}) = 1+T_r\mathrm{s}$$

Hence

$$\frac{\theta_0}{\theta_i}(\mathrm{s}) = \frac{1+T_i\mathrm{s}}{1+\left(T_r+\dfrac{f}{K_gK_m}\right)\mathrm{s}+\left(\dfrac{fT_m+J}{K_gK_m}\right)\mathrm{s}^2+\left(\dfrac{T_mJ}{K_gK_m}\right)\mathrm{s}^3} \qquad (5.23)$$

For zero velocity error, $b_1 = a_1$. Therefore

$$T_i = T_r + \frac{f}{K_g K_m} = 0{\cdot}13 + \frac{25}{250} = 0{\cdot}23$$

$$\omega_0 = \left(\frac{1}{a_3}\right)^{1/3} = \left(\frac{K_g K_m}{T_m J}\right)^{1/3} = \left(\frac{250}{0{\cdot}1 \times 2{\cdot}5}\right)^{1/3} = 10$$

$$b = a_2 \omega_0^2 = \left(\frac{f T_m + J}{K_g K_m}\right) \omega_0^2 = \left[\frac{(25 \times 0{\cdot}1) + 2{\cdot}5}{250}\right] 100 = 2{\cdot}0$$

and

$$c = a_1 \omega_0 = \left(T_r + \frac{f}{K_g K_m}\right) \omega_0 = \left(0{\cdot}13 + \frac{25}{250}\right) 10 = 2{\cdot}3$$

From Fig. 5.15, the percentage overshoot for the point $b = 2{\cdot}0$, $c = 2{\cdot}3$ is about 39%. From Fig. 5.16, for the point $b = 2{\cdot}0$, $c = 2{\cdot}3$, $M_p$ is about 1·55, and $\omega_B/\omega_0$ is about 1·85, hence the true bandwidth is approximately 18·5 rad/sec.

A disadvantage of this method of eliminating velocity lag by derivative of input shaping is that the time constant $T_i$ must be individually matched to the rate feedback time constant $T_r$ and to the viscous damping/feedforward gain ratio $(f/K_g K_m)$ if true elimination is to be achieved. Since the viscous damping torque at least is highly variable from one servo to the next, it is sometimes preferable to eliminate velocity lag by using a method which is independent of load characteristics.

## 5.14 Coefficient Plane Design of a System Using Integral Compensation for Eliminating Steady State Velocity Lag

Integral compensation for completely eliminating steady state lag is often preferred to the derivative methods outlined in Chapter 4 and in Section 5.13, since the perfect matching of the input derivative time constant $T_i$ to other system parameters is required for input derivative signal shaping, whilst the use of the error derivative signal shaping method can only be used in systems with zero load damping. Steady state errors in following ramp function inputs can be completely eliminated using a suitable integrating circuit, provided the $a_1$ and $b_1$ transfer function coefficients are directly controlled by the integrator time constant $T_p$. The introduction of the integrating network increases the order of the system transfer function by one, thus turning the second order systems discussed in Chapter 4 into appropriate third order systems. Neglecting the motor time constant $T_m$, but introducing a perfect integrating network in the feedforward path and acceleration feedback instead of rate feedback in the feedback path, the simple position control system studied in the previous section takes the form shown in Fig. 5.18.

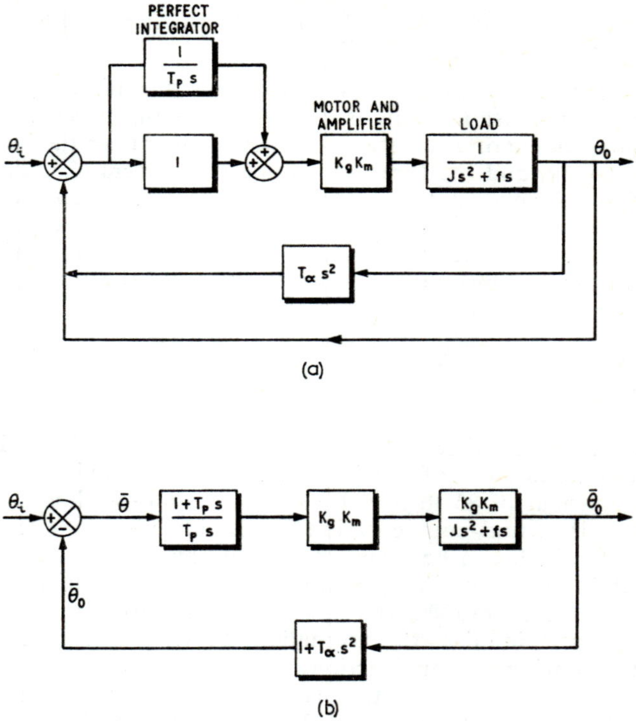


*Fig. 5.18. Block diagram of the third order zero velocity error servomechanism obtained by using integral compensation (see Example 5.14.1): (a) block diagram of servomechanism; (b) simplified block diagram*

The system transfer function is

$$\frac{\theta_0}{\theta_i}(s) = \frac{G(s)}{1+G(s)H(s)}$$

where

$$G(s) = \frac{K_g K_m(1+T_p s)}{(Js^2+fs)T_p s}$$

and

$$H(s) = 1 + T_\alpha s^2$$

Whence, in simplest form,

$$\frac{\theta_0}{\theta_i}(s) = \frac{1+T_p s}{1+T_p s+\left(T_\alpha+\dfrac{T_p f}{K_g K_m}\right)s^2+T_p\left(\dfrac{J}{K_g K_m}+T_\alpha\right)s^3} \qquad (5.24)$$

The reason for the choice of acceleration feedback $T_\alpha$ is now apparent, because not only is $a_1 = b_1$, but both these coefficients are independent of any system parameters other than the integrating network, and the three adjustable parameters $K_g$, $T_p$ and $T_\alpha$ provide independent adjustment of the transfer functions coefficients $a_1$, $a_2$ and $a_3$. Thus we can stipulate definite requirements for $b$, $c$ and $\omega_0$ and define the system behaviour uniquely and completely. An example will clearly illustrate the principles involved.

EXAMPLE 5.14.1

A position control servomechanism is to control a load characterised by a moment of inertia of 2·5 slugs ft² and a viscous damping coefficient of 25 lb ft rad$^{-1}$ sec. If the motor time constant is negligible, and compensation is provided by an integrating network which can be assumed perfect plus output acceleration feedback, determine the system parameters such that the system bandwidth is 20 rad/sec, the steady state velocity lag is zero, and the system minimises the normalised ITAE performance index when following a ramp function input.

The desired values of the normalised coefficients are given by Equation 5·22 as $b = 1{\cdot}30$, $c = 2{\cdot}94$. From Fig. 5.16 the normalised bandwidth $\omega_B/\omega_0$ for this combination of coefficients is 2·47. Thus

$$\omega_0 = \omega_B\left(\frac{\omega_0}{\omega_B}\right) = \frac{20}{2{\cdot}47} = 8{\cdot}10$$

Using Equation 5.24, $T_p$ is obtained directly from $a_1$ (or $b_1$), since

$$a_1 = \frac{c}{\omega_0} = T_p = \frac{2{\cdot}94}{8{\cdot}10} = 0{\cdot}363$$

$K_g K_m$ and $T_r$ are obtained by solving simultaneous equations.

From the $a_3$ coefficient we have

$$\frac{1}{a_3} = \frac{1}{\omega_0^3} = T_p\left(\frac{J}{K_gK_m}+T_\alpha\right) = \frac{1}{8\cdot 10^3}$$

Hence

$$\frac{2\cdot 5}{K_gK_m}+T_\alpha = \frac{1}{8\cdot 10^3\times 0\cdot 363}$$

and thus

$$\frac{2\cdot 5}{K_gK_m}+T_\alpha = \frac{1}{193} \tag{5.25}$$

From the $a_2$ coefficient of Equation 5.24

$$b = \omega_0^2\left(T_\alpha+\frac{T_pf}{K_gK_m}\right)$$

Thus

$$T_\alpha+\frac{25\times 0\cdot 363}{K_gK_m} = \frac{1\cdot 30}{8\cdot 1^2}$$

and, on simplification,

$$T_\alpha+\frac{9\cdot 10}{K_gK_m} = \frac{1}{50\cdot 4} \tag{5.26}$$

Equations 5.26 and 5.25 may be used to eliminate $T_\alpha$, leaving

$$K_gK_m = \frac{6\cdot 6}{\dfrac{1}{50\cdot 4}-\dfrac{1}{193}} = 451 \text{ lb ft/rad}$$

Substituting this value of $K_gK_m$ into Equation 5.26,

$$T_\alpha+\frac{9\cdot 1}{4\cdot 51} = \frac{1}{50\cdot 4}$$

giving $T_\alpha = -0\cdot 0004$, thus showing that a positive feedback of output acceleration is required in this system. The resultant block diagram is shown in Fig. 5.19, and the system transfer function in numerical form is

$$\frac{\theta_0}{\theta_i}(s) = \frac{1+0\cdot 36s}{1+0\cdot 363s+0\cdot 0198s^2+0\cdot 0019s^3} \tag{5.27}$$

Using Fig. 5.16, and $b = 1\cdot 30$, $c = 2\cdot 94$, we can determine $M_p \simeq 2\cdot 05$, and using Fig. 5.15 we find that the maximum percentage

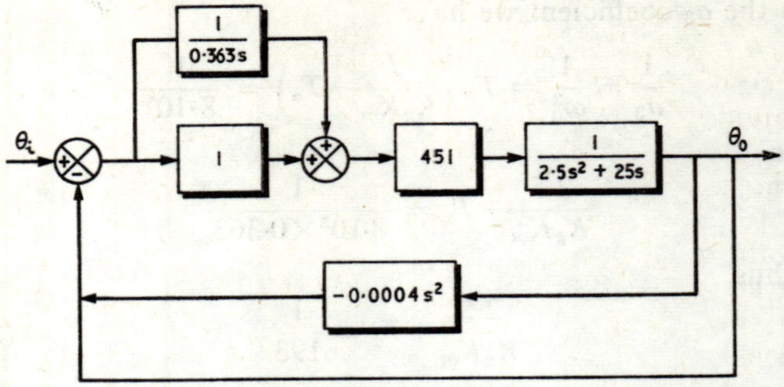


$$\frac{\theta_o}{\theta_i}(s) = \frac{1 + 0{\cdot}363s}{1 + 0{\cdot}363s + 0{\cdot}0198s^2 + 0{\cdot}0019s^3}$$

*Fig. 5.19. Block diagram of the third order zero velocity error servomechanism which has a 20 rad/sec bandwidth and a minimum ITAE following a ramp function input*

overshoot is approximately 50%. Both $M_p$ and the percentage overshoot are rather high, which is the penalty paid for minimising the normalised ITAE following the ramp function input, and the reader faced with the design of a zero velocity error system with three independent parameters might well choose different values of $b$ and $c$ to insert in the design process.

## 5.15 The Normalised Root Locus for the Third Order Characteristic Equation Bounded by $b = c = 3{\cdot}0$[43–45]

The design aids so far presented in this chapter greatly simplify the analysis and synthesis of third order unity numerator systems and third order zero velocity error systems provided that all three roots make a significant contribution to system behaviour. Although the synthesis techniques dealt with in Sections 5.7 and 5.8 were directed at unity numerator systems in particular, it follows that zero velocity error systems can be synthesised in exactly the same way. Two further problems remain: firstly we must consider the more general third order system with one or more system zeros present, and secondly we must consider the regions in the coefficient plane where the response is predominantly first order or second order because two or one poles respectively are insignificant in their effect upon system behaviour. These two problems are related, however, because the introduction of system zeros may cause the movement of the desired type of response outside the arbitrary boundaries of the coefficient plane defined by the ordinates $b = c = 3$, as we have already seen for the zero velocity error system for the case when severe restrictions are placed on the magnitudes of the percentage overshoot and $M_p$.

Unfortunately, the introduction of further variables such as the system zero locations makes the presentation of results difficult. However, since the third order characteristic equation is unchanged by the introduction of system zeros, it is still possible to avoid constructing the true root locus for even the most general third order system, since we can use the normalised root locus plot in the coefficient plane to determine the system poles very rapidly. The coefficient plane method will again be particularly time-saving when two design parameters are adjusted simultaneously, thus allowing the system pole-zero array to be determined very rapidly. Once the system pole-zero array has been constructed, the system behaviour is then examined using the conventional techniques of Chapter 3.

The normalised third order characteristic equation is defined by Equation 5.3 as

$$1+c\mathrm{S}+b\mathrm{S}^2+\mathrm{S}^3 = 0$$

and the equation may be factorised to yield

$$\left(\mathrm{S}+\frac{1}{T_1}\right)(\mathrm{S}^2+2\zeta\omega_n\mathrm{S}+\omega_n^2) = 0 \qquad (5.28)$$

where $T_1$ and $\omega_n$ refer specifically to the normalised S plane. Equating coefficients,

$$\left.\begin{aligned} \frac{\omega_n^2}{T_1} &= 1 \\ \frac{2\zeta\omega_n}{T_1}+\omega_n^2 &= c \\ \frac{1}{T_1}+2\zeta\omega_n &= b \end{aligned}\right\} \qquad (5.29)$$

When transformed to the true s plane, the factorised equation becomes

$$\left(s+\frac{\omega_0}{T_1}\right)(s^2+2\zeta\omega_n\omega_0 s+\omega_n^2\omega_0^2) \qquad (5.30)$$

thus yielding the true roots, which are geometrically similar to the normalised roots but multiplied by $\omega_0$. Should $\zeta$ in Equation 5.30 be greater than unity, the second order factor will factorise into

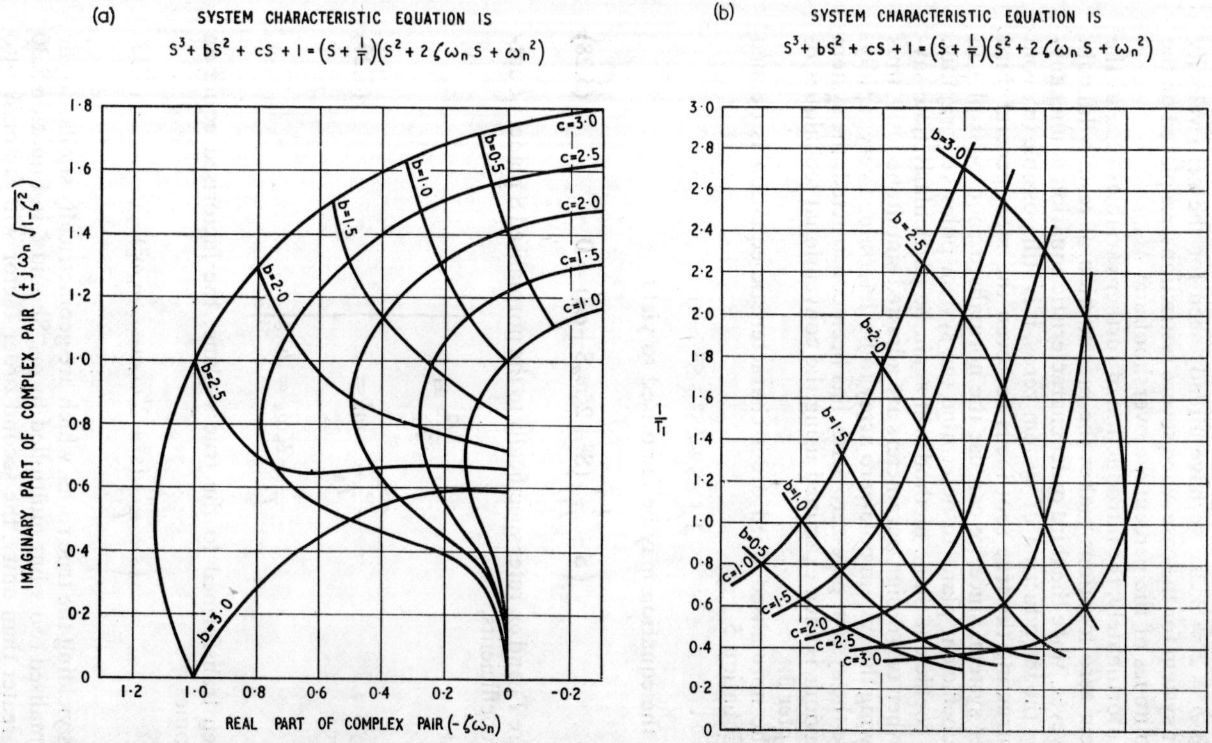


*Fig. 5.20. Roots of the normalised third order characteristic equation in the region of the coefficient plane wherein all three roots are significant in the response of a type (0,3) system: (a) complex roots; (b) real root*

two negative real roots, and the equation can be rewritten as

$$\left(s+\frac{\omega_0}{T_1}\right)\left[s+\omega_0\omega_n\left(\zeta+\sqrt{\zeta^2-1}\right)\right]\left[s+\omega_0\omega_n\left(\zeta-\sqrt{\zeta^2+1}\right)\right]=0 \tag{5.31}$$

In the interests of accuracy of resolution it is convenient to plot the roots of the normalised characteristic equation for two different regions of the coefficient plane quite separately. In the region bounded by $b = c = 3{\cdot}0$, $\zeta$ is never greater than one, and all roots are significant. Fig. 5·20(a) shows the root locus of the complex pole pair in terms of the normalised coefficients $b$ and $c$, thus allowing $\zeta$ and $\omega_n$ to be measured directly. To superimpose the negative real root $-1/T_1$ on this locus would confuse matters, and so this root is drawn separately in Fig. 5.20(b). As a simple illustration of the use of Fig. 5.20, let us consider the factorisation of the characteristic equation, with $b = 2{\cdot}0$, $c = 2{\cdot}5$. From Fig. 5.20(a), $\omega_n = 1{\cdot}29$ and the damping angle is 57°, giving $\zeta = 0{\cdot}54$, whilst from Fig. 5.20(b) the negative real root $-1/T_1 = -0{\cdot}60$. As a check, $\omega_n^2/T_1 = 1{\cdot}29^2 \times 0{\cdot}60 = 1{\cdot}00$ as given by Equation 5.29. The true roots are obtained simply by multiplication by $\omega_0$.

EXAMPLE 5.15.1

It is required to obtain the system pole-zero array for the third order autopilot shown in Fig. 5.21, which has a phase advance network in the feedforward path as the feedforward gain $K_g$ and the phase advance time constant $T_c$ are varied for the flight case corresponding to $T_c = 0{\cdot}035$ sec, $K_a = 335$ rad sec$^{-1}$ rad$^{-1}$ and $T_a = 0{\cdot}179$ sec. The system behaviour can then be obtained from the system pole-zero array as discussed in Chapter 3, thus allowing performance to be related back to the adjustable parameters $K_g$ and $T_e$.

The autopilot transfer function is given by

$$\frac{\phi_0}{\phi_D}(s) = \frac{G(s)}{1+G(s)H(s)}$$

where

$$G(s) = \frac{K_g K_a(1+T_e s)}{(1+T_c s)(1+T_a s)s}$$

and

$$H(s) = 1$$

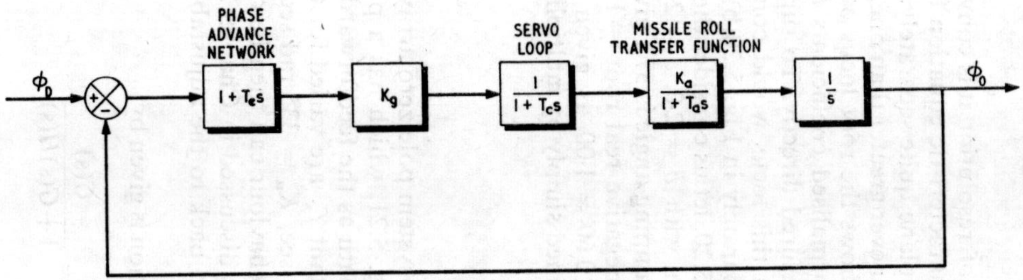


$$\frac{\phi_0}{\phi_D}(s) = \frac{1 + T_e s}{1 + \left(T_e + \frac{1}{K_g K_a}\right)s + \left(\frac{T_a + T_c}{K_g K_a}\right)s^2 + \left(\frac{T_a T_c}{K_g K_a}\right)s^3}$$

*Fig. 5.21. Block diagram of the missile autopilot with one system zero and three system poles used as Example 5.15.1*

On substitution and simplification, we obtain

$$\frac{\phi_o}{\phi_D}(s) = \frac{1+T_e s}{1+\left(T_e+\frac{1}{K_g K_a}\right)s+\left(\frac{T_a+T_c}{K_g K_a}\right)s^2+\left(\frac{T_a T_c}{K_g K_a}\right)s^3} \quad (5.32)$$

By inspection, the system zero is $-1/T_e$, and the system poles can be determined by the normalisation technique. Thus

$$\left.\begin{aligned} \omega_0 &= \left(\frac{K_g K_a}{T_a T_c}\right)^{1/3} \\ b &= \left(\frac{T_a+T_c}{K_g K_a}\right)\omega_0^2 \\ c &= \left(T_e+\frac{1}{K_g K_a}\right)\omega_0 \end{aligned}\right\} \quad (5.33)$$

After substituting the fixed parameters $T_a$, $T_c$ and $K_a$ into Equation 5.33 a suitable matrix of values of $T_e$ and $K_g$ is used to find

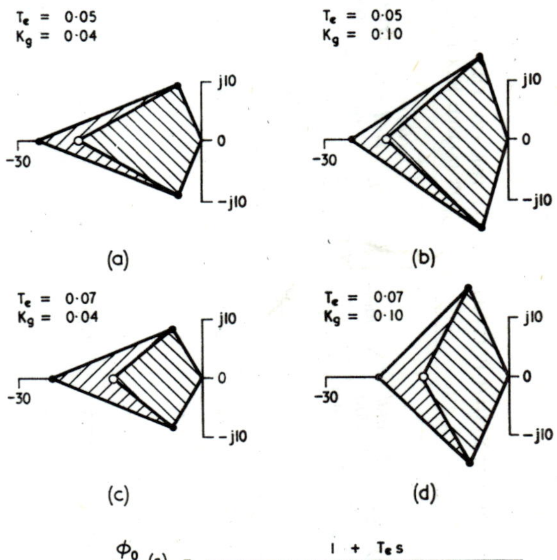


$$\frac{\phi_o}{\phi_D}(s) = \frac{1 + T_e s}{1 + \left(T_e + \frac{1}{K_g K_a}\right)s + \left(\frac{T_a+T_c}{K_g K_a}\right)s^2 + \left(\frac{T_a T_c}{K_g K_a}\right)s^3}$$

*Fig. 5.22. System pole-zero arrays for Example 5.15.1 obtained for the normalised root locus*

**Table 5.15.1**

DETERMINATION OF THE SYSTEM POLE-ZERO ARRAY FOR THE THIRD ORDER AUTOPILOT WITH TWO ADJUSTABLE PARAMETERS BY THE NORMALISED ROOT LOCUS METHOD

| *Design variable* $K_g$ | *Design variable* $T_e$ | *Normalisation frequency* $\omega_0$ | *Normalised coefficient b* | *Normalised coefficient c* | *Normalised real root* $-\frac{1}{T_1}$ | *True real root* $-\frac{\omega_0}{T_1}$ | *Damping angle* | *Damping ratio* | *Normalised undamped natural frequency* $\omega_n$ | *True undamped natural frequency* $\omega_0\omega_n$ | *True real system zero* $-\frac{1}{T_e}$ |
|---|---|---|---|---|---|---|---|---|---|---|---|
| 0·04 | 0·05 | 12·9 | 2·65 | 1·61 | −2·08 | −26·7 | 66° | 0·41 | 0·70 | 9·05 | −20 |
| 0·10 | 0·05 | 17·5 | 1·95 | 1·40 | −1·45 | −25·3 | 73° | 0·29 | 0·84 | 14·70 | −20 |
| 0·04 | 0·07 | 12·9 | 2·65 | 1·86 | −1·91 | −24·6 | 60° | 0·50 | 0·72 | 9·30 | −14·1 |
| 0·10 | 0·07 | 17·5 | 1·95 | 1·75 | −1·18 | −20·6 | 66° | 0·41 | 0·93 | 16·30 | −14·1 |

$\omega_0$, $b$ and $c$. The normalised roots are then obtained from Fig. 5.20, and multiplication by $\omega_0$ yields the true roots. Table 5.15.1 shows a typical layout for determining the system pole-zero arrays for a 2×2 matrix in $K_g$ and $T_e$, and the resultant system pole-zero arrays are shown in Fig. 5.22. This example shows that such a system may be rapidly synthesised using the normalised root locus to obtain the system poles in terms of two variables, and by using logarithmic approximations to the frequency response, a system satisfying, say, certain bandwidth and velocity constant requirements can be found very quickly without actually constructing a series of root loci with, say, $K_g$, varying along each locus for several particular values of $T_e$.

## 5.16 System Poles in the Far-off Regions of the Coefficient Plane

If attention is now turned to the areas of the coefficient plane which are outside the ordinates $b = c = 3{\cdot}0$, then if $\zeta = 1{\cdot}0$ in Equation 5.28 , two equal roots will result, and the boundaries corresponding to values of $b$ and $c$ which will result in three negative real roots will be uniquely defined. Substituting $\zeta = 1{\cdot}0$ and subsequently eliminating $T_1$ in Equation 5.29, the boundary equation in terms of $\omega_n$ is given by

$$\left.\begin{aligned} c &= \omega_n^2+\frac{2}{\omega_n} \\ b &= 2\omega_n+\frac{1}{\omega_n^2} \end{aligned}\right\} \qquad (5.34)$$

giving two boundaries corresponding to $\omega_n$ greater than and less than unity. The two boundaries meet at the cusp corresponding to three equal roots, which is the point in the coefficient plane defined by $b = 3$, $c = 3$, as shown in Fig. 5.23, stable complex roots existing in the region between these boundaries and the $bc = 1{\cdot}0$ line. Contours corresponding to constant values of $\omega_n$ and $\zeta$ in Equation 5.28 have been superimposed on Fig. 5.23 in the regions where these complex roots exist. It will be noted that the coefficient plane has been extended to the ordinates $b = c = 14$, which is certainly far enough to show the principles of pole dominance, and certainly would include virtually all practical control systems. From Fig. 5.23 it can be seen that the constant $\zeta$ contours are symmetrical about the $b = c$ line, whilst if the $\omega_n$ contours were extended $\omega_n$ would invert in value on meeting the $b = c$ line. Hence, if the coefficients $b$ and $c$ are interchanged, the damping ratio of the complex pole pair is unchanged but the undamped natural

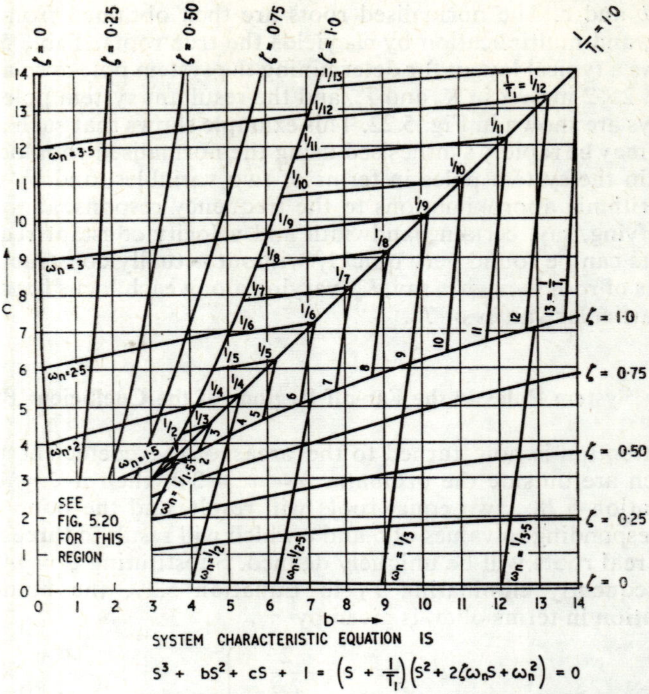


*Fig. 5.23. Coefficient plane plot of the roots of the normalised third order characteristic equation*

frequency is inverted. For example, the system defined by

$$S^3+4S^2+2S+1 = 0 \tag{5.35}$$

can be factorised, using Fig. 5.23 and the fact that $\omega_n^2/T_1 = 1$, into

$$(S+3{\cdot}51)[S^2+(2\times0{\cdot}47\times0{\cdot}53)S+0{\cdot}53^2] = 0$$

whilst the reversed symmetrical system

$$S^3+2S^2+4S+1 = 0 \tag{5.36}$$

can be similarly factorised into

$$\left(S+\frac{1}{3{\cdot}51}\right)[S^2+(2\times0{\cdot}47\times1{\cdot}89)S+1{\cdot}89^2] = 0$$

Fig. 5.24 shows the corresponding system pole arrays, and it can be seen that for conditions corresponding to Equation 5.35 the system will be dominated by the second order term, whilst for the

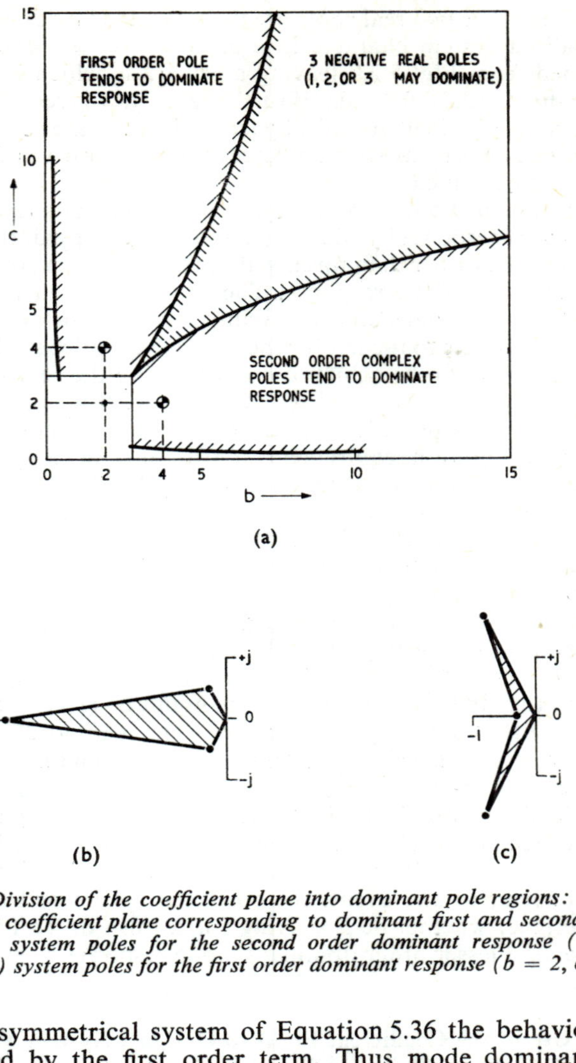


*Fig. 5.24. Division of the coefficient plane into dominant pole regions: (a) regions in the coefficient plane corresponding to dominant first and second order poles; (b) system poles for the second order dominant response ($b = 4$, $c = 2$); (c) system poles for the first order dominant response ($b = 2$, $c = 4$)*

reversed symmetrical system of Equation 5.36 the behaviour is dominated by the first order term. Thus mode dominance is achieved by a very small excursion away from the region bounded by $b = c = 3$, and the mode dominance effect becomes even more pronounced as the excursion becomes larger by making, say, $b = 6$ in Equation 5.35 or $c = 6$ in Equation 5.36.

For third order systems with three negative real poles, it is convenient to present one real root in chart form as shown in Fig.

5.23. The other two real roots are then obtained by division of the normalised system characteristic equation by the known factor obtained from Fig. 5.23, leaving a quadratic which is factorised using the standard formula. If further accuracy is required for any of the roots obtained graphically from Fig. 5.23, the roots so determined can be used as starting values for the iterative root-solving processes described in Chapter 6.

In many third order control systems it will be possible to obtain satisfactory response by allowing the complex pole pair to dominate the response, rather than forcing all three poles to make significant contributions to the response. The choice between the two approaches is essentially an engineering problem. For example, driving the third pole to the left for the missile autopilot studied earlier in the chapter can be achieved by decreasing the servo loop time constant $T_c$ from, say, 0·035 sec to 0·010 sec. If a suitable servo is available which provides this faster response without exceeding power consumption requirements, and at the same time is still physically small enough to fit into the space allocated for it, then it is fair enough to seek the dominant mode solution if this is capable of fulfilling the specification. Intuitively, however, it may be stated that often better use can be made of existing components by allowing more poles to make significant contributions to the system behaviour. This argument is partially borne out by the use of the Butterworth filter philosophy outlined in Section 1.21, which disposes uncompensated poles (those not near to a system zero) around the circle with radius $\omega_0$. It should be remembered, however, that the control system designer's aim is to produce a system which meets the specification laid down for it, and not necessarily in all cases to make the best use of components, and so unless the specification is relatively tight, there may be several possible systems which prove suitable. Economic and physical considerations, not to say ease of design, will therefore considerably influence the final choice; in any event, the coefficient plane will greatly aid the designer whether he is striving for a third order form of response or a lesser order dominant mode response.

## 5.17 Use of the Coefficient Plane to Design a System with a Load Disturbance Rejection Requirement

The coefficient plane can be used in the design of systems for which a feedforward stiffness requirement is included in the performance specification. For example, if feedforward gain, velocity constant and bandwidth are specified, then $b$, $c$ and $\omega_0$ can be determined as usual by solving simple algebraic equations. Such quantities as percentage overshoot and peak amplitude ratio can then be obtained from the coefficient plane to make sure that the specification can be met using three independent adjustable parameters without resulting in excessive values for percentage overshoot and peak amplitude ratio. In other words, the main specification of feedforward gain, velocity constant and bandwidth will require the provision of three independently adjustable parameters to meet these criteria alone, and if no other compensation is provided we shall have to accept whatever value of, say, percentage overshoot arises from the normalised coefficients $b$ and $c$ required to meet the main part of the specification. Although the percentage overshoot is not part of the fundamental specification in this case, a solution which would result in a percentage overshoot of 100% would in many cases be intuitively unacceptable, and this is immediately obvious from the coefficient plane. Specification of feedforward stiffness is met in applications where significant disturbance torques or forces are acting on the load and it is required to minimise the effect on load displacement.

In Fig. 2.24(c) it was seen that a disturbance torque acting on the missile roll channel could be considered as an equivalent deflection of the control surface for the purpose of determination of the effect of this disturbance torque on the missile roll position. A third order missile roll autopilot with three independently adjustable parameters ($K_g$, $T_r$ and $T_\alpha$), as shown in Fig. 5.9, was considered in detail in Example 3.5.1, where bandwidth and ITAE optimum form requirements were written into the specification. Fig. 5.25(a) shows the conventional block diagram for this autopilot with disturbance torque added, and Fig. 5.25(b) shows the block diagram redrawn assuming that the autopilot input $\phi_0$ and the disturbance torque $L_0$ are incorporated. From Fig. 5.25 (b) the transfer function relating disturbance torque and the output due to this disturbance torque may be written as

$$\frac{\phi_0}{L_0}(s) = \frac{G_p(s)G(s)}{1+G(s)H(s)} \tag{5.37}$$

where

$$G_p(s) = \frac{1}{L_\xi}$$

$$G(s) = \frac{K_a}{(1+T_a s)s}$$

and

$$H(s) = \frac{K_g(1+T_r s+T_\alpha s^2)}{1+T_c s}$$

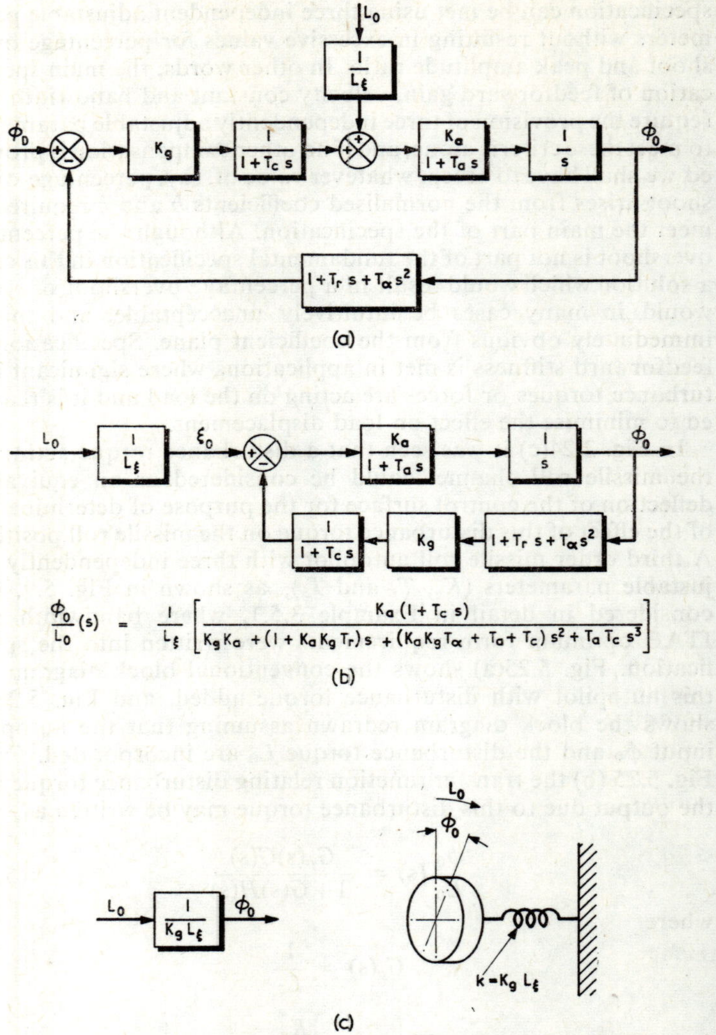


*Fig. 5.25. Stiffness requirement in a feedback control system for rejection of torque disturbances; (a) conventional block diagram; (b) redrawn block diagram assuming $\phi_0$ and $L_0$ to be uncorrelated; (c) static stiffness of the system* (s = *0*)

which, on manipulation, yields

$$\frac{\phi_0}{L_0}(s) = \frac{1}{L_\xi}\left[\frac{K_a(1+T_c s)}{K_g K_a+(1+K_a K_g T_r)s+(K_a K_g T_\alpha+T_a+T_c)s^2+T_a T_c s^3}\right] \tag{5.38}$$

where the denominator of Equation 5.38 is the system characteristic equation as expected. In the absence of further data it is customary at the design stage to assume that the disturbance torque is constant in magnitude (s = 0), and Equation 5.38 then takes the very simple form

$$\frac{\phi_0}{L_0} = \frac{1}{L_\xi K_g} \tag{5.39}$$

Now $L_\xi$ is a property of the load itself, and can therefore be assumed to be fixed, and hence to reduce the unwanted output displacement due to the disturbance torque $K_g$, the feedforward gain adjustment must be made as high as possible. Making the gain high will tend to result in an unstable system, and therefore under such circumstances considerable compensation is to be expected. Equation 5.39 can be thought of, in the static case, as a simple torsional spring as shown in Fig. 5.25 (c).

Example 5.17.1

For flight conditions corresponding to $K_a = 335$ rad sec$^{-1}$ rad$^{-1}$, $T_a = 0{\cdot}179$ sec, $T_c = 0{\cdot}035$ sec, and $L_\xi = 47{,}500$ lb ft disturbance torque per equivalent control surface deflection, design a roll autopilot to meet the following specification: bandwidth = 30 rad/sec, velocity constant = 20 sec$^{-1}$, and the static stiffness of the system to be at least 30,000 lb ft/rad in rejecting disturbance torques.

From the stiffness requirement we may write, from Equation 5.39,

$$\frac{\phi_0}{L_0} = \frac{1}{L_\xi K_g} = \frac{1}{30{,}000}$$

Hence

$$K_g = \frac{30{,}000}{47{,}500} \simeq \frac{1}{1{\cdot}60}$$

The autopilot transfer function is given by Equation 5.12 as

$$\frac{\phi_0}{\phi_D}(s) = \frac{1}{1+\left(T_r+\dfrac{1}{K_g K_a}\right)s+\left(T_\alpha+\dfrac{T_a+T_c}{K_g K_a}\right)s^2+\left(\dfrac{T_a T_c}{K_g K_a}\right)s^3}$$

Hence, in the usual way,

$$\left.\begin{aligned} a_1 &= \frac{c}{\omega_0} = T_r + \frac{1}{K_g K_a} \\ a_2 &= \frac{b}{\omega_0^2} = T_\alpha + \frac{T_a + T_c}{K_g K_a} \\ a_3 &= \frac{1}{\omega_0^3} = \frac{T_a T_c}{K_g K_a} \end{aligned}\right\} \qquad (5.40)$$

Fixing $K_g$ determines $\omega_0$ uniquely as can be seen by inspecting the $a_3$ coefficient; therefore, since $1/T_a T_c = 160$ for these flight conditions,

$$\omega_0 = \left(\frac{K_g K_a}{T_a T_c}\right)^{1/3} = \left(\frac{335 \times 160}{1{\cdot}6}\right)^{1/3} = 32{\cdot}2$$

For a unity feedback system, $C_v = 1/a_1$ when $a_0 = 1{\cdot}0$, and thus

$$a_1 = \frac{1}{C_v} = 0{\cdot}05$$

but from Equation 5.40 for $a_1$,

$$a_1 = T_r + \frac{1}{K_g K_a} = 0{\cdot}05$$

and therefore

$$T_r = 0{\cdot}05 - \frac{1}{K_g K_a} = 0{\cdot}05 - \frac{1{\cdot}6}{335} = 0{\cdot}0452$$

To determine $c$, we may write from Equation 5.40 for $a_1$

$$c = a_1 \omega_0 = 0{\cdot}05 \times 32{\cdot}2 = 1{\cdot}61$$

To determine the required value of $b$, the value of $\omega_B/\omega_0$ is determined as $30/32{\cdot}2 = 0{\cdot}93$, and $b$ can then be found from the intersection of the $\omega_B/\omega_0 = 0{\cdot}93$ contour and the $c = 1{\cdot}61$ line in Fig. 5.8, giving $b \doteq 2{\cdot}60$. Using Equation 5.40 for $a_2$, $T_\alpha$ is determined by

$$T_\alpha = a_2 - \frac{T_a + T_c}{K_g K_a} = \frac{b}{\omega_0^2} - \frac{T_a + T_c}{K_g K_a}$$

Hence

$$T_\alpha = 0{\cdot}0025 - \frac{0{\cdot}214 \times 1{\cdot}6}{335} = 0{\cdot}00148$$

Using the values of $b$ and $c$ determined from the system specification, we may write the system transfer function which satisfies the specification as either

$$\frac{\phi_o}{\phi_D}(s) = \frac{1}{1+1{\cdot}61\left(\dfrac{s}{32{\cdot}2}\right)+2{\cdot}60\left(\dfrac{s}{32{\cdot}2}\right)^2+\left(\dfrac{s}{32{\cdot}2}\right)^3} \tag{5.41}$$

or

$$\frac{\phi_o}{\phi_D}(s) = \frac{1}{1+0{\cdot}05s+0{\cdot}0025s^2+0{\cdot}0000299s^3} \tag{5.42}$$

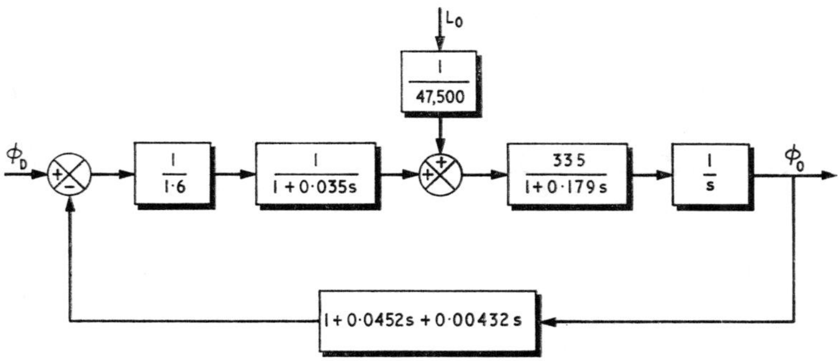


$$\frac{\phi_0}{\phi_D}(s) = \frac{1}{1+0{\cdot}05s+0{\cdot}00534s^2+0{\cdot}0000299s^2}$$

*Fig. 5.26. System meeting the specification of 30 rad/sec bandwidth, $C_v = 20$ sec$^{-1}$ and stiffness of 30,000 lb ft/rad (Example 5.17.1)*

From Fig. 5.6, the maximum percentage overshoot for this design ($b = 2{\cdot}60$, $c = 1{\cdot}61$) is found to be about 26%. The block diagram of this design is shown in Fig. 5.26.

The system pole-zero array is of some interest, and this can be determined from Fig. 5.20. Thus we may write in normalised factorised form

$$\frac{\phi_o}{\phi_D}(S) = \frac{1}{(S^2+0{\cdot}55S+0{\cdot}49)(S+2{\cdot}05)} \tag{5.43}$$

Converting to true factored form by multiplication of the roots by $\omega_0$, we obtain

$$\frac{\phi_o}{\phi_D}(s) = \frac{1}{(s^2+17{\cdot}7s+507)(s+66{\cdot}5)} \tag{5.44}$$

Thus for the quadratic term, $\omega_n = \sqrt{507} = 22{\cdot}5$, and $\zeta = 17{\cdot}7/2\omega_n = 0{\cdot}39$. Fig. 5.27 shows the corresponding system pole-zero array, and it can be seen that even for this design the third pole is situated such that it has little effect upon bandwidth and percentage overshoot, but will have a significant effect on $C_v$. The system pole-zero array is not necessary for the design process, but is included for comparison with the system pole-zero array which results from the next example, which is for the same stiffness and $C_v$ requirements as previously, but the bandwidth is

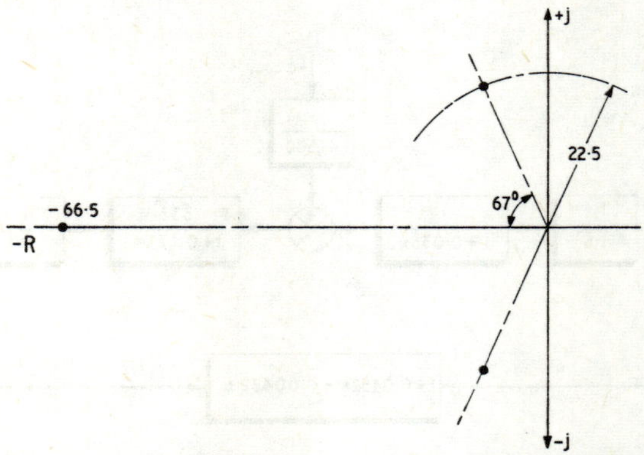


*Fig. 5.27. Pole-zero array for the system meeting the specification of 30 rad/sec bandwidth, $C_v$ = 20 sec$^{-1}$ and stiffness of 30,000 lb ft/rad (Example 5.17.1)*

reduced from 30 rad/sec to 20 rad/sec. Such a reduction in bandwidth may be considered a desirable compromise for systems where significant noise signals are present, and where a bandwidth of 20 rad/sec is such that good following of the command signal is achieved simultaneously with adequate filtering out of the noise signals. It will be seen that this moves the design point out of the third order dominant mode region bounded by $b = 3$, $c = 3$, into the region where the second order term completely dominates the response.

EXAMPLE 5.17.2

For flight conditions corresponding to $K_a = 335$ rad sec$^{-1}$ rad$^{-1}$, $T_a = 0{\cdot}179$ sec, $T_c = 0{\cdot}035$ sec and $L_\xi = 47{,}500$ lb ft disturbance torque per equivalent control surface deflection, design a roll autopilot to meet the following specification: bandwidth =

20 rad/sec, velocity constant = 20 sec$^{-1}$, and static stiffness of the system to be at least 30,000 lb ft/rad in rejecting disturbance torques.

Now, from Example 5·17.1, $\omega_0$ and $c$ are known since the feedforward gain and velocity constant requirements are unchanged. Thus $\omega_0 = 32{\cdot}2$ and $c = 1{\cdot}61$ as before, which means that $K_g$ and $T_r$ will also be unchanged. However, the normalised bandwidth is now $\omega_B/\omega_0 = 20/32{\cdot}2 = 0{\cdot}62$, which by inspection of Fig. 5.8 indicates that the design point will have moved outside the area defined by the coordinates $b = c = 3{\cdot}0$. The bandwidth is therefore derived from first principles for the known values of $c$ and $\omega_0$. In terms of the normalised coefficients we may write

$$\frac{\theta_0}{\theta_i}(S) = \frac{1}{S^3+bS^2+cS+1} \tag{5.45}$$

$\theta_0/\theta_i(S)$ has been used at this stage for generality. Substituting $S = j\omega$ in Equation 5.45, the amplitude ratio at frequency $\omega$ is defined by

$$\left|\frac{\theta_0}{\theta_i}\right|^2_\omega = \frac{1}{(1-b\omega^2)^2+(c\omega-\omega^3)^2} \tag{5.46}$$

The denominator of Equation 5.46 is equal to 2 for the bandwidth frequency, and in simplest form we may write, since $\omega = \omega_B/\omega_0$, the quadratic in $b$

$$b^2\left(\frac{\omega_B}{\omega_0}\right)^4 - 2b\left(\frac{\omega_B}{\omega_0}\right)^2 + \left(\frac{\omega_B}{\omega_0}\right)^6 + \left(\frac{\omega_B}{\omega_0}\right)^2 c^2 - 2\left(\frac{\omega_B}{\omega_0}\right)^4 c - 1{\cdot}0 = 0 \tag{5.47}$$

Substituting $\omega_B/\omega_0 = 0{\cdot}62$ and $c = 1{\cdot}61$ in Equation 5.47, the quadratic becomes

$$0{\cdot}149b^2 - 0{\cdot}772b - 0{\cdot}42 = 0$$

Hence

$$b = \frac{+5{\cdot}18 \pm \sqrt{27{\cdot}8+11{\cdot}3}}{2}$$

Since positive $b$ is required, $b = 5{\cdot}72$. The normalised equation meeting the specification is

$$\frac{\theta_0}{\theta_i}(S) = \frac{1}{S^3+5{\cdot}72S^2+1{\cdot}61S+1} \tag{5.48}$$

Now $a_2 = b/\omega_0^2 = 5{\cdot}72/32{\cdot}2^2 = 0{\cdot}00534$, and the new value of $T_\alpha$ is given by

$$T_\alpha = a_2 - \frac{T_a+T_c}{K_gK_a} = 0{\cdot}00534 - \frac{0{\cdot}214\times1{\cdot}6}{335} = 0{\cdot}00432$$

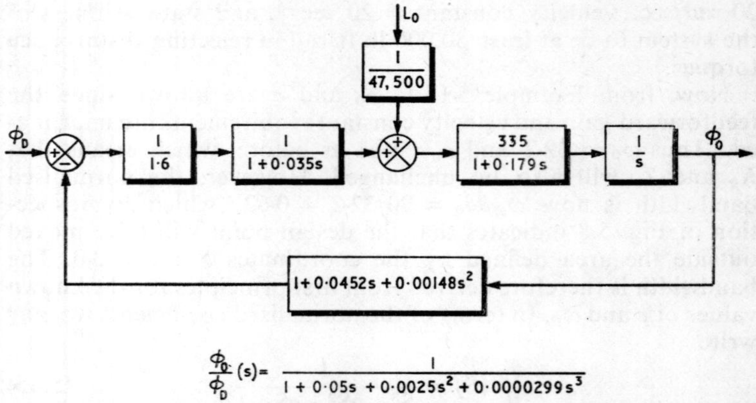


*Fig. 5.28. System meeting the specification of 20 rad/sec bandwidth, $C_v$ = 20 sec$^{-1}$ and stiffness of 30,000 lb ft/rad (Example 5.17.2)*

The system meeting this particular specification is shown in Fig. 5.28; the difference between this system and that shown in Fig. 5.26 is in the $T_\alpha$ term only. This difference in $T_\alpha$ is required to increase $b$ from 2·60 to 5·72. Using Fig. 5.23, Equation 5·48 may be factorised into the normalised system poles

$$\frac{\phi_0}{\phi_D}(S) = \frac{1}{(S^2+0{\cdot}26S+0{\cdot}18)(S+5{\cdot}46)} \qquad (5.49)$$

and, since $\omega_0 = 32{\cdot}2$, the true roots are defined by

$$\frac{\phi_0}{\phi_D}(s) = \frac{1}{(s^2+8{\cdot}32s+187)(s+176)} \qquad (5.50)$$

giving for the second order term $\omega_n = \sqrt{187} = 13{\cdot}7$, and $\zeta = 8{\cdot}32/2\omega_n = 0{\cdot}305$.

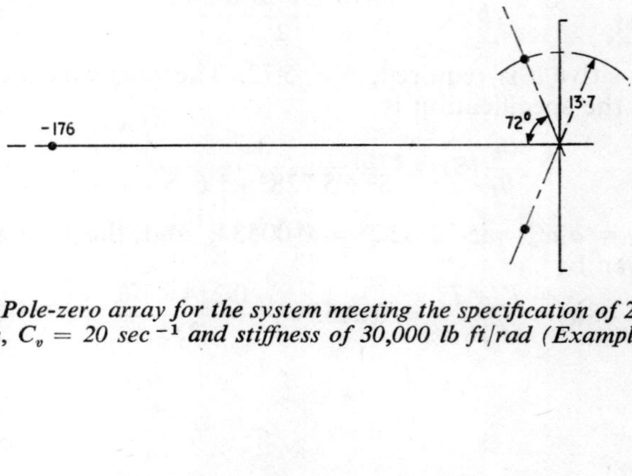


*Fig. 5.29. Pole-zero array for the system meeting the specification of 20 rad/sec bandwidth, $C_v$ = 20 sec$^{-1}$ and stiffness of 30,000 lb ft/rad (Example 5.17.2)*

The resultant system pole-zero array is shown in Fig. 5.29, and it can be seen that the third pole is well to the left of the s plane. Essentially the maximum percentage overshoot will be defined by the second order term with damping ratio 0·305, and this may be determined (Fig. 4.10 of Chapter 4) to be approximately 35%. We may similarly use Fig. 4.11 to check the system bandwidth. For a type (0,2) system with $\zeta = 0{\cdot}305$, $\omega_B/\omega_n = 1{\cdot}45$,

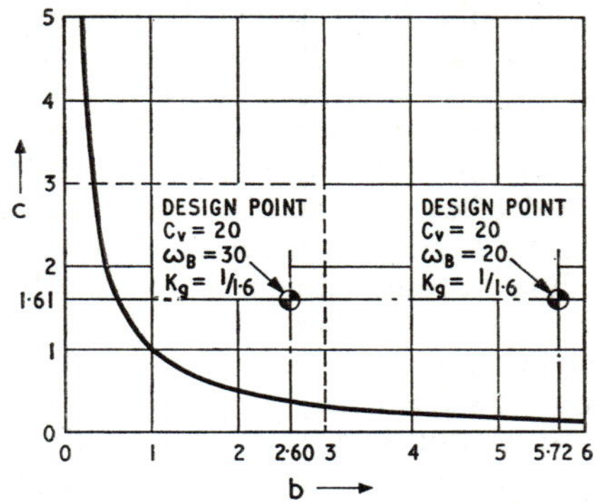


*Fig. 5.30. Comparison of design points for Examples 5.17.1 and 5.17.2*

hence $\omega_B = 1{\cdot}45\omega_n = 1{\cdot}45 \times 13{\cdot}7 = 20$ rad/sec, which is the required answer. The true transfer function for this example is

$$\frac{\phi_0}{\phi_D}(s) = \frac{1}{1+1{\cdot}61\left(\dfrac{s}{32{\cdot}2}\right)+5{\cdot}72\left(\dfrac{s}{32{\cdot}2}\right)^2+\left(\dfrac{s}{32{\cdot}2}\right)^3} \tag{5.51}$$

or

$$\frac{\phi_0}{\phi_D}(s) = \frac{1}{1+0{\cdot}05s+0{\cdot}00534s^2+0{\cdot}0000299s^3} \tag{5.52}$$

To compare the two designs corresponding to Example 5.17.1 and Example 5.17.2 in the coefficient planes, Fig. 5.30 has been drawn. The increase in percentage overshoot which occurs when the bandwidth is reduced from 30 rad/sec to 20 rad/sec arises because the specification of $C_v$ and feedforward stiffness determines $\omega_0$ and $c$ absolutely, and bandwidth can only be reduced by movement of the design point to the right of the coefficient plane along the $c$ = constant line, which movement in this region increases the maximum percentage overshoot.

From Equation 1.11 we would expect the hydraulic servo lag to form a system zero when considering load disturbance rejection, since the load disturbance rejection transfer function is the command signal following transfer function divided by the transfer function between the points of injection of the command and disturbance signals. In the terminology of Fig. 1.12, this transfer function is $G_1(s)G_2(s)$. Now, in general, $G_1(s)$ and/or $G_2(s)$ will be compensation terms added to obtain the desired system performance. If $G_1(s)$ and/or $G_2(s)$ contain complex zeros introduced to cancel with undesirable complex poles occurring in $G_3(s)$ (a design technique to be outlined in Section 6.11), these complex zeros will become complex poles in the load disturbance transfer function. We therefore expect the disturbance response to be highly oscillatory, since the compensation zeros are required to be lightly damped in many instances in order to achieve effective cancellation. This is indeed often so, and if a stringent disturbance response requirement is to be met under these conditions, the compensation procedure becomes much more complicated and it is necessary to use the special approach developed by Truxal.[50]

In both the above examples it was assumed that the disturbance signals were of very low frequency, thus allowing the performance specification to be written in terms of the stiffness requirement $K_g L_\xi$. It is of interest to compare the transient responses of the two systems in the disturbance rejection mode. From Equation 5.38 we may write the output to the disturbance transfer function as

$$L_\xi K_g \left[\frac{\phi_0}{L_0}(s)\right] = \frac{1+T_c s}{1+\left(T_r+\dfrac{1}{K_g K_a}\right)s+\left(T_\alpha+\dfrac{T_a+T_c}{K_g K_a}\right)s^2+\left(\dfrac{T_a T_c}{K_g K_a}\right)s^3} \quad (5.53)$$

This transfer function has the same poles as the output to command signal transfer function defined in Equation 5.11, but there is a system zero at $-1/T_c$. The effect of this zero is to liven up the transient response in the disturbance rejection mode. Since for both examples the third pole is well to the left of the root plane (Figs. 5.27 and 5.29), we may neglect the effect of this pole on the maximum percentage overshoot. We may therefore compute $\alpha$ for the remnant type (1,2) system and determine the maximum percentage overshoot directly from Fig. 4.10. The value of $\alpha$ is calculated from Equation 4.27 for the known pole-zero geometry. For Example 5.17.1, we have

$$\alpha = \frac{\omega_n}{2\zeta(-z_1)} = \frac{22{\cdot}5}{2\times 0{\cdot}39\times 28{\cdot}6} = 1{\cdot}00$$

and using Fig. 4.10, for $\zeta = 0{\cdot}39$, $\alpha = 1{\cdot}00$, the maximum percentage overshoot = 37%. Similarly, for Example 5.17.2,

$$\alpha = \frac{\omega_n}{2\zeta(-z_1)} = \frac{13{\cdot}7}{2\times 0{\cdot}305\times 28{\cdot}6} = 0{\cdot}79$$

whence the maximum percentage overshoot is 40%. Thus the system with lower bandwidth has a larger maximum percentage overshoot for both command signal following and load disturbance rejection, but will, in general, reject high-frequency noise more readily. Table 5.17.1 summarises the characteristics of the two systems.

**Table 5.17.1**

SUMMARY OF COMMAND SIGNAL FOLLOWING AND LOAD DISTURBANCE REJECTION CHARACTERISTICS FOR EXAMPLES 5.17.1 AND 5.17.2

| *System* | *Command signal following characteristics* | | | *Disturbance rejection characteristics* | | |
|---|---|---|---|---|---|---|
| | $\omega_B$ (rad/sec) | $C_v$ (sec$^{-1}$) | *P.O.* | $[\phi_0/L_0]_{s=0}$ (rad/lb ft) | $\alpha$ {*negl. 3rd pole* | *P.O.* |
| Example 5.17.1 | 30 | 20 | 26% | 1/30,000 | 1·00 | 37% |
| Example 5.17.2 | 20 | 20 | 35% | 1/30,000 | 0·79 | 42% |

## 5.18 Conclusion

The normalised coefficient plane is a very powerful tool, not only in the analysis but also in the synthesis of third order linear and quasilinear systems. Many third order systems can be designed simply by inspecting the coefficient plane in the region where the type of response is acceptable. For three-pole non-zero systems, the region where all three roots are more or less significant is approximately bounded by the coordinates $b = c = 3{\cdot}0$. Regions also exist where the response is dominated by a single pole, and where the response is dominated by a pair of complex poles. If the normalised coefficients are obtained for these systems, the approximations of third to first and third to second order systems become obvious. $M_p$, $\omega_B$ and percentage overshoot contours have been provided for both unity numerator and zero velocity error systems, whilst for the general third order system where $C_v$ is finite but is modified by the existence of a system zero, generalised root charts have been provided to assist in the determination of the system pole-zero array in terms of adjustable parameters.

An outstanding advantage of the coefficient plane is that for the unity numerator and zero velocity error systems it is unnecessary to draw the system root locus in order to study system performance. Also two parameter variations can be studied simultaneously, thus a *set* of loci is rendered redundant, since no approximations in $M_p$, percentage overshoot and $\omega_B$ are involved, and the system is not forced into a second order dominant system merely so that the designer can handle the design easily. For unity numerator systems it is found that for major loop applications the region around $b = c = 2{\cdot}0$ is useful, whilst for hydraulic servos acting inside a major loop, $bc = 1{\cdot}5$ in the region of $b = 0{\cdot}5$ will give an acceptable response, since the lightly damped oscillatory mode will be filtered out by the major loop itself.

Systems with three negative real poles can be designed very simply using the root locus method, since the only labour required is in the calibration of the locus. Often with such systems it is found that the requirement of three negative real poles means that one of the system poles does not migrate very far from the open loop position, and it is therefore possible to use the open loop pole as the first guess in an iteration process to evaluate the root numerically. Under such circumstances the solution may converge so rapidly that direct numerical methods are to be preferred; suitable numerical methods are discussed in Chapter 6. General methods applicable to the design of systems with three poles and one zero (not positioned so as to make $C_v = \infty$) involving s plane manipulation of the root locus will also be described in Chapter 6.

CHAPTER 6

# General Application of Transfer Function Techniques in the Analysis and Synthesis of Control Systems

## 6.1 Introduction

In this chapter the general problem of designing a system of any order to meet a given specification will be discussed in detail. Particular emphasis will be placed on those analysis and synthesis techniques which lead directly to the relationships between the system transfer function, the performance specification, and the fixed parameters of the control system. This chapter is therefore an extension of Chapter 3, which indicated how the root locus technique can be used for system design in the root or s plane.

Low-order control systems, particularly second and third order, can often be designed uniquely from three independent performance criteria such as velocity constant, bandwidth and percentage overshoot, since the techniques of Chapters 4 and 5 determine the system transfer function which can be interpreted in some preferred way to obtain the compensation terms required. For given load properties, a choice of compensation terms capable of forcing the system to have this transfer function may exist, and the designer may then choose between these alternative methods of compensation by considering such factors as noise rejection and load disturbance rejection in addition to the specified performance in following the command signal, or may altenatively decide on a particular design which may be preferable due to the availability of particular components. Of course, certain restrictions do apply to the type of system which is capable of meeting a particular specification, and if a low ratio of $\omega_B/C_v$ is sought, then it is normally necessary to make sure that one of the degrees of design freedom provided will result in the generation of a suitable system zero.

For higher order systems, the designer must achieve a reasonable balance between the complexity of system he can conveniently handle, and the best use of available components. The requirement for keeping the design process reasonably straightforward has led to the so-called 'dominant mode' approach in which the designer endeavours to make sure that system performance is dominated by a few system poles and zeros. Using this approach, the design problem is: *given the load characteristics and the system specification, to introduce suitable compensation in the feedforward, or feedback path, or both, to force the system pole-zero array into such a geometry that the dominant poles and zeros more or less fulfil the specification whilst the remaining system poles and zeros have been forced into such a position that they may be ignored, either due to their position in the* s *plane, or due to the relatively close proximity of a system pole and system zero.* As mentioned previously, and shown in Chapter 3, the far-off poles are more significant in their effect on $C_v$ than on $\omega_B$ or on the percentage overshoot and it may be necessary to allow for this when studying the dominant poles and zeros. For example, if a fifth order system has a dominant pair of system poles and a dominant system zero, and the desired $C_v$ is 50, then it may be considered necessary to ensure that $C_v$ of the dominant poles and zeros is at least 70 to counteract the effect of the far-off poles. However, one of the design techniques to be outlined will automatically constrain $C_v$ to have the desired value, at the same time fixing the dominant poles and zeros to meet approximately the bandwidth and percentage overshoot requirements.

An alternative design procedure is to endeavour, for a given order of system, to adjust the coefficients of the system transfer function to give a system which will minimise the ITAE criterion, provided that the form of response resulting from ITAE optimum transfer function is suitable for the particular application. However, such a design procedure requires that an excessive number of independently adjustable parameters are available, whereas in practice it is relatively rare to find a system with more than three independent parameters which are directly under the control of the system designer. Furthermore, the presence of non-linearities may impose limits on possible values of a least one of these parameters.

Provided the correct number of adjustable parameters is available, the design of an ITAE optimum system reduces to simple arithmetic once the system transfer function has been determined algebraically in terms of component values. Examples of the very simple design procedure involved are given in Chapters 4 and 5. The ITAE optimum system approach is therefore extremely useful for low-order systems, but for high-order systems it should be looked upon as an alternative to the dominant mode approach under circumstances where the latter approach would lead to a wasteful use of components, perhaps by choosing an excessively large or expensive motor. Even then, the designer should realise that the ITAE optimum pole-zero arrays are indications of suitable general areas in which the system poles and zeros may be placed as well as is practicable with the limited number of adjustable parameters at his disposal. For convenience, the ITAE optimum pole-zero arrays are always defined in terms of the normalised system transfer function of a particular order. Graham and Lathrop have shown that these system pole-zero arrays provide reasonably competitive systems when true time computation of the ITAE is undertaken. Nevertheless, for a given load configuration it may be possible to design a dominant mode system whose true-time ITAE is not much larger than that obtained from an alternative practical design with system poles and zeros satisfying ITAE optimum system geometry.

The design of high-order systems to achieve system specification, i.e. the determination of a suitable transfer function, is therefore a problem of deciding on the type of compensation which will force the root locus into one of perhaps a possible number of suitable geometries so that a satisfactory system pole-zero array can be chosen. This chapter is therefore mainly devoted to an exposition of various s plane compensation techniques which will allow the designer to choose the form of compensation most suited to his particular problem, but it will commence with a discussion on certain numerical methods which will provide a useful adjunct to the root locus technique. Emphasis will be placed on $C_v$, $\omega_B$ and percentage overshoot as performance criteria, but some consideration will also be given to load disturbance rejection.

## 6.2 Use of Synthetic Division to Find Real Roots of Polynomials

It is often desirable to evaluate rapidly the roots of specific polynomials. These numerically derived roots can then be used to check root loci, determine roots more accurately if the root locus values are given as the first trial, evaluate roots for simultaneous changes in two or more system parameters and, finally, given the actual system transfer function from a particular dynamic test procedure as the ratio of two polynomials, the system poles and zeros may be evaluated directly.

It will be found particularly convenient to solve polynomials which have been transformed using the normalisation technique already outlined in Chapters 1 and 5, as this procedure will in general avoid the use of very large, or very small, numbers in the computational process. The normalisation technique is not obligatory, of course, but we shall use the normalised equations when illustrating the use of synthetic division to determine real roots, and also when illustrating Lin's method for evaluating complex roots. A method of using synthetic division to evaluate complex roots has been evolved by Oldenburger,[51] but will not be discussed here.

The general $n^{\text{th}}$ order polynomial in s may be written as

$$a_n s^n + a_{n-1} s^{n-1} + \ \dots \ a_1 s + 1 = 0 \tag{6.1}$$

and, using the normalisation technique, we may write

$$s = \omega_0 S \tag{6.2}$$

where

$$\omega_0 = \left(\frac{1}{a_n}\right)^{1/n}$$

and transform Equation 6.1 into the normalised form

$$S^n + a'_{n-1} S^{n-1} + a'_{n-2} S^{n-2} + \ \dots \ a'_1 S + 1 = 0 \tag{6.3}$$

where

$$a'_{n-r} = \omega_0^{n-r} \, a_{n-r} \tag{6.4}$$

This leads to an alternative representation of Equation 6.3 as follows

$$S^n + a_{n-1}\omega_0{}^{n-1} S^{n-1} + a_{n-2}\omega_0{}^{n-2} S^{n-2} + \ \dots \ a_1\omega_0 S + 1 = 0 \tag{6.5}$$

i.e. the order of $\omega_0$ relating to a coefficient is the order of S.

If $n$ is odd, then Equation 6.3 must have at least one real root, and if this equation is the system characteristic equation, a necessary condition for stability is that every coefficient of the polynomial must be positive. Now, the $S^{n-1}$ coefficient represents the sum of the normalised roots taken one at a time, and hence if there is a large real root present it will contribute nearly all of the $S^{n-1}$ coefficient. Thus if a large real root exists, in normalised coefficients we may write

$$S^n + a'_{n-1} S^{n-1} \simeq 0$$

and hence

$$S \simeq -a'_{n-1} \tag{6.6}$$

In true coefficient form this would give a true root defined by

$$s \simeq -\frac{a_{n-1}}{a_n}$$

If a small root exists, then all terms involving $S^2$, etc. may be neglected as a first approximation, and the system normalised equation may be written as

$$a_1'S+1 \simeq 0$$

and hence

$$S \simeq -\frac{1}{a_1'} \tag{6.7}$$

In true coefficient form, assuming $a_0 = 1$, this would give a true root defined by

$$s \simeq -\frac{1}{a_1}$$

These values for large or small roots can be used as first trial values in any iteration process which is then used to determine the roots more accurately. One such iterative process involves the use of synthetic division, which is considerably quicker than the normal method of division. Two synthetic division formats are possible. The so-called left-to-right format is used to obtain small real roots, whilst the right-to-left format is used to obtain large roots.[51]

EXAMPLE 6.2.1

As an example of *left-to-right synthetic division*, let us consider the normalised polynomial

$$S^3+2S^2+4S+1 = 0 \tag{6.8}$$

which is the polynomial studied in Chapter 5 as Equation 5.36. By inspection of Equation 6.8 we see that the S coefficient is large, indicating a small root, and the first approximate root will be $-1/4 = -0{\cdot}25$. Note that the true magnitude of the root will be larger than $-0{\cdot}25$, and we could take our first trial root numerically larger than $-0{\cdot}25$, say $-0{\cdot}30$, if desired. However, for generality we shall develop the left-to-right synthetic division process from a trial root of $-0{\cdot}25$. The equation is written in descending order of S, and the first computation is as follows.

| $S^3$ | $S^2$ | $S^1$ | $S^0$ | |
|---|---|---|---|---|
| 1 | 2 | 4 | 1 | 0·25 |
| | $(0{\cdot}25\times 1)$ ↗ | $(0{\cdot}25\times 1{\cdot}75)$ ↗ | $(0{\cdot}25\times 3{\cdot}56)$ ↗ | |
| 1 | 1·75 | 3·56 | 0·11 | |
| | | ↑ Inverse of next trial value | ↑ Remainder | |

It is not necessary to write in the S line as shown, which is put in for clarity and plays no part in the computation. The root, with sign reversed, is written on the right-hand side as shown. The number 1 is brought down to the bottom line, and is then multiplied by 0·25 to form the entry in the second working line of the second column. The product $0{\cdot}25\times 1 = 0{\cdot}25$ is then subtracted from the $S^2$ coefficient, leaving $2-0{\cdot}25 = 1{\cdot}75$ to be inserted in the bottom line. 1·75 is then multiplied by 0·25 and the product 0·44 is inserted in the second line of the third column, whence it is subtracted from the S coefficient, leaving $4-0{\cdot}44 = 3{\cdot}56$ to be inserted below in the bottom line. Finally 3·56 is multiplied by 0·25, and the product 0·89 is inserted in the second line of the fourth column and is subtracted from the $S^0$ coefficient, leaving $1-0{\cdot}89 = 0{\cdot}11$ to be inserted in the fourth column, bottom line.

Points to note are that 0·11 is the remainder, and would, of course, be zero if the trial value chosen is indeed a root, and secondly that if the remainder is non-zero, a suitable next trial value is obtained by inverting the number found on the bottom line of the $S^1$ column. Thus in this case a suitable second trial value is given by $S = -1/3{\cdot}56 = -0{\cdot}28$, giving the second trial divisor as $+0{\cdot}28$. Considerable arithmetical rounding off is allowable, particularly in the early stages of the iteration process, and normally slide-rule accuracy is sufficient. If the roots are required particularly accurately, a hand calculating machine can be used for the last one or two iterations only. Using the new trial value, the second computation is

| | | | | |
|---|---|---|---|---|
| 1 | 2 | 4 | 1 | 0·28 |
| | 0·28 | 0·48 | 0·99 | |
| 1 | 1·72 | 3·52 | 0·01 | |

The remainder of 0·01 is considered negligible for this application, and therefore only two iterations are required for this particular example. Thus $S+0{\cdot}28$ is a factor of Equation 6.8 and, furthermore, the quadratic factor is given by the first three terms of the bottom line of the computation, i.e. by the expression $S^2+1{\cdot}72S+3{\cdot}52$. The factored terms of Equation 5.8 are therefore as follows:

$$(S+0{\cdot}28)(S^2+1{\cdot}72S+3{\cdot}52) = 0 \tag{6.9}$$

EXAMPLE 6.2.2

As an example of *right-to-left synthetic division*, let us consider the normalised polynomial

$$S^3+4S^2+2S+1 = 0 \tag{6.10}$$

which is the polynomial studied in Chapter 5 as Equation 5.35 and is, in fact, Equation 6.8 with the coefficients of $S^2$ and $S^1$ interchanged. By inspection of Equation 6.10 we see that the $S^2$ coefficient is relatively large compared with the $S^1$ coefficient and therefore a large root is suspected. The magnitude of this large root is given approximately by Equation 6.6 as $-4$. The true magnitude of this root will be numerically somewhat less than $-4$ as the other roots will make a small, but finite, contribution to the $S^2$ coefficient. However, again for generality, we shall develop the right-to-left synthetic division process from a trial root of $-4$. The equation is written in descending order of S, and the first computation is as follows:

| | $S^3$ | $S^2$ | $S^1$ | $S^0$ | 4 |
|---|---|---|---|---|---|
| | 1 | 4 | 2 | 1 | |
| Remainder → | $\frac{7}{64}$ Next trial → | $3\frac{9}{16}$ | $1\frac{3}{4}$ | 1 | |
| | $3\frac{9}{16}/4$ | $1\frac{3}{4}/4$ | $\frac{1}{4}$ | 0 | |

The fractions have been retained merely to show the method of computation from one column to the next. In practice, of course, decimals would be used straightaway. The procedure is started from the right, and unity is written in the second working line of the $S^0$ column. The trial root, with sign reversed, is written on the right-hand side, and this is divided into unity and written in the bottom line giving unity divided by 4, i.e. $\frac{1}{4}$. The $\frac{1}{4}$ is subtracted from the $S^1$ coefficient, giving $2-\frac{1}{4} = 1\frac{3}{4}$, and this is written in the second line of the second column. This number is itself divided by the trial root with sign reversed, giving $1\frac{3}{4}/4$, and this is written in the bottom line of the $S^2$ column, and then subtracted from the $S^2$ coefficient, giving $4-\frac{7}{16} = 3\frac{9}{16}$, which is written in the second line of the $S^2$ column. Finally $3\frac{9}{16}$ is divided by 4 and written in the bottom line of the $S^3$ column. The remainder is then found by subtracting $3\frac{9}{16}/4$ from the $S^3$ coefficient leaving, in this instance, $1-\frac{57}{64} = \frac{7}{64}$ which is inserted in the second line of the $S^3$ column.

For a true root value the remainder will be zero. If the remainder is non-zero, the next trial divisor is given by the number in the second line of the $S^2$ column. Thus it is not strictly necessary to evaluate the remainder when using the right-to-left synthetic division process as the number in the second line of the $S^{n-1}$ column will be equal to the trial divisor when the trial divisor is a true root.

Continuing with the iteration, the second trial divisor is $3\frac{9}{16} \triangleq 3{\cdot}56$, and the computation is

| | | | | |
|---|---|---|---|---|
| 1 | 4 | 2 | 1 | 3·56 |
| | 3·52 | 1·72 | 1 | |
| | 0·48 | 0·28 | 0 | |

The third trial divisor is thus 3·52, which leads to the following computation:

| | | | | |
|---|---|---|---|---|
| 1 | 4 | 2 | 1 | 3·52 |
| | 3·52 | 1·72 | 1 | |
| | 0·48 | 0·28 | 0 | |

Within the accuracy required for this particular application $-3{\cdot}52$ is a root of Equation 6.10, the second order term of which can be obtained directly from the bottom line of the synthetic division process as $S^2+0{\cdot}48S+0{\cdot}28$. Equation 6.10 may therefore be written in factored form as

$$(S+3{\cdot}52)(S^2+0{\cdot}48S+0{\cdot}28) = 0 \qquad (6.11)$$

In a higher order system the normal procedure is to remove the real roots one at a time using synthetic division, and then to use Lin's method to evaluate the remaining quadratic roots should two or more pairs of complex roots exist.

## 6.3 Lin's Method for Finding Complex Roots of Polynomials[52]

Lin's method can be used to determine real roots of a polynomial as an alternative to synthetic division, but the main virtue in Lin's method lies in its ability to determine complex roots. The further apart are the roots, the more quickly will the method converge, thus offering distinct advantages if the dominant mode philosophy is used.

Lin's method requires that the polynomial be written with $a_n = 1{\cdot}0$ at least, but we shall use the normalisation technique again, which will result in first and last coefficients equal to unity. To illustrate the method we shall apply it to the fourth order unity numerator ITAE optimum system which was initially discussed in Section 1.20. The roots form two complex pairs and are not particularly far apart, so that the solution will not converge particularly rapidly.

EXAMPLE 6.3.1

To determine the roots for the normalised ITAE optimum unity numerator fourth order system using Lin's method.

The normalised characteristic equation is, from Table 1.20.1,

$$S^4+2{\cdot}1S^3+3{\cdot}4S^2+2{\cdot}7S+1 = 0 \qquad (6.12)$$

The first trial using Lin's method, which is looking for the smallest complex pair, is to form a trial divisor from the last three terms of Equation 6.12. Thus the first trial divisor is the factor

$$3{\cdot}4S^2+2{\cdot}7S+1 = 0$$

which for our purpose is more conveniently written as

$$S^2+0{\cdot}8S+0{\cdot}30 = 0 \qquad (6.13)$$

since we can round off our trial divisors as required. This trial divisor defined by Equation 6.13 is divided into the polynomial equation in the standard manner as follows:

$$\begin{array}{rl}
 & S^2+1{\cdot}3S\ +2{\cdot}06 \\
S^2+0{\cdot}8S+0{\cdot}30) & S^4+2{\cdot}1S^3+3{\cdot}4S^2+2{\cdot}7S+1 \\
 & S^4+0{\cdot}8S^3+0{\cdot}30S^2 \\
 & \quad 1{\cdot}3S^3+3{\cdot}10S^2 \\
 & \quad 1{\cdot}3S^3+1{\cdot}04S^2+0{\cdot}39S \\
\text{Second trial divisor} \rightarrow & \qquad 2{\cdot}06S^2+2{\cdot}31S+1 \\
 & \qquad 2{\cdot}06S^2+1{\cdot}64S+0{\cdot}62 \\
\text{Remainder} \rightarrow & \qquad 0{\cdot}67S+0{\cdot}38
\end{array}$$

Since the remainder is non-zero, the second trial divisor is obtained from the $(n-1)$th remainder as shown. The second trial divisor is, therefore,

$$2{\cdot}06S^2+2{\cdot}31S+1 = 0$$

which is more conveniently written as

$$S^2+1{\cdot}12S+0{\cdot}49 = 0 \tag{6.14}$$

and the second computation is as follows:

$$\begin{array}{r}
S^2+0{\cdot}98S+1{\cdot}81 \\ \hline
S^2+1{\cdot}12S+0{\cdot}49)\,S^4+2{\cdot}1S^3+3{\cdot}4S^2+2{\cdot}7S+1 \\
S^4+1{\cdot}12S^3+0{\cdot}49S^2 \\ \hline
0{\cdot}98S^3+2{\cdot}91S^2+2{\cdot}7S \\
0{\cdot}98S^3+1{\cdot}10S^2+0{\cdot}48S \\ \hline
\text{Third trial divisor} \rightarrow 1{\cdot}81S^2+2{\cdot}22S+1 \\
1{\cdot}81S^2+2{\cdot}03S+0{\cdot}89 \\ \hline
0{\cdot}19S+0{\cdot}11
\end{array}$$

The third trial divisor is

$$1{\cdot}81S^2+2{\cdot}22S+1 = 0$$

which is rewritten as

$$S^2+1{\cdot}23S+0{\cdot}55 = 0 \tag{6.15}$$

and the third computation is

$$\begin{array}{r}
S^2+0{\cdot}87S+1{\cdot}78 \\ \hline
S^2+1{\cdot}23S+0{\cdot}55)\,S^4+2{\cdot}1S^3+3{\cdot}4S^2+2{\cdot}7S+1 \\
S^4+1{\cdot}23S^3+0{\cdot}55S^2 \\ \hline
0{\cdot}87S^3+2{\cdot}85S^2+2{\cdot}7S \\
0{\cdot}87S^3+1{\cdot}07S^2+0{\cdot}48S \\ \hline
\text{Fourth trial divisor} \rightarrow 1{\cdot}78S^2+2{\cdot}22S+1 \\
1{\cdot}78S^2+2{\cdot}19S+0{\cdot}98 \\ \hline
0{\cdot}03S+0{\cdot}02
\end{array}$$

The remainder is now relatively small, but carrying the iteration one stage further, we have the fourth trial divisor

$$1{\cdot}78S^2+2{\cdot}22S+1 = 0$$

which, on dividing through by 1·78, becomes

$$S^2+1{\cdot}25S+0{\cdot}56 = 0 \tag{6.16}$$

The fourth computation is therefore

$$
\begin{array}{r l}
 & S^2+0{\cdot}85S+1{\cdot}78 \\
\cline{2-2}
S^2+1{\cdot}25S+0{\cdot}56) & S^4+2{\cdot}1S^3+3{\cdot}4S^2+2{\cdot}7S+1 \\
 & S^4+1{\cdot}25S^3+0{\cdot}56S^2 \\
\cline{2-2}
 & \quad 0{\cdot}85S^3+2{\cdot}84S^2+2{\cdot}7S \\
 & \quad 0{\cdot}85S^3+1{\cdot}06S^2+0{\cdot}48S \\
\cline{2-2}
 & \qquad 1{\cdot}78S^2+2{\cdot}22S+1 \\
 & \qquad 1{\cdot}78S^2+2{\cdot}22S+1
\end{array}
$$

There is no remainder within the accurary required for this application, and hence the system poles are

$$(S^2+1{\cdot}25S+0{\cdot}56)(S^2+0{\cdot}85S+1{\cdot}78) = 0 \qquad (6.17)$$

giving for the first term an undamped natural frequency of 0·75 normalised rad/sec and a damping ratio of 0·83, and giving for the second term an undamped natural frequency of 1·33 normalised rad/sec and a damping ratio of 0·31. Being a unity numerator system there are no system zeros, and hence the system pole-zero array is that shown in Fig. 1.30.

As a further example of the use of Lin's method we shall consider the problem of checking a point on a root locus which has been constructed graphically. The example chosen is the fourth order missile autopilot studied in Chapter 3. For convenience the block diagram of the autopilot is repeated and shown in Fig. 6.1. Using the normal rules for transfer function manipulation, the system transfer function is written for convenience as

$$\frac{\phi_0}{\phi_D}(s) = \frac{a_0}{a_0+a_1s+a_2s^2+a_3s^3+a_4s^4} \qquad (6.18)$$

where

$$
\left.
\begin{aligned}
a_0 &= \frac{K_gK_a\omega_{ns}^2}{T_a} \\
a_1 &= \frac{K_gK_a\omega_{ns}^2T_r+\omega_{ns}^2}{T_a} \\
a_2 &= \frac{2\zeta_s\omega_{ns}}{T_a}+\omega_{ns}^2 \\
a_3 &= \frac{1}{T_a}+2\zeta_s\omega_{ns}
\end{aligned}
\right\} \qquad (6.19)
$$

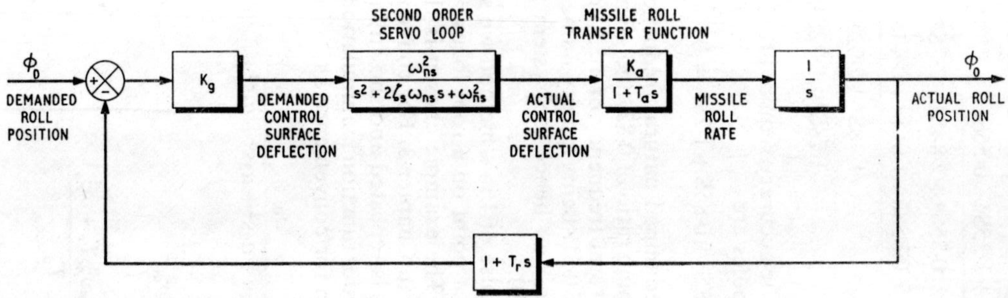


SYSTEM TRANSFER FUNCTION IS

$$\frac{\phi_0}{\phi_D}(s) = \frac{\dfrac{K_g K_a \omega_{ns}^2}{T_a}}{\dfrac{K_g K_a \omega_{ns}^2}{T_a} + \left(\dfrac{K_g K_a \omega_{ns}^2 T_r + \omega_{ns}^2}{T_a}\right)s + \left(\dfrac{2\zeta_s \omega_{ns}}{T_a} + \omega_{ns}^2\right)s^2 + \left(\dfrac{1}{T_a} + 2\zeta_s \omega_{ns}\right)s^3 + s^4}$$

*Fig. 6.1. Block diagram of the rate feedback stabilised fourth order missile roll autopilot used as Example 6.3.2*

Note that we have not made $a_0 = 1$ but this does not matter since we are going to normalise Equation 6.18 anyway. It is convenient for this example to deal with the relatively large numbers which arise when written in this form, rather than with the relatively small numbers which will arise if Equation 6.18 is divided top and bottom by $K_g K_a \omega_{ns}^2 / T_a$.

EXAMPLE 6.3.2

For the fourth order missile autopilot shown in Fig. 6.1, determine the system pole-zero array for the flight case corresponding to $K_a = 335$ rad sec$^{-1}$ rad$^{-1}$, $T_a = 0{\cdot}179$ sec, for conditions of rate feedback coefficient $T_r = \frac{1}{12}$, and feedforward gain $K_g = 0{\cdot}04$ if the servo loop undamped natural frequency is 30 rad/sec and the damping ratio is 0·5.

From the relationships defined in Equation 6.19 we evaluate the coefficients as

$$a_0 = \frac{K_g K_a \omega_{ns}^2}{T_a} = \frac{0{\cdot}04 \times 335 \times 30^2}{0{\cdot}179} = 67{,}400$$

$$a_1 = \frac{K_g K_a \omega_{ns}^2 T_r + \omega_{ns}^2}{T_a} = \frac{(335 \times 30^2 \times 0{\cdot}04)/12 + 30^2}{0{\cdot}179} = 10{,}660$$

$$a_2 = \frac{2\zeta_s \omega_{ns}}{T_a} + \omega_{ns}^2 = \frac{2 \times 0{\cdot}5 \times 30}{0{\cdot}179} + 900 = 1{,}068$$

$$a_3 = \frac{1}{T_a} + 2\zeta_s \omega_{ns} = \frac{1}{0{\cdot}179} + 2 \times 0{\cdot}5 \times 30 = 35{\cdot}6$$

and

$$a_4 = 1{\cdot}0$$

To apply the normalisation technique when $a_0 \neq 1{\cdot}0$ and $a_4 = 1{\cdot}0$, we define

$$\omega_0 = a_0^{1/4} = 67{,}400^{1/4} = 16{\cdot}1$$

and the normalised coefficients are

$$a_1' = \frac{a_1}{\omega_0^3} = \frac{10{,}660}{16{\cdot}1^3} = 2{\cdot}56$$

$$a_2' = \frac{a_2}{\omega_0^2} = \frac{1{,}068}{16{\cdot}1^2} = 4{\cdot}10$$

and

$$a_3' = \frac{a_3'}{\omega_0} = \frac{35{\cdot}6}{16{\cdot}1} = 2{\cdot}21$$

The normalised characteristic equation is

$$S^4+2{\cdot}21S^3+4{\cdot}10S^2+2{\cdot}56S+1 = 0 \tag{6.20}$$

and the convenient size of the coefficients of Equation 6.20 should certainly convince the reader of the advantages of using normalised equations in this numerical work.

The first trial solution is obtained from the last three terms of Equation 6.20, and when divided through by 4·10 and rounded off, the first trial divisor is

$$S^2+0{\cdot}62S+0{\cdot}24 \tag{6.21}$$

and the first computation is

$$\begin{array}{rl}
 & S^2+1{\cdot}59S+2{\cdot}87 \\ \hline
S^2+0{\cdot}62S+0{\cdot}24) & S^4+2{\cdot}21S^3+4{\cdot}10S^2+2{\cdot}56S+1 \\
 & S^4+0{\cdot}62S^3+0{\cdot}24S^2 \\ \hline
 & 1{\cdot}59S^3+3{\cdot}86S^2+2{\cdot}56S \\
 & 1{\cdot}59S^3+0{\cdot}99S^2+0{\cdot}38S \\ \hline
\text{Second trial divisor} \rightarrow & 2{\cdot}87S^2+2{\cdot}18S+1 \\
 & 2{\cdot}87S^2+1{\cdot}78S+0{\cdot}69 \\ \hline
 & 0{\cdot}40S+0{\cdot}31
\end{array}$$

Dividing the second trial divisor by 2·87 we obtain the divisor

$$S^2+0{\cdot}76S+0{\cdot}35 = 0 \tag{6.22}$$

and the second computation is then

$$\begin{array}{rl}
 & S^2+1{\cdot}45S+2{\cdot}56 \\ \hline
S^2+0{\cdot}76S+0{\cdot}35) & S^4+2{\cdot}21S^3+4{\cdot}10S^2+2{\cdot}56S+1 \\
 & S^4+0{\cdot}76S^3+0{\cdot}35S^2 \\ \hline
 & 1{\cdot}45S^3+3{\cdot}75S^2+2{\cdot}56S \\
 & 1{\cdot}45S^3+1{\cdot}10S^2+0{\cdot}51S \\ \hline
\text{Third trial divisor} \rightarrow & 2{\cdot}65S^2+2{\cdot}05S+1 \\
 & 2{\cdot}65S^2+2{\cdot}02S+0{\cdot}93 \\ \hline
 & 0{\cdot}03S+0{\cdot}07
\end{array}$$

Dividing the third trial divisor by 2·65, we obtain the divisor

$$S^2+0{\cdot}77S+0{\cdot}37 = 0 \tag{6.23}$$

and the third computation is

$$
\begin{array}{r l}
 & S^2+1{\cdot}44S\ +2{\cdot}62 \\
S^2+0{\cdot}77S+0{\cdot}37) & S^4+2{\cdot}21S^3+4{\cdot}10S^2+2{\cdot}56S+1 \\
 & S^4+0{\cdot}77S^3+0{\cdot}37S^2 \\
 & \quad 1{\cdot}44S^3+3{\cdot}73S^2+2{\cdot}56S \\
 & \quad 1{\cdot}44S^3+1{\cdot}11S^2+0{\cdot}53S \\
 & \qquad 2{\cdot}62S^2+2{\cdot}03S+1 \\
 & \qquad 2{\cdot}62S^2+2{\cdot}02S+0{\cdot}98 \\
 & \qquad\qquad 0{\cdot}01S+0{\cdot}02
\end{array}
$$

The remainder is now negligibly small and the normalised system poles are therefore given by

$$(S^2+1{\cdot}44S+2{\cdot}62)\,(S^2+0{\cdot}77S+0{\cdot}37) = 0 \qquad (6.24)$$

The damping ratio of the first term is 0·45, and the true undamped natural frequency is $16{\cdot}1\times\sqrt{2{\cdot}62} = 26{\cdot}1$ rad/sec, whilst the damping ratio of the second term is 0·63, and the true undamped natural frequency is $16{\cdot}1\times\sqrt{0{\cdot}37} = 9{\cdot}8$ rad/sec. If desired, Equation 6.24 may be rewritten in standard form, and the system transfer function for the stated conditions will then become

$$\frac{\phi_o}{\phi_D}(s) = \frac{1}{\left[1+\left(\frac{2\times0{\cdot}45}{26{\cdot}1}\right)s+\frac{s^2}{26{\cdot}1^2}\right]\left[1+\left(\frac{2\times0{\cdot}63}{9{\cdot}8}\right)s+\frac{s^2}{9{\cdot}8^2}\right]} \qquad (6.25)$$

since by inspection of Fig. 6.1 there are no system zeros. The poles defined by Equation 6.25 are found to lie on the root locus for the system, which has been previously drawn in Fig. 3.45, thus confirming the accuracy of the construction.

The solution has converged quite quickly in this particular application, and Lin's method allied to synthetic division can be a very useful aid for the control engineer, particularly under those circumstances where a digital computer is not available or for instances where the use of the digital computer is not justified, mainly due to the computer being uneconomic in terms of time when only a few equations have to be solved and the programming time is significant. It may also be argued that the use of hand-machine or slide-rule methods may lead to a quicker solution of the design problem, as the designer is always in contact with the problem. However, the control engineer requires many tools at his disposal, and root locus techniques, numerical methods, and analogue and digital computation should always be looked upon as complementary to each other, the ratio of the use of one to the other being determined by the particular application.

## 6.4 Fourth Order Transfer Functions

It is possible to develop certain transfer function techniques which are applicable directly to fourth order systems, particularly when the system has a second order dominant mode. In many cases it is possible to recognise the type of system by inspecting the normalised transfer function, which we shall define, for the fourth order system, as

$$S^4+aS^3+bS^2+cS+1 = 0 \tag{6.26}$$

where $a$, $b$ and $c$ are real and positive.

Absolute stability of the fourth order characteristic equation can be tested using the Routh Hurwitz criterion which, applied to Equation 6.26, yields the following array

$$\begin{array}{c|ccc} S^4 & 1 & b & 1 \\ S^3 & a & c & \\ S^2 & \dfrac{ab-c}{a} & 1 & \\ S^1 & \dfrac{[(ab-c)/a]c-a}{(ab-c)/a} & & \\ S^0 & 1 & & \end{array}$$

For absolute stability the numbers in the left-hand column of the array must be positive, hence we may write

$$\frac{ab-c}{a} > 0$$

and

$$\left(\frac{ab-c}{a}\right) c-a > 0$$

It follows that if the second inequality holds, the first must hold also; since we have defined $a$, $b$, and $c$ as positive, for the second inequality to hold, $ab-c$ must be positive, which is the meaning of the first inequality. Thus for absolute stability we need consider the factor $[(ab-c)/a]c-a$ only. Assuming now that $a$ and $c$ are given, and the problem is deciding on the critically stable value of $b$, then the second condition when equated to zero, gives us the relationship

$$abc-c^2-a^2 = 0$$

Hence

$$b = \frac{a^2+c^2}{ac} \tag{6.27}$$

and $b$ must be greater than that value defined by Equation 6.27. The absolute minimum value of $b$ in Equation 6.27 occurs when $a = c$, giving $b_{min} = 2{\cdot}0$

Graphical presentation of Equation 6.27 is possible, but for clarification of presentation it is better to determine constant $b$ lines in the $ac$ plane as follows. Rewriting Equation 6.27 in terms of the ratio $a/c$ gives

$$\left(\frac{a}{c}\right)^2 - b\left(\frac{a}{c}\right) + 1 = 0$$

Hence

$$\frac{a}{c} = \frac{b \pm \sqrt{b^2 - 4}}{2} \tag{6.28}$$

which shows immediately that the absolute minimum value of $b$ for a stable system is $b = 2$. Equation 6.28 is plotted in Fig. 6.2, a log-log scale having been chosen for ease of presentation. The stability diagram in the $ac$ plane is shown in Fig. 6.2. To interpret the

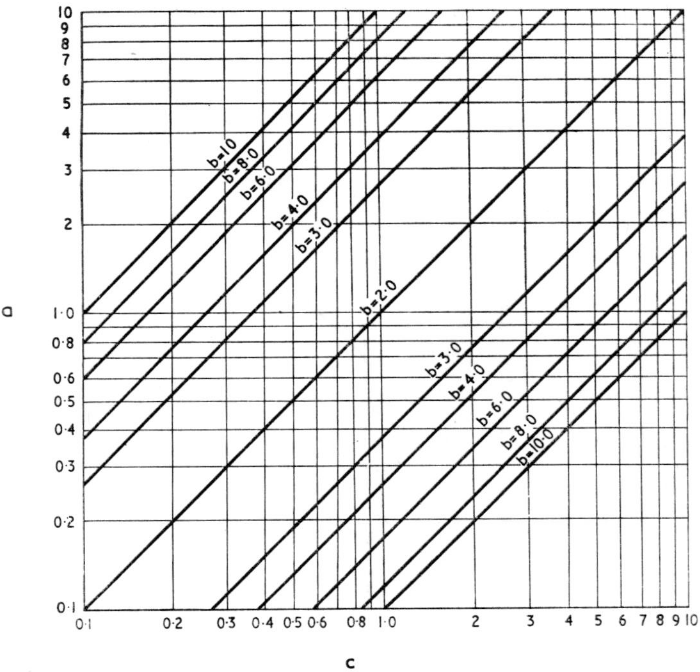


*Fig. 6.2. Routh Hurwitz stability conditions for the fourth order normalised characteristic equation* $S^4 + aS^3 + bS^2 + cS + 1 = 0$

diagram it is necessary to consider the three possible cases which may arise.

If $a$ and $c$ are given, the value of $b$ required for stability must be greater than the value of $b$ at that coordinate $(a, c)$ as shown in Fig. 6.3(a). For example, if $a = 1{\cdot}5$, $c = 0{\cdot}4$, the critical value of $b$ is 4·0, and hence a stable system will require that $b > 4{\cdot}0$.

If $a$ and $b$ are given, then there are upper and lower limits of $c$ which will provide a stable system. The critical values of $c$ lie at the intersection points of the respective $a$ and $b$ lines as shown in Fig. 6.3(b). For example, given $a = 1{\cdot}0$, $b = 3{\cdot}0$, the critical values of $c$ are 0·37 and 2·60, and therefore values of $0{\cdot}37 < c < 2{\cdot}60$ will provide a stable system.

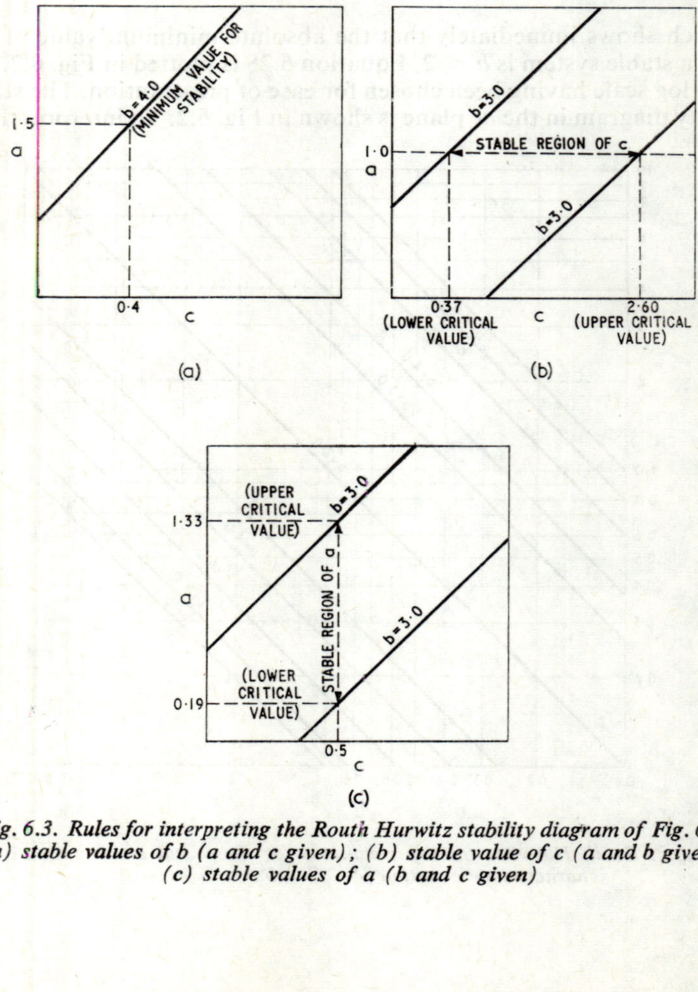


*Fig. 6.3. Rules for interpreting the Routh Hurwitz stability diagram of Fig. 6.2: (a) stable values of b (a and c given); (b) stable value of c (a and b given); (c) stable values of a (b and c given)*

If $b$ and $c$ are given, then there are upper and lower limits of $a$ which will provide a stable system. The critical values of $a$ lie at the intersection points of the respective $b$ and $c$ lines as shown in Fig. 6.3(c). For example, given $c = 0{\cdot}5$, $b = 3{\cdot}0$, the critical values of $a$ are 0·19 and 1·33, and therefore values of $0{\cdot}19 < a < 1{\cdot}33$ will result in a stable system.

Turning our attention now to fourth order linear systems with a dominant second order mode, it will be remembered from Equation 3.60 that a fourth order characteristic equation can be written in the form

$$1 + a_1 s + a_2 s^2 + a_3 s^3 + a_4 s^4 = 0 \tag{6.29}$$

or

$$\left[1 + \left(\frac{2\zeta_1}{\omega_n}\right) s + \frac{s^2}{\omega_n^2}\right]\left[1 + \left(\frac{2\zeta_2}{\lambda\omega_n}\right) s + \frac{s^2}{\lambda^2\omega_n^2}\right] = 0 \tag{6.30}$$

where, in Equation 6.30, the dominant mode has a damping ratio $\zeta_1$ and undamped natural frequency $\omega_n$, whilst the secondary mode has a damping ratio $\zeta_2$ and undamped natural frequency $\lambda\omega_n$. The ratio $\lambda$ of the undamped natural frequencies is thus an indication of mode dominance, since, in general, the larger is $\lambda$, the less important will be the secondary mode. No restriction is placed on damping ratios at this stage, but in general at least the dominant poles will be underdamped. Now, Equation 6.30 may be expanded to give

$$1 + 2\left(\zeta_1 + \frac{\zeta_2}{\lambda}\right)\frac{s}{\omega_n} + \left(1 + \frac{1}{\lambda^2} + \frac{4\zeta_1\zeta_2}{\lambda}\right)\frac{s^2}{\omega_n^2} + 2\left(\frac{\zeta_1}{\lambda^2} + \frac{\zeta_2}{\lambda}\right)\frac{s^3}{\omega_n^3} + \frac{s^4}{\lambda^2\omega_n^4} = 0 \tag{6.31}$$

Hence

$$a_0 = 1$$

$$a_1 = \frac{2}{\omega_n}\left(\zeta_1 + \frac{\zeta_2}{\lambda}\right)$$

$$a_2 = \frac{1}{\omega_n^2}\left(1 + \frac{1}{\lambda^2} + \frac{4\zeta_1\zeta_2}{\lambda}\right)$$

$$a_3 = \frac{1}{\omega_n^3}\left(\frac{\zeta_1}{\lambda^2} + \frac{\zeta_2}{\lambda}\right)$$

and

$$a_4 = \frac{1}{\lambda^2\omega_n^4}$$

Introducing the normalisation frequency $\omega_0 = (1/a_4)^{1/4} = \omega_n\sqrt{\lambda}$, the normalised coefficients are

$$c = a_1\omega_0 = 2\sqrt{\lambda}\left(\zeta_1+\frac{\zeta_2}{\lambda}\right)$$

$$b = \lambda\left(1+\frac{1}{\lambda^2}+\frac{4\zeta_1\zeta_2}{\lambda}\right)$$

and

$$a = 2\sqrt{\lambda}\left(\frac{\zeta_1}{\lambda}+\zeta_2\right)$$

and the normalised fourth order characteristic equation becomes

$$S^4+2\sqrt{\lambda}\left(\frac{\zeta_1}{\lambda}+\zeta_2\right)S^3+\lambda\left(1+\frac{1}{\lambda^2}+\frac{4\zeta_1\zeta_2}{\lambda}\right)S^2+$$
$$2\sqrt{\lambda}\left(\frac{\zeta_2}{\lambda}+\zeta_1\right)S+1 = 0 \qquad (6.32)$$

By inspecting Equation 6.32 we see that as mode dominance increases ($\lambda \rightarrow \infty$), we may write the following limiting values for the coefficients

$$\left.\begin{aligned} a &\simeq 2\sqrt{\lambda}\zeta_2 \\ b &\simeq \lambda \\ c &\simeq 2\sqrt{\lambda}\zeta_1 \end{aligned}\right\} \qquad (6.33)$$

i.e. $b$ is approximately equal to the ratio of the undamped natural frequencies, $a$ is proportional to the secondary mode damping, whilst $c$ is proportional to the dominant mode damping.

An approximate, but very useful, technique for solving fourth order equations may be developed from Equations 6.32 and 6.33. Reasonable approximations to $\lambda_1$, $\zeta_1$, and $\zeta_2$, may be found from the following simple formulae which neglect $1/\lambda^2$ as being insignificant:

$$\left.\begin{aligned} \lambda &\simeq b-\frac{ac}{b} \\ \zeta_1 &\simeq \frac{c}{2\sqrt{\lambda}}-\frac{a}{2\lambda^{3/2}} \\ \zeta_2 &\simeq \frac{a}{2\sqrt{\lambda}}-\frac{c}{2\lambda^{3/2}} \end{aligned}\right\} \qquad (6.34)$$

To use these formulae $\lambda$ must, of course, be calculated before $\zeta_1$ and $\zeta_2$ are determined. The accuracy of parameters determined by Equation 6.34 depends on the relative position in the s plane of the dominant and secondary poles, but the approximations are reasonably good for many applications, as we shall now show by means of two examples.

EXAMPLE 6.4.1

Determine, using the approximate method, the factors of the equation

$$S^4+2{\cdot}68S^3+6{\cdot}20S^2+2{\cdot}68S+1 = 0 \quad (6.35)$$

From this equation, $a = 2{\cdot}68$, $b = 6{\cdot}20$ and $c = 2{\cdot}68$. Hence

$$\lambda \simeq b-\frac{ac}{b} = 6{\cdot}20-\frac{2{\cdot}68^2}{6{\cdot}20} = 5{\cdot}04$$

and $\sqrt{\lambda} \simeq 2{\cdot}24$, $\lambda^{3/2} \simeq 11{\cdot}3$. Therefore, substituting for $\lambda$ in $\zeta_1$ and $\zeta_2$, we have

$$\zeta_1 \simeq \frac{c}{2\sqrt{\lambda}}-\frac{a}{2\lambda^{3/2}} = \frac{2{\cdot}68}{2\times2{\cdot}24}-\frac{2{\cdot}68}{2\times11{\cdot}3} = 0{\cdot}60-0{\cdot}12 = 0{\cdot}48$$

and

$$\zeta_2 = \frac{a}{2\sqrt{\lambda}}-\frac{c}{2\lambda^{3/2}} = \frac{2{\cdot}68}{2\times2{\cdot}24}-\frac{2{\cdot}68}{2\times11{\cdot}3} = 0{\cdot}60-0{\cdot}12 = 0{\cdot}48$$

The true system poles may be written down by comparison with Equation 6.30 to give

$$\left[1+\left(\frac{2\times0{\cdot}48}{\omega_n}\right)s+\frac{s^2}{\omega_n^2}\right]\left[1+\left(\frac{2\times0{\cdot}48}{5{\cdot}04\omega_n}\right)s+\frac{s^2}{5{\cdot}04^2\omega_n^2}\right] = 0 \quad (6.36)$$

Exact values are $\lambda = 5{\cdot}0$, $\zeta_1 = \zeta_2 = 0{\cdot}5$, which shows the sort of accuracy which may be expected from this very rapid method of solving fourth order equations.

EXAMPLE 6.4.2

Determine, using the approximate method, the factors of the equation

$$S^4+2{\cdot}34S^3+10{\cdot}9S^2+4{\cdot}62S+1 = 0$$

From this equation

$$\lambda \simeq b-\frac{ac}{b} = 10{\cdot}9-\frac{2{\cdot}34\times4{\cdot}62}{10{\cdot}9} = 10{\cdot}9-1{\cdot}0 = 9{\cdot}90$$

**Table 6.4.1**

COEFFICIENTS OF NORMALISED FOURTH ORDER EQUATION WITH SECOND ORDER DOMINANT MODE

($\zeta_1 = 0{\cdot}5$, $\lambda = 5{\cdot}0$, $\zeta_2$ varied)

| $\zeta_2$ | $a$ | $b$ | $c$ |
|---|---|---|---|
| 0·3 | 1·79 | 5·80 | 2·50 |
| 0·5 | 2·68 | 6·20 | 2·68 |
| 0·7 | 3·58 | 6·60 | 2·86 |
| 1·0 | 4·92 | 7·20 | 3·13 |

and $\sqrt{\lambda} \simeq 3{\cdot}15$, $\lambda^{3/2} \simeq 31{\cdot}2$. Therefore, substituting for $\lambda$,

$$\zeta_1 = \frac{4{\cdot}62}{2\times 3{\cdot}15} - \frac{2{\cdot}34}{31{\cdot}2\times 2} = 0{\cdot}74 - 0{\cdot}04 = 0{\cdot}70$$

and

$$\zeta_2 = \frac{2{\cdot}34}{2\times 3{\cdot}15} - \frac{4{\cdot}62}{31{\cdot}2\times 2} = 0{\cdot}37 - 0{\cdot}075 = 0{\cdot}295$$

The true system poles may be written down by comparison with Equation 6.30 to give

$$\left[1+\left(\frac{2\times 0{\cdot}70}{\omega_n}\right)s+\frac{s^2}{\omega_n^2}\right]\left[1+\left(\frac{2\times 0{\cdot}295}{9{\cdot}9\omega_n}\right)s+\frac{s^2}{9{\cdot}9^2\omega_n^2}\right] = 0 \quad (6.37)$$

Exact values are $\lambda = 10$, $\zeta_1 = 0{\cdot}70$, $\zeta_2 = 0{\cdot}30$, thus showing how accuracy is increased as $\lambda$ is increased.

The method does not give particularly accurate results for the fourth order autopilot studied as Example 6.3.2, since $\lambda$ is only about 2·7. Approximate values obtained are $\zeta_1 = 0{\cdot}53$, $\zeta_2 = 0{\cdot}39$, whereas the exact values are $\zeta_1 = 0{\cdot}63$, $\zeta_2 = 0{\cdot}45$. It would be

**Table 6.4.2**

COEFFICIENTS OF NORMALISED FOURTH ORDER EQUATION WITH SECOND ORDER DOMINANT MODE

($\zeta_1 = 0{\cdot}5$, $\zeta_2 = 0{\cdot}5$, $\lambda$ varied)

| $\lambda$ | $a$ | $b$ | $c$ |
|---|---|---|---|
| 3 | 2·31 | 4·33 | 2·31 |
| 5 | 2·68 | 6·20 | 2·68 |
| 10 | 3·48 | 11·10 | 3·48 |

possible to use the approximate value of $\zeta_1 = 0{\cdot}53$ as part of the first trial divisor in Lin's method if desired, and this might make the solution converge in fewer steps than would be the case when using the standard routine.

As a guide to the magnitude of the normalised coefficients to be found in dominant mode fourth order systems, Table 6.4.1 shows the variation as $\zeta_2$ is altered with $\zeta_1$ and $\lambda$ constant, whilst Table 6.4.2 shows the variation as $\lambda$ is altered with $\zeta_1$ and $\zeta_2$ constant. Further normalised coefficients are shown in graphical form in Fig. 6.4, and these coefficients should assist the reader in

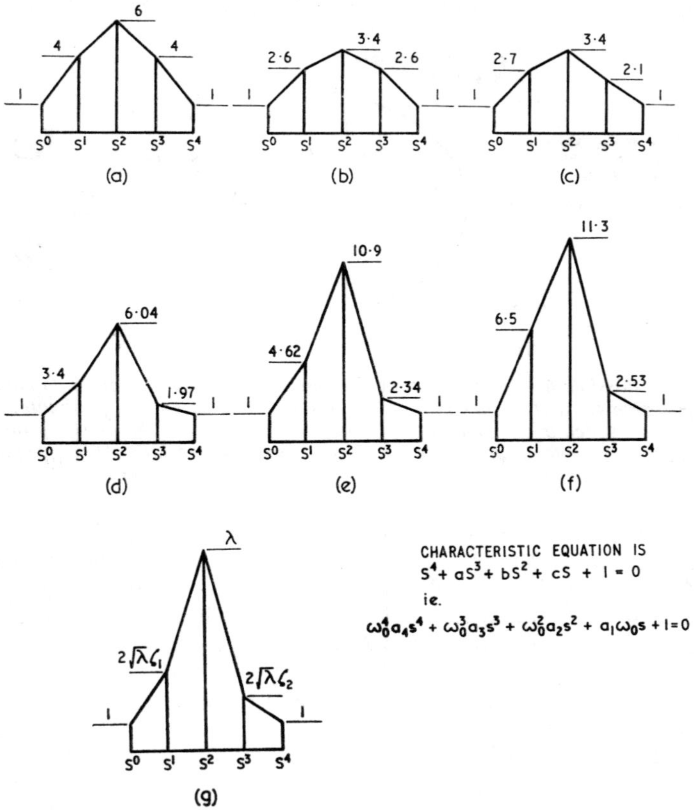


*Fig. 6.4. Normalised coefficients for typical fourth order characteristic equations: (a) binomial; (b) Butterworth; (c) unity numerator ITAE; (d) dominant mode ($\lambda = 5$, $\zeta_1 = 0{\cdot}7$, $\zeta_2 = 0{\cdot}3$); (e) dominant mode ($\lambda = 10$, $\zeta_1 = 1{\cdot}0$, $\zeta_2 = 0{\cdot}3$); (f) dominant mode ($\lambda = 10$, $\zeta_1 = 10$, $\zeta_2 = 0{\cdot}3$); (g) dominant mode limiting value of coefficients for $\lambda \gg 1{\cdot}0$*

recognising the basic types, since Figs. 6.4 (a), (b) and (c) represent systems where all four poles are significant, whilst the remaining systems are mode dominant to a greater or lesser degree.

## 6.5 ITAE Optimum System Pole-zero Arrays

The concept of suitable system pole-zero arrays for a given order of system was introduced in Chapter 1, and it was concluded that the ITAE performance index defined an acceptable system pole-zero array for applications where a particular number of poles and zeros might be significant in determining system performance. It

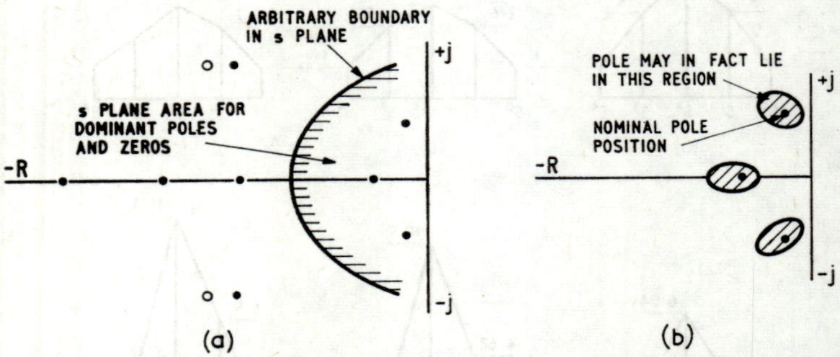


*Fig. 6.5. Possible use of the ITAE performance index in the design of dominant mode systems: (a) typical eighth order system considered to be dominated by three poles which approximately fulfil the pole geometry of the ITAE optimum third order unity numerator system; (b) an example of the latitude in pole positions which would still result in acceptable performance*

has been emphasised that the ITAE pole and zero positions are an *indication*, and no more, of suitable regions in the s plane for positioning the significant poles and zeros. Thus, for example, in an eight order system, if five poles are effectively at infinity, then a satisfactory design may be achieved by placing the remaining three poles (and any zeros required) in approximately the positions defined by the ITAE index for the dominant mode system, as shown in Fig. 6.5.

Certain ITAE optimum pole-zero arrays have already been discussed in Chapters 1, 4 and 5 on an *ad hoc* basis. Susini[53] has published the optimum system pole-zero arrays for systems with up to five poles for conditions of (a) zero displacement error ($a_0 = b_0$), (b) zero displacement error and zero velocity error ($a_0 = b_0$, $C_v = \infty$), (c) zero displacement error, zero velocity error and zero acceleration error ($a_0 = b_0$, $C_v = \infty$, $C_a = \infty$).

The minor discrepancies in the coefficients quoted by Susini on the one hand, and Lathrop and Graham[20, 21] on the other hand, are not significant. Fig. 6.6 shows the third order systems resulting

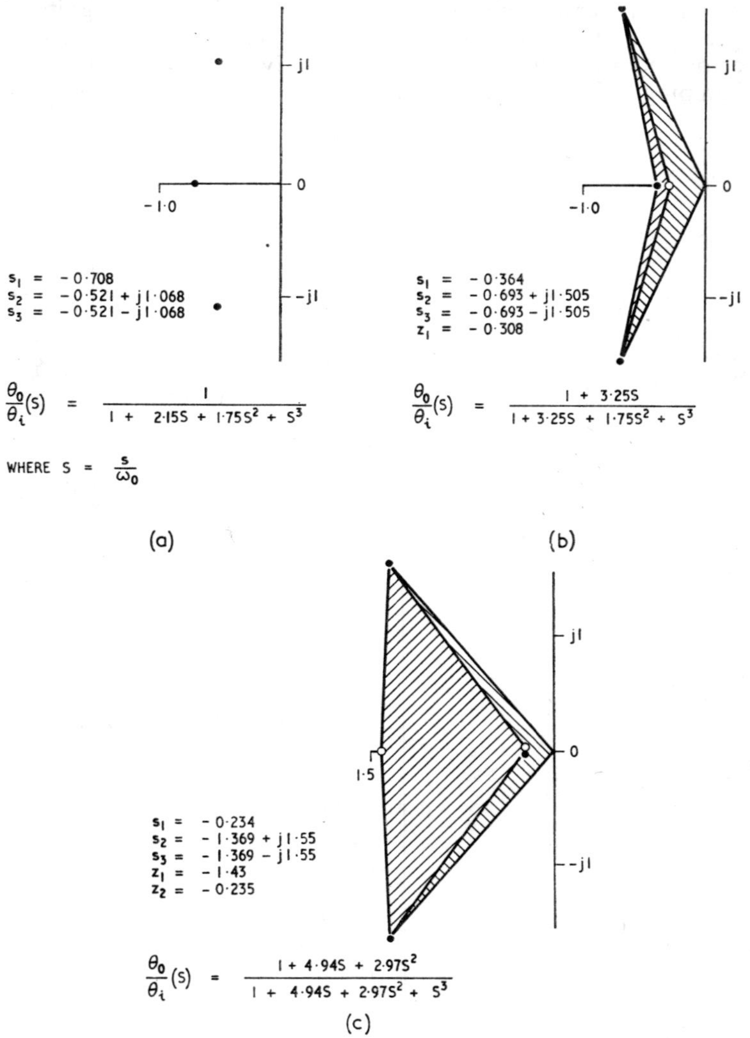


*Fig. 6.6. ITAE optimum third order systems optimised following a step function input (Susini's results): (a) zero displacement error system ($a_0 = b_0$); (b) zero displacement error and zero velocity error system ($a_0 = b_0$, $C_v = \infty$); (c) zero displacement error, zero velocity error and zero acceleration error system ($a_0 = b_0$, $C_v = \infty$, $C_a = \infty$)*

from the minimisation of the ITAE index following a step function input. For the zero displacement error system of Fig. 6.6(a) the complex poles have a damping ratio of about 0·42, but the maximum percentage overshoot is reduced by the significant presence of the third pole. The introduction of the zero velocity error requirement ($C_v = \infty$) has relatively little effect on the damping ratio of the complex poles, as shown in Fig. 6.6(b), but

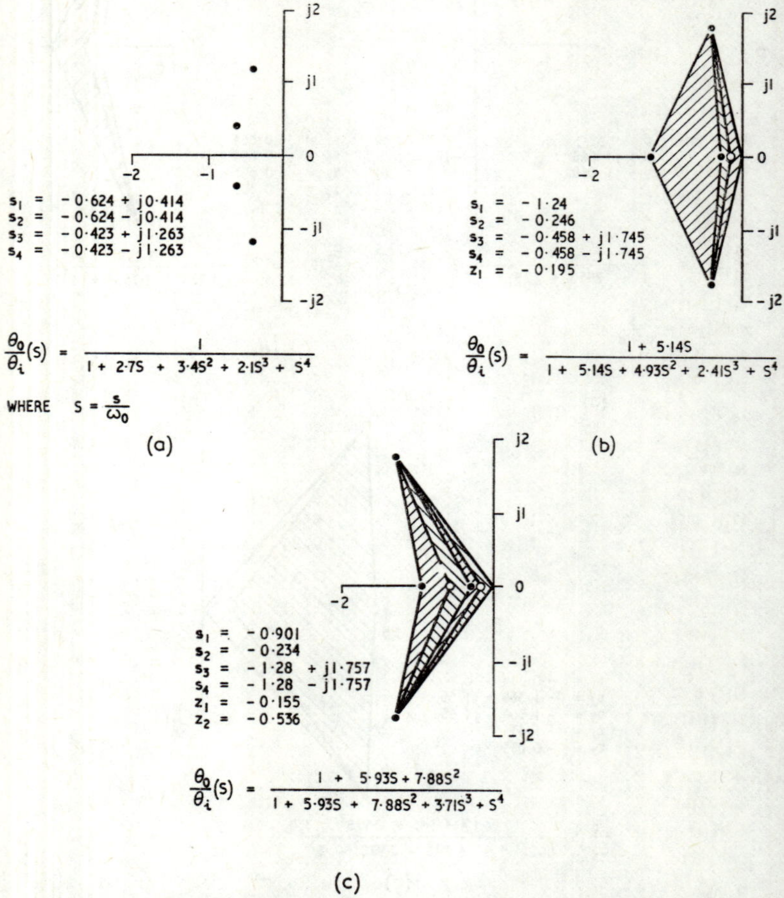


*Fig. 6.7. ITAE optimum fourth order systems optimised following a step function input (Susini's results); (a) zero displacement error system ($a_0 = b_0$); (b) zero displacement error and zero velocity error system ($a_0 = b_0$, $C_v = \infty$); (c) zero displacement error, zero velocity error and zero acceleration error system ($a_0 = b_0$, $C_v = \infty$, $C_a = \infty$)*

the natural frequency has increased. The third pole has moved in to just over half its former value, whilst the system zero is introduced near this third pole, and on the origin side of it, as would be expected as this zero has to eliminate the effect of all three poles on $C_v$. Introducing the $C_a = \infty$ requirement in addition to $C_v = \infty$, the complex poles increase slightly in undamped natural frequency and damping ratio, but the third pole and zero move in relatively close to the origin, and the second zero is introduced on the negative real axis more or less in line with the complex poles.

Fig. 6.7 shows the corresponding fourth order system pole-zero arrays, and from Fig. 6.7(a) we see that the zero displacement error system has two pairs of complex poles, the more lightly damped pair having a damping ratio of about 0·31. Introducing the zero velocity error requirement in addition, we see that the damping ratio of the lightly damped poles is little changed, although their undamped natural frequency increases by about 50%. The other two poles no longer form a complex pair, but have both migrated to the negative real axis, one pole moving very close to the origin where it is joined, on the origin side, by the system zero, whilst the other pole has moved away from the origin but to a position where this pole still considerably affects the maximum percentage overshoot. When, in addition, zero acceleration error is specified, the damping ratio of the complex pole pair increases, as do also their natural frequency, the pole and zero near the origin move closer to the origin, the zero much more so than the pole, whilst the fourth pole moves nearer the origin to about 75% of the value for the zero velocity error case.

Fifth order systems are considered in Fig. 6.8. For the zero displacement error system shown in Fig. 6.8(a), it can be seen that there are two complex pole pairs, one pair of which is lightly damped with a damping ratio of about 0·28. The negative real pole is further from the origin than the complex poles, but will still affect the maximum percentage overshoot. Adding the zero velocity lag requirement causes the negative real pole to move towards the origin as shown in Fig. 6.8(b). The lightly damped complex poles increase in undamped natural frequency by about 50%, and the damping ratio decreases slightly. An increase in the undamped natural frequency of the more heavily damped poles can be observed, and there is a definite decrease in their damping ratio. Finally, a system zero is added near the negative real pole, and on the origin side of it. The specification of zero acceleration error in addition to the previous two requirements drives the negative real pole even further towards the origin, and two system zeros, identical in value, are added near this pole, but on the far side from the origin.

The damping ratio of both complex pole pairs is increased as shown in Fig. 6.8(c), in fact the pair with the lower natural frequency has the damping ratio increased almost to unity, i.e. the pole pair almost forms a double pole on the negative real axis.

These nine system pole-zero arrays indicate how the pole-zero array is affected on the one hand by the specification, and on the other hand by the order of the system. In theory, for a specification

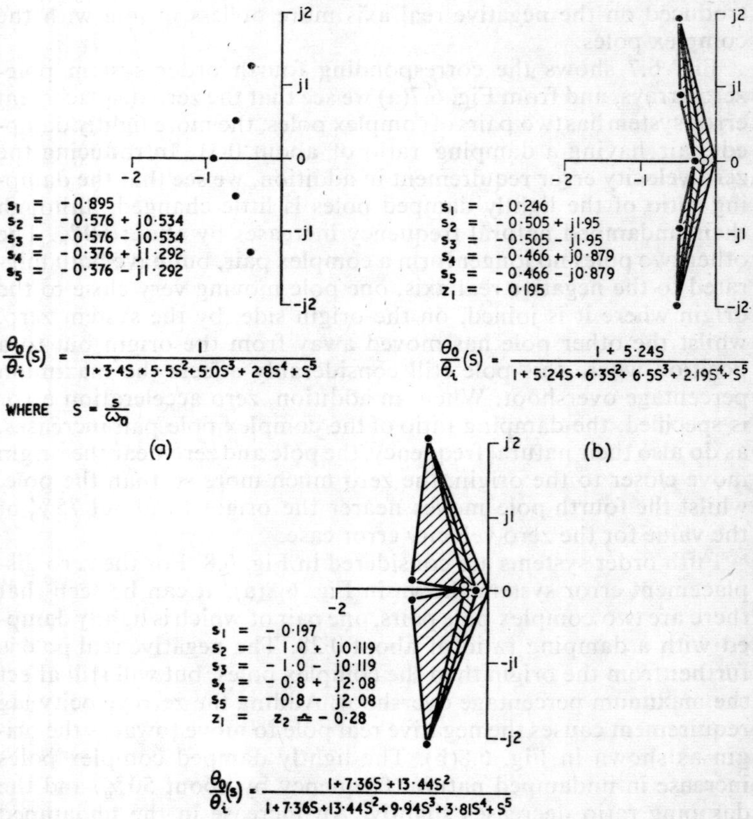


*Fig. 6.8. ITAE optimum fifth order systems optimised following a step function input (Susini's results): (a) zero displacement error system ($a_0 = b_0$); (b) zero displacement error and zero velocity error system ($a_0 = b_0$, $C_v = \infty$); (c) zero displacement error, zero velocity error and zero acceleration error system ($a_0 = b_0$, $C_v = \infty$, $C_a = \infty$)*

which included intermediate values of $C_v$ and $C_a$, it would be necessary to simulate the system on an analogue computer to determine the best pole-zero array to fit the specification, and at the same time to minimise the ITAE performance criterion. In practice, however, this degree of refinement is not normally necessary, as sufficient general guidance is really incorporated in the systems studied in Figs. 6.6, 6.7 and 6.8.

## 6.6 Alternative s Plane Design for Meeting Specific Requirements

As further evidence that several possible solutions may exist for specific design problems, consider the two system pole-zero arrays shown in Fig. 6.9. The second order system is a simple unity numerator system with a damping ratio of 0·70, whilst the third order

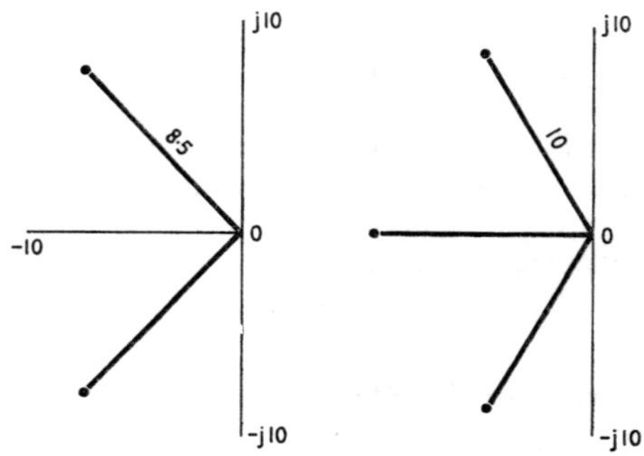


*Fig. 6.9. Second and third order system pole-zero arrays which have similar transient responses*

system is also a simple unity numerator system with the complex pole pair damping ratio equal to 0·5. However, the presence of the third pole in the position shown results in a transient response which is not dissimilar to that obtained for the second order system with higher damping ratio, as can be seen in Fig. 6.10. There are, of course, certain distinct differences between the responses, such as the rise time, however this is defined, but by and large the shape of the response is similar enough for either system to be acceptable for many applications. Of course, there are other differences between the systems, and the second order system has the lower bandwidth, but higher $C_v$, which might influence the final choice of pole geometry. On the other hand it may well be that the third order pole-zero array is a much more practicable proposition for given load characteristics, and therefore the designer should

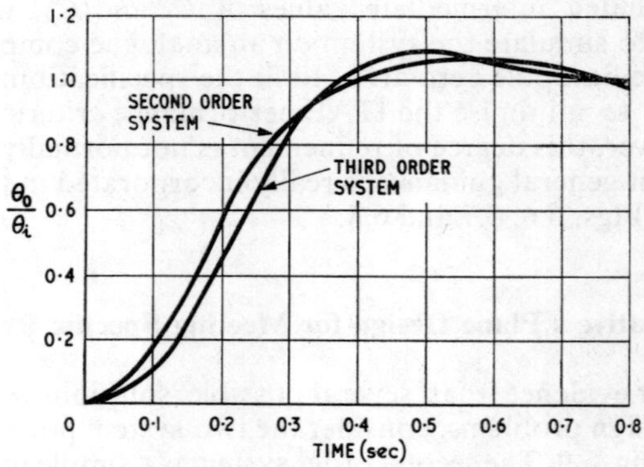


*Fig. 6.10. Transient responses of the second and third order systems with similar behaviour*

always be prepared, where necessary, to accept the existence of three or even four dominant poles plus possibly one or two dominant zeros. The difficulty facing the designer is how to make best use of the extra freedom resulting from departure from a necessarily two-pole dominant system, and this can only be overcome by intuition and experience, although a thorough understanding of Chapters 3, 5 and 6 will be of great assistance.

## 6.7 Design in Terms of the Root Locus

Design in terms of the root locus is essentially based on the use of the s plane for the evaluation of various compensation techniques. Thus given the fixed elements of the system, the problem is to introduce, often on an *ad hoc* basis, compensatory open loop* poles and zeros to force the root locus to pass through certain regions of the s plane so that specific system poles and zeros are possible. If a dominant mode system is required, then we must force the dominant poles into certain regions, as shown, for example, in Fig. 6.5(b), and at the same time make sure that the other branches of the locus are of such a geometry that all other poles are effectively at infinity, or approximately cancel with a system zero. Clearly the dominant mode system poles and zeros are chosen to meet such specifications as bandwidth and precentage overshoot. Some allowance on $C_v$ may have to be made to compensate for the neglected poles since, in general, these have more

* We conform with the open loop terminology here, but the reader is reminded of the physical significance of this term as developed in Chapter 2. Thus open is short for opened, i.e. broken.

effect on $C_v$ than on the other common specifications, as was discussed in detail in Chapter 3.

An understanding of the mechanism of changing root locus shapes by the addition of compensatory open loop poles and zeros to the basic fixed open loop poles and zeros can greatly simplify the s plane design problem by indicating a suitable type of compensation for a given application. This understanding is best obtained by the examination of basic types of root locus, starting with a simple second order system and building up gradually to fourth order systems. The arguments put forward for changing the locus of the simple systems can be extended to higher order systems by splitting the s plane into regions of three or four poles and zeros, and crudely sketching the locus according to that shown for the corresponding basic type of locus. Refinements if necessary to improve the accuracy of the locus can then be made once it is certain that the proposed design is sufficiently promising.

Yeh,[54] and Bhattacharyya,[55] amongst others,[56] have suggested specific basic types of locus, and these are included in Figs. 6.11 to 6.18 inclusive, together with certain additions which have been found necessary to form a logical development of the principles of s plane compensation. All the loci shown are for systems with an open loop pole at the origin, since this book is mainly concerned with such systems. The original literature does include systems without a pure integrator in the loop.

Fig. 6.11 shows the modification to the root locus of a second order system with an open loop pole at the origin when a single open loop zero is added. The basic locus is shown in Fig. 6.11(a), and it can be seen that the locus either lies on the negative real axis, or on the high-gain asymptotes, which are at angles $\pm\pi/2$, and which intersect the negative real axis midway between the open loop poles. There is no limit on the damping ratio which may be obtained from this basic system, but there is a definite limit to the undamped natural frequency which can be obtained for, say, a damping ratio of 0·70. Adding an open loop zero between the open loop poles constrains the locus to lie only on the negative real axis, as shown in Fig. 6.11(b), whilst adding an open loop zero to the left of both open loop poles allows the system loop poles to lie either on the negative real axis or on the circular locus shown in Fig. 6.11(c). Thus complex system poles are possible in the latter case, and for a given position of the compensation zero there is a minimum value of damping ratio possible, as was seen in Example 3.4.1. By judicious choice of the position of the compensation zero, the system poles can be forced into positions in the s plane with satisfactory combinations of damping ratio and undamped natural frequency.

The basic third order system with three negative real poles, one at the origin, is shown in Fig. 6.12(a), and in the absence of a compensation zero there will be three high-gain asymptotes at angles $\pm\pi/3$ and $\pi$. Clearly the system will become unstable if the gain is made sufficiently high, and the natural frequency of the complex poles corresponding to a reasonable value of damping ratio

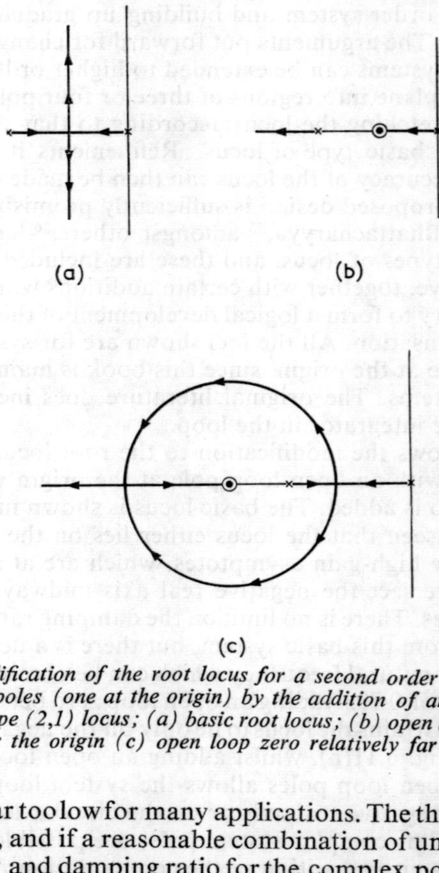


*Fig. 6.11. Modification of the root locus for a second order system with two real open loop poles (one at the origin) by the addition of an open loop zero, resulting in a type (2,1) locus; (a) basic root locus; (b) open loop zero situated relatively near the origin (c) open loop zero relatively far from the origin*

is probably far too low for many applications. The third pole moves off to infinity, and if a reasonable combination of undamped natural frequency and damping ratio for the complex pole pair can be obtained without further compensation, then the third pole effect can often be neglected. If an open loop zero is added to the left of all three open loop poles, as shown in Fig. 6.12(b), then the third pole will migrate to this zero, and the other two poles will form a locus with asymptotes at $\pm\pi/2$, and will be similar to the locus of the uncompensated second order system shown in Fig. 6.11(a). A suitable choice of zero position will bring the intersection of the high-gain asmyptotes (at angles $\pm\pi/2$) with the negative real axis

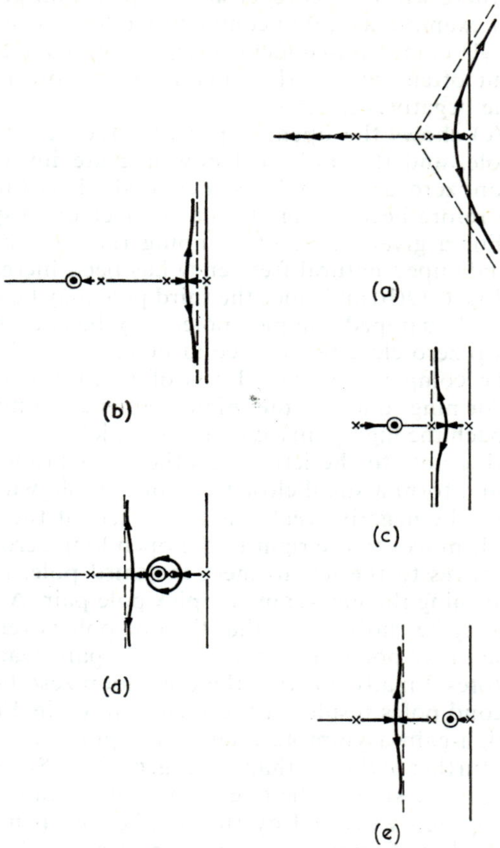


*Fig. 6.12. Modification of the root locus for a third order system with three real open loop poles (one at the origin) by the addition of an open loop zero, resulting in a type (3,1) locus: (a) basic root locus; (b) addition of a zero far to the left; (c) addition of a zero between the second and third poles; (d) addition of a zero between the second and third poles but close to the second pole; (e) addition of a zero near the imaginary axis*

to the left of the origin, thus resulting in a stable system for all values of gain. Note that the addition of an open loop zero always reduces the number of high-gain asymptotes by one, thus turning a (3,0) locus into a (3,1) locus, the first number in the bracket being the number of open loop poles, and the second number in the bracket being the number of open loop zeros. Of course, as we shall see later, if the compensation zero is generated by a passive network, then there will be a compensation open loop pole as well, and thus, for example, a (2,0) uncompensated locus would be turned into at (3,1) compensated locus still with high-gain asymptotes at $\pm\pi/2$, but often with vastly different asymptote intersection points on the negative real axis.

Fig. 6.12(c) shows the open loop zero added to the right of the third pole, and this pole will now migrate inwards to the compensation zero as the gain is increased. This third system pole will therefore become significant in effect upon system performance. For a given value of damping ratio of the complex poles, the undamped natural frequency has been increased compared with Fig. 6.12(b) and since the third pole may be significant, relatively lightly damped complex poles may be chosen. Moving the open loop zero close to the second pole, but to the left of it, results in the compensated root locus of Fig. 6.12(d), the third pole now forming one of the high-frequency complex roots which approach the high-gain asymptotes, which themselves have moved considerably to the left. The other two branches of the locus meet and form a small closed contour, as shown, which, of course, meets the negative real axis to the left of the open loop zero; one pole moves to the right to the open loop zero whilst the other pole moves to the left to meet the third pole, these latter poles then forming the high-gain complex pole pair. A gain value would normally be chosen such that the one pole is very near the zero, and the other poles are a complex pole pair near the high-gain asymptotes. Finally, moving the open loop zero between the first and second poles results in the locus shown in Fig. 6.12(e), where the high-gain asymptote intersection point, for given pole positions, is further to the left than for Fig. 6.12(d). Since the intersection point is the sum of the open loop poles minus the sum of the open loop zeros divided by the number of open loop poles minus the number of open loop zeros, movement of the zero to the right, all other conditions being unchanged, must result in a movement of the intersection point to the left. Of course this movement of the zero close to the origin means that a system pole will be near the origin also, but if the open loop zero is a system zero also, the effects will tend to cancel (except on $C_v$) provided the system pole and system zero are very close together.

A third order system with complex open loop poles is shown in Fig. 6.13(a): there will be high-gain asymptotes at angles $\pm\pi/3$ and $\pi$. The pole at the origin simply moves along the negative real axis towards infinity, and the complex poles move into the right-hand plane as the gain is made sufficiently high. Adding a zero on the negative real axis can stabilise the system, as shown in Figs. 6.13(b) and 6.13(c), but if the complex open loop poles are lightly damped, then this mode of stabilisation is rarely satisfactory as, in general, there will be a system pole pair with a damping ratio smaller than the damping ratio of these complex open loop poles. Methods of overcoming this difficulty will become apparent shortly, when the addition of complex zeros is discussed.

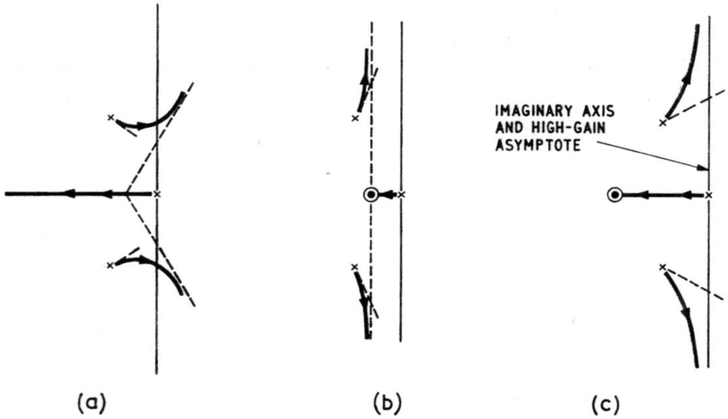


*Fig. 6.13. Modification of the root locus for a third order system with complex open loop poles and an open loop pole at the origin by the addition of a single open loop zero, resulting in a type (3,1) locus: (a) basic root locus; (b) modification of the root locus caused by the addition of an open loop zero near the origin; (c) modification of the root locus caused by the addition of an open loop zero away from the origin*

The effect on the root locus of adding two real open loop zeros to a system with two real open loop poles and a pole at the origin is shown in Fig. 6.14. Several types of contours can be formed. As far as a dominant mode approach is concerned, the advantage of the system shown in Fig. 6.14(a) over that of Fig. 6.14(b) is that a high natural frequency may be obtained for a given damping ratio of the dominant complex system poles. For the dominant roots of the system shown in Fig. 6.14(c) to be complex, say, with a damping ratio of 0·70, the third pole has moved very close to the smaller open loop zero, and if this is also a system zero then the system will be very close to a genuine second order system. Finally, the system shown in Fig. 6.14(d) can also be made to have a response (except for $C_v$) which is dominated by the second order complex poles, as the pole at the origin will be very close to the open loop zero, which must also be a closed loop zero for the complex system poles to dominate the response.

Modifications possible to the root locus of a third order system with complex open loop poles which may be achieved by the addition of two open loop zeros are shown in Fig. 6.15. The uncompensated locus is shown in Fig. 6.15(a), and additions of a pair of

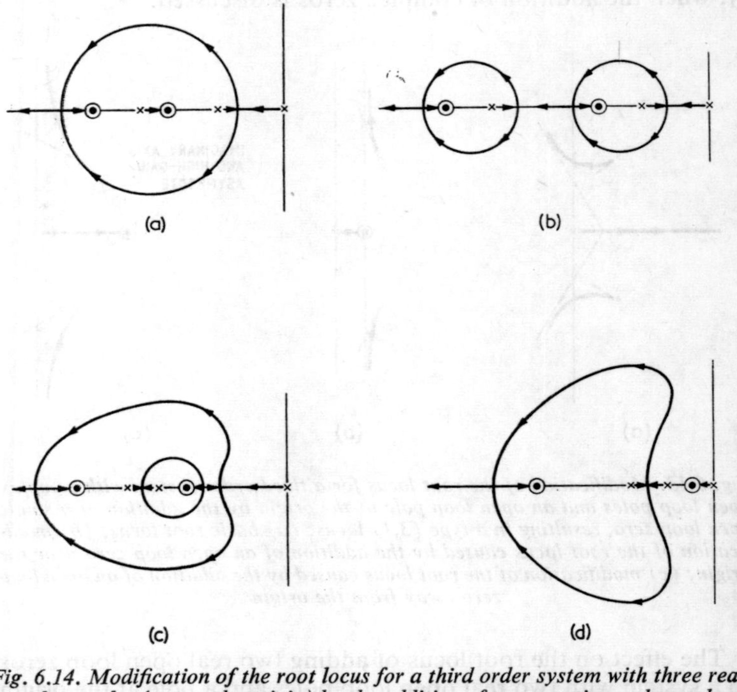


*Fig. 6.14. Modification of the root locus for a third order system with three real open loop poles (one at the origin) by the addition of two negative real open loop zeros, resulting in a type (3,2) locus: (a) largest pole close to smallest zero, forming a single large closed contour; (b) largest pole and largest zero moved further to the left, forming two small closed contours; (c) smallest zero moved to the right beyond the second pole, forming two concentric closed contours; (d) smallest zero close to the origin, forming a single large closed contour (a slow real root is present which might be turned into a dipole)*

complex open loop zeros near the complex open loop poles are shown in Fig. 6.15(b) and 6.15(c). It can be seen that in both cases instability is avoided, and the third pole at the origin marches off to infinity; the position of this third pole will be very significant indeed in determining system response. If the open loop poles and open loop zeros are of the same value, their effect on the locus is cancelled, and if the open loop zeros are system zeros also, the effect on system performance is also cancelled. If the open loop

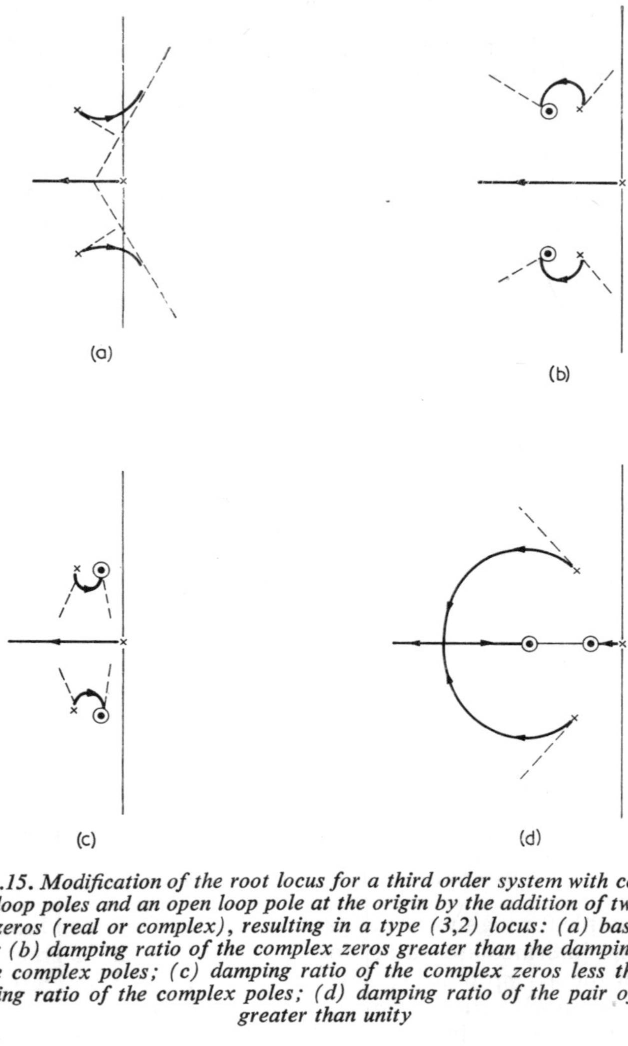


*Fig. 6.15. Modification of the root locus for a third order system with complex open loop poles and an open loop pole at the origin by the addition of two open loop zeros (real or complex), resulting in a type (3,2) locus: (a) basic root locus; (b) damping ratio of the complex zeros greater than the damping ratio of the complex poles; (c) damping ratio of the complex zeros less than the damping ratio of the complex poles; (d) damping ratio of the pair of zeros greater than unity*

zeros are both real, as shown in Fig. 6.15(d), then the system can be easily stabilised, reasonable system complex poles can be determined, but a pole near the origin is introduced, which means that this pole may dominate the response unless the open loop zero nearby is also a system zero.

Fourth order systems with four real open loop poles, one at the origin, and compensated by an open loop zero, making (4,1) type loci are shown in Fig. 6.16. The uncompensated locus is shown in Fig. 6.16(a), the four high-gain asymptotes being at angles of $\pm\pi/4$ and $\pm 3\pi/4$, thus ensuring that the uncompensated system will

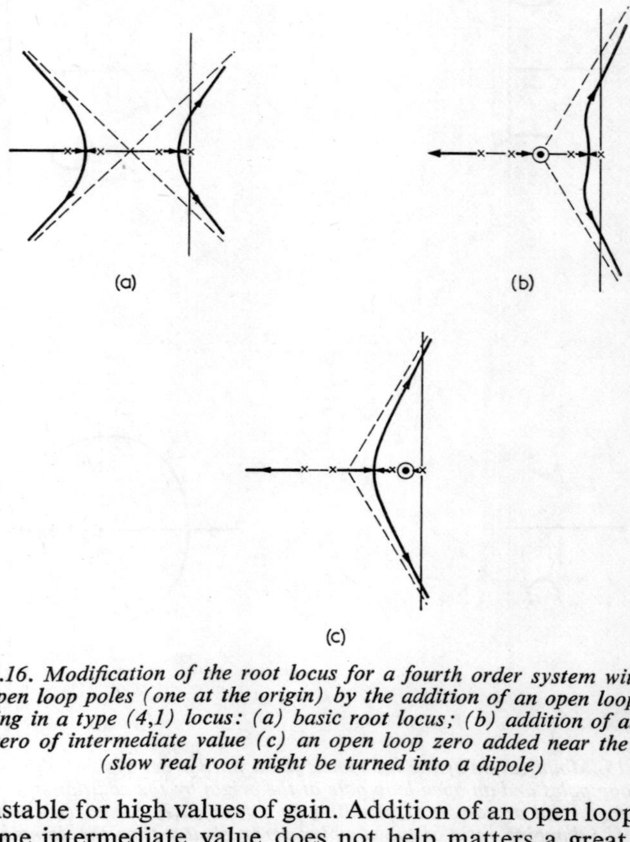


*Fig. 6.16. Modification of the root locus for a fourth order system with four real open loop poles (one at the origin) by the addition of an open loop zero, resulting in a type (4,1) locus: (a) basic root locus; (b) addition of an open loop zero of intermediate value (c) an open loop zero added near the origin (slow real root might be turned into a dipole)*

be unstable for high values of gain. Addition of an open loop zero at some intermediate value does not help matters a great deal, even if the asymptotes are now at angles of $\pm\pi/3$ and $\pi$, since a high undamped natural frequency coupled with reasonable damping ratio is not possible for the dominant complex poles as shown in Fig. 6.16(b). Adding the open loop zero near the origin is of considerable benefit, as a reasonable combination of damping ratio and undamped natural frequency for the complex poles is now possible. However, once again the pole near the origin will considerably affect the system response unless the open loop zero nearby is also a system zero.

Fourth order systems with a pair of complex open loop poles are shown in Fig. 6.17. In the basic locus shown in Fig. 6.17(a), it is the system poles emanating from the negative real open loop

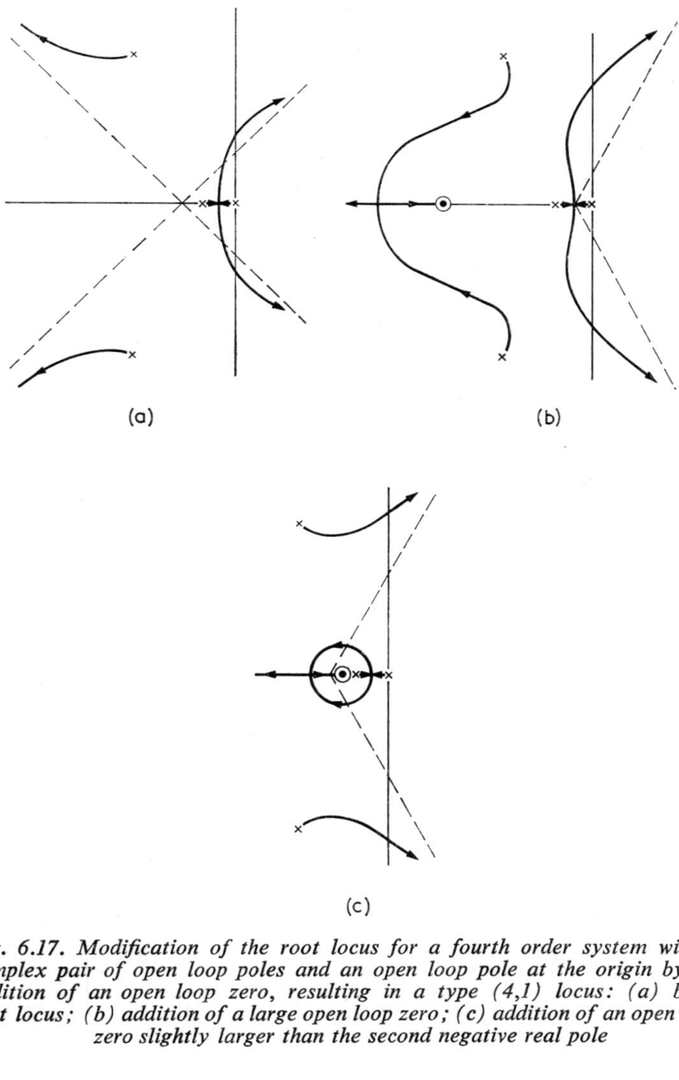

*Fig. 6.17. Modification of the root locus for a fourth order system with a complex pair of open loop poles and an open loop pole at the origin by the addition of an open loop zero, resulting in a type (4,1) locus: (a) basic root locus; (b) addition of a large open loop zero; (c) addition of an open loop zero slightly larger than the second negative real pole*

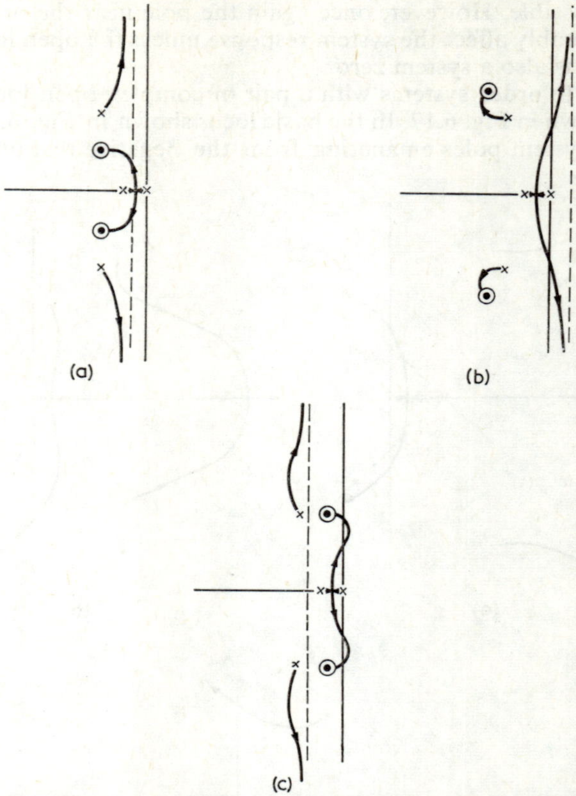


*Fig. 6.18. Modification of the root locus for a fourth order system with a complex pair of open loop poles and an open loop pole at the origin by the addition of a complex pair of open loop zeros, resulting in a type (4,2) locus: (a) open loop zeros with the same $\omega_n$ as the complex open loop poles but with larger $\zeta$; (b) open loop zeros with the same $\zeta$ as the complex open loop poles but with larger $\omega_n$; (c) open loop zeros with the same $\omega_n$ as the complex open loop poles but with lower $\zeta$*

poles which meet and form complex poles which eventually migrate into the right-hand plane. If the open loop complex poles are moved nearer the imaginary axis, and the second open loop pole on the negative real axis moved far enough to the left, it will be the system poles emanating from the complex open loop poles which will migrate into the right-hand plane. As can be seen from Fig. 6.17(b), the addition of an open loop zero far to the left will improve matters slightly, but with the particular open loop pole geometry shown, will still result in a very low damping ratio of the dominant complex system poles for a reasonable value of undamped natural frequency. More improvement is obtained by placing the open loop zero close to the second real pole, but to the left of it as shown in Fig. 6.17(c), since it is now the complex poles resulting from the complex open loop poles which migrate into the right-hand plane, whilst the poles near the origin form a locus very similar to that shown for the compensated second order system of Fig. 6.11(c).

Finally, to conclude this brief review of root locus modification by the addition of open loop poles and zeros, the compensation of a fourth order system with a pair of complex open loop poles by the the addition of a pair of complex open loop zeros is shown in Fig. 6.18. The high-gain asymptotes are at $\pm\pi/2$ and, depending on the position of the compensation zeros, either the poles emanating from the complex open loop poles or those emanating from the real open loop poles can migrate to the high-gain asymptotes, as can be seen by comparing Fig. 6.18(a) and 6.18(b). Finally, the system can still become unstable for certain positions of the complex open loop zeros as shown in Fig. 6.18(b) and 6.18(c). Fig. 6.18(a) appears quite promising in an attempt to obtain a reasonable combination of damping ratio and undamped natural frequency for the dominant complex system poles.

The problem of introducing compensation elements to change the shape of the root locus so that a particular branch of the locus passes through a specific point in the s plane is most easily understood by considering constant phase angle loci and constant gain contours. Chu[57] developed the concept that the root locus is but one of an infinite number of constant phase angle loci in the s plane. Thus given specific open loop poles and zeros, the root locus is the locus of all points in the s plane which result in nett angles of $180°\pm N360°$ measured from the open loop poles and zeros in accordance with Rule 9 of the root locus construction described in Section 3.3. The s plane may now be explored for other nett phase angles such as $\pm 135°$, $\pm 90°$, etc. A set of such phase angle loci for a simple uncompensated third order system is shown in Fig. 6.19, where the conventional root locus is shown as a full line. For a linear system the positive and negative values of the phase angle are symmetrical about the real axis as shown. These phase angle contours may be readily determined using a Spirule, and each contour may be calibrated in terms of 'gain', i.e. the conventional product of the vectors from the point of interest to the respective open loop poles divided by the product of the vectors from the point of interest to the respective open loop zeros. However, the construction of constant gain contours can be greatly simplified by using potential analogy which yields the significant result that gain contours intersect the constant phase angle contours orthogonally. Thus given the conventional root locus, calibrated in terms of gain, the gain contours can be sketched freehand once the phase angle loci are available. In practice it is rarely necessary to construct the phase angle loci and gain margin contours to determine the type of compensation required since it is often sufficient to evaluate at the desired point in the s plane the nett phase angle and gain at that point due to the open loop poles and zeros of the uncompensated system. For this point in the s plane to be on the root locus,

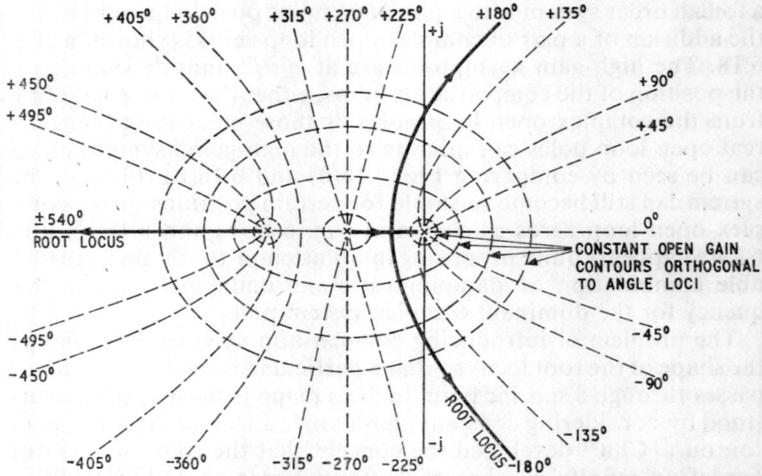


*Fig. 6.19. Set of phase angle loci and constant gain contours for a typical third order system*

the nett angle must be $180° \pm N360°$, which determines the phase contribution from the compensation elements, and if there is a specific gain requirement at this point the attenuation or amplification of gain at the point required from the compensation elements is also known. Thus the general s plane design problem involves the selection of compensation elements which will meet the phase and gain requirements determined in this way.

So far we have only considered compensation in very general terms, and have studied some interesting pictures showing how the locus is changed by the addition of open loop poles and zeros. Before considering specific examples of compensation, it is pertinent to describe briefly the more common methods of achieving this compensation. At the present time, integral (or phase lag), derivative (or phase lead), cancellation, and feedback are the four simple types of compensation which are normally considered before more complex schemes are investigated. The incorporation of an integral network in the feedforward path introduces an open loop pole and an open loop zero relatively close together, with the pole nearest the origin. It is a convenient form of compensation for increasing $C_v$ without materially affecting the system transient or frequency response. Derivative compensation, when achieved by a passive network, introduces an open loop pole and an open loop zero, the zero being very much closer to the origin, often by a factor of 5 or 10. Several stages of lead compensation may be introduced if required. Cancellation compensation involves the insertion of a compensation network, the zeros of which approximately cancel with unwanted open loop poles, leaving much more desirably placed poles of the compensation network as the significant open loop poles.

From Fig. 3.1 it can be seen that the system zeros are the feedback path zeros plus the feedback path poles, the open loop poles are the feedforward path poles plus the feedback path poles, and the open loop zeros are the feedforward path zeros plus the feedback path zeros. Cancellation, derivative, and integral compensation using passive networks generally involve the introduction of a compensation network with equal numbers of poles and zeros, the zeros of which, being feedforward zeros, will be system zeros. If the order of the compensation network is $q$, and the order of the uncompensated system is $r$, then the use of series compensation using passive lead, passive lag, or passive imperfect cancellation, will transform the $(r, 0)$ locus into an $(r+q, q)$ locus and will transform the $(r, 0)$ system into an $(r+q, q)$ system provided the basic system consists of feedforward poles only (similar simple rules can be developed for other basic systems). For these three forms of compensation, where system poles have migrated almost to the compensation open loop zeros, an appropriate dipole is formed which will have little effect on transient or frequency response. In contrast, the use of feedback compensation need not necessarily introduce system zeros. For example, pure rate feedback will introduce an open loop zero, but as the compensation involves the introduction of a feedback path zero only, a system zero will not result. Therefore with pure feedback compensation (i.e. no measurement lags), an $(r, 0)$ locus will transform to an $(r, q)$ locus, but the system will remain an $(r, 0)$ system. The introduction of pure feedback compensation therefore considerably affects the shape of the locus but does not introduce near cancelling system dipoles, as was the case with series compensation. Imperfect feedback compensation involving measurement lags will result in feedback path poles and hence in system zeros. If the feedback terms are wholly obtained from passive networks, then an $(r, 0)$ locus will transform to an $(r+q, q)$ locus and the system will transform from an $(r, 0)$ system to an $(r+q, q)$ system. The $q$ system poles thus introduced can be used to advantage if required.

It is hoped that the foregoing discussion will clarify the various principles involved in the different forms of elementary compensation which will be discussed individually in Sections 6.9 to 6.12. As previously stated, cancellation compensation involves the approximate cancellation of unwanted open loop poles and their substitution by more suitable open loop poles. If a second order compensation network is used, then an *RC* network is feasible, provided that the denominator of the compensation transfer function factorises into real poles. Thus using passive networks we may reasonably expect to cancel two unwanted complex open loop poles by complex zeros, but additionally, we must expect to replace them with two negative real poles which are much more conveniently placed. Exact cancellation tends to be wasteful of feedforward gain, and may give rise to non-linear problems, but in general exact cancellation is not necessary. If non-linear problems do arise due to the use of the cancellation compensation method, it might be argued that the power element and signal amplifier chosen are not adequate for the particular application.

## 6.8 $C_v$ in Terms of Open Loop Gain

In view of the fact that $C_v$ is commonly specified as a definite performance requirement, and that s plane compensation is often carried out on the root locus drawn in terms of feedforward gain, it is helpful at this stage to evaluate the relationship between the two. Neglecting any shaping which takes place outside the loop itself, i.e. input signal shaping, the general single loop control system with a basic integrator in the feedforward path appears as shown in Fig. 6.20(a). The feedforward gain $K$ is the scalar gain obtained when the factors are written in the form $1+T\mathrm{s}$. If terms such as $T_\alpha$ represent feedforward path zeros, terms such as $T_\beta$ represent feedforward path poles, terms such as $T_\gamma$ represent feedback path zeros, and terms such as $T_\delta$ represent feedback poles, we may write the system transfer function as

$$\frac{\theta_0}{\theta_i}(\mathrm{s}) = \frac{\prod(1+T_\alpha \mathrm{s})\times\prod(1+T_\delta \mathrm{s})}{\prod(1+T_\alpha \mathrm{s})\times\prod(1+T_\gamma \mathrm{s})+\dfrac{\mathrm{s}}{K}\prod(1+T_\beta \mathrm{s})\times\prod(1+T_\delta \mathrm{s})} \tag{6.38}$$

and the transfer function coefficients of interest are

$$a_0 = b_0 = 1 \tag{6.39}$$

$$b_1 = \sum T_\alpha + \sum T_\delta \tag{6.40}$$

and

$$a_1 = \frac{1}{K} + \sum T_\alpha + \sum T_\gamma \tag{6.41}$$

Hence, $C_v$, according to Equation 1.27, is defined by

$$\frac{1}{C_v} = a_1 - b_1 = \frac{1}{K} + \sum T_\gamma - \sum T_\delta \tag{6.42}$$

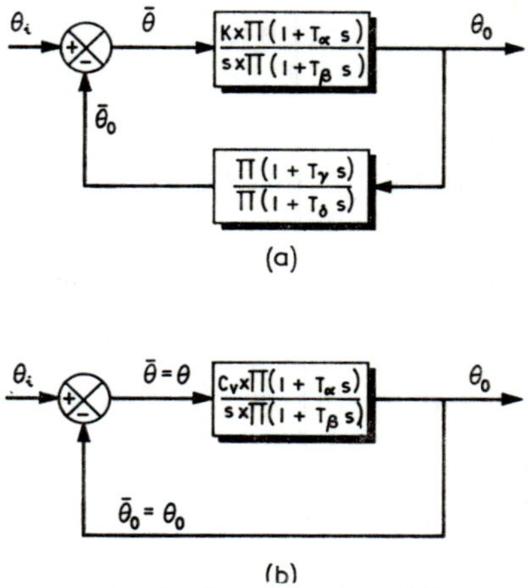


*Fig. 6.20. Block diagrams for the determination of $C_v$ in terms of the feedforward gain for single loop control systems with a basic integrator in the feedforward path; (a) block diagram for determining the relationship between $C_v$ and feedforward gain for the general single loop control system; (b) special case for the determination of $C_v$ from feedforward gain and unity feedback control systems*

Thus from Equation 6.42 we have the very important result that for unity feedback path systems $C_v$ = feedforward gain $K$, and thus for such systems we may obtain the value of $C_v$ directly from the root locus plot. The root locus gain obtained from the root locus plot is, of course, written in terms not of $1+Ts$, but in terms of $s+1/T$. For *unity feedback systems only* we must therefore write

$$C_v = \frac{\text{root locus gain} \times \prod T_\beta}{\prod T_\alpha}$$

or, alternatively,

$$\frac{\bar{\theta}_0}{\bar{\theta}}(s) = \frac{\theta_0}{\theta}(s) = \frac{C_v \times \prod(1+T_\alpha s)}{s \times \prod(1+T_\beta s)} \tag{6.43}$$

where $\prod T_\beta$ is the product of the time constants of the feedforward poles and $\prod T_\alpha$ is the product of the time constants of the feedforward zeros. Equation 6.43 is in a particularly easy form to remember, and the corresponding block diagram is shown in Fig. 6.20(b).

## 6.9 Lag Compensation in the s Plane

A lag compensation network with transfer function $(1+\tau s)/(1+\alpha\tau s)$ with $\alpha$ larger than unity may be inserted in the feedforward path of a system with dominant poles so placed that the transient and frequency responses are satisfactory but whose $C_v$ is too low, for the purpose of increasing $C_v$ without materially affecting the position of the dominant poles. As can be seen in the next example, from the system pole-zero array the effect is observable as the addition of a dipole quite near to the origin and the system zero is nearest the origin, hence giving rise to the description 'integrating dipole'. The system pole thus introduced is normally *much* closer to the origin than are the conventional dominant poles. This additional pole introduces a low-amplitude long-tailed component to the response, and if this component is not to affect settling time, the dipole pole and zero must be very close indeed.

EXAMPLE 6.9.1.

A unity feedback control system with open loop transfer function $18/[s(s+6)]$ has a satisfactory transient and frequency response, but $C_v$ should be 20, and is only 3·0 for the basic system. Design a lag network which will raise $C_v$ to 20 and yet will maintain the dominant poles in materially the same place in the s plane.

The block diagram of the uncompensated system is shown in Fig. 6.21(a), and the basic root locus is shown in Fig. 6.21(b). The uncompensated system transfer function is

$$\frac{\theta_0}{\theta_i}(s) = \frac{1}{1+\left(\frac{6}{18}\right)s+\frac{s^2}{18}} \tag{6.44}$$

Hence $\omega_n^2 = 18$ and $\omega_n = 4{\cdot}25$ rad/sec. Thus

$$\zeta = \frac{0{\cdot}33\,\omega_n}{2} = 0{\cdot}71$$

For this basic system,

$$C_v = \frac{1}{a_1} = 3{\cdot}0$$

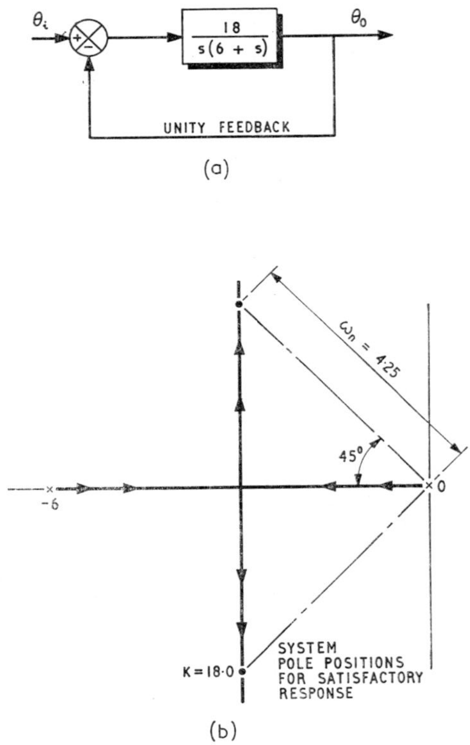


*Fig. 6.21. A system with satisfactory transient response but a low $C_v = 3{\cdot}0$ (Example 6.9.1): (a) block diagram for a system with satisfactory transient response but low $C_v$; (b) root locus for the uncompensated system with $C_v = 3{\cdot}0$*

Based on the consideration of 95% settling time, Truxal[58] has stated that for an integrating dipole to have an insignificant effect on transient response, the system pole must be no further away from the origin then 1·05 times the distance from the system zero. Let us evaluate the dipole geometry required to meet the $C_v$ requirement and this 1·05 restriction. Assuming the dominant poles do not move significantly by the introduction of the compensation network, let the distance of the added system zero from the origin be z, and the distance of the added system pole is then 1·05z. For the resultant three-pole one-zero system we may write

$$-\frac{1}{C_v} = \sum \frac{1}{\text{system poles}} - \sum \frac{1}{\text{system zeros}}$$

Hence

$$-\frac{1}{C_v} = -\frac{2\zeta}{\omega_n} + \frac{1}{1{\cdot}05\,z} - \frac{1}{z} \tag{6.45}$$

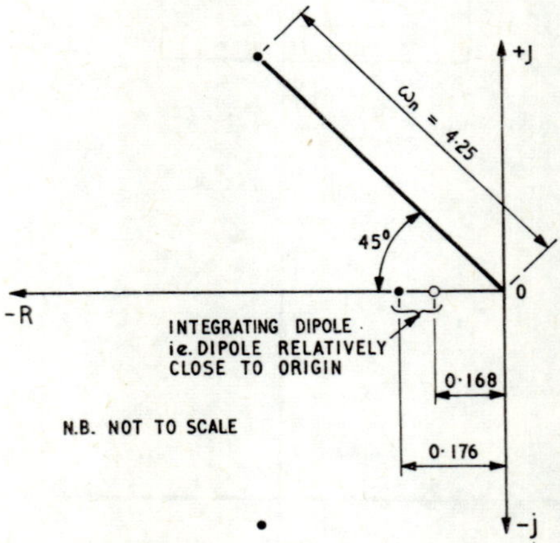


*Fig. 6.22. System pole-zero array for increasing $C_v$ from 3·0 to 20 but maintaining an unchanged transient response using the integrating dipole technique (Example 6.9.1)*

Substituting for $C_v$, $\zeta$ and $\omega_n$, we have, on reversing signs,

$$\frac{1}{20} = \frac{1}{3} + \frac{1}{z}\left(1 - \frac{1}{1{\cdot}05}\right)$$

and, on simplification,

$$z = -\frac{60}{17} \times \frac{0{\cdot}05}{1{\cdot}05} = -0{\cdot}168$$

and the corresponding system pole is $1{\cdot}05z = -0{\cdot}176$. The required system pole-zero array is shown in Fig. 6.22.

We have now to determine a suitable network to force the system into this pole-zero array. The mechanism can be explained by pointing out that we are going to add to the open loop configuration a lag network near the origin in the s plane to turn the basic (2,0) locus shown in Fig. 6.11(a) into the (3,1) locus shown in Fig. 6.12(d), with the third pole tucking in closely behind the open loop (and system) zero. A block diagram of the proposed compensated system is shown in Fig. 6.23(a). Because of the

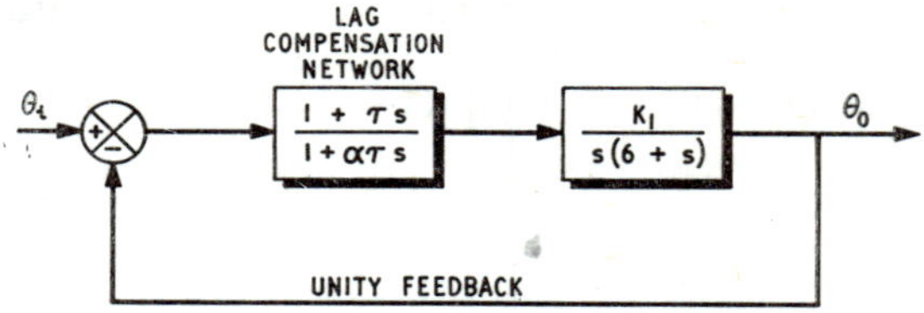


(a)

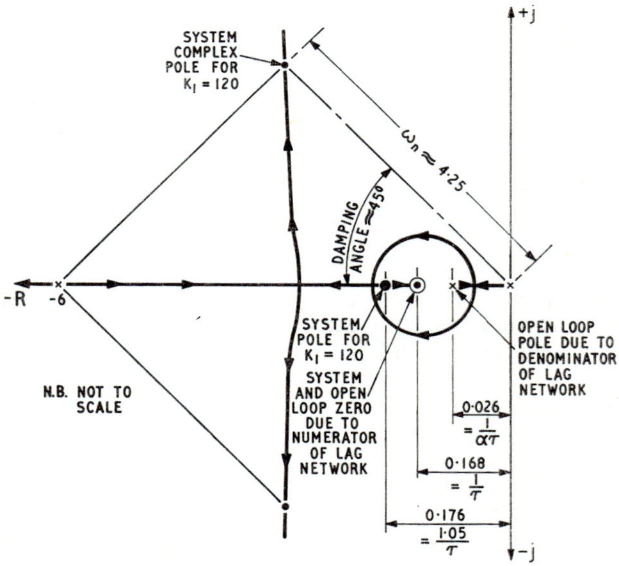


(b)

*Fig. 6.23. Introduction of a lag network to create an integrating dipole which will maintain the existing satisfactory transient response but will increase $C_v$ from 3·0 to 20 (Example 6.9.1): (a) block diagram of the system with a series lag compensation network added to form an integrating dipole in the system pole-zero array; (b) root locus of the compensated system showing the mechanism of creating the integrating dipole with a lag network*

presence of the compensation network, with transfer function

$$G_c(s) = \frac{1+\tau s}{1+\alpha\tau s}$$

the feedforward gain assumes some new value $K_1$, and we know from Equation 6.42 that $C_v$ is related to the feedforward gain in a straightforward manner. The open loop equation for the compensated system, written in the $\prod(1+Ts)$ form, is

$$\frac{\bar{\theta}_0}{\bar{\theta}}(s) = \frac{K_1(1+\tau s)}{6\left(1+\frac{s}{6}\right)(1+\alpha\tau s)s} = \frac{C_v(1+\tau s)}{s\left(1+\frac{s}{6}\right)(1+\alpha\tau s)} \tag{6.46}$$

which is the factored form required by Equation 6.43 and hence $C_v = K = K_1/6$. Since $C_v = 20$, we have immediately $K_1 = 120$.

The system transfer function for the compensated system is

$$\frac{\theta_0}{\theta_i}(s) = \frac{\dfrac{K_1(1+\tau s)}{6(1+s/6)(1+\alpha\tau s)s}}{1+\dfrac{K_1(1+\tau s)}{6(1+s/6)(1+\alpha\tau s)s}} \tag{6.47}$$

where $K_1 = 120$, and $\tau$ is known from the system zero position; i.e. $\tau = 1/0{\cdot}168 = 5{\cdot}95$. Simplifying Equation 6.47, we have

$$\frac{\theta_0}{\theta_i}(s) = \frac{1+5{\cdot}95s}{1+6{\cdot}0s+\left(\dfrac{1+6\times 5{\cdot}95\alpha}{120}\right)s^2+\left(\dfrac{5{\cdot}95\alpha}{120}\right)s^3} \tag{6.48}$$

but if the complex poles are still in approximately the same place as for the uncompensated system, we may write, from the system pole-zero array of Fig. 6.22,

$$\frac{\theta_0}{\theta_i}(s) \doteqdot \frac{1+5{\cdot}95s}{(1+5{\cdot}67s)\left(1+\dfrac{s}{3}+\dfrac{s^2}{18}\right)}$$

and hence a numerical transfer function can be obtained:

$$\frac{\theta_0}{\theta_i}(s) \doteqdot \frac{1+5{\cdot}95s}{1+6{\cdot}0s+1{\cdot}95s^2+0{\cdot}315s^3} \tag{6.49}$$

Comparing Equations 6.48 and 6.49 we see that we have one unknown, $\alpha$, and two coefficients from which $\alpha$ may be obtained. The two values of $\alpha$ will not quite agree, as the addition of the lag network will have slightly moved the complex system poles. However, the magnitude of the difference between these two values of $\alpha$ will give some indication of the degree of approximation involved.

From the $a_3$ coefficient we have

$$\alpha = \frac{120\times0{\cdot}315}{5{\cdot}95} = 6{\cdot}40$$

and from the $a_2$ coefficient we have

$$\alpha = \frac{(1{\cdot}95\times120)-1}{6\times5{\cdot}95} = 6{\cdot}50$$

The good agreement between these two values of $\alpha$ shows that the assumption of little movement of the complex system poles is reasonable. Finally, the pole of the compensating network is situated at

$$-\frac{1}{\alpha\tau} = -\frac{1}{5{\cdot}95\times6{\cdot}40} = -0{\cdot}026$$

The compensating network transfer function is therefore

$$G_c(s) = \frac{1+\tau s}{1+\alpha\tau s} = \frac{1+5{\cdot}95s}{1+38{\cdot}0s}$$

Fig. 6.23(b) shows the actual root locus, not drawn to scale, and indicates clearly how the introduction of a lag network with a pole nearest the origin results in a dipole with a system zero nearest the origin. In practice, once the mechanism of the compensation is understood, it would not be necessary to draw the root locus for this type of problem, the good agreement in $\alpha$ being sufficient evidence that the $C_v$ requirement is met with little change in the transient or frequency response which has been guaranteed by the factor of 1·05 on the system dipole.

## 6.10 Passive Lead Compensation in the s Plane

A single-stage passive derivative, or lead network, will have a transfer function

$$G_c(s) = \frac{1}{\alpha}\left(\frac{1+\alpha\tau s}{1+\tau s}\right)$$

and unlike the use of the lag compensation network, where it was found that both the compensation network pole and zero were relatively close to the origin, the addition of a lead compensation network can result in the network zero being relatively close to the origin and the network pole being far to the left. For example, if

$\tau = 0{\cdot}025$ and $\alpha = 10$, then the network zero is at $-4{\cdot}0$, whilst the network pole is at $-40$. The introduction of the lead network allows the salient branch of the root locus to be driven to the left, thus permitting an increase in the undamped natural frequency obtainable for a particular damping ratio of the dominant complex pole. However, the movement of the branch of the locus to the left is sometimes achieved at the expense of introducing a dipole to the system pole-zero array somewhat near the origin, with the system pole nearest the origin.

To illustrate the principles involved in lead compensation we shall firstly investigate the effect of addition of a known compensation network and then, secondly, we shall determine the compensation network required to allow the system to meet a given specification. In both examples we shall consider the addition of a single-stage lead network to the basic third order uncompensated system, thus turning a (3,0) system shown in Fig. 6.12(a) into a (4,1) system shown in Fig. 6.16(c). A comparison of these two figures immediately shows how the dominant complex poles may be driven to the left.

EXAMPLE 6.10.1

The third order roll autopilot loop is compensated by the incorporation of a lead network with transfer function

$$G_c(s) = \frac{1}{10}\left(\frac{1+0{\cdot}25s}{1+0{\cdot}025s}\right)$$

as shown in Fig. 6.24. Sketch the root locus and determine the system pole-zero arrays corresponding to complex system poles with damping ratios of 0·707 and 0·50.

The open loop transfer function for Fig. 6.24 is

$$\frac{\overline{\phi_0}}{\overline{\phi}}(s) = \frac{335\,K_g(1+0{\cdot}25s)}{10(1+0{\cdot}025s)(1+0{\cdot}179s)(1+0{\cdot}035s)s} \qquad (6.50)$$

and in the root locus form

$$\frac{\overline{\phi_0}}{\overline{\phi}}(s) = \frac{K(s+4)}{(s+40)\,(s+5{\cdot}6)\,(s+28{\cdot}6)s} \qquad (6.51)$$

where

$$K = \frac{335\times 0{\cdot}25K_g}{10\times 0{\cdot}025\times 0{\cdot}179\times 0{\cdot}035} = 53{,}600K_g$$

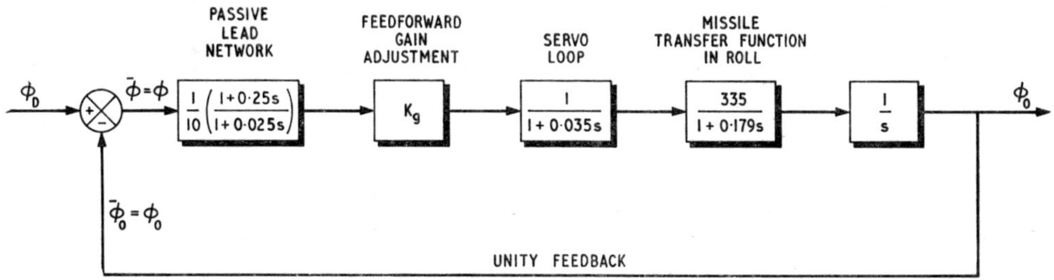


*Fig. 6.24. Block diagram of the basic third order autopilot compensated with a passive lead network to drive dominant poles to the left so that bandwidth, etc., is increased; a dipole with a system pole nearer the origin is introduced as a consequence (Example 6.10.1)*

From Equation 6.51 we may write down the open loop poles and zeros as

$$s_1 = 0$$
$$s_2 = -40$$
$$s_3 = -5{\cdot}6$$
$$s_4 = -28{\cdot}6$$

and

$$z_1 = -4{\cdot}0$$

Loci will exist on the negative real axis at points where the total number of open loop poles plus open loop zeros to the right is odd, hence there will be a branch of the locus between 0 and $-4{\cdot}0$, between $-5{\cdot}6$ and $-28{\cdot}6$, and between $-40$ and $-\infty$. There are $4-1$ high-gain asymptotes, and these will be at angles $\pm\pi/3$ and $\pm\pi$, and the asymptote intersection point on the negative real axis will be given by

$$\frac{\sum \text{open loop poles} - \sum \text{open loop zeros}}{n-m} = -\frac{(28{\cdot}6+40+5{\cdot}6)-4}{3}$$
$$= -23{\cdot}4$$

Clearly there must be a breakaway point on the negative real axis between $-5{\cdot}6$ and $-28{\cdot}6$, and using the small-angle approximation technique outlined in Section 3.3 *et seq.*, this is found to be at a distance of $-12{\cdot}85$. Further points on the locus are then found using a Spirule, and the complete locus is shown in Fig. 6.25.

The vectors required for the computation of $K_g$ corresponding to the damping ratio of 0·707 are shown in Fig. 6.26(a). Hence the root locus gain is given by

$$K = \frac{\prod \text{vectors to open loop poles}}{\prod \text{vectors to open loop zeros}}$$
$$= \frac{15{\cdot}3 \times 12 \times 21{\cdot}6 \times 31{\cdot}2}{12{\cdot}7}$$
$$= 9{,}750$$

Now, for this example, $K = 53{,}600\ K_g$; hence, for $\zeta = 0{\cdot}707$,

$$K = \frac{9{,}750}{53{,}600} = \frac{1}{5{\cdot}5}$$

Similarly the vectors required for the computation of $K_g$ corresponding to a damping ratio of 0·50 are shown in Fig. 6.26(b);

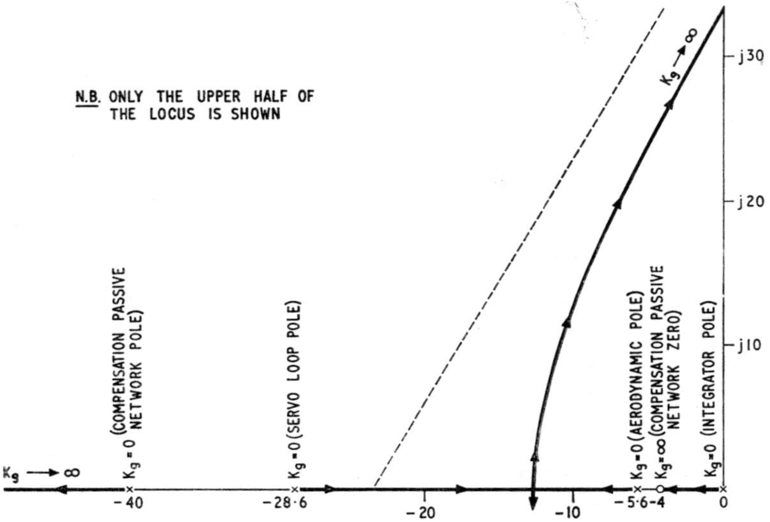


*Fig. 6.25. Compensated (4,1) root locus for the third order autopilot with passive lead network added (Example 6.10.1)*

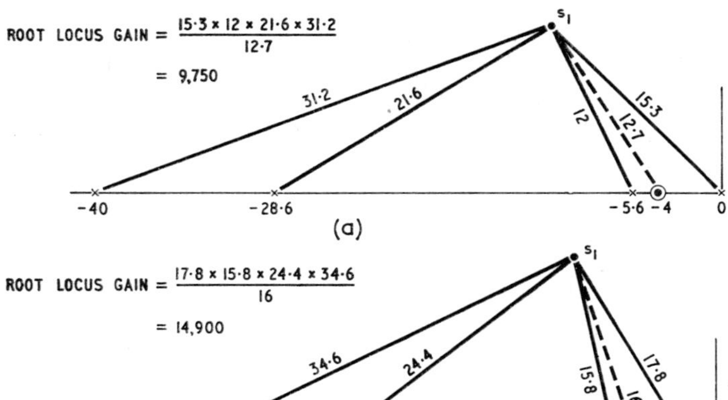


*Fig. 6.26. Vector diagrams for the computation of feedforward gain for specific values of ζ of complex poles (Example 6.10.1): (a) ζ = 0·707; (b) ζ = 0·50*

hence the root locus gain is given by

$$K = \frac{17{\cdot}8 \times 15{\cdot}8 \times 24{\cdot}4 \times 34{\cdot}6}{16} = 14{,}900$$

Hence, for $\zeta = 0{\cdot}50$,

$$K = \frac{14{,}900}{53{,}600} = \frac{1}{3{\cdot}60}$$

The other roots corresponding to these particular values of $K_g$ may be determined in a number of ways, since it may reasonably be assumed that one pole is near the open loop zero and an iteration technique can be used to find this pole. The fourth pole can then be found using the invariant centre of gravity theorem which is applicable in this case, since $n-m > 1{\cdot}0$. Alternatively, the locus may be calibrated in terms of gain in the usual way, or the system transfer function coefficients may be evaluated and the two remaining roots obtained by dividing the known quadratic factor into the characteristic equation.

Using any of these methods, for $K_g = 1/5{\cdot}50$, the third pole is at $-3{\cdot}30$, and the fourth pole is at $-50$. The corresponding system pole-zero array is shown in Fig. 6.27(a), the system zero at $-4$

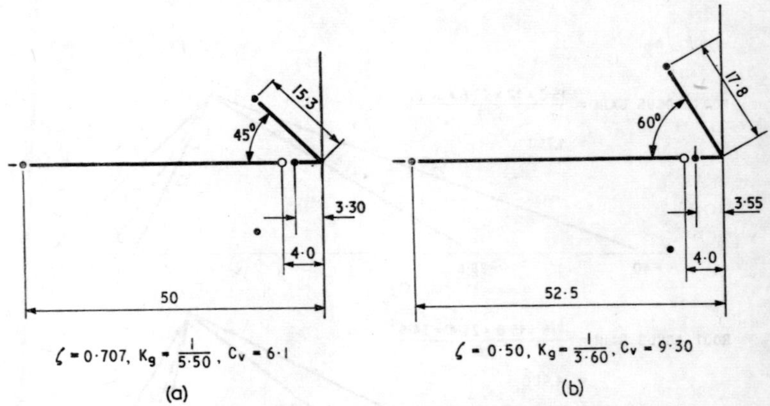


*Fig. 6.27. System pole-zero arrays for the basic third order autopilot transformed into a (4,1) system by the addition of a passive lead network (Example 6.10.1): (a) $\zeta = 0{\cdot}707$, $K_g = 1/5{\cdot}50$, $C_v = 6{\cdot}1$; (b) $\zeta = 0{\cdot}50$, $K_g = 1/3{\cdot}60$, $C_v = 9{\cdot}30$*

being obtained by writing down the system transfer function or simply remembering that feedforward path zeros are system zeros. Since the system has unity feedback path, $C_v$ may be obtained from the open loop gain very simply, it being remembered from Equation 6.43 that

$$\frac{\bar{\phi}_0}{\bar{\phi}}(s) = \frac{C_v \times \prod(1+T_\alpha s)}{s \times \prod(1+T_\beta s)} \tag{6.52}$$

and from Equation 6.50, by direct comparison with Equation 6.52, we have

$$C_v = \frac{335\ K_g}{10} = \frac{335}{5{\cdot}5 \times 10} = 6{\cdot}10$$

For the case where the damping ratio of the complex poles is 0·50, i.e. $K_g = 1/3{\cdot}60$, the third pole is at $-3{\cdot}55$ and the fourth pole is at $-52{\cdot}5$. The corresponding system pole-zero array is shown in Fig. 6.27(b). The new value of $C_v$ is

$$C_v = \frac{335\ K_g}{10} = \frac{335}{3{\cdot}60 \times 10} = 9{\cdot}30$$

To check the validity of our working it is possible to check $C_v$ directly from the system pole-zero array in accordance with Equation 3·32. For $K_g = 1/5{\cdot}50$, the value of $C_v$ from the system poles and zeros is 6·0, and for $K_g = 1/3{\cdot}6$ the value of $C_v$ from the system poles and zeros is 9·40.

Note from Fig. 6.27 the presence of the dipole, which is nothing like so near the origin as when lag compensation was used. In both cases the transient and frequency responses are slightly affected by the presence of the dipole. However, since the complex system poles are much more suitably placed than for the uncompensated system, the overall benefit is still very considerable.

We are now in a position to consider the choice of a lead compensation to enable a system to meet a given performance specification. In Fig. 6.28, $s_1$ represents a point in the s plane through which it is desired that a branch of the root locus shall pass. If a single-stage lead network of transfer function

$$G_c(s) = \frac{1}{\alpha}\left(\frac{1+\alpha\tau s}{1+\tau s}\right)$$

is inserted in the feedforward path, then an open loop zero is added at $-1/\alpha\tau$, and an open loop pole is added at $-1/\tau$. In general $\alpha$ will be considerably greater than unity, say 5 or 10. From the vector diagram shown in Fig. 6.28, the nett angular effect of adding the lead network will be a clockwise contribution of $\delta_1-\gamma_1$. Since in Chapter 3 for convenience we defined anticlockwise angles as positive, the nett contribution at $s_1$ will be $-(\delta_1-\gamma_1)$. The absolute limiting value for the phase contribution of a single-stage lead network occurs when the compensation open loop zero is at the origin, and the compensation open loop pole is at infinity, as shown in Fig. 6.29.

Note that practical limitations on network components will reduce the maximum phase contribution possible at a point $s_1$ due to a single-stage lead compensation network to a value somewhat

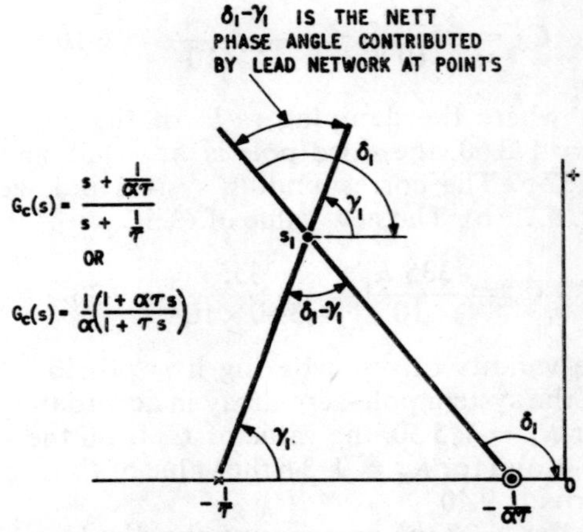


*Fig. 6.28. Vector diagram showing the phase angle contributed at a point in the* s *plane by a single-stage passive lead network*

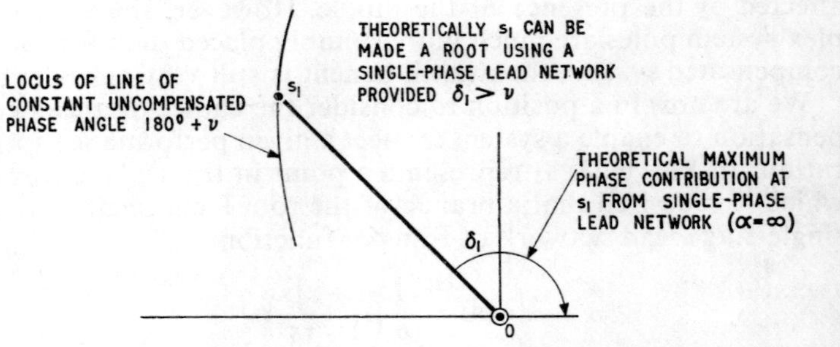


*Fig. 6.29. Absolute maximum phase contribution at* s *due to a single-stage passive lead network*

below the angle $\delta_1$ defined in Fig. 6.29. Thus if the phase angle locus of the uncompensated system, as defined, for example, in Fig. 6.19, passing through the point $s_1$ is greater than $180°+\delta_1$, a two-stage compensation network at least is required. A rough and ready guide to a practical nett phase contribution from a single-stage lead network is about 90°, but this is considerably a function of the point $s_1$; however one might reasonably expect to resort to double-stage compensation of the form

$$G_c(s) = \frac{1}{\alpha^2}\left(\frac{1+\alpha\tau s}{1+\tau s}\right)^2$$

if the phase angle contour passing through the $s_1$ point is in the region of 270°. Providing the limiting angle criterion is met, then the root locus can be made to pass through $s_1$ and, furthermore, a wide variety of compensation networks can be chosen which will provide the necessary phase contribution. However, the system specification may require that not only does the locus pass through $s_1$, but that the feedforward gain at that point is a certain value, thus meeting the $C_v$ requirement in a unity feedback system. It is then necessary to evaluate the gain requirement as well as the angle requirement, and the problem is to determine a network which meets both these requirements simultaneously. Fortunately a graphical technique is available which makes the design procedure quite straightforward. The procedure is best illustrated by means of an example.

EXAMPLE 6.10.2

It is required that the roll autopilot loop shown in Fig. 6.30 should meet the specification of a pair of complex poles with $\omega_n = 15$ and $\zeta = 0{\cdot}5$ (for acceptable transient and frequency response), and at the same time $C_v$ is required to be 8·0. Determine the transfer function of a suitable single-stage lead network.

The basic uncompensated open loop pole-zero array is shown in Fig. 6.31, and the desired root location $s_1$ is shown at $\omega_n = 15$, $\zeta = 0{\cdot}5$. By direct measurement the phase angle contour passing through the point $s_1$ is $120°+100°+32° = 252°$. For the root locus to pass through the point $s_1$ the phase angle at this point must be 180°, hence the nett lead angle must be $252°-180° = 72°$, which dictates one requirement for the compensation network. The uncompensated root locus gain at this point is simply the product of the vectors from $s_1$ to the open loop poles, i.e. the root locus gain $= 25{\cdot}6\times 13\times 15 = 5{,}000$.

Adding the compensation network to the s plane is shown in Fig. 6.32, where at this stage $c$ and $\alpha$ (or $\alpha$ and $\tau$) are unknown. The correction to the root locus gain due to the addition of the compensation network is $b/a$, and the new root locus gain will be 5,000 $b/a$. Referring to Fig. 6.30, the open loop transfer function is

$$\frac{\bar{\phi}_0}{\bar{\phi}}(s) = \frac{K_g K_a(1+\alpha\tau s)}{\alpha(1+\tau s)(1+T_c s)(1+T_a s)s} \tag{6.53}$$

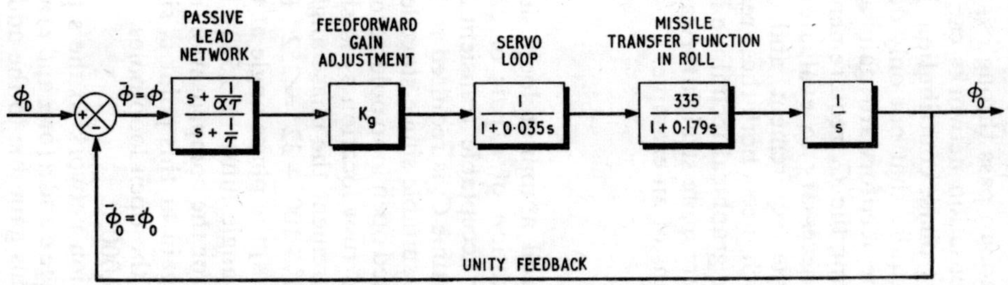


*Fig. 6.30. Passive lead compensated autopilot required to meet specification of $C_v = 8$ and dominant pole characteristics of $\omega_n = 15$, $\zeta = 0{\cdot}5$ (Example 6.10.2)*

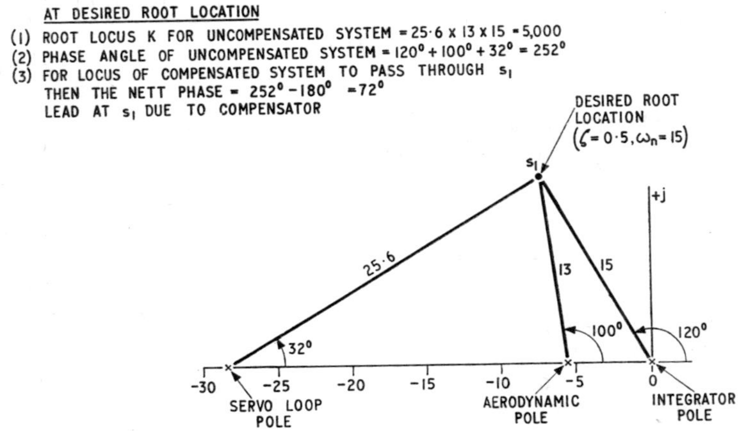


*Fig. 6.31. Vector diagram showing the lead angle required from the phase advance network to achieve the desired conjugate system pole pair location of $\zeta = 0{\cdot}5$, $\omega_n = 15$ (Example 6.10.2)*

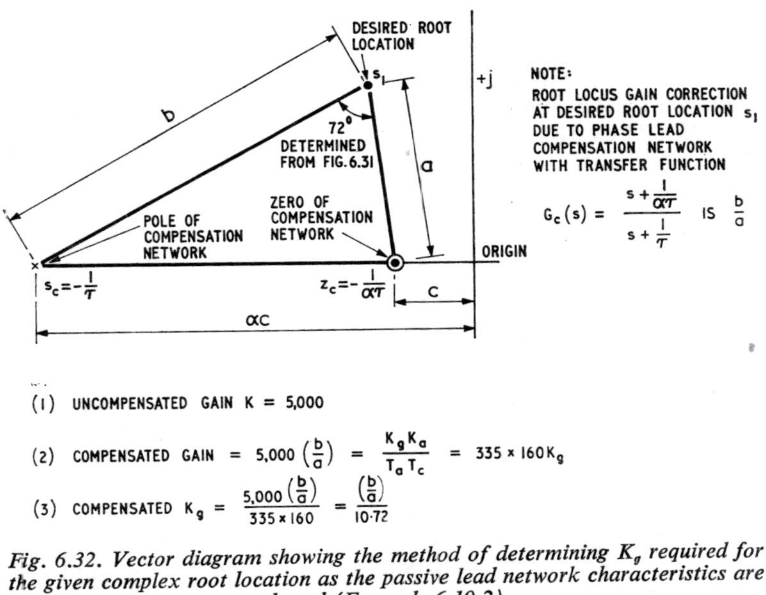


*Fig. 6.32. Vector diagram showing the method of determining $K_g$ required for the given complex root location as the passive lead network characteristics are altered (Example 6.10.2)*

and since the system is unity feedback we may write, from Equation 6.43,

$$C_v = \frac{K_g K_a}{\alpha} \tag{6.54}$$

Now the root locus gain $K$ is given when the open loop equation is written in the particular form

$$\frac{\bar{\phi}_0}{\bar{\phi}}(\mathrm{s}) = \frac{K \times \prod\left(\mathrm{s}+\frac{1}{T_\alpha}\right)}{\mathrm{s} \times \prod\left(\mathrm{s}+\frac{1}{T_\beta}\right)}$$

where $T_\alpha$ is the general feedforward zero time constant defined in Section 6.8, and is unrelated to the same symbol used elsewhere in the book as the acceleration feedback constant. From Fig. 6.30, we manipulate to obtain the form

$$\frac{\bar{\phi}_0}{\bar{\phi}}(\mathrm{s}) = \frac{K_g K_a\left(\mathrm{s}+\frac{1}{\alpha\tau}\right)}{T_a T_c\left(\mathrm{s}+\frac{1}{T_c}\right)\left(\mathrm{s}+\frac{1}{T_a}\right)\left(\mathrm{s}+\frac{1}{\tau}\right)} \tag{6.55}$$

Substituting known values, we determine from Equation 6.55 the specific relationship

$$K = 335 \times 160 K_g = 53{,}600 K_g$$

But the compensated root locus gain, in terms of the vectors introduced by the compensation network is 5,000 $b/a$, hence $b/a = 10{\cdot}72 K_g$. Thus two simultaneous equations can be determined for $K_g$. From the $C_v$ requirement of Equation 6.54, we may write

$$K_g = \frac{8\alpha}{335} \tag{6.56}$$

and from the root location requirement, we may write

$$K_g = \frac{b/a}{10{\cdot}72} \tag{6.57}$$

Both $\alpha$ and $b/a$ are functions of the s plane location of the compensation open loop pole and zero, which must contribute 72° phase angle at $\mathrm{s}_1$. The procedure is now to assume a position $\mathrm{z}_c$ for the compensation open loop zero, draw the vector to $\mathrm{s}_1$, construct the 72° angle, and thus determine the position $\mathrm{s}_c$ of the compensation open loop pole as shown in Fig. 6.32. For the zero position chosen, $\alpha$ is determined simply by the ratio $\mathrm{s}_c/\mathrm{z}_c$, and $b/a$ can be measured directly. The values of $K_g$ to meet the $C_v$ and root

TABLE REQUIRED FOR CONSTRUCTION OF SUBSIDIARY GRAPHS (FIGS. 6.34 AND 6.35)

| ASSUMED ZERO POSITION $z_c$ | CORRESPONDING POLE POSITION P | $\alpha = \frac{P}{z_c}$ | b | a | $\frac{b}{a}$ | $1/K_g$ FROM ROOT LOCATION REQUIREMENT $= 10{\cdot}72/\left(\frac{b}{a}\right)$ | $1/K_g$ FROM $C_v$ REQUIREMENT $= 41{\cdot}8/\alpha$ |
|---|---|---|---|---|---|---|---|
| 2 | 22·7 | 11·35 | 20 | 14·2 | 1·41 | 7·60 | 3·59 |
| 3 | 24·8 | 8·26 | 21·6 | 13·8 | 1·56 | 6·85 | 5·06 |
| 4 | 27·6 | 6·90 | 23·8 | 13·5 | 1·76 | 6·08 | 6·06 |
| 5 | 30·9 | 6·20 | 26·7 | 13·3 | 2·00 | 5·35 | 6·75 |
| 6 | 35·8 | 5·96 | 31 | 13·1 | 2·36 | 4·54 | 7·00 |
| 7 | 42·8 | 6·11 | 37·6 | 13·0 | 2·89 | 3·71 | 6·84 |
| 8 | 54·6 | 7·84 | 48·6 | 13·0 | 3·74 | 2·86 | 5·34 |

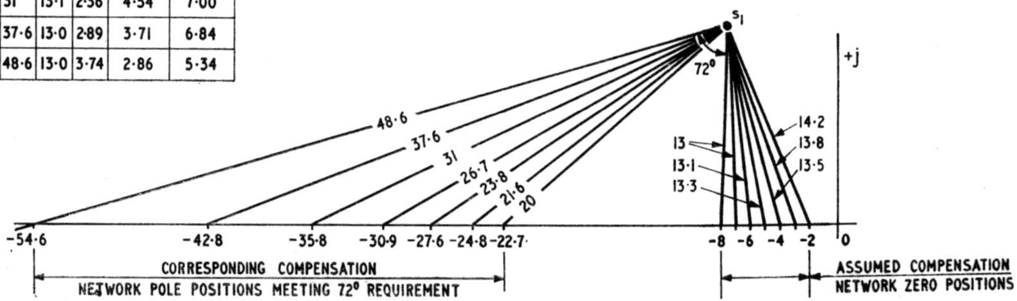


*Fig. 6.33. Determination of data for the construction of subsidiary graphs required for the design of the system to meet the requirements $C_v = 8{\cdot}0$, $\omega_n = 15$ and $\zeta = 0{\cdot}5$ using passive lead compensation (Example 6.10.2)*

requirements can then be evaluated from Equations 6.56 and 6.57 respectively. When the two values of $K_g$ thus evaluated are equal, the compensation network will be completely determined.

If the compensation zero $z_c$ is allowed to move along the negative real axis, the construction can be repeated for a number of values of $z_c$ as shown in Fig. 6.33. Computations involved in the determination of $K_g$ are also tabulated in Fig. 6.33. Due to the size of the numbers involved it is convenient to plot $1/K_g$ in preference to $K_g$ in the subsidiary graphs required to determine $\alpha$ and $\tau$. $1/K_g$ is plotted against the zero position in Fig. 6.34,

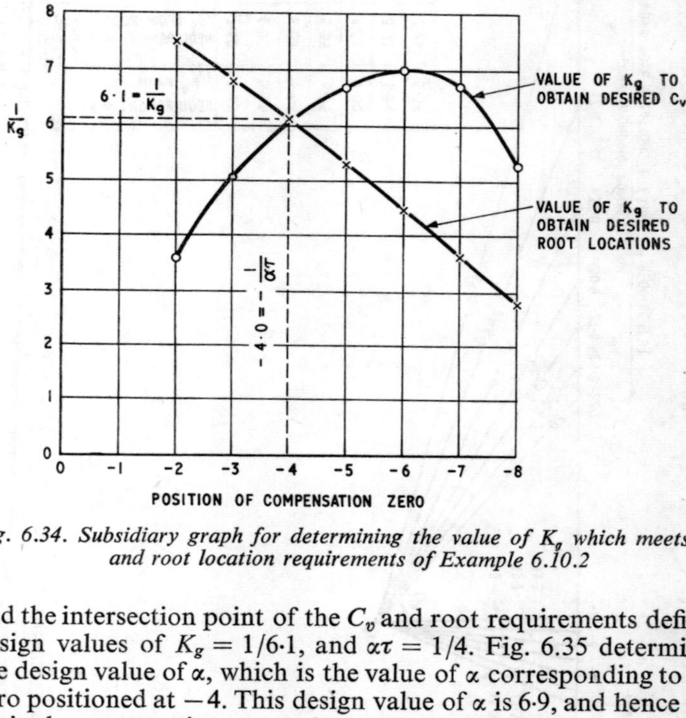


*Fig. 6.34. Subsidiary graph for determining the value of $K_g$ which meets $C_v$ and root location requirements of Example 6.10.2*

and the intersection point of the $C_v$ and root requirements defines design values of $K_g = 1/6{\cdot}1$, and $\alpha\tau = 1/4$. Fig. 6.35 determines the design value of $\alpha$, which is the value of $\alpha$ corresponding to the zero positioned at $-4$. This design value of $\alpha$ is 6·9, and hence the desired compensation network transfer function is

$$G_c(s) = \frac{1}{6{\cdot}9}\left(\frac{1+0{\cdot}25s}{1+0{\cdot}036s}\right)$$

giving the compensation open loop zero at $-4{\cdot}0$, and the compensation open loop pole at $-27{\cdot}6$. The block diagram of the final design is shown in Fig. 6.36.

Determination of the system pole near the zero at $-4{\cdot}0$ is simplified by using the vector diagram shown in Fig. 6.37. The design value of $b/a$ from the table of Fig. 6.33 is 1·76, hence the root locus gain is $5{,}000b/a = 8{,}800$. From Fig. 6.37, the root locus gain is given by

$$K = \frac{(24{\cdot}6-x)(23{\cdot}6-x)(1{\cdot}6+x)(4-x)}{x} = 8{,}800$$

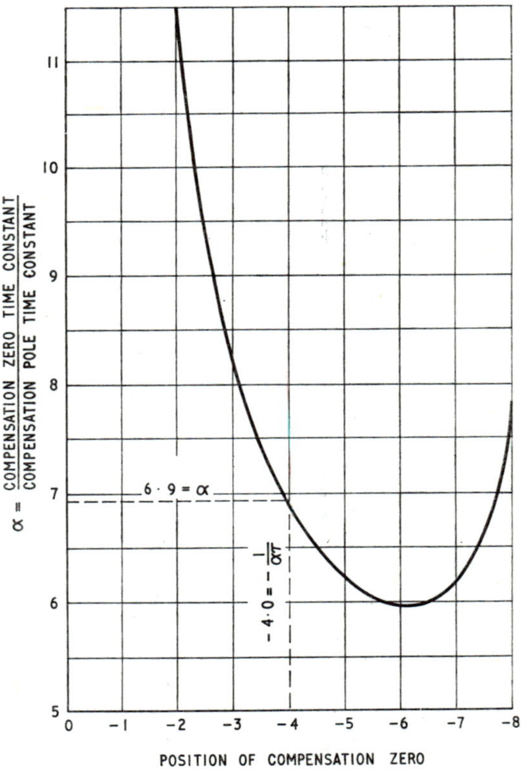


*Fig. 6.35. Subsidiary graph for determining the passive lead network which meets $C_v$ and root location requirements of Example 6.10.1*

which may be solved using an iteration process giving $x = 0{\cdot}46$, and the pole position at $-3{\cdot}54$. To find the fourth pole position for this value of gain the centre of gravity theorem is used. For gain = 0, the poles are given by $-(28{\cdot}6+27{\cdot}6+5{\cdot}6+0) = -61{\cdot}8$.

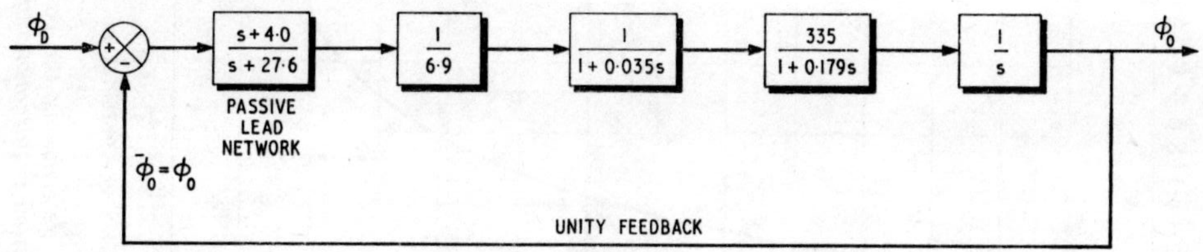


*Fig. 6.36. Block diagram of the passive lead compensated system which meets the $C_v = 8{\cdot}0$, $\omega_n = 15$ and $\zeta = 0{\cdot}50$ requirements of Example 6.10.2*

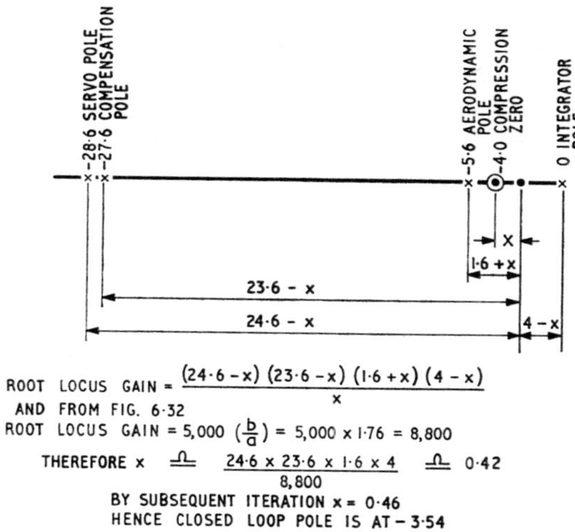


*Fig. 6.37. Vector diagram for the determination of the closed loop pole near the phase advance zero for the passive lead compensated system of Example 6.10.2*

Let the fourth pole be at distance $y$ from the origin, then $-(3{\cdot}54+y+7{\cdot}50+7{\cdot}50) = -61{\cdot}8$, and thus $y = 43{\cdot}26$. The final system pole-zero array is shown in Fig. 6.38. Obviously the dipole will slightly affect the transient and frequency response, but basically these responses will be dominated by the complex poles, and the $C_v$ requirement is definitely met. The dipole will tend to increase the settling time as a result of the system pole being nearest the origin.

Use of this graphical technique is not restricted to low-order systems; it is a very powerful design tool where a dominant mode

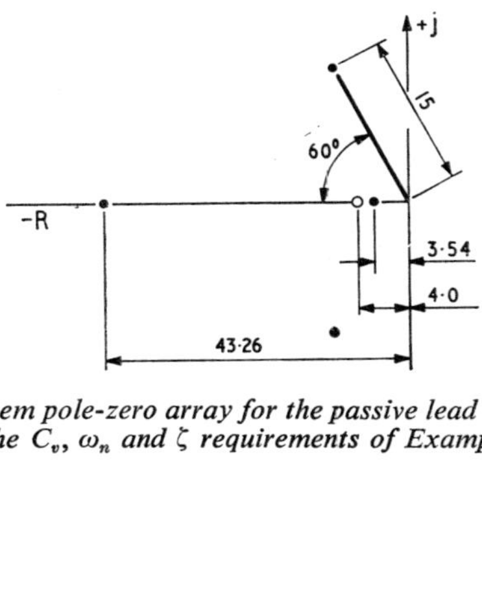


*Fig. 6.38. Final system pole-zero array for the passive lead compensated system meeting the $C_v$, $\omega_n$ and $\zeta$ requirements of Example 6.10.2*

complex pair plus a $C_v$ requirement has to be met, provided the main problem is to drive two poles to the left to obtain the desired root locations. If more than two poles have to be driven to the left, additional compensation is required, and cancellation compensation may be used as an adjunct to lead compensation, provided serious non-linear problems are not encountered as a result of the compensation.

## 6.11 Cancellation Compensation in the s Plane

Cancellation compensation involves the introduction of networks such that unwanted poles in the feedforward path are cancelled approximately by the zeros of the compensating network, leaving the poles of the compensating network as replacements. If the unwanted poles are cancelled exactly, then that branch of the locus disappears completely, and the system pole is therefore exactly cancelled by the system zero at the same point, since zeros in the feedforward path are system zeros also. For inexact cancellation, that branch of the locus will be very short (in general), and for the value of gain chosen the system pole will have migrated from the open loop position towards the open loop (and system) zero. Thus the odds are that the system pole and zero will be closer together, and more nearly cancelled out, than for the corresponding open loop pole and open loop zero. Even with approximate cancellation, therefore, the deleterious effect of particular open loop poles can be removed from the system performance.

As an illustration of the technique, consider the fourth order autopilot shown in Fig. 6.39, where, if a relatively high bandwidth is required, the open loop pole at $-1/T_a$ and the complex open loop poles with undamped natural frequency $\omega_{ns}$ and damping ratio $\zeta_s$ will prove a source of embarrassment, as can be seen from the basic root locus for the (4,0) system shown in Fig. 6.17(a). The negative real pole can be cancelled by the insertion of a phase advance network with transfer function

$$G_{c1}(s) = \frac{1}{\alpha}\left[\frac{1+T_a s}{1+\left(\dfrac{T_a}{\alpha}\right)s}\right]$$

thus replacing the pole at $-1/T$ by a pole at $-\alpha/T_a$. Cancellation of the complex open loop poles can be achieved by the insertion of a second order compensation network with transfer function

$$G_{c2}(s) = \frac{s^2+2\zeta_s\,\omega_{ns}s+\omega_{ns}{}^2}{(s+a)(s+b)}$$

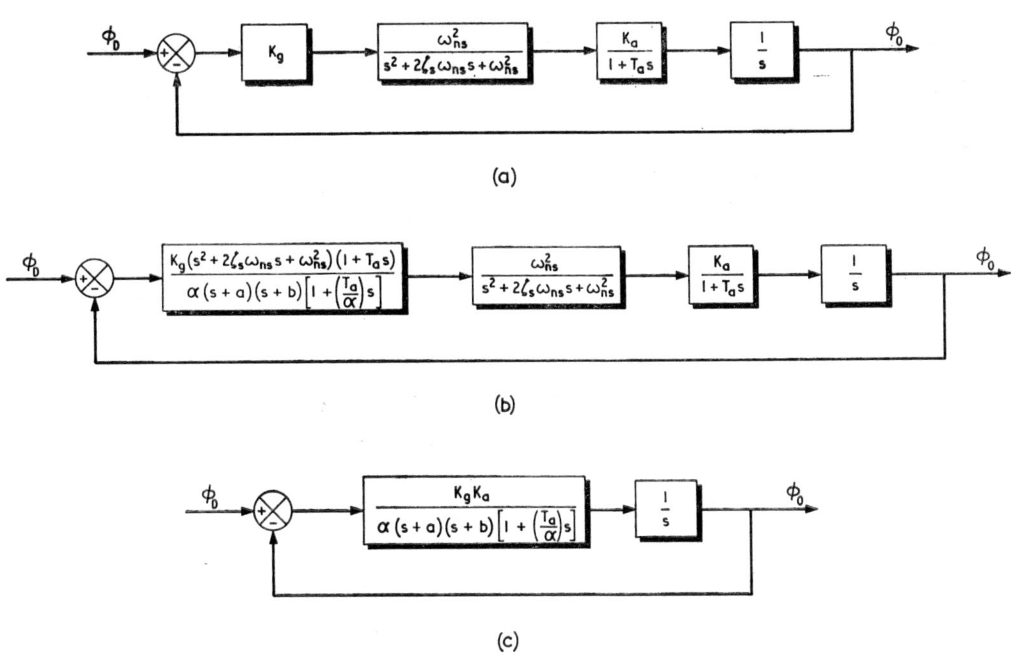


*Fig. 6.39. Use of the cancellation type of compensation to remove the effect of unwanted open loop poles from the system response: (a) block diagram of basic uncompensated autopilot; (b) block diagram of compensated autopilot; (c) simplified block diagram of compensated autopilot assuming perfect cancellation of unwanted open loop poles*

which involves the introduction of two new open loop poles at $-a$ and $-b$ on the negative real axis. Fig. 6.39(b) shows the compensated autopilot with the two compensation networks included. Assuming perfect cancellation, the uncompensated block diagram can be replaced by the simplified compensated block diagram of Fig. 6.39(c). Corresponding manipulation in the s plane is shown in Fig. 6.40. Uncompensated open loop poles are shown in Fig. 6.40(a), and adding the phase advance network

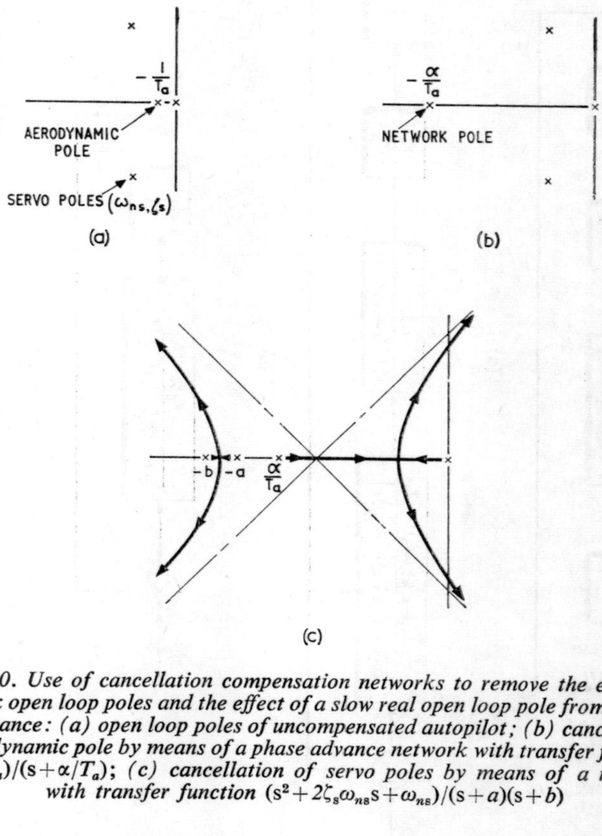


*Fig. 6.40. Use of cancellation compensation networks to remove the effect of complex open loop poles and the effect of a slow real open loop pole from system performance: (a) open loop poles of uncompensated autopilot; (b) cancellation of aerodynamic pole by means of a phase advance network with transfer function* $(s+1/T_a)/(s+\alpha/T_a)$; *(c) cancellation of servo poles by means of a network with transfer function* $(s^2+2\zeta_s\omega_{ns}s+\omega_{ns})/(s+a)(s+b)$

to cancel the aerodynamic pole results in the modified open loop pole array of Fig. 6.40(b). Cancellation of the complex open loop poles by the insertion of the second order compensation network and subsequent replacement by real open loop poles is shown in Fig. 6.40(c). The resultant root locus for the system with four open loop poles (at 0, $-\alpha/T_a$, $-a$ and $-b$) is similar to the locus shown in Fig. 6.16(a), but the dominant poles have been forced much further to the left compared with Fig. 6.17(a), and with judicious choice of the new open loop poles a great increase in system bandwidth is possible compared with the uncompensated system. As stated previously, the drawback with the cancellation compensation method is that feedforward gain (of amplifiers and power elements) may be used wastefully, and saturation-type non-linearities may cause trouble due to the high signal levels that may be present because of the amplification caused by the compensation networks.

In general, cancellation compensation, passive lag compensation, and passive lead compensation are all characterised by the introduction in the feedforward path of a network with equal numbers of poles and zeros. The open loop zeros thus introduced become system zeros, which are required approximately to cancel system poles by forming dipoles.

## 6.12 Feedback Compensation in the s Plane

Feedback compensation has already been met in Chapters 3, 4 and 5, and in this section we shall assume that the compensation is achieved by feeding back to the main comparator signals which are proportional to the various output derivatives. If, in practice, the feedback compensation is achieved by feeding back derivatives from some point other than the system output, the block diagram can be manipulated to include the compensation terms in the main feedback path using the techniques outlined in Section 3.11. Similar manipulation may be undertaken if the compensation signals are fed back to points other than the main comparator. If it is assumed that pure derivative signals can be obtained by instantaneous direct measurement of the physical quantities involved, the compensation is known as pure feedback compensation, and we may write the feedback transfer function in the form

$$H(s) = 1+T_r s+T_\alpha s^2+\dots \qquad (6.58)$$

i.e. there are feedback path zeros but there are no feedback path poles. We know from past work that feedback path zeros *do not* form system zeros, so the technique of using pure feedback compensation forces the root locus to take up a certain geometry, but system dipoles will not be formed because of the absence of the system zero which is present when lead or lag compensation is introduced. Specifically, if the open loop equation is of order $(r, 0)$, the introduction of a rate feedback term will result in a compensated root locus of order $(r, 1)$, but the system transfer function will still be of order $(r, 0)$, as was seen in Example 3.5.1.

If the derivative signals are obtained by using passive networks, or if significant lags are present in the instruments, the compensation will no longer be pure, and the result will be the introduction of poles in the feedback path, which will give rise to system zeros. For example, suppose the measurement of rate feedback is accompanied by an exponential lag, then we may write

$$H(s) = 1+\frac{T_r s}{1+T_{rL}s} \tag{6.59}$$

which may be simplified to

$$H(s) = \frac{1+(T_r+T_{rL})s}{1+T_{rL}s} \tag{6.60}$$

i.e. we have introduced an open loop pole at $-1/T_{rL}$ and an open loop zero at $-1/(T_r+T_{rL})$, but we have also introduced a system zero at $-1/T_{rL}$. Under some circumstances impure feedback compensation may be more advantageous than pure feedback compensation, due to the introduction of the system zero. Impure rate feedback compensation therefore transforms an $(r, 0)$ uncompensated locus into an $(r+1, 1)$ locus, and also transforms the $(r, 0)$ uncompensated system into an $(r+1,1)$ compensated system.

Extensive contributions to the literature concerning feedback compensation have been made by Thaler *et al.*[59, 60] and the deliberate use of a feedback path pole has been described there in detail. We shall concentrate on the manipulation of the root locus using pure feedback compensation only, since the extension of the techniques described hereafter to the impure feedback compensation case is relatively straightforward.

Using one of the techniques suggested by Thaler, the feedforward gain may be fixed by threshold or stiffness (load disturbance rejection) considerations. This will determine the system transfer function for the basic uncompensated system, and the system poles and zeros can be used as the starting and terminating points for the root locus calibrated in terms of the compensating element. As an example of the application of the feedback compensation technique, let us consider the problem of stabilising a third order system for a particular value of feedforward gain. This is the same example as was solved so beautifully in Chapter 5 using the coefficient plane method. In fact, the coefficient plane method is ideal for solving the design problem if the basic system is third order, and pure rate and acceleration feedback compensation are used to achieve the desired system performance. However, the example illustrates feedback compensation from first principles.

EXAMPLE 6.12.1

The third order system autopilot shown in Fig. 6.41(a) is to be stabilised using pure rate feedback such that the damping ratio of the complex system poles is to be a maximum for the case when the feedforward gain, fixed by stiffness requirements is $K_g = 1/1{\cdot}6$.

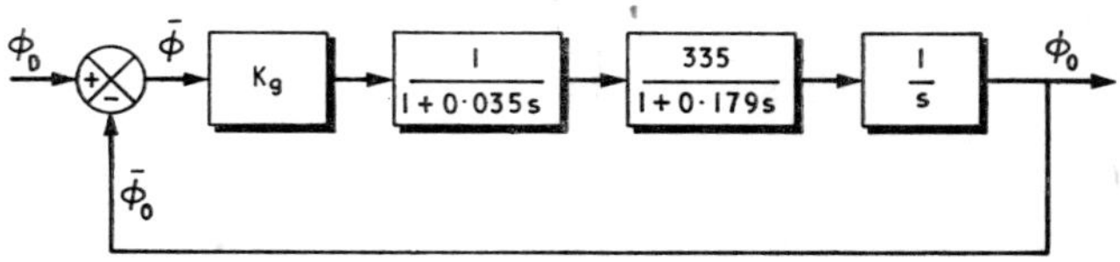


(a)

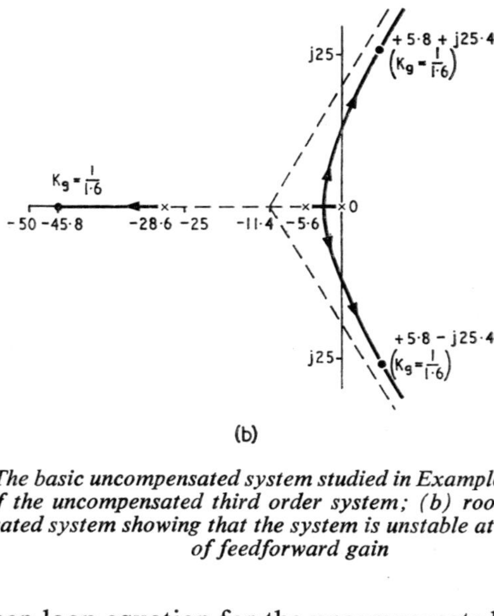


(b)

*Fig. 6.41. The basic uncompensated system studied in Example 6.12.1: (a) block diagram of the uncompensated third order system; (b) root locus plot of the uncompensated system showing that the system is unstable at the specified value of feedforward gain*

The open loop equation for the uncompensated system is

$$\frac{\bar{\phi}_0}{\bar{\phi}}(s) = \frac{335K_g}{(1+0{\cdot}035s)(1+0{\cdot}179s)s} \tag{6.61}$$

giving three open loop poles at 0, $-28{\cdot}6$ and $-5{\cdot}6$. The asymptotes intersect at a distance $-(28{\cdot}6+5{\cdot}6)/3 = -11{\cdot}4$ from the origin and, of course, the asymptotes are at angles $\pm\pi/3$ and $\pi$ as shown in Fig. 6.41(b). The breakaway point on the negative real axis is approximately $-2{\cdot}7$. The locus is then sketched using a Spirule to determine the intermediate points, and when the locus is calibrated in terms of feedforward gain, the uncompensated system poles are found to be at $s = -45{\cdot}8$ and $s = 5{\cdot}8 \pm j25{\cdot}4$, thus showing that the uncompensated system is unstable at the specified value of gain.

We may write the system transfer function of the uncompensated system symbolically as

$$\frac{\phi_0}{\phi_D}(s) = \frac{1}{1+\dfrac{s}{K_gK_a}+\left(\dfrac{T_a+T_c}{K_gK_a}\right)s^2+\left(\dfrac{T_aT_c}{K_gK_a}\right)s^3} \qquad (6.62)$$

and for the chosen values of $K_g$, $K_a$, $T_a$ and $T_c$ Equation 6.62 may be written numerically in the form $a_0 \neq 1$:

$$\frac{\phi_0}{\phi_D}(s) = \frac{33{,}500}{33{,}500+160s+34{\cdot}2s^2+s^3} \qquad (6.63)$$

Using the pole positions indicated in Fig. 6.41(b), the factored equation is

$$\frac{\phi_0}{\phi_D}(s) = \frac{33{,}500}{(s+45{\cdot}8)(s-5{\cdot}8+j25{\cdot}4)(s-5{\cdot}8-j25{\cdot}4)} \qquad (6.64)$$

Considering now the compensated system, the actual block diagram is shown in Fig. 6.42(a). Using the normal block diagram manipulation techniques outlined in Chapter 3, the compensation term is isolated as shown in Fig. 6.42(b) and finally, in Fig. 6.42(c), the simplest form for root locus determination in terms of $T_r$ is shown. It will be noted that the dummy open loop equation thus generated is

$$\frac{\overline{\phi}_0}{\overline{\phi}}(s) = \frac{33{,}500\,T_rs}{(s+45{\cdot}8)\,(s-5{\cdot}8+j25{\cdot}4)\,(s-5{\cdot}8-j25{\cdot}4)} \qquad (6.65)$$

Equation 6.65 has three open loop poles, which are the system poles of the uncompensated system for the desired value of feedforward gain. The starting point of the compensated locus is therefore the uncompensated system poles, i.e. $s = -45{\cdot}8$ and $s = 5{\cdot}8 \pm j25{\cdot}4$. There is also an open loop zero at the origin since $T_r = 0$ is a zero of Equation 6.65. The compensated root locus is thus of (3,1) type, and the high-gain asymptotes are at $\pm\pi/2$. The intersection point of the asymptotes with the negative real axis is

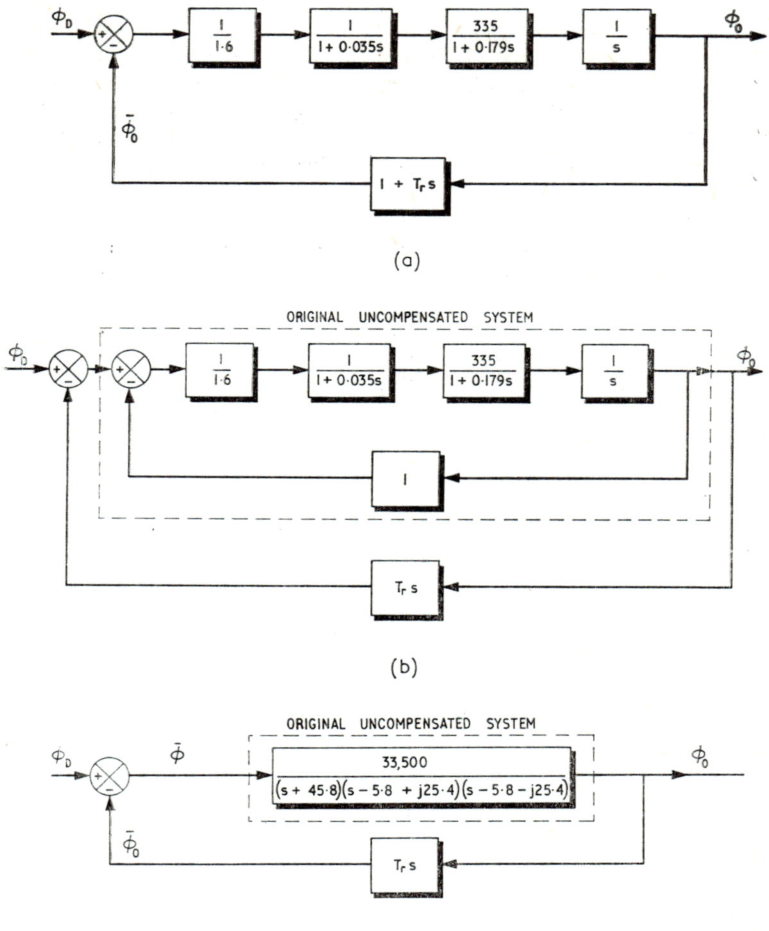


*Fig. 6.42. Block diagram manipulation for the pure rate feedback compensated third order system of Example 6.12.1: (a) conventional block diagram showing rate feedback compensation; (b) block diagram manipulation to free the compensation term; (c) block diagram in final form for construction of root locus in terms of the rate feedback compensation coefficient $T_r$*

given by

$$\frac{-(45{\cdot}8-5{\cdot}8-5{\cdot}8)}{2} = -17{\cdot}10$$

The locus leaves the open loop pole at $+5{\cdot}8+j25{\cdot}4$ at an angle of $+143°$, and by symmetry leaves the open loop pole at $+5{\cdot}8-j25{\cdot}4$ at an angle of $-143°$. These two branches of the locus approach the high-gain asymptotes, whilst the third pole migrates from the open loop pole at $-45{\cdot}8$ to the open loop zero at the origin, as shown in Fig. 6.43.

From the compensated root locus it can be seen that the damping ratio of the complex poles is a maximum at the position defined by $s = -10 \pm j48{\cdot}0$. The root locus gain $K$ is in this case

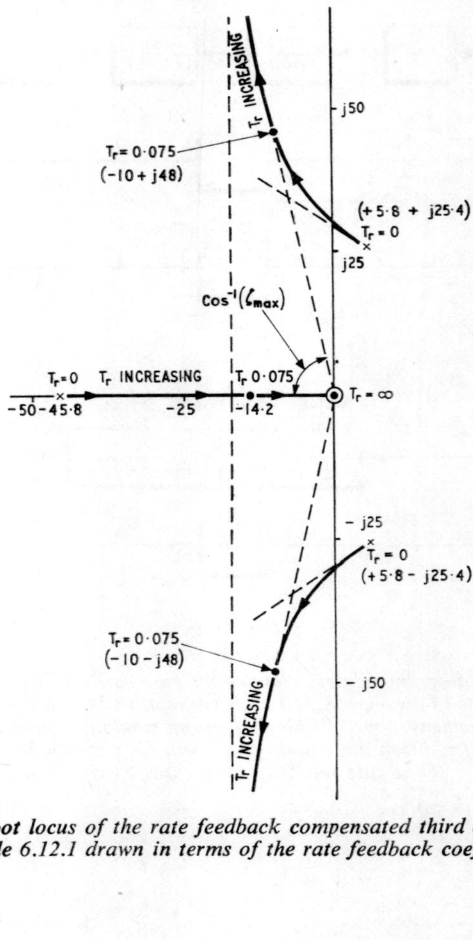


*Fig. 6.43. Root locus of the rate feedback compensated third order system of Example 6.12.1 drawn in terms of the rate feedback coefficient $T_r$*

equal to 33,500 $T_r$, as can be seen from Equation 6.65. Vectors for the evaluation of $T_r$ at the point $s = -10+j48{\cdot}0$ are shown in Fig. 6.44. Now, root locus gain is the product of the vectors from

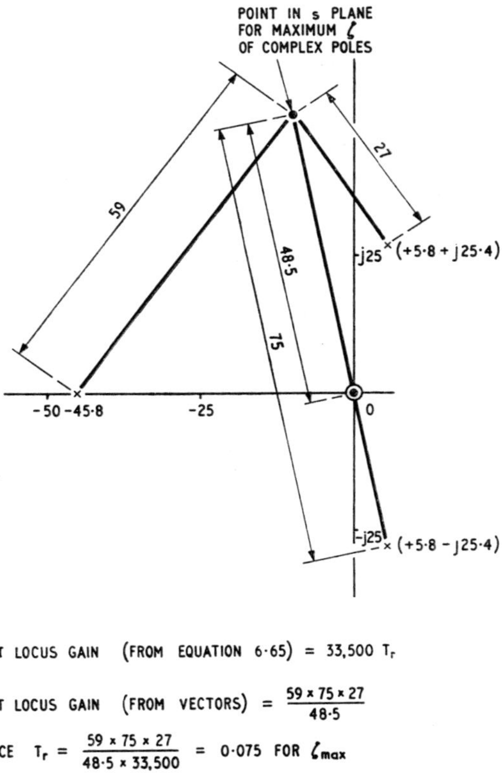


*Fig. 6.44. Vector diagram for the computation of $T_r$ for the maximum possible damping ratio of complex system poles for the pure rate feedback compensated third order system of Example 6.12.1*

the chosen point to the open loop poles divided by the product of the vectors from the chosen point to the open loop zeros. Hence

$$33{,}500T_r = \frac{27\times 59\times 75}{48{\cdot}5}$$

giving $T_r = 0{\cdot}075$. The third pole for this value of $T_r$ is given by the centre of gravity theorem and is at $-14{\cdot}2$. Hence the system pole-zero array for maximum damping ratio of the complex poles is given by

$$s_1 = -10+j48{\cdot}0$$
$$s_2 = -10-j48{\cdot}0$$

and

$$s_3 = -14{\cdot}2$$

The damping ratio of the complex poles is about 0·20, and the response obtained for this system may be unacceptable for many applications. It is therefore necessary to consider the provision of further compensation which will not only meet the stiffness requirement, but will also give a satisfactory response. We still have the problem of interpreting the response of a three-pole system: the coefficient plane method of Chapter 5 does allow us to work directly in terms of such quantities as bandwidth or $M_p$, whereas using the root locus technique we must either use a dominant mode approach, the approximate ITAE approach, or evaluate the performance for sample system pole arrays.

A method of introducing further compensation is to introduce acceleration feedback in addition to the rate feedback compensation already introduced. The effect on the root locus is again best studied by means of an example. Using pure rate and pure acceleration feedback, the locus of a (3,0) uncompensated system will be transformed into a (3,2) compensated locus, resulting in a (3,0) compensated system.

EXAMPLE 6.12.2

Sketch the root locus for the third order system of Example 6.12.1 for the case where rate feedback $T_r s$ and acceleration feedback $T_\alpha s$ are available for compensation. A suitable ratio of $T_r/T_\alpha$ is 30. (This choice of ratio is necessary at this stage because we are plotting the root locus for one variable only, which is $T_r$ in this case.)

Figs. 6.45(a), (b) and (c) show the block diagrams for the uncompensated, compensated and manipulated systems. The compensation term $T_r s + T_\alpha s^2$ is rewritten in the form $T_r s[1 + (T_\alpha/T_r)s]$ so that provided we have fixed $T_\alpha/T_r$ the locus can be drawn with $T_r$ variable. The dummy open loop equation used for the compensated root locus is, from Fig. 6.45(d),

$$\frac{\bar{\phi}_0}{\bar{\phi}}(s) = \frac{T_r s\left(1+\dfrac{s}{30}\right) 33{,}500}{(s+45{\cdot}8)(s-5{\cdot}8+j25{\cdot}4)(s-5{\cdot}8-j25{\cdot}4)} \qquad (6.66)$$

since $T_\alpha/T_r = 1/30$.

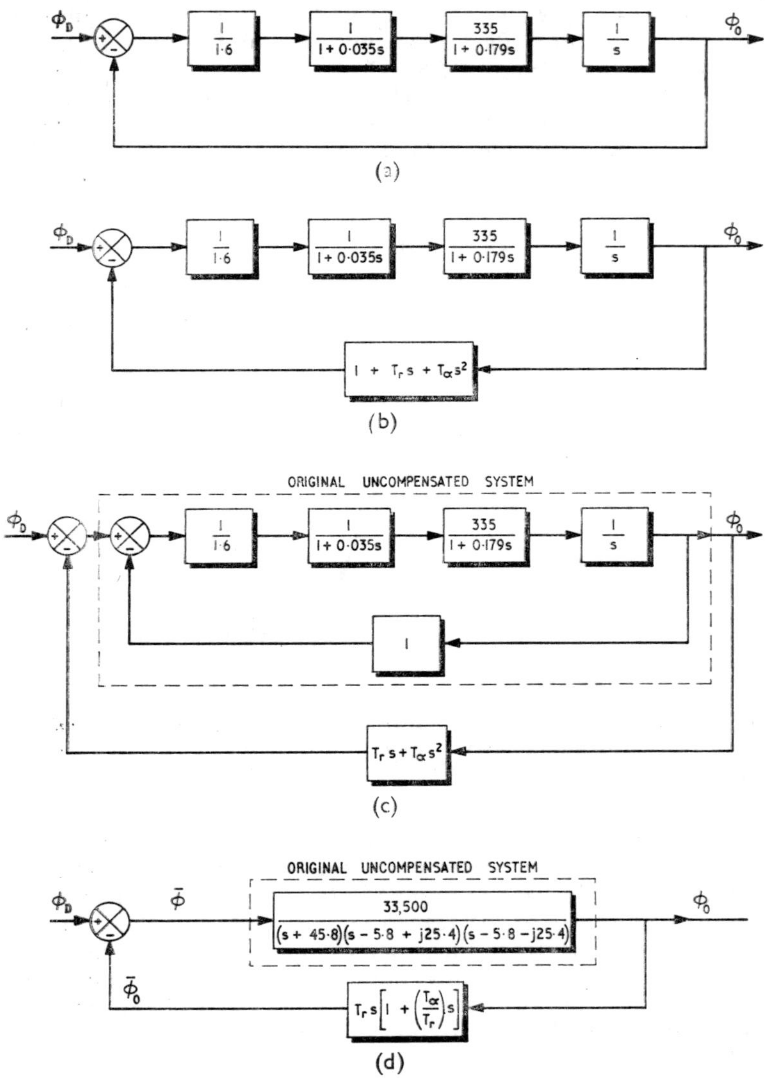


*Fig. 6.45. Block diagrams for the pure rate feedback and pure acceleration feedback compensated system of Example 6.12.2: (a) basic uncompensated third order system; (b) conventional block diagram for the system with pure rate feedback and pure acceleration feedback compensation; (c) manipulation of the block diagram to free the compensation terms; (d) block diagram in final form for the construction of the root locus in terms of the rate feedback coefficient $T_r$*

There are two open loop zeros of Equation 6.66 at the origin and at $-30$, and three open loop poles as for Example 6.12.1 at $-45{\cdot}8$, $+5{\cdot}8+j25{\cdot}4$ and $+5{\cdot}8-j25{\cdot}4$. The locus leaves the open loop pole at $+5{\cdot}8+j25{\cdot}4$ at an angle of 176°, and by symmetry the locus leaves the open loop pole at $+5{\cdot}8-j25{\cdot}4$ at an angle of $-176°$, as shown in Fig. 6.46. Using the method shown in the Appendix for taking the effect of the complex open loop poles

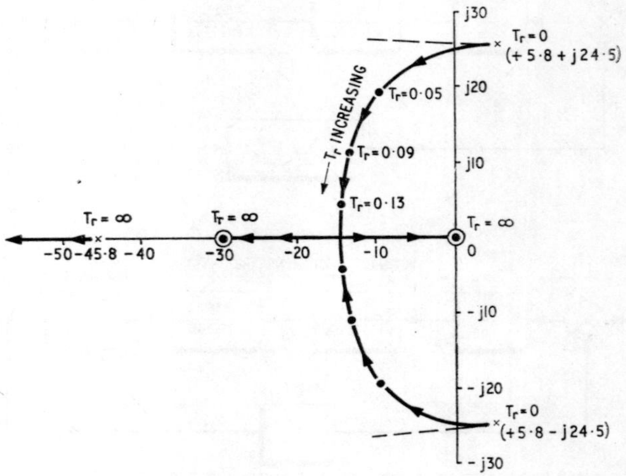


*Fig. 6.46. Root locus for the pure rate feedback and pure acceleration feedback compensated third order system of Example 6.12.2, drawn for the rate feedback coefficient $T_r$ being variable and $T_r/T_\alpha = 30$*

into account, the meeting point on the negative real axis is found to be at $-14{\cdot}5$. The intermediate points are found by using the Spirule and the locus is calibrated in terms of $T_r$ using the normal vector method. It can be seen from Fig. 6.46 that the complex pole pair migrate from the complex open loop poles into the left-hand plane, turn towards the negative real axis and meet at $-14{\cdot}5$ to split into two real roots, one of which migrates to the origin and the other to the open loop zero at $-30$. The third pole migrates towards infinity from $-45{\cdot}8$.

Obviously a much better system has resulted from the introduction of the second compensation term. Since the third pole is moving away from $-45{\cdot}8$ towards infinity, the effect on system performance is soon negligible. For a damping ratio of 0·707, the undamped natural frequency of the complex poles is about 18 rad/sec, and for a damping ratio of 0·5, the undamped natural frequency has increased to 21 rad/sec. In both cases quoted the third pole is, in fact, well to the left of the s plane, and has no effect on system performance. Clearly we now have the basis of a good system, which was obtained by choosing an arbitrary value of $T_r/T_\alpha$, and there is the possibility of improving performance even more by varying $T_r/T_\alpha$; but what aspect of performance are we most interested in? We have already specified the feedforward gain. Suppose $C_v$ is also of interest. For the pure rate and pure acceleration feedback compensated autopilot we have the system transfer function

$$\frac{\phi_0}{\phi_D}(s) = \frac{1}{1+\left(T_r+\dfrac{1}{K_g K_a}\right)s+\left(T_\alpha+\dfrac{T_a+T_c}{K_g K_a}\right)s^2+\left(\dfrac{T_a T_c}{K_g K_a}\right)s^3} \tag{6.67}$$

as was shown in Equation 5.12, Section 5.5.

Now, for Equation 6.67, $a_0 = b_0 = 1$, $b_1 = 0$ and

$$a_1 = T_r + \frac{1}{K_g K_a}$$

Hence we may write

$$C_v = \frac{1}{T_r+\dfrac{1}{K_g K_a}} \tag{6.68}$$

a result which could have been obtained equally well from Equation 6.42. $1/K_g K_a$ has been fixed from the stiffness requirement, and is 0·0048. The absolute maximum value of $C_v$ is therefore (in the absence of other compensation) equal to 210, but remember that this would be obtained for an unstable system. Clearly, realistic figures for $C_v$ consistent with the stiffness requirement will be much lower than 200. Equation 6.68 does allow the locus to be calibrated in terms of $C_v$ as well as $T_r$, and the root locus grid for four values of $C_v$ and three ratios of $T_r/T_\alpha$ has been constructed for the system studied in Example 6·12.2 and is shown in Fig. 6.47. $C_v = 25$ corresponds to $T_r = 0{\cdot}035$, $C_v = 20$ corresponds to $T_r = 0{\cdot}045$, $C_v = 15$ corresponds to $T_r = 0{\cdot}062$, whilst $C_v = 10$ corresponds to $T_r = 0{\cdot}095$. Fig. 6.47 shows that we can meet the stiffness and $C_v$ requirements, but cannot also choose specific definite values of $\omega_n$ and $\zeta$ for the dominant poles. This we expect, since we cannot meet four independent specifications with only three adjustable parameters, $K_g$, $T_r$ and $T_\alpha$. However, a study of the dominant poles does show that a reasonable compromise may be obtained.

The third pole positions are omitted from Fig. 6.47, but they are shown in Fig. 6.48, which shows the system pole-zero arrays for the various values of $T_\alpha$ and $T_r$ meeting $C_v$ specifications of 15, 20 and 25 only. It can be seen that the higher the value of $C_v$ required, the more lightly damped are the complex poles. Note the complete absence of system zeros in Fig. 6.48, highlighting the fact that feedback compensation is a technique adjusting pole positions only. Specifically, the inclusion of simple compensation using

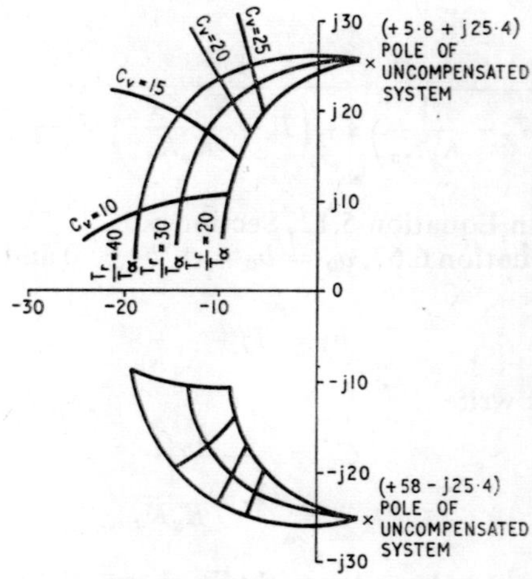


*Fig. 6.47. Root locus grid for dominant poles of the third order system compensated by pure rate feedback and pure acceleration feedback to achieve $C_v$ and stiffness specifications together with a reasonable transient response*

the $i^{\text{th}}$ output derivative will allow independent adjustment of the $i^{\text{th}}$ coefficient of the system characteristic equation. If rate feedback is included, then direct adjustment of the $a_1$ coefficient is possible, and if acceleration feedback is included, then direct adjustment of the $a_2$ coefficient is possible. Basically, feedback compensation is an orderly method of compensation, usually involving the lower feedback derivatives only, which can be used to stabilise the system and drive the dominant poles into a suitable position.

The absence of a $C_v$ requirement for the third order system compensated by rate and acceleration feedback would enable a stiffness specification and particular values of the complex pole damping ratio and the undamped natural frequency to be met directly.

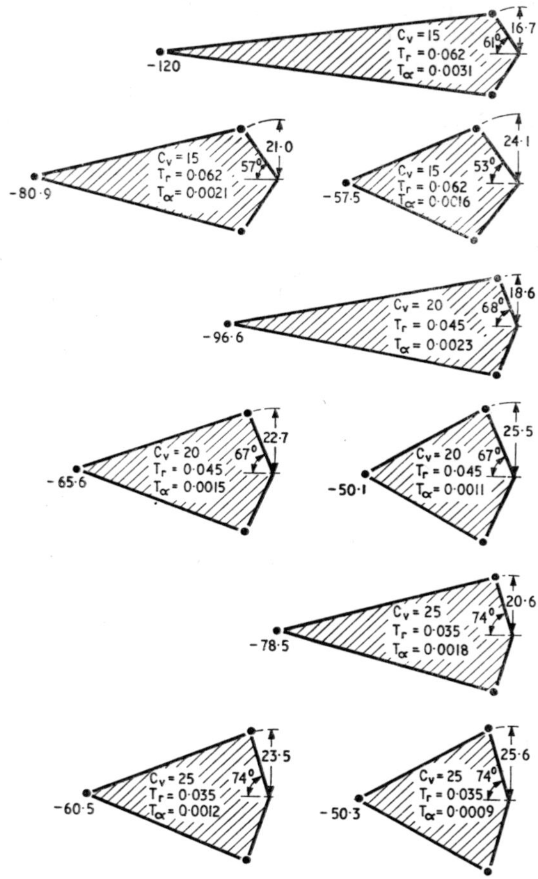


*Fig. 6.48. System pole-zero arrays for the pure rate feedback and pure acceleration feedback compensated third order system meeting a given stiffness requirement*

## 6.13 Method of Obtaining System Poles Directly from the Open Loop Bode Plot

In Section 6.8 it was observed that for unity feedback systems, with a basic integrator in the feedforward path, $C_v$ is proportional to feedforward gain, and when the feedforward gain is written in the form $[K\times\prod(1+T_\alpha s)]/[s\times\prod(1+T_\beta s)]$, the proportionality factor is unity, as can be observed from Fig. 6.20(b). Thus for unity feedback systems the $C_v$ specification can be met by inspecting the open loop response, and the Bode plot is particularly useful in this respect as $C_v$ is then simply the intersection point of the basic 20 db/decade slope extended to the 0 db line as was shown in Fig. 1.27. Open loop Bode plots are therefore helpful design aids when passive lead or passive lag networks are used for compensation in unity feedback systems, as changes in $C_v$ are readily observable by inspecting the behaviour of the previously mentioned intersection point. An approximate method which determines the system poles, and hence the approximate system transfer function, directly from the open loop Bode plot is therefore a valuable tool for the designer to have available, and such a method has been developed by Chen,[11] following initial work in this field by Biernson.[61] If desired, the method can be looked upon as a crude alternative construction for the root locus in place of the conventional construction shown in Chapter 3. An elementary knowledge of the conventional root locus technique is useful in understanding Chen's method, which is best illustrated by reference to a specific open loop Bode plot. Only the logarithmic straight-line approximation to the open loop amplitude ratio is required.

Fig. 6.49 shows a typical Bode plot for a unity feedback control system with feedforward transfer function

$$G(s) = \frac{C_v\left(1+\dfrac{s}{\omega_z}\right)}{s\left(1+\dfrac{s}{\omega_1}\right)\left(1+\dfrac{s}{\omega_2}\right)\left(1+\dfrac{s}{\omega_3}\right)} \tag{6.69}$$

Chen's method is to divide arbitrarily the Bode plane into three regions corresponding to a high $|G(j\omega)|$ region (Region I) where the gain at a particular frequency is greater than $+15$ db, a low $|G(j\omega)|$ region (Region III) where the gain at a particular frequency is less than $-15$ db, and a region of intermediate $|G(j\omega)|$, Region II, which is bounded by the $\pm 15$ db ordinates. The $\pm 15$ db boundaries have been chosen by Chen on the basis of experience with various problems. System poles are now determined quite separately for each of the three regions, it being assumed that the poles in any region are dependent only on the Bode plot for that region. In terms of the root locus the results may be summarised as follows:

*Region I* corresponds to those system poles which have migrated almost to the open loop zeros. There will be an approximate system pole for every breakpoint in Region I where the Bode

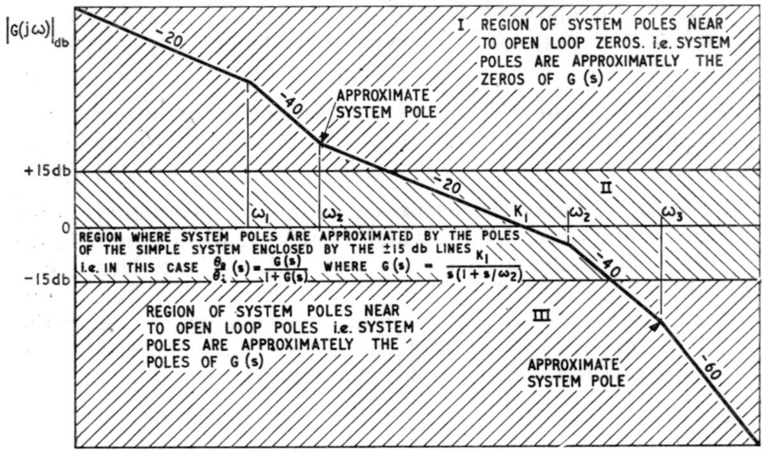


*Fig. 6.49. The basis of Chen's method for obtaining system poles from the open loop Bode plot*

plot slope *decreases* by 20 db/decade. Hence any open loop zero with a breakpoint observable in the Bode plot in Region I is an approximate system pole.

*Region III* corresponds to those system poles which have migrated only a little way from the open loop poles. There will be an approximate system pole for every breakpoint in Region III where the Bode plot slope *increases* by 20 db/decade. Hence any open loop pole with a breakpoint observable in the Bode plot Region III is an approximate system pole.

*Region II* corresponds to poles which are at some intermediate point between the appropriate open loop pole and open loop zero. The relatively low order system which results in the open loop Bode plot enclosed between the $\pm 15$ db ordinates, completely neglecting all happenings outside this region, is then analysed by whatever means are available to determine the system poles in this region.

Finally, the approximate system transfer function can be written down by taking the system poles evaluated for all three regions, and then adding the system zeros, which in this case are the open loop zeros since the system is of unity feedback form with lead or lag compensation resulting in feedforward path zeros. Applying these rules to the open loop Bode plot shown in Fig. 6.49, we obtain the following results:

*Region I.* An open loop zero breakpoint occurs at frequency $\omega_z$, which is an approximate system pole, hence we may write

$$s_1 \doteq -\omega_z \tag{6.70}$$

*Region III.* An open loop pole breakpoint occurs at frequency $\omega_3$, which is an approximate system pole, hence we may write

$$s_2 \doteq -\omega_3 \tag{6.71}$$

*Region II.* We may use whatever method is convenient to solve the second order system which is defined within the $\pm 15$ db lines. The open loop plot consists of a 20 db/decade line which intersects the 0 db line at the point $K_1$, and subsequently there is a breakpoint at frequency $\omega_2$ where the slope increases to 40 db/decade. The much-simplified open loop equation which fits this characteristic is

$$G(s) = \frac{K_1}{s\left(1+\dfrac{s}{\omega_2}\right)} \tag{6.72}$$

and treating this simple system as a closed loop system to determine the intermediate values of poles, we have

$$\frac{\theta_0}{\theta_i}(s) = \frac{G(s)}{1+G(s)} = \frac{1}{1+\dfrac{s}{K_1}+\dfrac{s^2}{K_1\omega_2}} \tag{6.73}$$

and the intermediate poles are thus defined by $\omega_n = \sqrt{K_1\omega_2}$, $\zeta = \sqrt{\omega_2/4K_1}$.

Reverting now to the root locus for the system with the Bode plot shown in Fig. 6.49, the locus appears as shown in Fig. 6.50,

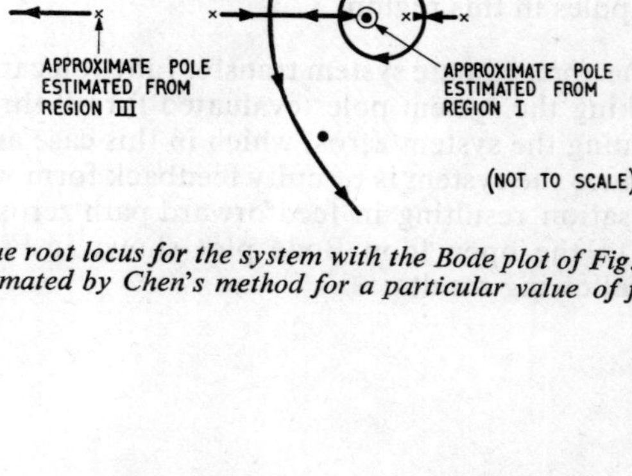


*Fig. 6.50. True root locus for the system with the Bode plot of Fig. 6.49 showing the poles estimated by Chen's method for a particular value of feedback gain*

the approximate poles obtained using Chen's method being superimposed on the conventional locus. We have already seen that, for dipoles occurring near the origin, even small displacements between pole and zero can be significant, and it is therefore necessary to determine a more accurate value for the system pole near

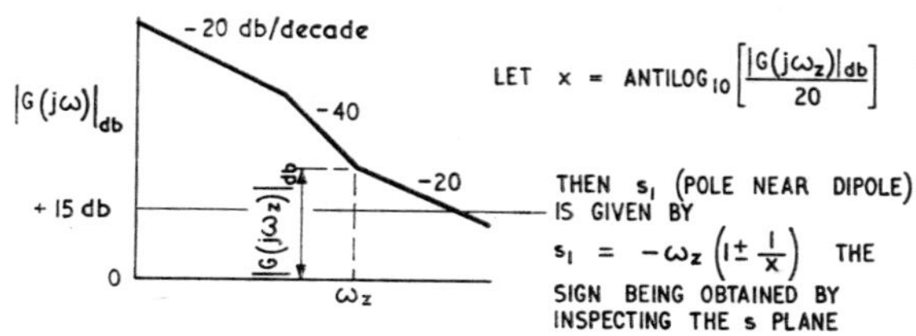


*Fig. 6.51. Demonstration of the method of obtaining the correction factor for the position of the system pole forming the significant dipole directly from the Bode plot*

the open loop zero. A simple method of determining a more accurate value for this pole is shown in Chestnut and Mayer,[62] and is illustrated in Fig. 6.51. The parameter $x$ is defined by

$$x = \text{antilog}_{10}\left[\frac{|G(j\omega_z)|_{\text{db}}}{20}\right] \tag{6.74}$$

In this particular example, $|G(j\omega_z)|$ is about 23 db, giving

$$x = \text{antilog}_{10}\left(\frac{23}{20}\right) = 13{\cdot}5$$

The more exact value for the pole near the open loop zero is given by

$$s_1 = -\omega_z\left(1 \pm \frac{1}{x}\right) \tag{6.75}$$

the sign of $x$ being determined by inspecting the s plane to see whether the pole is larger or smaller than the open loop zero. From the root locus plot we see in this particular case that the pole is larger than the zero, hence

$$s_1 = -\omega_z\left(1 + \frac{1}{13{\cdot}5}\right) = -1{\cdot}075\omega_z \tag{6.76}$$

It is not necessary to sketch the locus to determine the sign of $x$, since this can be obtained by reference to the appropriate standard root locus shown in Figs. 6.11 to 6.19 inclusive. Using Equations 6.71, 6.73 and 6.76, together with the known system zero at $-\omega_z$,

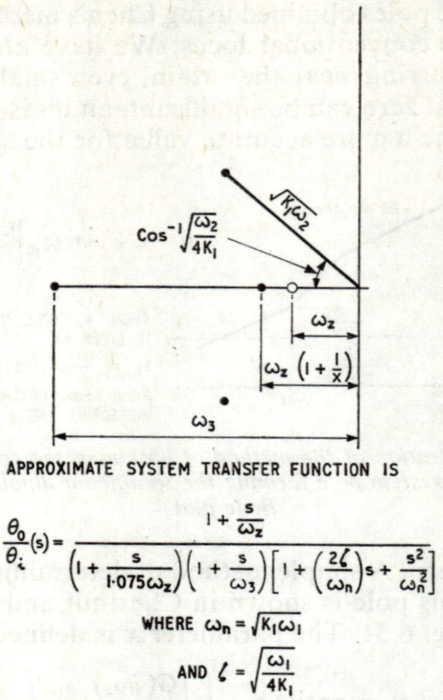


*Fig. 6.52. Final system pole-zero array including estimate of the dipole obtained using Chen's method on the system with open loop Bode plot shown in Fig. 6.49*

the approximate system transfer function in factored form is

$$\frac{\theta_0}{\theta_i}(s) \simeq \frac{1+\dfrac{s}{\omega_z}}{\left(1+\dfrac{s}{1\cdot075\omega_z}\right)\left(1+\dfrac{s}{\omega_3}\right)\left[1+\left(\dfrac{2\zeta}{\omega_n}\right)s+\dfrac{s^2}{\omega_n^2}\right]} \tag{6.77}$$

where $\omega_n = \sqrt{K_1\omega_2}$ and $\zeta = \sqrt{\omega_2/4K_1}$. Fig. 6.52 shows the system pole-zero array as obtained by Chen's method. As a numerical example of the method we shall now consider Chen's solution[63] to a problem posed by Oldenburger in a paper on the use of algebraic methods in the design of control systems (using numerical root-solving techniques, etc.).

EXAMPLE 6.13.1

The block diagram for an aircraft gas turbine control system is shown in Fig. 6.53. Determine the system pole-zero array for the

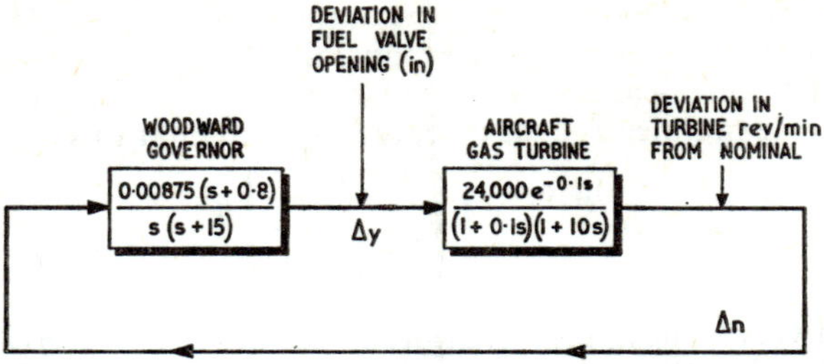


*Fig. 6.53. Block diagram of aircraft gas turbine control systems used as an example for estimating system poles directly from the open loop Bode plot*

system using the Woodward governor with a feedforward gain of 0·00875 in of valve closure per rev/min increase in turbine speed.

It will be noticed that the gas turbine has a transportation lag of $e^{-0\cdot1s}$, i.e. no change in speed occurs until 0·1 sec after the valve movement demanding the change in speed. For sinusoidal excitation the transportation lag contributes a phase lag of $0\cdot1\omega$ rad.

Chen's method requires that the transportation lag be represented, in the s plane, by a minimum phase transfer function. Thus, using the first term in a binomial expansion, we may write the expression

$$e^{-0\cdot1s} = \frac{1}{e^{0\cdot1s}} \simeq \frac{1}{1+0\cdot1s} \tag{6.78}$$

and the s plane representation is shown in Fig. 6.54 (a). The corresponding non-minimum phase approximation is shown in Fig.

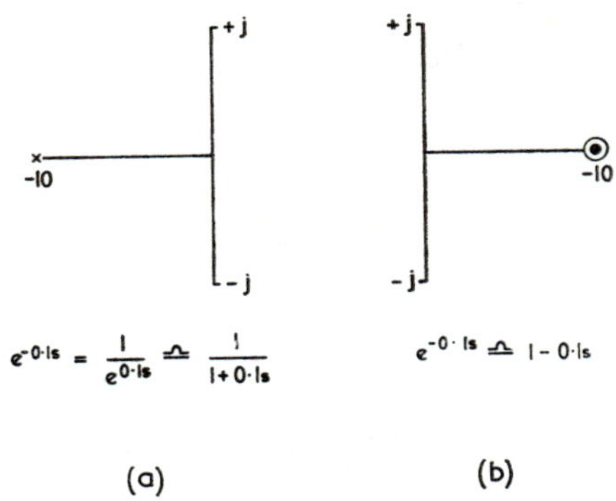


*Fig. 6.54. Crude* s *plane representation of deadtime: (a) simple minimum phase* s *plane representation of deadtime necessary for use with Chen's method; (b) simple non-minimum phase* s *plane representation of deadtime*

6.54(b). More exact representation of the transportation lag involves taking more terms in the expansion. Other representations of transportation lag are possible which avoid zeros in the right-hand plane; for example, Truxal discusses approximations of the form

$$e^{-0\cdot1\,s} = \left[\frac{1}{1+\left(\frac{0\cdot1}{n}\right)s}\right]n \tag{6.79}$$

which involves the approximation of the deadtime by $n$ open loop poles at $n/0{\cdot}1$. Such an approximation could be used if the first order minimum phase approximation is not good enough.

From Fig. 6.53, the open loop equation for the speed control system is

$$G(s) = \frac{0{\cdot}00875\times24{,}000(s+0{\cdot}8)e^{-0\cdot1s}}{s(s+15)(1+0{\cdot}1s)(1+10s)} \tag{6.80}$$

and using the deadtime approximation of Equation 6.78 we re-write the open loop equation in Bode form

$$G(s) = \frac{11{\cdot}2\left(1+\frac{s}{0{\cdot}8}\right)}{s\left(1+\frac{s}{0{\cdot}1}\right)\left(1+\frac{s}{15}\right)\left(1+\frac{s}{10}\right)^2} \tag{6.81}$$

giving $C_v = 11{\cdot}2$ directly from Equation 6.43.

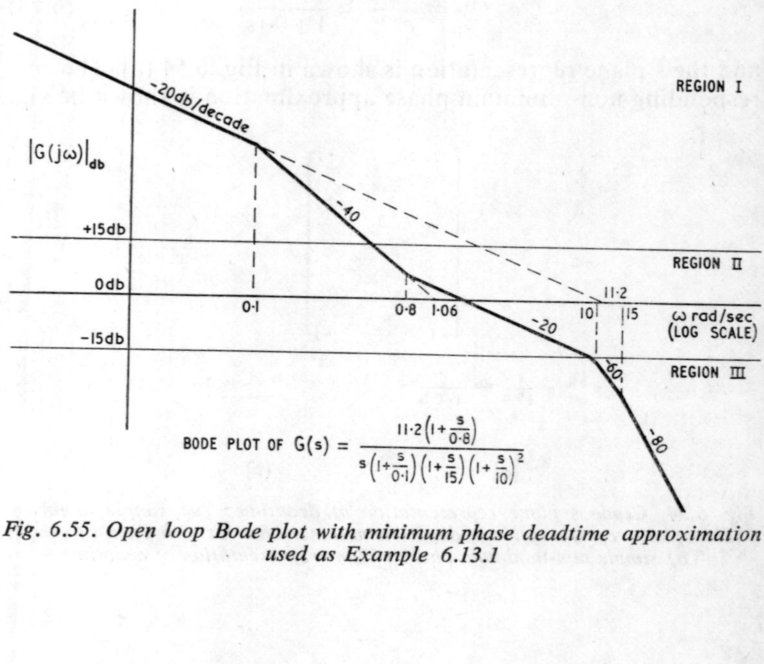


*Fig. 6.55. Open loop Bode plot with minimum phase deadtime approximation used as Example 6.13.1*

The Bode plot of Equation 6.81 is shown in Fig. 6.55, and it can be seen that there are no open loop zeros with breakpoints in Region I, hence there are no system poles which have migrated near to an open loop zero. In Region III there is a double breakpoint due to a pair of open loop poles at $-10$, and a single breakpoint due to an open loop pole at $-15$. Thus there are three system poles which may be considered to have moved only a little from the open loop poles. The three system poles are therefore defined by $(1+s/15)(1+s/10)^2$. Inside the $\pm 15$ db lines we may write down a simplified open loop transfer function which is independent of the Bode plot outside these limits. From Fig. 6.55 the simplified open loop transfer function is

$$G(s) = \frac{1{\cdot}06^2}{s^2}\left(1+\frac{s}{0{\cdot}8}\right) \tag{6.82}$$

where the $1{\cdot}06^2$ term arises because of the basic $-$ 40 db/decade slope. In this region we may therefore write

$$\frac{G(s)}{1+G(s)} = \frac{1+\dfrac{s}{0{\cdot}8}}{1+\dfrac{s}{0{\cdot}8}+\dfrac{s^2}{1{\cdot}06^2}} \tag{6.83}$$

Hence $\omega_n = \sqrt{1{\cdot}06^2} = 1{\cdot}06$ and $\zeta = \omega_n/2\times 0{\cdot}8 = 0{\cdot}66$. In Cartesian coordinates the quadratic is defined by $s = -0{\cdot}70 \pm j0{\cdot}80$. The final solution using Chen's method is therefore obtained by adding Equation 6.83 to the three roots found in Region III:

$$\frac{G(s)}{1+G(s)} \triangleq \frac{1+\dfrac{s}{0{\cdot}8}}{\left(1+\dfrac{s}{15}\right)\left(1+\dfrac{s}{10}\right)^2\left[1+\left(\dfrac{2\times 0{\cdot}66}{1{\cdot}06}\right)s+\dfrac{s^2}{1{\cdot}06^2}\right]} \tag{6.84}$$

Fig. 6.56 shows the system pole-zero array corresponding to Equation 6.84, and it will be noted that because of the nature of the transportation lag representation there are five system poles. The solution using the non-minimum representation of $e^{-0{\cdot}1s} = 1-0{\cdot}1s$ is shown in Fig. 6.57; it will be noted that this is a four-pole system. Both pole-zero arrays, of course, possess the system zero at $-4$, which is also an open loop zero appearing in the feedforward path. Although at first sight the pole-zero arrays of Figs. 6.56 and 6.57 appear somewhat different, the transient responses obtained from the arrays are, for engineering purposes, quite similar. Example 6.13.1 has illustrated the usefulness of the method of obtaining the system transfer function directly from the open loop Bode plot. This method is specially useful for synthesis of dominant mode systems for which $C_v$ or feedforward gain is specified and feedforward compensation is used. Since the method utilises only the information available in the logarithmic straight-line approximation to the open loop amplitude ratio, it follows that there will be some restrictions imposed on the method when significant open loop complex poles are present, since the Bode plot used will be independent of the damping ratio of these complex poles.

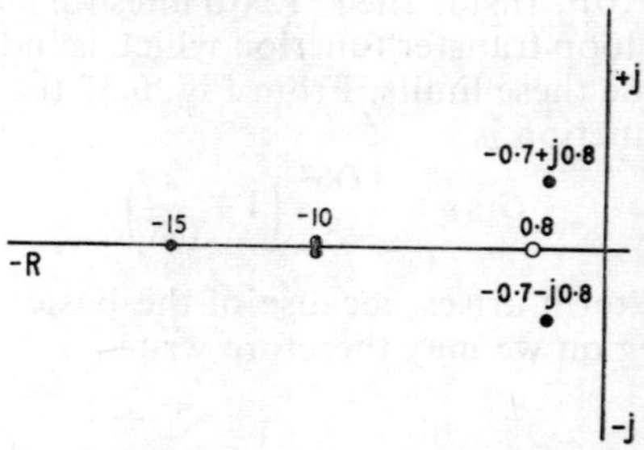


*Fig. 6.56. System pole-zero array for Example 6.13.1 obtained from the open loop Bode plot using the method of Chen*

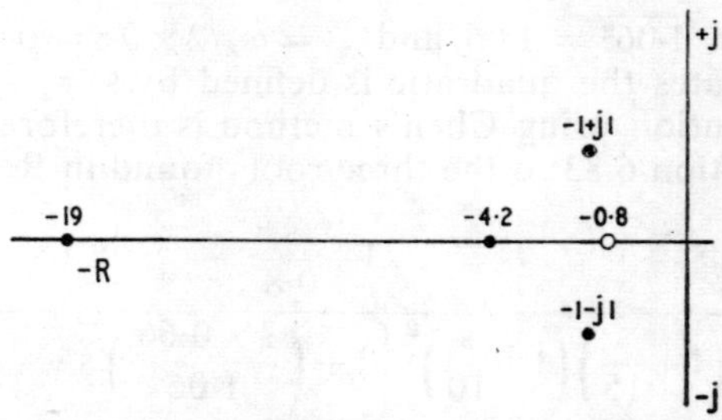


*Fig. 6.57. Exact system pole-zero array for Example 6.13.1*

When the system defined within the $\pm 15$ db ordinates is a third order system, the cubic root locus charts of Chapter 5 may be used with great advantage. For systems where a double breakpoint exists in the $\pm 15$ db region, the system may be simplified without too much loss in accuracy by replacing a double pole at frequency $\omega_1$ by a pole at $a\omega_1$ plus a pole at $\omega_1/a$ where $a$ is an arbitrarily chosen number less than 2. By judicious choice of $a$ it is possible to replace the double pole inside the $\pm 15$ db region with one pole inside this region and a pole outside the region, as shown in Fig. 6.58, thus reducing the order of the simplified system of Region II by one which will increase the speed of computation. Chen's method can be used to approximate rapidly the root locus of the dominant poles, as a function of the compensation parameters. The positions of these poles can then be refined by using Lin's method or synthetic division. The method is restricted to systems of the unity feedback form because of the assumptions made in Region I, which revolve around the use of the error/input transfer function to determine the poles in this region. Taking normal precautions,

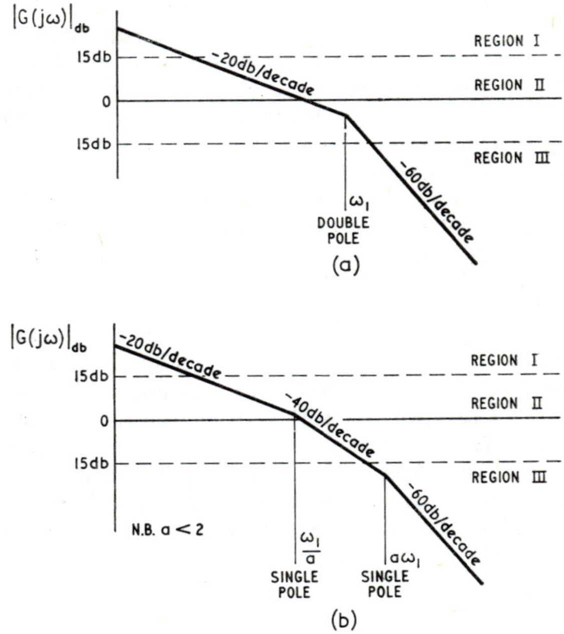


*Fig. 6.58. Approximate method for replacing a double pole in Region II prior to using Chen's method: (a) actual Bode plot; (b) approximate method for replacing a double pole in Region II by two single poles, one in Region II and one in Region III*

the method can be applied to non-unity feedback systems rewritten in the unity feedback path form. Extensive applications of Chen's method are quoted by Canfield, and include the *ad hoc* design of multi-loop systems solving the series of loops in succession. As will now be described, it is, however, possible to combine the principles of Chen with the approach adopted in Chapter 4 in order to produce an elegant synthesis technique which defines the compensation elements required without recourse to trial and error procedures.

## 6.14 A Rapid Synthesis Technique for the Direct Determination of Compensation Elements

A recent paper[64] has suggested a rapid technique whereby the necessary and sufficient compensation elements required in a feedback control system can be obtained. In particular, under conditions where the specification demands that deterministic performance criteria in different domains must be satisfied simultaneously, the new technique has tremendous advantages. It is an established engineering fact that most feedback control systems are dominated by a few poles and zeros, whilst in Chapter 4 we saw that a specification involving bandwidth, percentage overshoot, and velocity constant uniquely determined the system pole-zero geometry of a type (1,2) system. In Chapter 4 it was assumed that the system zero could be generated without increasing the number of system poles. In practice this is rarely possible, but there is no reason why the system zero cannot be generated without increasing the number of *dominant* system poles.

Therefore, when designing higher order systems we position the dominant two poles and one zero such that the specification is nominally met by this array. We then determine the compensation elements such that the dominant pole-zero array is only slightly affected, and also make sure that any additional system poles and zeros resulting from the compensation elements do not significantly affect performance criteria. From Chapter 3 it is known that of the three criteria of interest, only $C_v$ is affected by the far-off poles. By working from the open loop Bode plot, $C_v$ is constrained to be correct whatever the system pole-zero geometry.

EXAMPLE 6.14.1

Determine suitable compensation elements such that an inertia load of 2·5 slugs ft$^2$ and associated viscous damping of 2·5 lb ft rad$^{-1}$ sec, is controlled in accordance with the specification: percentage overshoot = 15%, $\omega_B$ = 15 rad/sec and $C_v$ = 30 sec$^{-1}$.

In Example 4.11.1 it was shown that a two-pole one-zero transfer function with $a_0 = b_0 = 1$, $a_1 = 0{\cdot}142$, $b_1 = 0{\cdot}109$, $a_2 = 0{\cdot}012$, would satisfy this specification. To obtain the equivalent feedforward transfer function of a unity feedback path system, we write

$$\left[\frac{\theta_o}{\theta_i}(s)\right]_{1,2} = \frac{G(s)}{1+G(s)}$$

Hence

$$G(s) = \frac{\left[\frac{\theta_0}{\theta_i}(s)\right]_{1,2}}{1-\left[\frac{\theta_0}{\theta_i}(s)\right]_{1,2}} = \frac{30\left(1+\frac{s}{9\cdot 16}\right)}{s\left(1+\frac{s}{2\cdot 79}\right)}$$

where by inspection it is seen that the $C_v = 30$ requirement is inherently met.

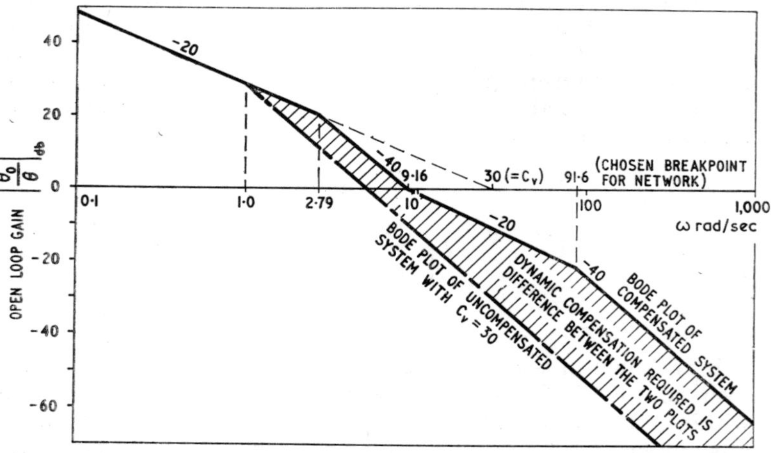


*Fig. 6.59. Open loop Bode plot of the system designed to meet the specifications of $\omega_B$ = 15 rad/sec, $C_v$ = 30 sec$^{-1}$ and 15% overshoot simultaneously*

In assessing dominance we now interpret the dominant pole-zero array as resulting from the open loop Bode plot in the 0 db region. Therefore the 0 db region must contain only the elements $[30(1+s/9\cdot 16)]/[s(1+s/2\cdot 79)]$ as shown in Fig. 6.59. Now, the load transfer function is

$$G_L(s) = \frac{1}{2\cdot 5s(1+s)}$$

and, clearly, to meet the $C_v$ requirement a scalar gain of $30\times 2\cdot 5$ = 75 is necessary. The uncompensated Bode plot of correct gain is therefore $75/[s(1+s)]$, and this is also shown in Fig. 6.59. Clearly the difference between the compensated and uncompensated plots must be provided by the compensation elements. We see that the basic compensation must be

$$G_c(s) = \frac{(1+s)\left(1+\dfrac{s}{9{\cdot}16}\right)}{1+\dfrac{s}{2{\cdot}79}}$$

but this is not physically realisable, in general, and if passive network compensation is used, another compensation pole must be chosen. If this is at $s = -91{\cdot}6$, then the network is easily realisable physically, and the breakpoint will occur outside the $-15$ db limit, thus resulting in a system pole well to the left and not significantly affecting bandwidth or percentage overshoot. The final compensation chosen is

$$G_c(s) = 75\left(\frac{1+s}{1+\dfrac{s}{2{\cdot}79}}\right)\left(\frac{1+\dfrac{s}{9{\cdot}16}}{1+\dfrac{s}{91{\cdot}6}}\right)$$

It is shown in the reference quoted that the actual percentage overshoot is 17·2% and the actual bandwidth is 16·2 rad/sec, well within practical tolerances. Note that if the Chen rules are literally applied to the resultant Bode plot, the negative breakpoint at $\omega = 2{\cdot}79$ is neglected, and this leads to considerable error in determination of the system pole-zero array. The present writer therefore places some flexibility on the $+15$ db limit to the extent of increasing it when there are negative breakpoints just outside the Chen boundary.

$G_c(s)$ may be interpreted as a double lead network as shown in Fig. 6.60 (a), but the $(1+s)/(1+s/2{\cdot}79)$ part can be achieved in a variety of ways using rate feedback around the load and motor. The gains in various parts of the system are then at the discretion of the designer, to match with component gains in the best possible way. Fig. 6.60(b) shows the resultant block diagram for the case when the transfer function prior to the minor loop is simply a single-stage lead network.

The complexity of the compensation elements is a measure of the practicality of forcing the system response to be dominated by two poles and one zero. The example chosen results in a type (1,3) system, but there is no fundamental restriction on the order of system which may be considered.

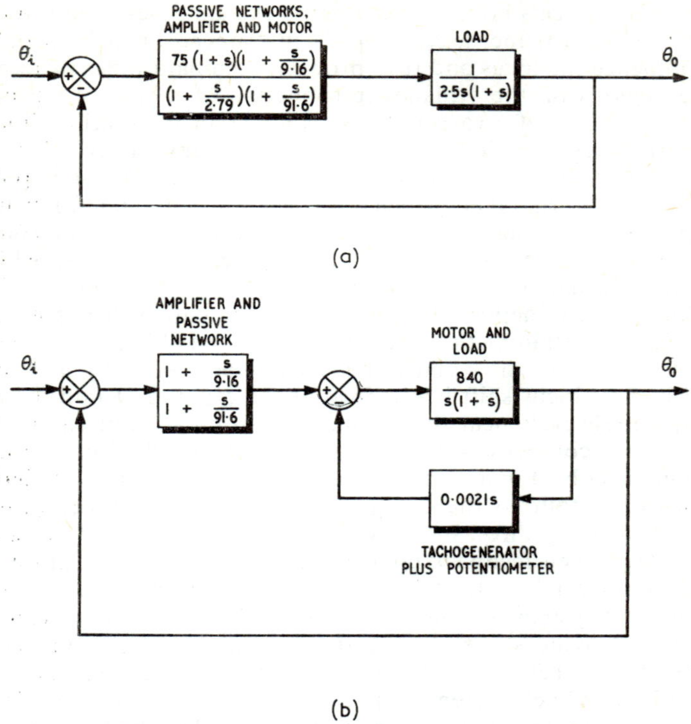


*Fig. 6.60. Alternative designs meeting the Bode plot requirement of Fig. 6.59 and hence resulting in 15% overshoot, 15 rad/sec bandwidth and 30 sec$^{-1}$ velocity constant: (a) two-stage lead compensated servomechanism; (b) rate feedback plus single-stage lead compensated servomechanism*

## 6.15 Conclusion

Compensation of linear control systems to meet specifications including such criteria as velocity constant, percentage overshoot, bandwidth and feedforward stiffness can be interpreted in the s plane as the the addition of open loop poles and zeros to force the root locus into an acceptable shape. The choice of the form of compensation depends on whether it is sufficient merely to force the system poles into specific parts of the s plane or whether it is necessary to provide significant system zeros, either to contribute directly to the system response or to cancel approximately the effect of a system pole.

Relatively simple standard forms of compensation can be used to improve system performance. Cancellation compensation effectively cancels both open loop and system poles in unwanted parts of the s plane, leaving newly introduced open loop poles to influence the locus and thus produce dominant poles in acceptable regions of the s plane. Introduction of a passive lag network in the feedforward path can induce an integrating dipole into the system pole-zero array without significantly affecting the position of the other system poles. Therefore if a system pole-zero array is obtained which satisfies transient and frequency response requirements, but which does not meet $C_v$ requirements, then the lag network will result in an integrating dipole which can be adjusted in the s plane to meet the $C_v$ specification. The addition of a lead network in the feedforward path will enable the root locus containing the dominant poles to be shifted considerably to the left of the s plane, thus greatly increasing system bandwidth. This locus shift is obtained at the expense of introducing a dipole relatively near the origin. There is great similarity in the use of lead compensation and in the use of cancellation compensation. Feedback compensation involves the feedback of derivatives of output to the comparison element. If the derivatives are obtained by direct measurement, and there are no measurement lags, then feedback compensation involves the addition of open loop zeros only, thus modifying the root locus but not generating system zeros as well. Feedback compensation can be used to provide satisfactory transient and frequency responses for systems meeting $C_v$ and feedforward gain requirements.

Unity feedback systems with a basic integrator in the feedforward path possess the important property that the feedforward gain is equal to $C_v$ when the open loop transfer function is written in the $1+Ts$ form. In particular, $C_v$ is given by the intersection of the initial $-20$ db/decade slope, extended to meet the 0 db line on the open loop Bode plot. Therefore Chen's method for determining system poles and hence the system transfer function from the open loop Bode plot is a useful design aid, since $C_v$ or the feedforward gain can be met by inspection, and the compensation network adjusted until, using the approximate method, a promising system pole-zero array is determined. The system pole positions may then be determined more accurately using Lin's numerical method or synthetic division as appropriate. Some special techniques which can be used on fourth order systems have been included to enable the reader to recognise standard forms of normalised transfer functions simply by inspecting the coefficients. For dominant mode systems of fourth order with a dominant pole pair, the system poles may be obtained very rapidly indeed using extremely simple formulae.

CHAPTER 7

# Transfer Function Techniques Applied to Non-linear Control Systems

## 7.1 The Meaning of Non-linearity

In Chapter 1 a linear system was defined in several ways, one of which was that a system could be considered linear if its performance could reasonably be described for all values of time by a single linear differential equation with constant coefficients. Any system which cannot be described in this way is automatically defined as a non-linear system, and the principle of superposition cannot be applied to the system, nor will an absolutely unique system transfer function exist. Note that this rigorous definition of non-linearity classifies as non-linear a system whose performance can be defined by a set of linear differential equations, whose coefficients are functions of time or signal amplitude.

The existence of non-linearities not only makes system analysis and synthesis much more difficult, but may also give rise to the occurrence of any one of a number of purely non-linear phenomena. Two common non-linear phenomena are limit cycling and jump resonances; other non-linear phenomena such as subharmonic generation are outside the scope of this book. The limit cycle phenomenon is shown in Fig. 7.1, and may occur continually or spasmodically in certain types of non-linear systems. For certain applications such a limit cycle may be perfectly acceptable provided that the limit cycle amplitude and frequency are at the right level.[65,66] Generally speaking it is found that

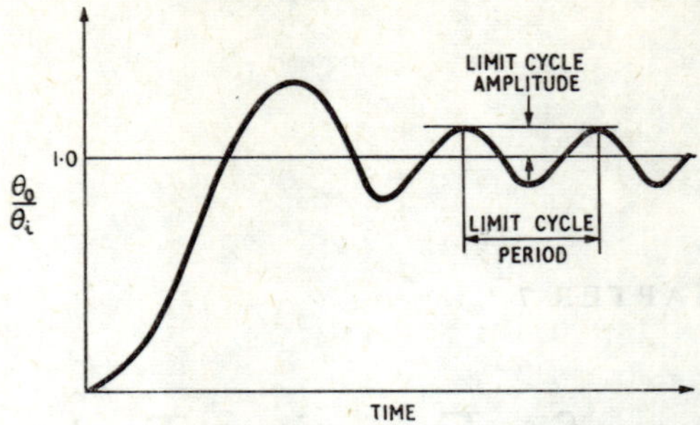


*Fig. 7.1. Limit cycle phenomenon found in certain types of non-linear systems*

the limit cycle frequency must be at least ten times the bandwidth frequency, and the amplitude must be very small indeed. If a high limit cycle frequency is obtained the low amplitude requirement is normally met as a matter of course. Limit cycling systems have certain practical advantages over linear systems, since stiction of mechanical components is continuously overcome and the effects of deadzone due to finite manufacturing tolerances may be minimised by using the limit cycle as a carrier signal, as in an electrical a.c. system. A typical example of the jump resonance phenomenon revealed by steady state frequency excitation is shown in Fig. 7.2. In this particular case, as the excitation frequency is increased the system behaves as normal until the point $B$ (corresponding to frequency $\omega_1$) is reached, when there will be a sudden increase in the steady state amplitude ratio to the point $C$. Further increases in excitation frequency will result in the amplitude ratio decreasing continuously along the line $CD$. If the frequency response test is now repeated but

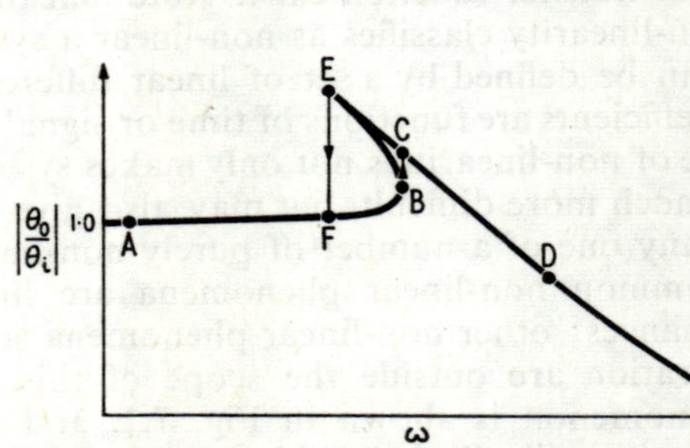


***Fig. 7.2. Jump phenomenon found in the frequency response characteristics of certain types of non-linear systems: as frequency of excitation is increased, jump discontinuity is the line BC; as frequency of excitation is decreased, jump discontinuity is the line EF***

with the frequency decreasing, the amplitude ratio will increase continuously along the line *DCE*, where *E* is at a lower frequency than *C*. At the point *E* (corresponding to frequency $\omega_2$) there will be a sudden reduction in amplitude ratio to the value given by point *F*. The system behaviour down to zero frequency is then perfectly normal. $\omega_1$ is the jump frequency for increasing frequency, and $\omega_2$ is that for decreasing frequency.

Quite apart from the possible existence of non-linear phenomena, the variation in performance as the input signal amplitude is varied may be considerable. Fig. 7.3 shows the experimental frequency responses obtained for an electrohydraulic

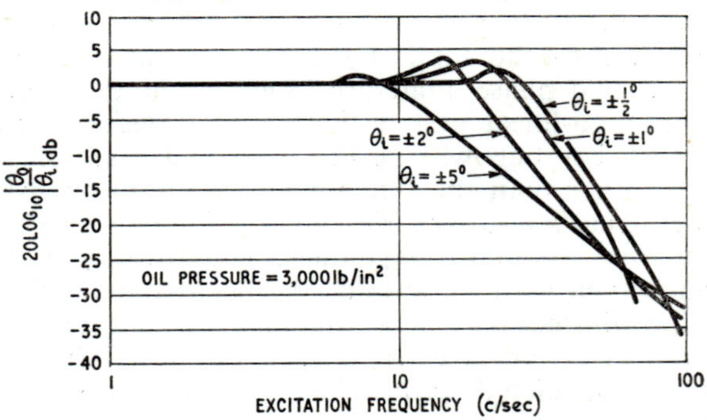


*Fig. 7.3. A set of experimental frequency responses obtained for an electrohydraulic servo showing the effect of the magnitude of the input signal*

servo for four different values of input amplitude. It will be noted that for this particular servo system bandwidth decreases as input amplitude is increased, whilst the peak amplitude ratio reaches a maximum value of $+3\frac{1}{2}$ db (1·50), which occurs for an input amplitude of about $\pm 2°$. The reader is warned that the effect of non-linearities on system performance can be much more spectacular or, alternatively, far less spectacular, than that exhibited in this fairly common example, and furthermore the nature of the effect may be unexpected in situations where a number of non-linearities combine to affect system behaviour.

Linear system response to any input stimulus is uniquely defined by the system transfer function or synonymously by the system pole-zero array. It therefore follows that for linear systems, responses to any input can be uniquely correlated to any other input. For example, the step function response may be transformed to the frequency domain and vice versa. In non-linear systems no such unique transformation exists, although under specific conditions it is possible to infer certain relationships. The existence of a limit cycle may not be apparent from studying the system frequency response, nor may the existence of a jump resonance be apparent from the system step function response.

## 7.2 Transfer Functions of Non-linear Systems

A sine wave applied to a linear system will result, in the steady state, in a pure sinusoidal output waveform. The output waveform will be of exactly the same frequency as the input, but with a different peak value and with the waveform displaced relative to the input by a certain phase angle. Both output peak and phase angle are functions of the applied frequency. In a non-linear system the output waveform will generally be repetitive but will be distorted and therefore contain frequencies other than the fundamental. The output waveform may be written as a Fourier series, and the fundamental harmonic calculated using standard techniques. This fundamental output waveform is then used to describe the behaviour of the system, the relationship between this fundamental output waveform and the input waveform being termed the describing function, i.e. the function which describes the behaviour of the output fundamental waveform in the frequency domain. Strictly speaking the describing function has a meaning only for sinusoidal excitation. Thus, although the system pole-zero array based on the describing function may be determined, the array is only valid near the $j\omega$ axis and can only be used to estimate transient response under conditions where the transient input to the non-linearity can be adequately represented by a piecewise set of sine waves or damped sine waves.

For small perturbations of the system about specific operating points, many non-linear systems may be linearised and the transfer functions obtained are valid in the linear sense, since the small signal response to any type of input may be determined from the pole-zero array, and the small signal frequency response can be correlated directly with the small signal transient response. A system pole-zero array will exist for every steady state operating point under consideration. If a root locus can be constructed for a particular system, the difference between these two methods of assessing the transfer function of a non-linear system is that the locus variable for the describing method will be the signal amplitude, whilst in the small perturbation method the locus variables will be steady state operating points such as chamber pressures and valve openings in a hydraulic servo. The situations in which two methods will provide meaningful information relating to system performance will be discussed when the methods themselves are evaluated in detail later in the chapter. It is necessary first to discuss the more common types of non-linearity in order to understand the type of problem which will be met when the non-linear systems are studied using these two non-linear transfer function techniques.

## 7.3 Classification of Non-linearities

It is particularly convenient at this stage to assume that the system has just one non-linearity present, and that this non-linearity is frequency independent. These assumptions will simplify the classification of the various types of non-linearities, although frequency dependent non-linearities and multiple non-linearities can be allowed for in certain design techniques. The block diagram for a system with a single frequency independent

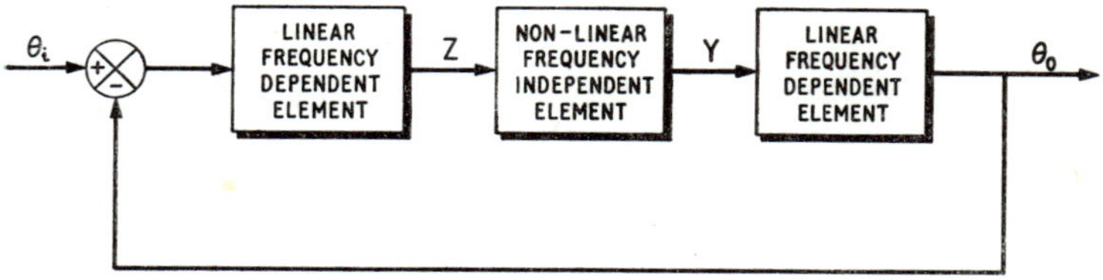


*Fig. 7.4. Block diagram of a non-linear control system wherein the non-linearity is frequency independent and can be easily isolated*

non-linearity is shown in Fig. 7.4, and in practice the non-linearity can occur anywhere in the system and in either the feed-forward or feedback path. Thus the non-linearity may occur in such places as the signal amplifier or the torque motor of a position servomechanism. Using the notation that $Z$ is the input to the non-linearity and $Y$ is the output from the non-linearity, typical frequency independent non-linearities are shown in Fig. 7.5. The linear element, soft spring, and hard spring characteristics of Fig. 7.5(a) are all continuous and single-valued, whilst the bang-bang, saturation, and deadzone characteristics of Fig. 7.5(b) are single-valued but discontinuous, since there are sharp changes in slope. The hysteresis characteristic of Fig. 7.5(c) is continuous but is not single-valued, since for any value of $Z$ there is no unique value of $Y$, whilst the mechanical backlash characteristic shown in Fig. 7.5(d) is similar to the hysteresis characteristic except that it is discontinuous at the end of the stroke. Non-single-valued non-linearities cannot be handled by the small perturbation method, and discontinuous non-linearities can only be handled within the continuous region, which really negates the use of the method under these circumstances. On the

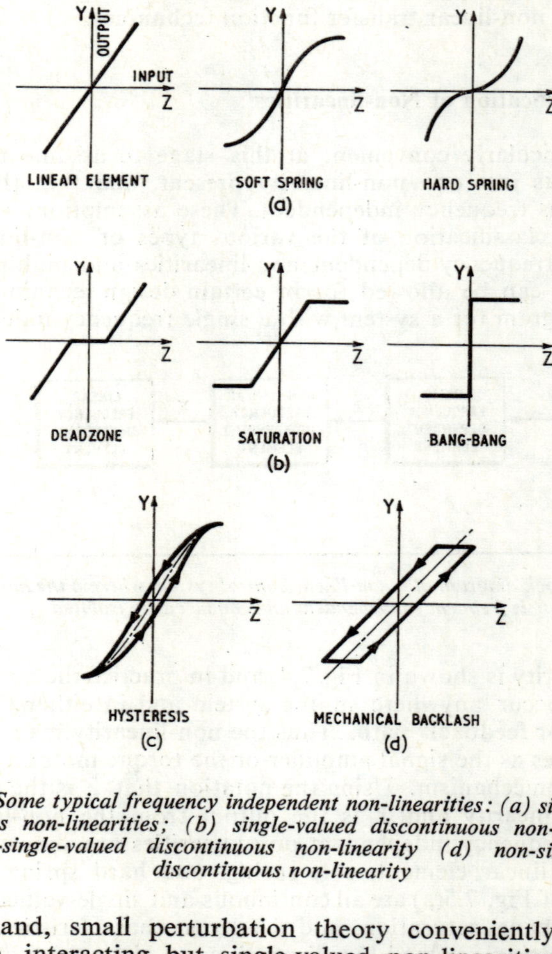


*Fig. 7.5. Some typical frequency independent non-linearities: (a) single-valued continuous non-linearities; (b) single-valued discontinuous non-linearities; (c) non-single-valued discontinuous non-linearity (d) non-single-valued discontinuous non-linearity*

other hand, small perturbation theory conveniently handles multiple, interacting but single-valued non-linearities. All the non-linearities shown in Fig. 7.5 are symmetrical in the sense that the forward characteristics ($Z$ positive) are the same as the reverse characteristics ($Z$ negative). Characteristics which are

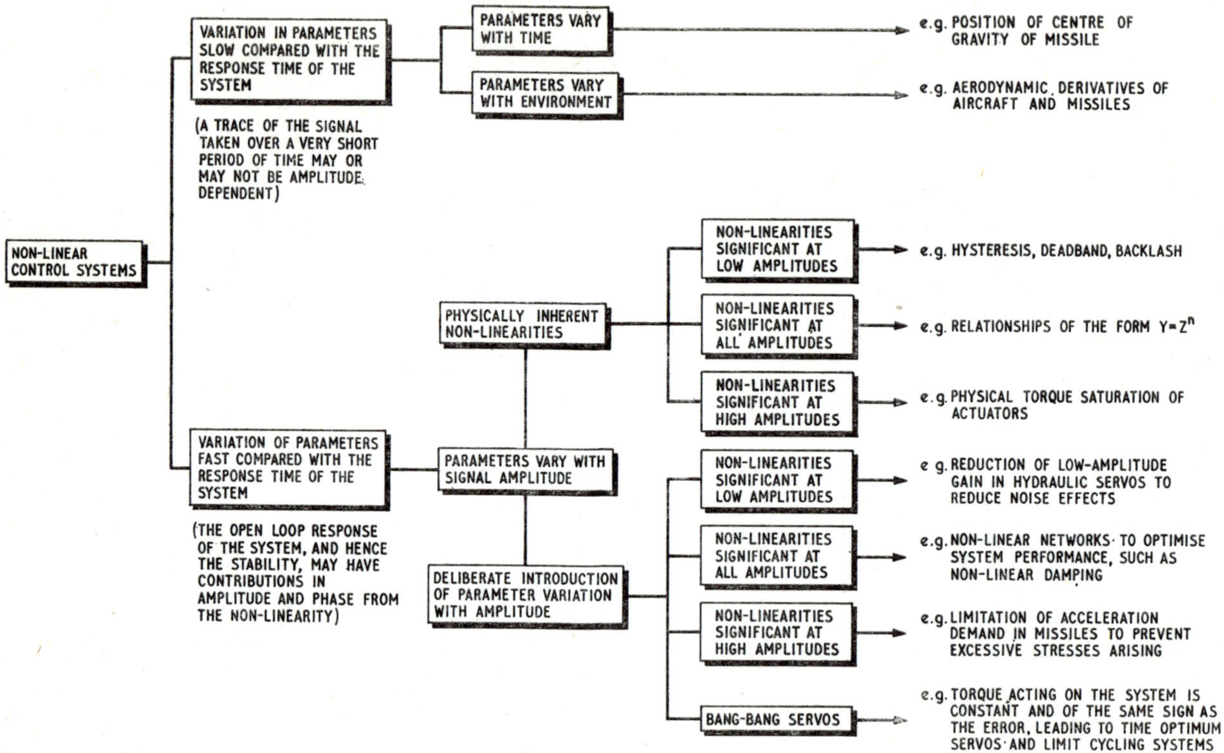


*Fig. 7.6. A simple classification of non-linear control systems*

asymmetrical can occur in feedback control systems, particularly in electronic components.

Owing to the lack of a unified theory capable of handling non-linear control systems, the treatment of, and attitude towards, non-linearities has become extremely parochial, and people working in one particular field are often completely unaware that the particular problems facing them may have already been studied and substantially solved sometime previously in another field, the only but very necessary common ground being the differential equation in s governing the system performance. To illustrate and emphasise the various attitudes of control engineers towards the non-linear system the chart of Fig. 7.6 has been drawn up. We can see immediately that there are several branches to the family tree, the first major division being the breakdown into amplitude dependent non-linearities and time dependent non-linearities (which may or may not be amplitude dependent at any particular instant of time). In transfer function terminology the first category will have a transfer function with coefficients which are a function of $\theta_i$ (and $\omega$ if the transfer function is a describing function involving a frequency dependent element), whilst for the second category the coefficients of the transfer function will also be affected by the time state and environmental state of the system. As an example of the second category the behaviour of a fixed-wing aircraft to a demand for pitch acceleration is shown in Fig. 7.7. The wide variation in response as a function of the

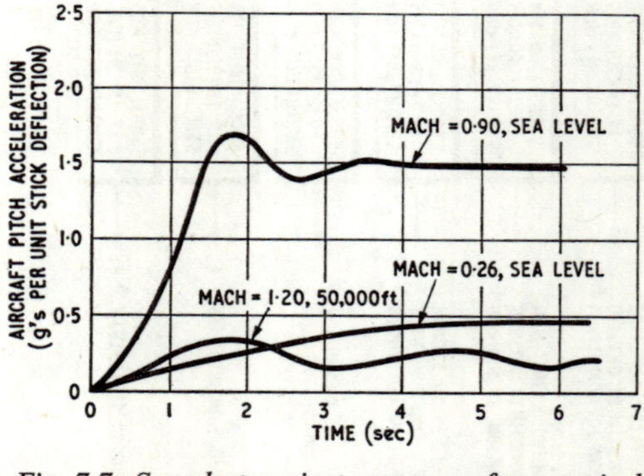


*Fig. 7.7. Sample transient responses for a typical environmental type of non-linearity*

environmental state is clearly shown, but it would appear that each response could be reasonably approximated by a suitable linearised transfer function, although further evidence would be required to establish whether the response is amplitude dependent for any of these flight cases, since only one test result is available in each case.

Turning now to the amplitude dependent group of non-linearities, we see that there are two vastly different sub-groups depending on whether the source of the non-linearity is physically inherent due to physical laws or production manufacturing tolerances, or whether the non-linearity is deliberately introduced to improve the system performance in some way. The physically inherent group are self-evident, and generally the attitude is to reduce the effects of the non-linearities as much as possible by good design and the highest standard of manufacture compatible with the economics of the situation.

Non-linearities may be deliberately introduced for a variety of reasons. Probably the best known example is the bang-bang form of control where the limitation in torque, for example, is recognised, and this torque is always demanded whenever there is a system error, however large or small. Instead of demanding a restoring torque proportional to the system error, a constant torque of the sign required to reduce the error is demanded. For example, if a step function input is applied to a second order system, the torque available initially accelerates the load, and provided the torque is switched to decelerate at the right instant the load will come to rest without overshoot in the minimum possible time. Generation of the right switching line involves the use of a non-linear network; it may be shown that the bang-bang system with the correct non-linear network will always have a faster response than any linear system with the same motor, and hence it is known as the time optimum system. Because of the non-linear operation the response time will be a function of the input amplitude, as shown in Fig. 7.8. With such systems it is necessary to modify the design to match the type of input signal expected, and this is one reason why the transfer function concept is inapplicable to this type of system since the design and test procedure should be based on the particular input or inputs of interest. For an $n^{th}$ order system $n-1$ torque switching points are required for the time optimum response,[67] and the optimum switching points are found by the use of the phase-space technique wherein an $n^{th}$ order differential equation is reduced to $n$ simultaneous first order differential equations. Unfortunately the lack of scope for the application of transfer function techniques to this type of system means that a detailed study of them is outside the scope of this book, but a brief mention has been made at this stage to keep things in perspective since the bang-bang mode of control is the obvious choice for certain projects.

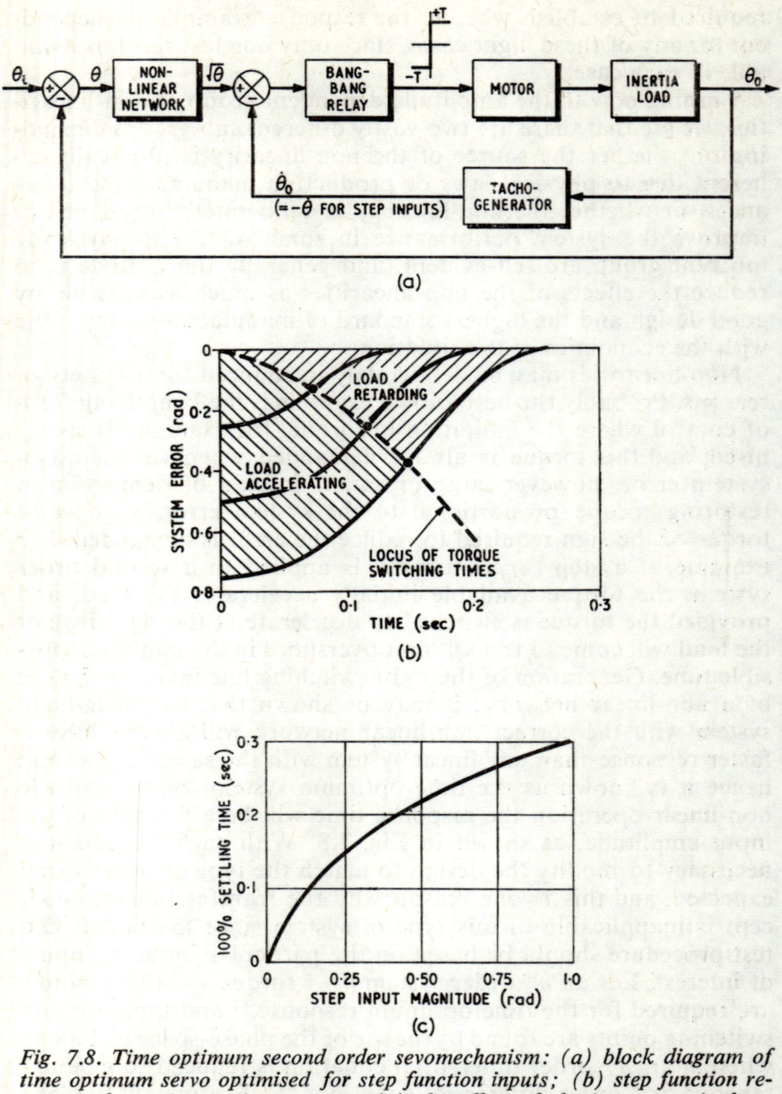


*Fig. 7.8. Time optimum second order sevomechanism: (a) block diagram of time optimum servo optimised for step function inputs; (b) step function responses for a time optimum servo: (c) the effect of the input magnitude on 100% settling time*

Other examples of the deliberate use of non-linear elements include the incorporation of saturation-type limits in missile autopilots to restrict the lateral acceleration of the missile to a value compatible with structural strength of the wing joints, irrespective of the lateral acceleration demanded by the guidance system; and the incorporation of a deadzone in an electro-hydraulic servo to reduce the power consumption due to low-amplitude noise which cannot be removed by means of frequency domain filtering. Having presented a fairly wide appraisal of non-linearities and their uses and abuses, we are now in a position to discuss the two specific non-linear techniques of interest in the transfer function approach, namely small perturbation theory and the describing function approach, and relate these to specific systems and non-linearities.

## 7.4 Introduction to Small Perturbation Theory (Small Signal Behaviour about a Steady State Operating Point)

Using small perturbation theory it is possible to linearise certain types of non-linear control elements and thus obtain system transfer functions with constant coefficients. These transfer functions are then treated in exactly the same way as normal linear transfer functions. Because of the change in coefficients of the transfer function as the steady state operating point is changed, it is necessary to study a number of such transfer functions corresponding to a suitable number of operating points. Such criteria as bandwidth, peak amplitude ratio and so on, can be determined from the linearised transfer function, and these criteria can then be plotted against the value of the steady state operating point. For example, the oil flow through a hydraulic spool valve is a non-linear function of valve opening, and it is possible to evaluate the linearised transfer function for small movements about selected values of valve opening. These linearised transfer functions are then used to evaluate, say, the peak amplitude ratio and the bandwidth, and these can be plotted against the steady state valve opening. Similarly a root locus plot can be drawn, and the variable along the locus will then be the steady state operating point.

The linearisation process involved in the small perturbation approach is such that departures from a steady state operating point are small enough to produce transfer function coefficients which are sufficiently constant to be relatively independent of the amplitude of the applied excitation signal *provided this is small*. We are thus replacing a non-linear differential equation by a set of linearised differential equations with coefficients dependent upon the steady state operating point. It follows, since we are considering a set of linearised differential equations, that only phenomena peculiar to linear systems can be investigated. Thus the existence of purely non-linear phenomena such as limit cycles, sub-harmonic oscillations and jump resonances are not predicted with the small perturbation method, although sometimes it is possible to infer the existence of a limit cycle under certain circumstances when the linearised theory predicts instability in the linear sense at a particular steady state operating point.

Small perturbation theory is applicable to non-linear control elements where the form of non-linearity is continuous and single-valued. Thus non-linearities of the type $Y = Z^{\mu}$ may be studied by the method, but hysteresis and relay-type non-linearities cannot be linearised in this particular way. There is no limitation imposed by the small perturbation method on the order of the system differential equation which may be handled, and the technique is thus equally suitable for both low and high order systems. An outstanding advantage of the small perturbation method is the ability to decouple multiple interacting non-linearities, and for very complex interacting systems such as jet engine fuel control systems it is often the only feasible method of studying behaviour prior to using an analogue computer to obtain a more exact solution. As a design technique, small perturbation theory is used to predict system performance for small perturbations about judiciously chosen steady state operating points, and if small signal behaviour about these points is satisfactory it is concluded that the system performance is also satisfactory for large departures from the steady state operating points. Naturally, the advent of the analogue computer as a design tool does mean that the foregoing hypothesis can be easily proved or disproved once a trial design has been established.

## 7.5 Linearising Equations by the Small Perturbation Method

If the output-input relationship for the non-linear element is available in graphical form the small perturbation method is particularly straightforward in application. Suppose the input to the non-linear element is $Z$, and the output from the non-linear element is $Y$, such that $Y = f(Z)$ as shown in Fig. 7.9. Let the steady state operating point of interest be defined by the coordinates $(Y_0, Z_0)$ then clearly $Y_0 = f(Z_0)$. The general relationship between $Y$ and $Z$ is thus

$$Y = f(Z) \tag{7.1}$$

If we now consider small increments about the steady state operating point ($Y_0$, $Z_0$) we may write

$$Y = Y_0 + \Delta Y \tag{7.2}$$

and

$$Z = Z_0 + \Delta Z \tag{7.3}$$

Substituting these new values of $Y$ and $Z$ in Equation 7.1, we obtain

$$Y_0 + \Delta Y = f(Z_0 + \Delta Z) \tag{7.4}$$

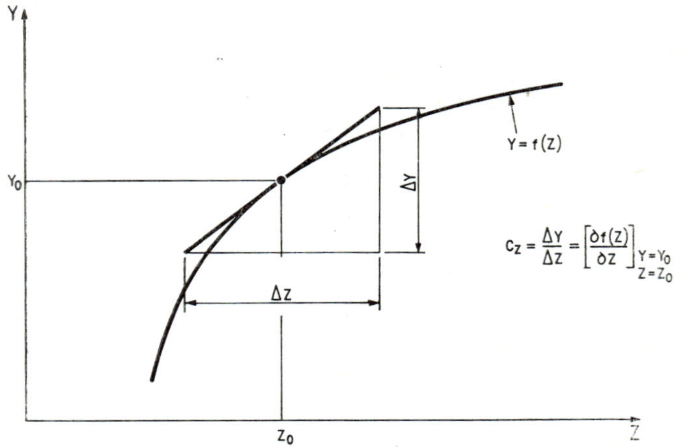


*Fig. 7.9. Application of small perturbation theory to a single continuous non-linearity*

Equation 7.4 may be expanded in a Taylor series about ($Y_0$, $Z_0$) to give

$$Y_0 + \Delta Y = f(Z_0) + \left[\frac{\partial f(Z)}{\partial Z}\right]_{Z=Z_0} \Delta Z \tag{7.5}$$

where second order and higher terms in the series are neglected. Since $f(Z_0) = Y_0$, Equation 7.5 may be simplified to

$$\Delta Y = \left[\frac{\partial f(Z)}{\partial Z}\right]_{Z=Z_0} \Delta Z \tag{7.6}$$

Introducing the general notation

$$C_Z = \left[\frac{\partial f(Z)}{\partial Z}\right]_{Z=Z_0} \tag{7.7}$$

we may write the linearised gain of the non-linear element as

$$\Delta Y = \left[\frac{\partial Y}{\partial Z}\right]_{Z=Z_0} \Delta Z = C_Z \Delta Z \tag{7.8}$$

Equation 7.8 is the result we would obtain by intuition, since $[\partial Y/\partial Z]_{Z=Z_0}$ is simply the slope of the $Y-Z$ graph at $Z = Z_0$.

The linearisation technique may be applied to a function of any number of variables. For example, if the inputs to an element are $Y$ and $Z$, and the output is $X$, then the output-input relationship for the non-linear element may be written

$$X = f(Y, Z) \tag{7.9}$$

If $X_0$, $Y_0$ and $Z_0$ are the steady state operating points, then for small displacements about the operating point the new values of $X$, $Y$ and $Z$ are, respectively,

$$X = X_0 + \Delta X \tag{7.10}$$

$$Y = Y_0 + \Delta Y \tag{7.11}$$

and

$$Z = Z_0 + \Delta Z \tag{7.12}$$

Substituting the new values into Equation 7.9, we have

$$X_0 + \Delta X = f(Y_0 + \Delta Y,\ X_0 + \Delta X) \tag{7.13}$$

The Taylor series expansion, neglecting terms higher than the first, is

$$X_0 + \Delta X = f(Y_0, Z_0) + \left[\frac{\partial f(Y_0, Z_0)}{\partial Y}\right]_{\substack{Z=Z_0\\ Y=Y_0}} \Delta Y + \left[\frac{\partial f(Y_0, Z_0)}{\partial Z}\right]_{\substack{Z=Z_0\\ Y=Y_0}} \Delta Z \tag{7.14}$$

Hence the output may be written

$$\Delta X = \left[\frac{\partial X}{\partial Y}\right]_{\substack{Z=Z_0\\ Y=Y_0}} \Delta Y + \left[\frac{\partial X}{\partial Z}\right]_{\substack{Z=Z_0\\ Y=Y_0}} \Delta Z \tag{7.15}$$

or, in general notation,

$$\Delta X = C_Y \Delta Y + C_Z \Delta Z \tag{7.16}$$

If graphs relating $X$ and $Y$ with $Z$ constant, and graphs relating $X$ and $Z$ with $Y$ constant, are available, then $C_Y$ and $C_Z$ are obtained by direct measurement as shown in Fig. 7.10. Every combination of $Y$ and $Z$ chosen will define a steady state operating

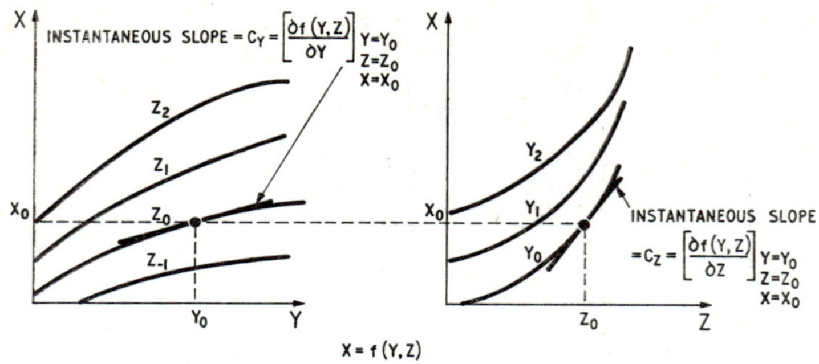


*Fig. 7.10. The application of small perturbation theory to a double continuous non-linearity*

point, and hence will automatically define a small perturbation differential equation. It follows that at the design stage there will be the same number of differential equations to be solved as there are steady state operating points.

## 7.6 Application of Small Perturbation Theory to the Mechanical-Hydraulic Jack Servomechanism

Small perturbation theory is often used to linearise the differential equations describing the performance of hydraulic and pneumatic systems, since the efficient use of fluid power often results in non-linear operation of such systems. The non-linearity in the main arises from two sources: firstly it is convenient to use circular ports in spool valves, thus resulting in flow through the port being a non-linear function of valve opening even with the valve pressure drop constant, and secondly the flow is affected by the valve pressure drop which is in turn affected by the motion of the controlled load, thus leading to a cross-coupling effect between fluid flow and load dynamics. The first non-linearity may also be aggravated or smoothed, depending on circumstances, by the variation in the coefficient of discharge of the valve as the valve is opened.

A typical mechanical-hydraulic jack servo is shown in Fig. 7.11 and is representative of one particular form of aircraft power control. It will be noted from Fig. 7.11 that the actuator piston is earthed, and the load moves with the actuator body. With the arrangement shown, a movement of the spool valve to the right will allow high-pressure oil to flow into the right-hand side of the actuator, and low-pressure oil will exhaust from the left-hand side of the actuator back to the storage tank, where the oil will

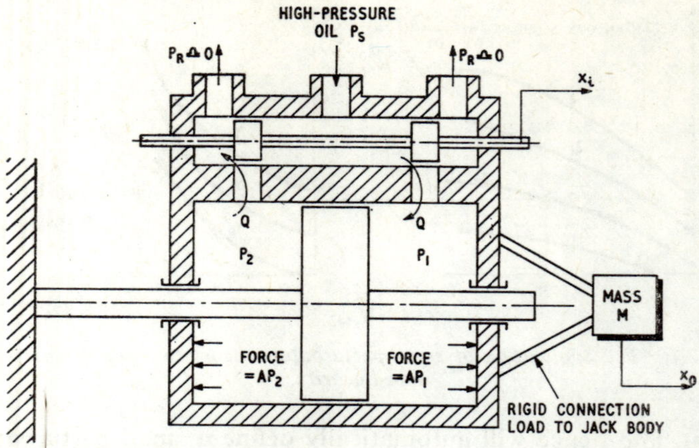


*Fig. 7.11. Typical mechanical-hydraulic jack servomechanism*

be stored until such time as it is pumped up to supply pressure and transferred to the accumulator. Because of these oil flows into and out of the actuator, a pressure drop will be generated across the actuator and will result in a nett force acting to the right which will accelerate the load until the load and body have moved into the position where the valve will be closed and no further oil flow will take place. The mechanical-hydraulic jack servo is thus a simple follow-up device with a very large power amplification, since the input power level at the valve input is of the order of a few watts whilst the power controlled by the valve may be of the order of several horsepower.

The application of small perturbation theory to the mechanical-hydraulic jack servo is made in terms of perturbations of valve movement and perturbations of chamber pressures about suitable steady state operating points.[47, 68] For this study we shall adopt the convention of capital letters for quantities relating to the steady state values, and lower case letters for quantities relating to perturbations about the steady state values. Thus $P$ is the steady state pressure in a particular part of the system, whilst $p$ is the perturbed pressure for small increments about the steady state pressure. We shall assume that flow conditions do not lead to cavitation in either side of the actuator chamber, i.e. $P_1 > 0$ and $P_2 > 0$.

If $P_v$ is the steady state pressure drop across the valve inlet port, $P_c$ is the steady state pressure drop across the actuator piston, and $P_1$ and $P_2$ are the respective chamber pressures, we may write

$$P_c = P_1 - P_2 \tag{7.17}$$

For a jack with perfect flow symmetry, the pressure drop across the exhaust port will be equal to the pressure drop across the inlet port and thus, provided the exhaust pressure is negligible,

$$P_v = P_s - P_1 = P_2 \tag{7.18}$$

where $P_s$ is the system supply pressure, which may be several thousand pounds per square inch. $P_1$ and $P_2$ may be eliminated between Equations 7.17 and 7.18, leaving

$$P_v = \frac{P_s - P_c}{2} \tag{7.19}$$

If the steady state load on the jack is zero, then the pressure drop across the valve is $P_s/2$, and the steady state chamber pressures are $P_1 = P_2 = P_s/2$.

The flow of oil through the valve inlet port and into the actuator will equal the flow of oil out of the actuator through the valve exhaust port for conditions of perfect flow symmetry, and under these circumstances we can consider the flow conditions through the inlet port only in order to determine the system dynamics. Flow through spool valves is turbulent in nature, and is therefore a function of the square root of the valve pressure drop. In general, therefore, for a valve opening $X_v$, the valve flow is

$$Q = f(X_v, \sqrt{P_v}) \tag{7.20}$$

If the coefficient of discharge of the valve orifice is constant, $Q$ will be proportional to $X_v$ for rectangular ported valves. Eliminating $P_v$ from Equation 7.20 by using the relationship in Equation 7.19, we may write

$$Q = f\left(X_v, \sqrt{\frac{P_s - P_c}{2}}\right) \tag{7.21}$$

Using the first terms of the Taylor series expansion, Equation 7.21 may be expanded about steady state operating points $Q_0$, $X_{v0}$, $P_{c0}$, to give

$$Q_0 + q = f\left(X_{v0}, \sqrt{\frac{P_s - P_{c0}}{2}}\right) + \left[\frac{\partial Q}{\partial X_v}\right]_{\substack{X_v = X_{v0} \\ P_c = P_{c0}}} x_v + \left[\frac{\partial Q}{\partial P_c}\right]_{\substack{X_v = X_{v0} \\ P_c = P_{c0}}} p_c \tag{7.22}$$

where the partial derivatives aie taken in the requisite direction at the steady state operating point. Thus $[\partial Q/\partial X_v]_{\substack{X_v = X_{v0} \\ P_c = P_{c0}}}$ is the slope of the $Q$–$X_v$ graph for $P_c$ constant $= P_{c0}$ evaluated at

$X_v = X_{v0}$ and $[\partial Q/\partial P_c]_{\substack{X_v=X_{v0}\\P_c=P_{c0}}}$ is the slope of the $Q$–$P_c$ graph for $X_v$ constant $= X_{v0}$ evaluated at $P_c = P_{c0}$. Defining

$$\left[\frac{\partial Q}{\partial X_v}\right]_{\substack{X_v=X_{v0}\\P_c=P_{c0}}} = C_e$$

and

$$\left[\frac{\partial Q}{\partial P_c}\right]_{\substack{X_v=X_{v0}\\P_c=P_{c0}}} = -C_p$$

Equation 7.22 may be written in standard form

$$q = C_e x_v - C_p p_c \tag{7.23}$$

where $x_v$ and $p_c$ are the small perturbations about the operating point.

Note how the small perturbation method has uncoupled the two general non-linearities in $X_v$ and $P_c$ which were previously being multiplied together. In fact, since $P_c$ is a function of the instantaneous load behaviour, as will be shown in Equation 7.27, the true non-linear equation has cross-coupling between the valve motion and the load motion which is particularly difficult to handle except by the small perturbation method or by automatic computation, whether by analogue or digital means.

The flow into the right-hand side of the actuator will be composed of a direct flow term plus a compressibility flow term. The compressibility flow term arises because, under hydraulic servo operating conditions, the oil cannot be considered incompressible, and this flow will be proportional to the rate of change of pressure in the right-hand side of the actuator. If $A$ is the ram net cross-sectional area, $V$ is the volume of oil trapped between the valve and the actuator piston in the steady state, and $N$ is the bulk modulus of the oil (between $10^5$ and $2\times10^5$ lb/in$^2$, depending on the type of oil and the amount of entrained air), the perturbed flow into the right-hand side of the jack is

$$q = A\mathrm{s}x_0 + \left(\frac{V}{N}\right)\mathrm{s}p_1 \tag{7.24}$$

$$\underset{\substack{\text{Swept volume}\\\text{term}}}{\uparrow} \quad + \quad \underset{\substack{\text{Compressibility}\\\text{term}}}{\uparrow}$$

Owing to flow symmetry, the rate of change of pressure in the right-hand chamber will be one half of the rate of change of pressure drop across the actuator piston. Hence

$$\mathrm{s}p_1 = \frac{\mathrm{s}p_c}{2} \tag{7.25}$$

and substituting in Equation 7.24, we have

$$q = Asx_0 + \left(\frac{V}{2N}\right) sp_c \tag{7.26}$$

If the load is rigidly connected to the actuator body then, from Newton's Law, the perturbed nett force is

$$Ap_c = Ms^2x_0 \tag{7.27}$$

and the flow into the right-hand side of the actuator becomes

$$q = Asx_0 + \left(\frac{VM}{2NA}\right) s^3x_0 \tag{7.28}$$

Equating the valve flow to the flow into the right-hand side of the actuator and transposing, the relationship between load motion and valve motion is

$$\frac{x_0}{x_v}(s) = \frac{1}{\left(\frac{A}{C_e}\right)s\left[1+\left(\frac{C_pM}{A^2}\right)s+\left(\frac{VM}{2NA^2}\right)s^2\right]} \tag{7.29}$$

The second order term in Equation 7.29 represents the closed-valve oscillatory mode of the mass on the oil column, and is normally of high frequency but of low damping ratio. For the particular design of servo shown in Fig. 7.11, the valve opening is defined as

$$x_v = x_i - x_0 \tag{7.30}$$

and thus the small perturbation system transfer function is

$$\frac{x_0}{x_i}(s) = \frac{1}{1+\left(\frac{A}{C_e}\right)s+\left(\frac{C_pM}{AC_e}\right)s^2+\left(\frac{VM}{2C_eNA}\right)s^3} \tag{7.31}$$

If the Routh Hurwitz criterion is applied to Equation 7.31, the resultant array is

$$\begin{array}{c|cc}
s^3 & \dfrac{VM}{2C_eNA} & \dfrac{A}{C_e} \\[2ex]
s^2 & \dfrac{C_pM}{AC_e} & 1 \\[2ex]
s^1 & \dfrac{\dfrac{C_pM}{AC_e}\times\dfrac{A}{C_e}-\dfrac{VM}{2C_eNA}}{\dfrac{C_pM}{AC_e}} & \\[2ex]
s^0 & 1 &
\end{array}$$

For critical stability,

$$\frac{C_p M}{AC_e} \times \frac{A}{C_e} = \frac{VM}{2C_e NA}$$

and, in the simplest form,

$$\frac{C_p}{C_e} = \frac{V}{2NA} \tag{7.32}$$

Note that the left-hand side of Equation 7.32 is a function of the valve characteristics only. Typical valve flow curves are shown

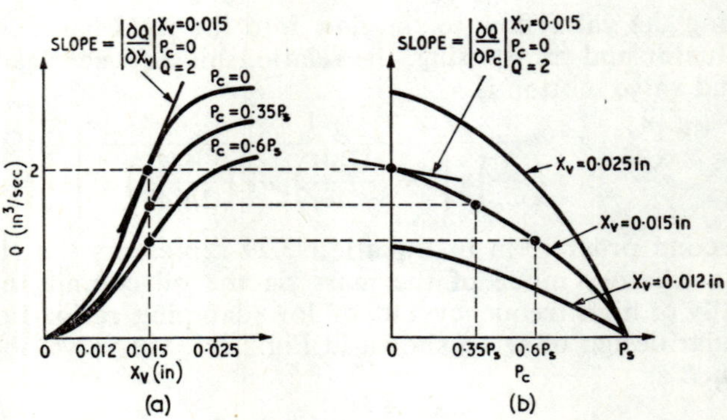


*Fig. 7.12. Valve flow curves for a typical mechanical-hydraulic jack servomechanism: (a) valve flow versus valve opening (load pressure drop constant); (b) valve flow versus load pressure drop (valve opening constant)*

in Fig. 7.12, and it can be seen that $C_p$ tends to be relatively small, particularly if the load pressure drop $P_c$ is small. This is so for most hydraulic valves, particularly in the region of zero valve opening and so, theoretically, hydraulic servos are unstable for small perturbations about the zero error steady state operating point. For absolute stability of the servomechanism in the linear sense about all steady state operating points in $X_v$ and $P_c$, the ratio $C_p/C_e$ must *always* be greater than $V/2NA$.

One common method of stabilising hydraulic servos is to introduce a small bleed hole across the actuator piston. The leakage flow through the bleed hole may often, but not always, be considered as laminar flow and therefore proportional to the pressure drop across the piston. Alternatively if experimental leakage results for the bleed hole are available, small perturbation coefficients can be obtained for this term too. If at the design stage the damping term $C_p$ is considered negligible, Equation 7.23 reduces to

$$q = C_e x_v \tag{7.33}$$

and the flow into the right-hand side in the presence of leakage flow defined by $Lp_c$ becomes

$$q = As x_0 + Lp_c + \left(\frac{V}{2N}\right) sp_c \tag{7.34}$$

When the process of elimination of $q$, $x_v$ and $p_c$ is repeated, the system transfer function becomes

$$\frac{x_0}{x_i}(s) = \frac{1}{1+\left(\frac{A}{C_e}\right)s+\left(\frac{LM}{AC_e}\right)s^2+\left(\frac{VM}{2C_eNA}\right)s^3} \tag{7.35}$$

Thus the leakage flow term is additive to $C_p$ since, by comparison with Equation 7.31, the system for both flow components significant would become $[(L+C_p)/AC_e]\,M$.

Application of the Routh Hurwitz criterion to Equation 7.35 results in the simple stability relationship

$$\frac{L}{C_e} \geqslant \frac{V}{2NA} \tag{7.36}$$

By adjusting the leakage flow term $L$, the servo may in theory be suitably stabilised. Using the flow Equations 7.33 and 7.34 it is possible to construct a small perturbation block diagram showing the mechanics of operation of the hydraulic servomechanism. Thus if the pressure drop flow term $C_p p_c$ is neglected, we may interpret the equations as follows. A valve-opening $x_v$ produces a perturbed flow $q$ which is available to move the actuator body, but which in addition has to provide the leakage flow and compressibility flow as well. Thus the nett flow available to displace the actuator body is the valve flow minus the compressibility flow minus the leakage flow, and this may be interpreted as a comparator situated immediately after the valve. The $p_c$ term necessary to develop the leakage and compressibility flows can be thought of as generated from the output position using the relationship defined by Newton's Law in Equation 7.27. The resulting block diagram is shown in Fig. 7.13, and shows that the hydraulic servo is basically a multi-loop system with physically inherent minor loops. If compressibility, leakage and load effects are negligible, the system transfer function for this type

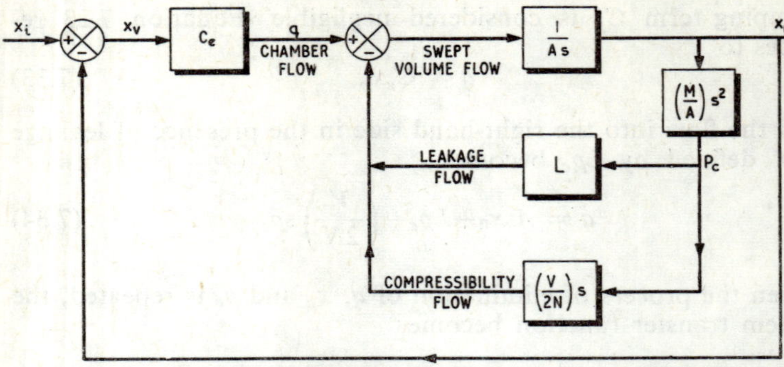


*Fig. 7.13. Small perturbation theory block diagram for the mechanical-hydraulic jack servo*

of hydraulic servo can be reduced to the simple lag form

$$\frac{x_0}{x_i}(s) = \frac{1}{1+\left(\frac{A}{C_e}\right)s} \tag{7.37}$$

In fact, it is often found that even when the system is genuinely third order the bandwith can be determined reasonably accurately from Equation 7.37, since the complex roots due to the oil compressibility mode will be of relatively high frequency and will not significantly affect the response in the region of bandwidth frequency.

## 7.7 Determination of Hydraulic Servo Transfer Functions and Root Loci in Terms of the Steady State Operating Points

EXAMPLE 7.7.1

A hydraulic jack-type servomechanism has the following characteristics:

$L = 0{\cdot}0022\ \text{in}^5\ \text{lb}^{-1}\ \text{sec}^{-1}$

$M = 50/386 = 0{\cdot}13\ \text{lb sec}^2\ \text{in}^{-1}$

$V = 1{\cdot}5\ \text{in}^3$

$A = 1\ \text{in}^2$

$N = 10^5\ \text{lb/in}^2$

$C_e$ varies between 30 in²/sec and 190 in²/sec for the zero load

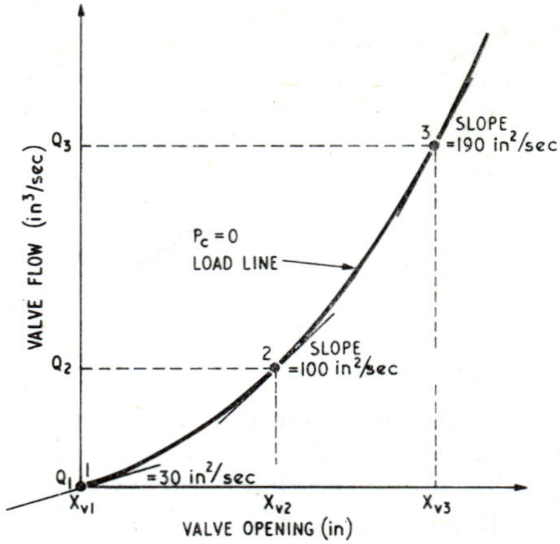


*Fig. 7.14. Valve flow versus valve opening characteristics ($C_p = 0$) for the mechanical-hydraulic jack servo of Example 7.7.1*

pressure drop case as shown in Fig. 7.14
$C_p \simeq 0$

If it can be assumed that the servo operates in the lightly loaded condition, i.e. $P_c = 0$ is a suitable load line, determine the small perturbation transfer function for the three operating points shown on the flow versus valve opening curve. Hence determine the root locus as $C_e$ is varied, i.e. as the steady state operating point is changed.

The transfer function coefficients are, from Equation 7.35,

$$a_0 = 1$$

$$a_1 = \frac{A}{C_e}$$

$$a_2 = \frac{LM}{AC_e}$$

and

$$a_3 = \frac{VM}{2NAC_e}$$

The numerical values of the transfer function coefficients are therefore as shown in Table 7.7.1.

**Table 7.7.1**

TRANSFER FUNCTION COEFFICIENTS AS A FUNCTION OF STEADY STATE OPERATING POINTS FOR THE MECHANICAL-HYDRAULIC SERVO OF EXAMPLE 7.7.1

| *Steady state operating point* | $C_e$ | $a_0$ | $a_1$ | $a_2$ | $a_3$ |
|---|---|---|---|---|---|
| 1 | 30 | 1 | $33\times10^{-3}$ | $9{\cdot}5\times10^{-6}$ | $32{\cdot}4\times10^{-9}$ |
| 2 | 100 | 1 | $10\times10^{-3}$ | $2{\cdot}85\times10^{-6}$ | $9{\cdot}7\times10^{-9}$ |
| 3 | 190 | 1 | $5{\cdot}3\times10^{-3}$ | $1{\cdot}50\times10^{-6}$ | $5{\cdot}1\times10^{-9}$ |

It is enlightening to evaluate $\omega_0$, $b$ and $c$ for these steady state operating points so that the coefficient plane discussed in Chapter 5 may be used to study the change in system poles as the operating point is changed. The required relationships are

$$\omega_0 = \left(\frac{1}{a_3}\right)^{1/3}$$

$$b = a_2\omega_0^2$$

and

$$c = a_1\omega_0$$

The results are shown in Table 7.7.2.

Note that Point 3 gives the product $bc = 1{\cdot}55$, thus confirming that a suitable hydraulic servo design point would lie on the $bc = 1{\cdot}5$ line in the region of $b = 0{\cdot}5$. As the valve gain is reduced the design point moves into the region where the first order mode tends to dominate the response in accordance with the results shown in Fig. 5.24. The open loop poles may be determined from Equation 7.29 with $L$ substituted for $C_p$. There will be an open loop pole at nought and a complex, lightly damped pole pair due to the oscillatory mode of the mass on the oil column.

**Table 7.7.2**

NORMALISATION FREQUENCY AND NORMALISED COEFFICIENTS AS FUNCTIONS OF STEADY STATE OPERATING POINTS FOR THE MECHANICAL-HYRAULIC SERVO OF EXAMPLE 7.7.1

| *Steady state operating point* | $C_e$ | $\omega_0$ | $b$ | $c$ |
|---|---|---|---|---|
| 1 | 30 | 313 | 0.98 | 10.3 |
| 2 | 100 | 470 | 0.63 | 4.7 |
| 3 | 190 | 580 | 0.505 | 3.07 |

The undamped natural frequency of this pole pair equals $\sqrt{2NA^2/VM}$ as determined from the second order factor of Equation 7.29, and substituting for the values for this particular servo, we obtain the undamped natural frequency

$$\omega_{nc} = \sqrt{\frac{2NA^2}{VM}} = \sqrt{\frac{10^5 \times 2 \times 386}{50 \times 1{\cdot}5}} = 1{,}015 \text{ rad/sec}$$

and, since $C_p \simeq 0$,

$$\frac{2\zeta_c}{\omega_{nc}} = \frac{LM}{A^2} = \frac{0{\cdot}0022 \times 50}{1^2 \times 386}$$

Hence

$$\zeta_c = \frac{0{\cdot}0022 \times 50}{1^2 \times 386} \times \frac{1{\cdot}015}{2} = 0{\cdot}145$$

The open loop poles are shown in Fig. 7.15, and the root locus shown in the same figure has been constructed using Table

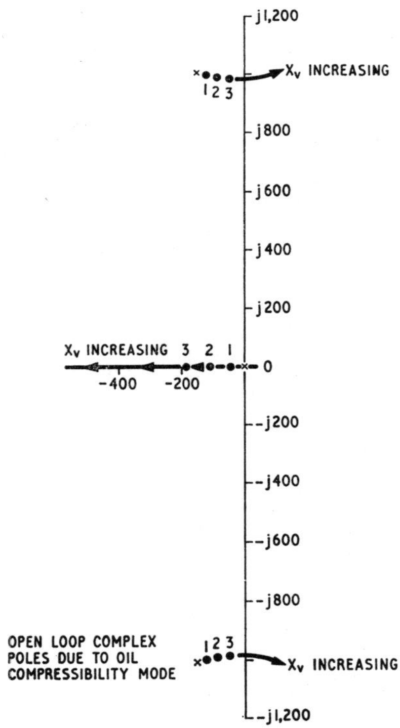


*Fig. 7.15. Small perturbation root locus in terms of the steady state operating points on the $X_v$ curve for the mechanical-hydraulic servo of Example 7.7.1*

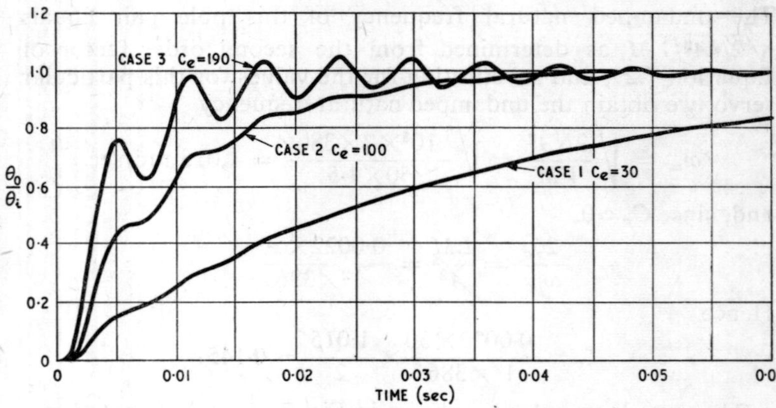


*Fig. 7.16. Small perturbation transient response for three steady state operating points on the flow valve opening curve ($P_c = 0$) for the mechanical-hydraulic servo of Example 7.7.1*

7.7.2 and Fig. 5.23 to determine the complex pair of system poles and then dividing the system characteristic equation by this complex pair to determine the third pole, since on this occasion the root locus is constructed much more rapidly by using the coefficient plane method of Chapter 5; for higher order systems the normal root locus method would be used and the points on the locus corresponding to the various steady state operating points would be determined by interpolation in the usual way. Systems with multiple non-linearities would require a set of root loci; for example in the hydraulic servo, if the pressure drop is likely to be significant, then it would be necessary to repeat the root locus of Fig. 7.15 for those values of $P_c$ necessary to obtain reasonable coverage of the problem. The effect of significant changes in $P_c$ will be to multiply the unloaded value of $C_e$ by the factor $\sqrt{1-P_c/P_s}$. It is this square-rooting effect which makes the hydraulic servo performance relatively insensitive to small changes in load dynamics. However, since $C_p$ is a function of the load pressure drop, as shown in Fig. 7.12(b), this term will not normally be negligible if the load pressure drop becomes large, and will have to be taken into account in the system characteristic equation.

The transient responses for small displacements about the three steady state operating points are shown in Fig. 7.16. For Case 1 the system response is very much dominated by the first order mode, although the oscillatory mode is apparent in the early stages of the response. In Case 2 the response is still very much dominated by the first order mode, although the oscillatory mode is more clearly visible, whilst for Case 3 the oscillatory mode is so lightly damped that the settling time is longer than for Case 2. The response for Case 3 represents the worst case normally allowable in practice, since a relatively small alteration in the position of the complex open loop poles would result in an unstable system. These responses are for small displacements about three steady state operating points for a lightly loaded hydraulic servo. From the responses we might conclude that the design is acceptable, knowing that individual responses might take any of the forms shown. Since the function of the servo is to keep the error, and hence the valve opening, small, an experimental trace in response to a true input signal might well show that on average the system behaviour is best expressed by the small perturbation transfer function for Case 1. For a large step input covering the slopes corresponding to Cases 1, 2 and 3, we might well find that the response in the early stages might look like the response for small perturbations about point 2, since this is roughly the average slope for the large opening, whilst in the later stages of the response it might well become quite deadbeat as the Point 1 behaviour becomes significant. Such inferences can only be drawn on the basis of experience gained from analogue computer and experimental results. The small perturbation method should generally be used to fix roughly certain system parameters, such as choosing a particular value of valve flow gain, prior to a more detailed study on an analogue computer.

## 7.8 The Describing Function Concept (Linearisation of Frequency Response for Large Departures into the Non-linear Operating Region)

Linearisation of non-linear systems for large departures into the non-linear operating region can often be achieved by using the describing function method, which is essentially a frequency domain approximation. Owing to the nature of non-linear systems it is rarely possible to infer behaviour in response to any other type of excitation from a knowledge of the describing function. The term describing function may be used to refer to the behaviour of an individual element, or to a complete system response. For a given excitation frequency, a linear system is described, as we have seen in Chapter 1, by the amplitude ratio of the input and output harmonic waveforms, and the phase lag between the two waveforms, both characteristics being measured in the steady state. When non-linear systems are excited by a harmonic waveform, the output waveform will contain multiples of the excitation frequency. In other words, there will be a distortion of the waveform in addition to phase shift and amplitude scaling. Fig.

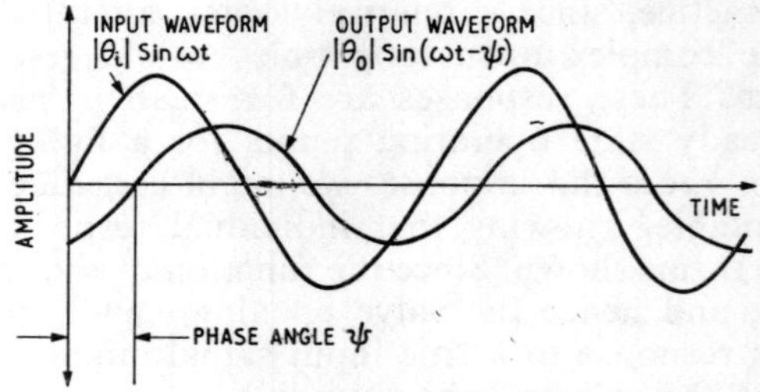


(a)

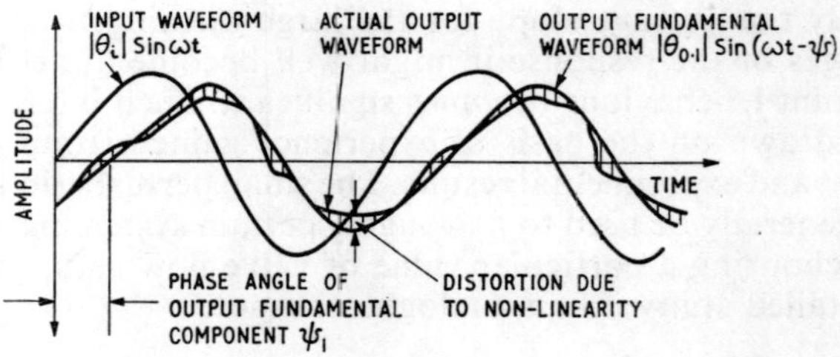


(b)

*Fig. 7.17. Comparison of steady state response to a sinusoidal input of linear and non-linear control systems leading to a definition of the describing function: (a) steady state response of a linear control system to a sinusoidal input for a given input frequency; (b) steady state response of a non-linear control system to a sinusoidal input for a given input frequency and amplitude*

7.17(a) shows the situation for the linear system, and Fig. 7.17(b) shows the same situation for the non-linear system. The describing function for the non-linear system is obtained by determining by measurement or by calculation the fundamental harmonic of the non-linear output waveform, and treating the amplitude and phase data relating to this fundamental harmonic in exactly the same way as if the system were linear.

Clearly, from Fig. 7.17(b) it is possible to determine a wide range of fundamental harmonic waveforms of the right frequency. How should we choose the correct waveform? Fortunately, a consistent answer can be obtained fairly readily, and the conventional describing function is based on the fundamental output harmonic which is defined by the first sine and cosine term in the standard Fourier series. This particular fundamental harmonic is reasonably easy to calculate for a wide range of frequency independent non-linearities, and it is equally fortuitous that the

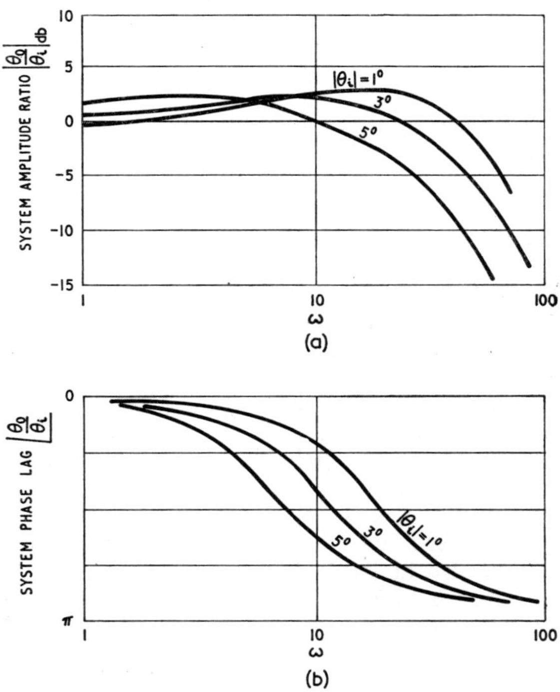


*Fig. 7.18. Typical describing function for system response based on the output fundamental harmonic: (a) amplitude contribution; (b) phase lag contribution*

integration processes inherent in a wide range of test equipment force most transfer function analysers to display the amplitude and phase lag associated with this harmonic. Mathematically it may be shown that this choice of fundamental harmonic minimises the root mean square error between the chosen waveform and the actual non-linear waveform, and is therefore a 'best fit' in the mean squares sense.

Many non-linear elements to be found inside control systems are dependent only on signal amplitude at the input to the non-linearity, and thus element describing functions are often independent of frequency. For system describing functions, $|\theta_0/\theta_i|$ will be a function of $|\theta_i|$ and excitation frequency, even if the element inside the loop is not frequency dependent, and the same applies to the system phase lag also. A typical system describing function is shown in Fig. 7.18, both amplitude and phase components being necessary for a complete description of the system, although sometimes only the amplitude component is of direct interest.

## 7.9 Some Uses and Limitations of the Describing Function Technique

For large amplitude excitation the describing function is often the only feasible way of crystallising system behaviour, and is by far the most popular method of synthesising non-linear systems which are outside the scope of small perturbation theory and the phase plane technique. The basic assumption made in the use of the describing function technique is that the performance can be meaningfully described by reference to the fundamental harmonic of output only. This assumption therefore requires that suitable linear filtering exists, preferably both before and after the non-linearity, to attenuate the higher harmonics. In fact, the basic describing function theory, developed independently by, amongst others, Kochenburger,[69] Greif,[70] and Tustin,[71] assumes that the non-linearity in a feedback control system is frequency independent, and is preceded and followed by a suitable linear filter. On the whole, these requirements are unnecessarily restrictive and, as we shall see in Section 7.14, it is possible to obtain reasonable results in a system with very little filtering present. The describing function method has also been applied, with some success, to systems with multiple non-linearities and with frequency dependent non-linearities. It should be pointed out that the describing function technique can give unexpectedly good, and sometimes misleadingly bad, results, and therefore the describing function technique in this day and age should be looked upon as a means of rapidly synthesising an approximate answer to the design problem prior to studying the system on an analogue computer to verify not only the design, but occasionally the validity of the describing function approximations also.

At the synthesis stage the stability of the system can be determined using the theoretical or experimental describing functions for individual non-linear elements. Stability is best assessed by the use of broken loop techniques, but the root locus method can also be used in certain circumstances. For stable systems, a variety of techniques exist for determination of the system describing function given the element describing functions and the linear transfer functions encountered inside the loop. It is customary to require the describing function to be determined as a function of frequency with $|\theta_i|$ fixed. Some of the methods for determination of the system describing function are laborious if a definite $|\theta_i|$ is required.

A meaningful root locus can be drawn which can be calibrated in terms of the *input to the non-linearity*, and the resultant poles, together with the system zeros, can be used to determine the system describing function by evaluating the frequency responses in a scalar or logarithmic form. These frequency responses are also calibrated in terms of the input to the non-linearity. However, some difficulty arises in interpreting the root locus in terms of *system input magnitude*, and it is therefore generally more convenient to interpret the system describing function itself as the graphical representation of the system transfer function plotted as a function of frequency and amplitude of excitation. Theoretically it is possible, for any given input amplitude $|\theta_i|$, to use this amplitude ratio and phase lag data in order to obtain a reasonable fit approximate linear transfer function (and hence reasonable system poles and zeros). This quasi-linear transfer function can then be used to describe the system behaviour in the frequency domain for the prescribed value of input only.

For example, a given set of data may yield two approximate transfer functions covering the frequency range of interest. At $|\theta_i| = 1°$,

$$\left[\frac{\theta_0}{\theta_i}(j\omega)\right]_{\theta_i=1°} = \frac{1}{1+0{\cdot}10(j\omega)+0{\cdot}03(j\omega)^2+0{\cdot}001(j\omega)^3} \tag{7.38}$$

whilst at $|\theta_i| = 3°$,

$$\left[\frac{\theta_0}{\theta_i}(j\omega)\right]_{\theta_i=3°} = \frac{1}{1+0{\cdot}20(j\omega)+0{\cdot}04(j\omega)^2+0{\cdot}002(j\omega)^3} \tag{7.39}$$

Generally speaking the graphical representation of the system transfer functions for non-linear systems is perfectly adequate. The use of best-fit transfer functions may be necessary in circumstances where test equipment (on-line or otherwise) measures a set of quasi-linear transfer function coefficients directly, such as achieved by the Wayne Kerr analogue modelling technique.

## 7.10 Use of Element Describing Functions to Determine System Stability

The general block diagram for a feedback control system with a single frequency independent non-linearity is shown in Fig. 7.19. Classical describing function theory requirements of pre- and post-linear filtering are observed by the inclusion of the terms $G_1(j\omega)$ and $G_2(j\omega)$. The input to the non-linearity is $I_M$, and it is assumed that in the steady state the filtering action is sufficient to ensure that the input to the non-linearity is a perfect sine wave,

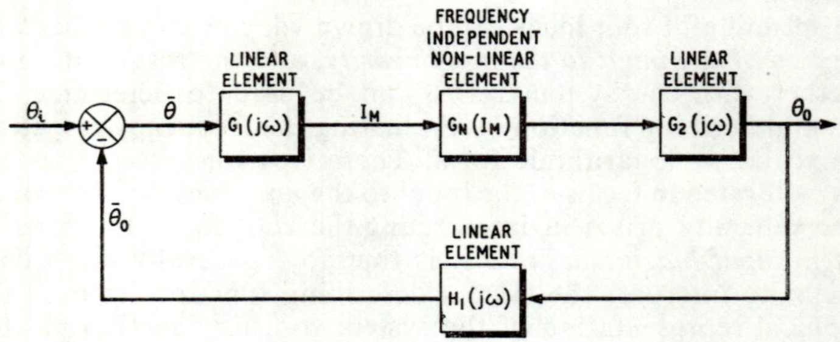


*Fig. 7.19. Block diagram for a feedback control system with a single frequency independent non-linearity*

which may therefore be written in the form $I_M \sin \omega t$. In the steady state the non-linear element is replaced by the describing function $G_N(I_M)$, which will have an amplitude contribution only for single-valued non-linearities such as deadzone and saturation, but will have amplitude and phase contributions for non-single-valued non-linearities such as backlash and hysteresis. For deadzone the interpretation of the describing function is essentially that of a variable gain which increases with $I_M$ and reaches the linear value of gain as $I_M \to \infty$, whilst for saturation the interpretation of the describing function is also that of a variable gain, but a gain which decreases as $I_M$ exceeds the level of input required for saturation. Simple interpretations can generally be made for any amplitude dependent non-linearity.

When using element describing functions to determine system stability it is possible to use open loop techniques or the root locus techniques. It is generally preferable to use open loop techniques, since the element describing function is easily expressed in terms of signal levels occurring inside the loop. The broken loop equation corresponding to Fig. 7.19 can be written in the standard form, which in this case is

$$\frac{\bar{\theta}_0}{\bar{\theta}}(I_M, j\omega) = G_1(j\omega)\, G_N(I_M)\, G_2(j\omega)\, H_1(j\omega) = -1 \qquad (7.40)$$

For fixed linear elements, Equation 7.40 can be plotted on a Nyquist diagram in the usual way for various values of $I_M$ and

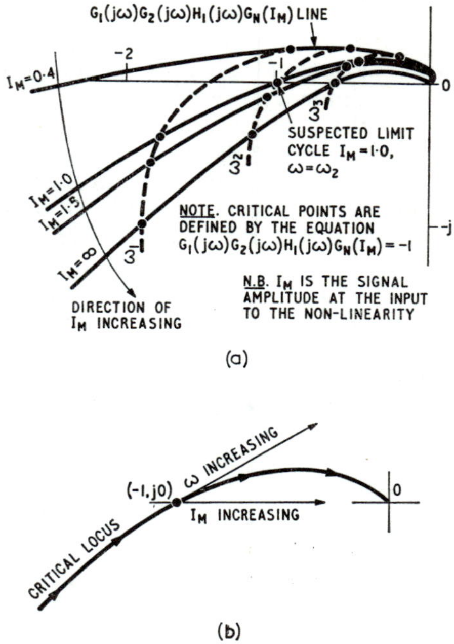


*Fig. 7.20. Use of the Nyquist plot together with the describing function to predict the existence of limit cycles in a system with backlash: (a) Nyquist plot for a third order system with backlash showing encirclement of the Nyquist point for certain signal amplitudes; (b) Nyquist form of Loeb criterion for the existence of a stable limit cycle in a system where the non-linearity is independent of frequency*

$\omega$. To illustrate a typical Nyquist plot, let us consider the case where the non-linearity is backlash. The backlash describing function will be examined in detail in Section 7.11; at this stage it is sufficient to know that $G_N(I_M)$ will have an attenuation effect which will decrease as $I_M$ is increased, and a phase lag which will decrease as $I_M$ is increased. Fig. 7.20(a) shows a Nyquist plot for a third order system with backlash; $I_M = \infty$ is the standard linear open loop response. It can be seen from Fig.

7.20(a) that if the input to the non-linearity is small enough, the Nyquist point is encircled. The values of amplitude and frequency corresponding to the critical locus as it passes through the critical point are, respectively, $I_M = 1{\cdot}0$, $\omega = \omega_2$. It will be noted by comparison with the linear response that at this value of $I_M$ the backlash describing function is contributing 16° phase lag and sufficient attenuation to drag the $\omega_2$ point into the Nyquist point as shown.

Encirclement of the Nyquist point in this manner can be interpreted as showing the *possible* existence of a limit cycle of amplitude $I_M = 1{\cdot}0$ and frequency $\omega = \omega_2$. Non-encirclement of the Nyquist point over the possible operating range of $I_M$ can be interpreted, within the limits of the describing function technique, as meaning that the system is both stable in the linear sense and free from limit cycles. Suspected limit cycles may be stable or unstable in their own right, and it is therefore necessary to test the limit cycle for stability. An unstable limit cycle will just oscillate to infinity, or to the next stable limit cycle condition. Loeb[72] has suggested a simple limit cycle stability criterion which may be applied to a system where the non-linearity is frequency independent, and this is shown in Fig. 7.20(b). If the non-linearity is frequency dependent, the limit cycle stability test becomes much more complicated. In the simple case shown the requirement for a stable limit cycle is that as the variable-frequency constant-amplitude locus passing through the critical point is traversed in the direction of increasing frequency, the amplitude must be increasing to the right as shown in Fig. 7.20(b). A similar interpretation of limit cycle stability can be made using either Bode plots or Nichols plots based on the broken loop response, since both these methods of analysis hinge around the concept of the Nyquist point as was discussed in Chapter 1.

When examining non-linear feedback control systems for the existence of limit cycles it is often found convenient to rewrite Equation 7.40 in the following form[73]

$$G_1(j\omega)G_2(j\omega)H_1(j\omega) = -\frac{1}{G_N(I_M)} \qquad (7.41)$$

The locus in terms of frequency corresponding to the linear frequency dependent elements of the left-hand side of Equation 7.41 is plotted separately from the amplitude dependent inverse describing function shown on the right-hand side of the equation. Possible limit cycles are then indicated by intersection points of the two curves as shown in Fig. 7.21.

Although Bode and Nichols plots are viable alternatives to the Nyquist plot for studying the stability of non-linear feedback

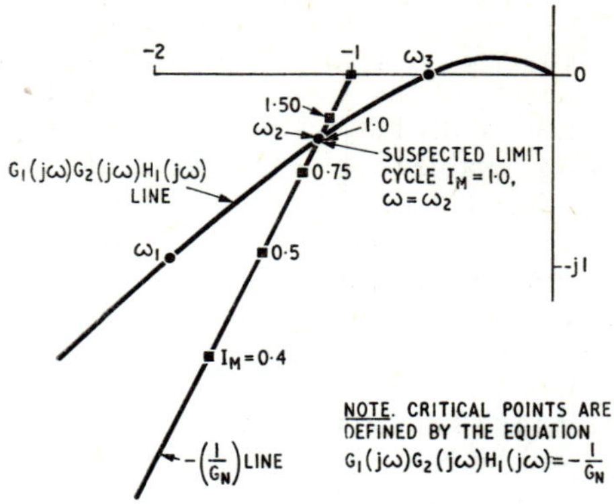


*Fig. 7.21. An alternative method of using the Nyquist plot for the prediction of limit cycles: the linear response and the inverse describing function are plotted separately, and any intersections are suspected limit cycles*

control systems, the Nichols plot is the more common of the two. This is partially due to the comparative clarity of data presentation, and partially due to the possibility of using Nichols contours in the rather laborious process of determining the system response. The Nichols chart representation of the third order system with backlash is shown in Fig. 7.22, and for this example the gain-phase presentation is particularly clear. To remove the limit cycle indicated by the intersection of the curves representing the left-hand side and right-hand side of Equation 7.41, plotted in decibels, the linear part of the response on the Nichols chart must

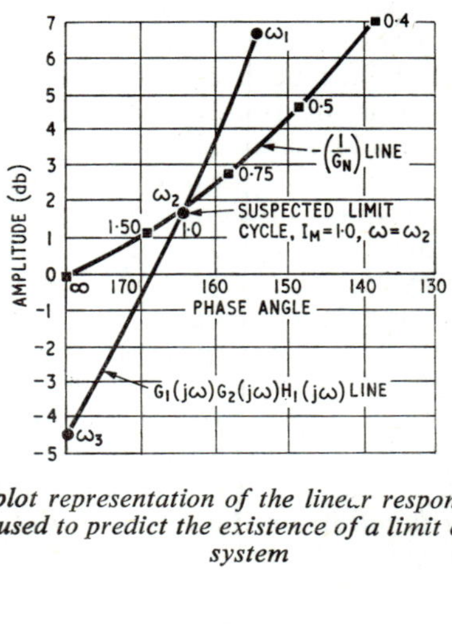


*Fig. 7.22. Nichols plot representation of the linear response and the backlash describing function used to predict the existence of a limit cycle in a third order system*

be moved to the right and/or downwards, whilst the corresponding movement required on the Nyquist plot is the attenuation or anticlockwise rotation of the linear curve away from the inverse describing function plot. The whole range of linear compensation techniques is available to solve this problem. Quite clearly the limit cycle is removed if one of the following alternatives is adopted: (a) reduction of feedforward gain (this will drop the linear line downwards on the Nichols chart and attenuate the radii on the Nyquist plot); (b) phase-advance shaping or rate feedback shaping. The latter compensation will move the linear line mainly to the right on the Nichols chart but will also provide some amplification which has to be allowed for, whilst a similar amplification also results on the Nyquist plot but the locus can be swung round in an anticlockwise direction as required. A first approximation to the magnitude of phase advance required would be about 35° at frequency $\omega'$, since this would probably be enough to make the linear line clear the inverse backlash describing function line for all values of input amplitude, but this figure could only be confirmed or otherwise by a more thorough study of the low-frequency characteristics of the linear part of the system. Suitably compensated Nyquist and Nichols plots are shown in Figs. 7.23(a) and (b) and, since intersection is avoided, it may be concluded, within the limitation of describing function theory, that limit cycles are unlikely in the compensated system.

In practice, because of the neglect of the higher harmonics, a limit cycle may exist in practice even though no such limit cycle is predicted using describing function theory. This can occur when the phase margin in the worst case is small and the rounding-off errors due to the describing function method become significant. Such non-linear systems should be checked on an analogue computer once the design has been established to make sure that these rounding-off errors do not invalidate the proposed design. Limit cycles can always be removed by providing perfectly linear components, but this solution is normally impracticable from engineering and economic considerations.

The root locus method may also be used to determine the existence of limit cycles, since these will be indicated by the locus crossing the imaginary axis. When the non-linearity is amplitude dependent only, and possesses no phase shift, the non-linearity may be treated as a variable feedforward gain, and the root locus for a single-loop system is therefore identical to that for a conventional linear locus drawn in terms of feedforward gain, except that the locus variable will be either the element describing function or the input amplitude at the non-linearity, since these two quantities are dependent. Because the describing function is a frequency domain concept, the root locus for a non-linear system can only yield stability information based on the $j\omega$ axis, and system frequency response data in terms of the input amplitude at the non-linearity. The stability test on the $j\omega$ axis is directly equivalent to an examination of the Nyquist point in broken loop methods. Considerable effort is required to translate into a useful form the system frequency response which can be obtained from the root locus, since this is

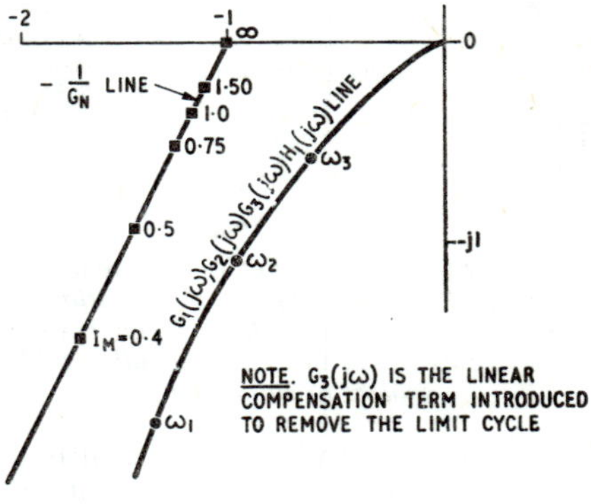


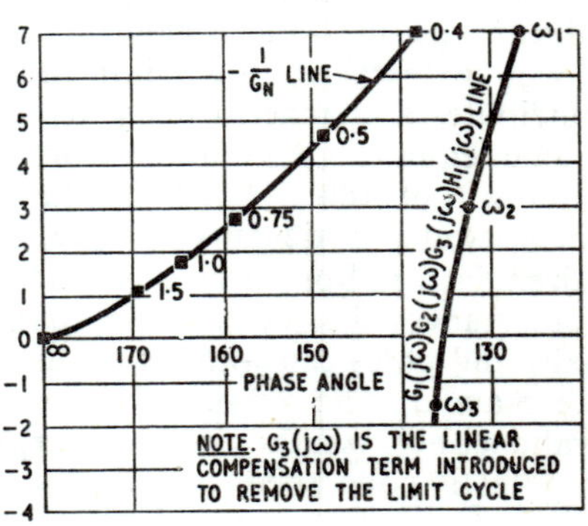


*Fig. 7.23. Example of removal of the limit cycle using a passive phase advance network: (a) Nyquist plot of the inverse describing function and linear response for a compensated system with a backlash-type non-linearity (no crossing points exist and therefore no limit cycles are suspected); (b) Nichols plot of the inverse describing function and linear response for a compensated system with a backlash-type non-linearity (no crossing points exist and therefore no limit cycles are suspected)*

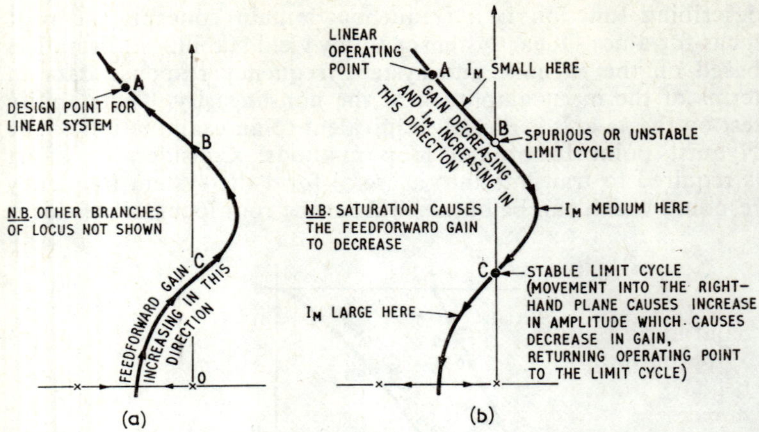


*Fig. 7.24. Use of the root locus to predict the existence of limit cycles in a conditionally stable control system subject to saturation in the feedforward path: (a) root locus for a conditionally stable linear system; (b) use of root locus to predict limit cycle existence in a conditionally stable system which is subject to saturation*

in terms of the input amplitude at the non-linearity, whereas the response normally required is in terms of the system input amplitude $|\theta_i|$. However, the root locus method is the basis of McAllister's method[74] for obtaining the system describing function given the linear and non-linear elements inside the loop, and this method will be described in detail in Section 7.16. When non-linearity contributes a change in phase as well as a change in gain, the root locus equation is modified from the conventional angular requirement of $180° \pm N360°$ by the phase contribution of the non-linearity at a particular value of $I_M$. In other words, the angular requirement will appear as a function of the s plane. Under these circumstances the locus is no longer symmetrical about the real axis, and poles can wander around the s plane in an indiscriminate manner. The majority of the root locus rules are no longer valid either. As far as the interpretation of the resultant system poles is concerned, no difficulty should arise if it is remembered that the results are valid for harmonic excitation only.

As an example of the interpretation of a root locus plot in terms of stable and unstable limit cycles, let us consider the root locus plot for a conditionally stable control system as shown in Fig. 7.24(a). Such a system would normally be designed to possess a pair of complex poles in the position $A$ shown near the imaginary axis, and the poles will be in this position during normal linear operation. If, for some reason, such as during warm-up time, the feedforward gain is reduced, the system will become unstable by virtue of this pole pair moving into the right-hand half-plane through point *B*. In a non-linear system such a reduction in gain can be effectively achieved by the application of a large demand signal which will saturate the system, as will become apparent when the describing function for a saturating element is determined in the next section. Having passed through point *B* a further reduction in gain due to the input amplitude at the non-linearity increasing as a consequence of this instability will drive the complex poles to the point *C* on the imaginary axis, and any further gain reduction beyond this point will stabilise the system because the poles will have moved back into the left-hand half-plane. We therefore conclude that, if the system input amplitude is such that the system does saturate, the system can become unstable for certain modes of operation due to the movement of the poles into the right-hand half-plane.

Note that although for a linear system the locus as the feed-forward gain is increased will be in the direction *CBA*, the locus for the non-linear system for increasing input amplitude will be in the reverse direction *ABC*. Thus, when the poles migrate into the right-hand half-plane, the system will be unstable in the linear sense, and the amplitude will therefore increase. In turn the increased amplitude $I_M$ will effectively reduce the gain, thus driving the poles to the point *C* whence a limit cycle of frequency $\omega_c$ and amplitude $I_{Mc}$ at the non-linearity will be set up. When limit cycling at this point, any slight tendency for the system poles to move into the right-hand half-plane will result in an increase in signal amplitude which will decrease the gain and drive the poles back to the limit cycle point *C*, which is therefore known as a stable limit cycle point. No such restoring action occurs at point *B*, which is thus an unstable limit cycle point through which the response oscillates with increasing magnitude in the same way as would happen with a linear system with the requisite value of gain. Although the describing function technique will often predict the existence of limit cycles, it does not indicate the transient system input conditions which might generate the limit cycle.

## 7.11 Derivation of the Describing Function for Frequency Independent Non-linearities

The calculation of the conventional describing function for frequency independent non-linearities is quite straightforward if the method of Fourier series is used. Let the input to the non-linear device be defined as $I_M \sin \omega t$, then if the output waveform from the device is $f(t)$, the output waveform may be written as a Fourier series,

$$f(t) = \frac{A_0}{2} + A_1 \sin \omega t + B_1 \cos \omega t + A_2 \sin 2\omega t + B_2 \cos 2\omega t + \cdots \tag{7.42}$$

where the Fourier series coefficients are

$$A_r = \frac{2}{\pi} \int_0^{\pi} f(t) \sin r\omega t \, d\omega t \tag{7.43}$$

and

$$B_r = \frac{2}{\pi} \int_0^{\pi} f(t) \cos r\omega t \, d\omega t \tag{7.44}$$

If the non-linearity is single valued, i.e. contains none of the memory properties found in hysteresis and backlash, $B_r = 0$, which considerably simplifies the computation. The describing function assumes that only $A_1$ and $B_1$ are required to describe adequately the element and the higher harmonics are therefore discarded, but they can be calculated from $A_r$ and $B_r$ if required. Under these circumstances the describing function is written

$$G_N(I_M) = \left| \frac{O_1}{I_M} \right| \underline{/-\psi} \tag{7.45}$$

Assuming that the phase relationship is, in fact, a phase lag, and substituting $A_1$ and $B_1$, the describing function becomes

$$G_N(I_M) = \left| \frac{\sqrt{A_1^2 + B_1^2}}{I_M} \right| \underline{/\tan^{-1} \frac{B_1}{A_1}} \tag{7.46}$$

Thus if the element is single valued, $B_1 = 0$ and the phase lag of the element will be zero. To illustrate the method, consider the saturating element shown in Fig. 7.25(a) which is a simple output limiter. The output waveform is shown in Fig. 7.25(b) for a complete cycle, but the symmetry property may be used and the integration need only be carried out in the range 0 to $\pi$ as defined in Equations 7.43 and 7.44. By inspection $f(t)$ may be written

$$\left.\begin{aligned} f(t) &= K_N I_M \sin \omega t & 0 \leq \omega t \leq \alpha \\ f(t) &= K_N I_L = K_N I_M \sin \alpha & \alpha \leq \omega t \leq \pi - \alpha \\ f(t) &= K_N I_M \sin \omega t & \pi - \alpha \leq \omega t \leq \pi \end{aligned}\right\} \tag{7.47}$$

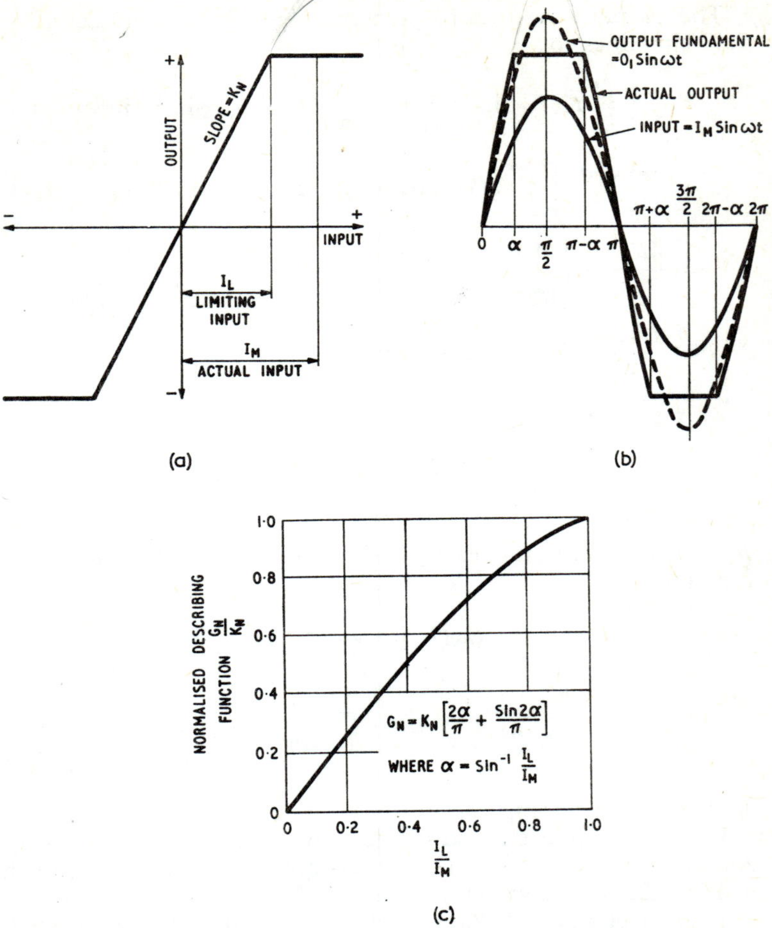


*Fig. 7.25. Describing function for saturation: (a) output-input relationship; (b) input and output waveforms; (c) normalised describing function*

where $\alpha = \sin^{-1}(I_L/I_M)$. Hence, substituting Equation 7.47 in Equation 7.43, we have

$$A_1 = \frac{2}{\pi}\int_0^{\alpha} K_N I_M \sin^2 \omega t \, d\omega t + \frac{2}{\pi}\int_{\alpha}^{\pi-\alpha} K_N I_L \sin \omega t \sin \alpha \, d\omega t + \frac{2}{\pi}\int_{\pi-\alpha}^{\pi} K_N I_M \sin^2 \omega t \, d\omega t \qquad (7.48)$$

The output waveform is symmetrical about $\pi/2$, and Equation 7.48 simplifies to

$$A_1 = \left[\frac{4}{\pi}\int_0^{\alpha} K_N I_M \sin^2 \omega t \, d\omega t + \frac{4}{\pi}\int_{\alpha}^{\pi/2} K_N I_M \sin\alpha \sin\omega t \, d\omega t\right]$$

$$= \frac{4K_N I_M}{\pi}\left[\int_0^{\alpha} \tfrac{1}{2}(1-\cos 2\omega t)\, d\omega t + \int_{\alpha}^{\pi/2} \sin\alpha \sin\omega t \, d\omega t\right]$$

$$= \frac{4K_N I_M}{\pi}\left\{\left[\frac{1}{2}\left(1-\frac{\sin 2\omega t}{2}\right)\right]_0^{\alpha} + \sin\alpha\left[-\cos\omega t\right]_{\alpha}^{\pi/2}\right\}$$

$$= \frac{K_N I_M}{\pi}(2\alpha - \sin 2\alpha + 4\sin\alpha\cos\alpha)$$

and, in simplest form,

$$A_1 = \frac{K_N I_M}{\pi}(2\alpha + \sin 2\alpha) \qquad (7.49)$$

$B_1 = 0$ since the non-linearity is single valued, and the describing function is

$$G_N(I_M) = \frac{A_1}{I_M} \underline{/-\psi}$$

i.e.

$$G_N(I_M) = \frac{K_N}{\pi}(2\alpha + \sin 2\alpha) \underline{/0^\circ} \qquad (7.50)$$

This is shown in Fig. 7.25(c).

Using the same method the describing function for the deadzone may be obtained. The output-input relationship is shown in Fig. 7.26(a), the waveforms in Fig. 7.26(b) and the describing function itself in Fig. 7.26(c). Since the deadzone is single valued, there is again no phase lag contributed by the non-linearity. The describing function for the deadzone will be zero for small inputs and will approach the linear value of gain, $K_N$, for large inputs. Backlash is not single valued, however, and the describing function will have a phase contribution which will be a function of the input amplitude. The physical meaning of backlash is shown in Fig. 7.27(a) and the resultant input-output relationship in Fig. 7.27(b). Where the backlash is in a pair of spur gears, the distance $2D$ represents the distance between the mating teeth when measured along the pitch circle. The phase lag caused by the backlash is apparent from a study of the waveforms shown

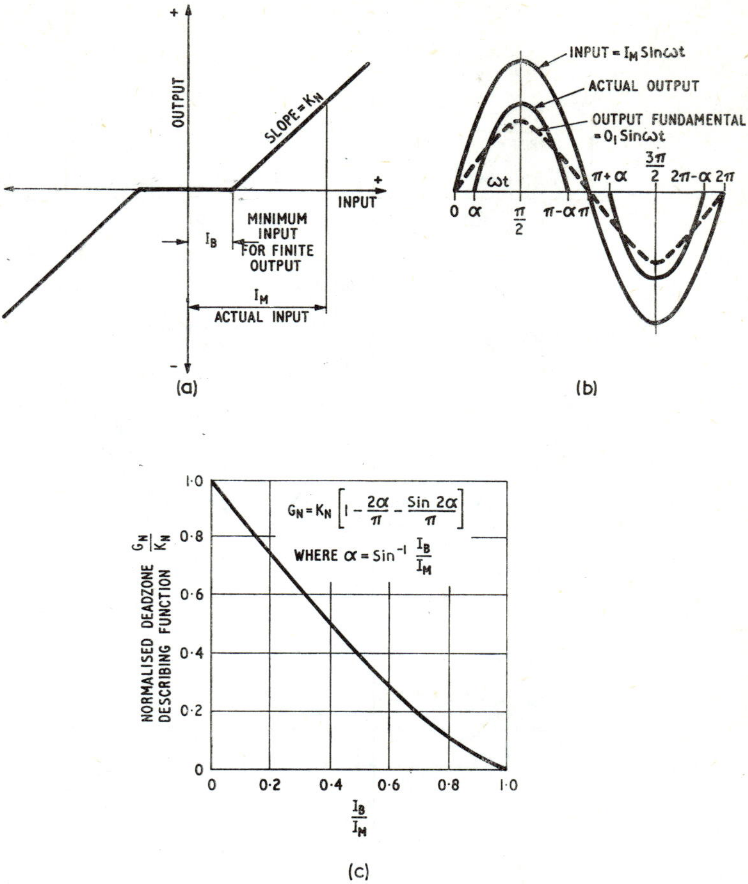


*Fig. 7.26. Describing function for deadzone: (a) output-input relationship; (b) input and output waveforms; (c) normalised describing function*

in Fig. 7.27(c), and both phase and amplitude parts of the describing function are shown in Fig. 7.27(d), the phase lag approaching 90° as the input signal is reduced to the same magnitude as the semi-clearance *D*. Nyquist and Nichols forms of the backlash describing function have already been met in Figs. 7.21 and 7.22, respectively. The relatively large phase lags obtainable as a result of the presence of backlash explains the great trouble with limit cycling sometimes caused in feedback control systems prone to this type of non-linearity.

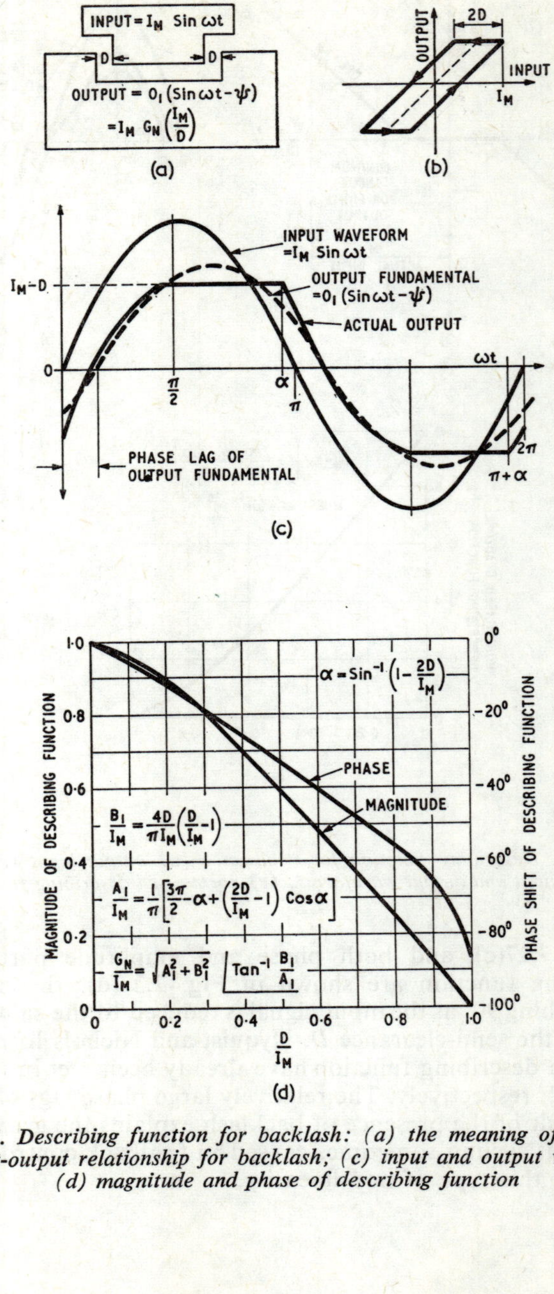


*Fig. 7.27. Describing function for backlash: (a) the meaning of backlash, (b) input-output relationship for backlash; (c) input and output waveforms; (d) magnitude and phase of describing function*

## 7.12 Describing Functions for Variable-gain Elements and Dynamic Components

Often the input-output relationship of an element may be expressed in the form of a variable-gain equation, and it is then usually possible to derive a describing function by assuming a sinusoidal input and manipulating the resulting relationship to obtain the output fundamental harmonic. Let us consider the cubic gain relationship

$$O = K_N I^3 \tag{7.51}$$

where $O$ is the output, and assume, as before, that the input is a pure sine wave, $I_M \sin \omega t$. Substituting in Equation 7.51, we obtain

$$O(\omega t) = K_N I_M^3 \sin^3 \omega t \tag{7.52}$$

Now, $\sin^3 \omega t$ may be expanded to give

$$\sin^3 \omega t = \tfrac{1}{4}(3 \sin \omega t - \sin 3\omega t) \tag{7.53}$$

and if the third harmonic is neglected we may write the describing function as the output fundamental divided by the input; hence

$$G(I_M) = \frac{K_N I_M^3 \frac{3}{4} \sin \omega t}{I_M \sin \omega t} \tag{7.54}$$

and, in simplest form,

$$G_N(I_M) = \tfrac{3}{4} K_N I_M^2 \tag{7.55}$$

Greif[70] has evaluated the describing function for the general case as

$$O = K_N I^{\mu} \tag{7.56}$$

for both integer and non-integer values of $\mu$. For integer values, the solution takes the same form as for $\mu = 3{\cdot}0$, whilst for non-integer $\mu$ the solution is evaluated using a set of gamma-function tables. The describing function does not contribute a phase lag as the non-linearity is single valued. A specimen output-input relationship is shown in Fig. 7.28(a), and input and output waveforms are shown in Fig. 7.28(b), both for $\mu > 1{\cdot}0$, whilst the normalised describing function for sample integer and decimal values is shown in Fig. 7.28(c). For $\mu = 1{\cdot}0$ the normalised describing function, is, of course, equal to unity for all values of input.

The term 'dynamic component' is normally applied to elements where the output is a function of the input and input derivatives and integrals, which may form cross-coupling terms.

Dynamic component characteristics may arise from the component differential equations directly, or from fitting an output-input relationship to a set of experimental data. The procedure for obtaining the describing function for a dynamic component is similar to that used for obtaining the describing function for the variable-gain element.[77] For example, let the input be $I$ as before, and let the output be $O = f(I, \dot{I}, \text{etc.})$ such that

$$O = a\ddot{I}+b\int I\,dt+cI^3 \tag{7.57}$$

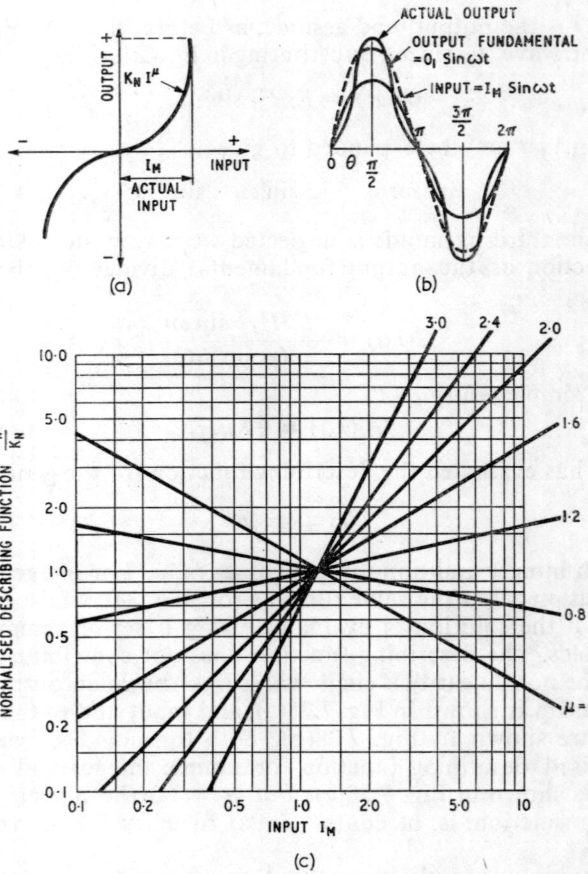


*Fig. 7.28. Describing function for a continuous variable-gain element as evaluated by Greif: (a) output-input relationship; (b) input and output waveforms; normalised describing function*

Assuming $I = I_M \sin \omega t$, we have, on substitution and appropriate differentiation and integration,

$$O = -(aI_M\omega^2) \sin \omega t - \left(\frac{bI_M}{\omega}\right) \cos \omega t + \left(\frac{cI_M^3}{4}\right) (3 \sin \omega t - \sin 3\omega t)$$

Ignoring the third harmonic, and collecting like terms, we have

$$O_1 = (\tfrac{3}{4}cI_M^3 - aI_M\omega^2) \sin \omega t - \left(\frac{bI_M}{\omega}\right) \cos \omega t \qquad (7.58)$$

hence the describing function is

$$G_N(I_M,\omega) = \frac{O_1}{I_M}$$

$$= \left[(\tfrac{3}{4}cI_M^2 - a\omega^2)^2 + \frac{b^2}{\omega^2}\right]^{1/2} \Bigg/ \tan^{-1} \frac{-(b/\omega)}{\tfrac{3}{4}cI_M^2 - a\omega^2} \qquad (7.59)$$

and is quite clearly a function of input amplitude and excitation frequency. A word of warning is necessary concerning the validity of the describing function method for dynamic components, which may affect the interpretation of any results on stability which may be obtained using these describing functions. The output waveform from a symmetrical frequency independent element will contain only odd harmonics, i.e. no second harmonic can cloud the issue, which is very beneficial since, in general, the amplitude of the $n^{\text{th}}$ harmonic is smaller than the $(n-1)^{\text{th}}$ harmonic and so on, quite apart from the beneficial effect of the linear filtering in the loop which would reduce the magnitude of the third harmonic (at frequency $3\omega$) much more than the magnitude of the second harmonic (at frequency $2\omega$). Absence of the second harmonic from the response due to the type of non-linearity present therefore assists in the validity of the describing function technique, whilst existence of the second harmonic due to the type of non-linear element will reduce the validity of the describing function technique. Although not evident from the specific example chosen, it is in fact possible to have an irreducible second order harmonic output from the dynamic component due to a term involving $(I, \dot{I})$, which may be of sufficient magnitude to completely invalidate the describing function approach even to the extent of being very misleading. Much more care is therefore required in the interpretation of systems containing such dynamic components or, indeed, any system containing frequency dependent non-linearities.

## 7.13 Graphical Method for Determination of the Describing Function for a Frequency Independent Non-linearity

Several graphical methods for the determination of describing functions for frequency independent non-linearities have appeared in the literature. A graphical method is useful when the describing function does not exist as standard knowledge, or when the output-input relationship has been obtained experimentally, or when the non-linearities are straightforward but possess a number of discontinuities. Graphical methods may thus either supplement or replace the conventional mathematical computation of the describing function. The method to be described here is mathematically exact, and is particularly suitable for evaluation on a drawing-board if available. This method is due to Ku and Chen,[76] and can be applied to single-valued and non-single-valued non-linearities. In-phase and quadrative components of the describing function are determined independently, and the physical significance of the situation is obvious throughout the construction. Perhaps the greatest advantage of this particular method of describing function determination over the mathematical solution of $A_1$ and $B_1$ in Equations 7.43 and 7.44 respectively, is that the solution is no more difficult whatever the shape and number of discontinuities of the output-input relationship. No proof will be offerred for the method. The interested reader should refer to the original literature for complete details; we shall concentrate on the construction itself, and compare several sample answers with those which have already been quoted in Section 7.12.

If the output-input relationship for a frequency independent non-linearity is single valued, then the locus traversed during one cycle of steady state operation is merely along the curve relating output to input. No area is enclosed in the output-input plane as would be the case with the hysteresis type of non-linearity and there will be no phase lag due to the non-linearity. To evaluate $A_1$ for such a non-linearity, consider the single-valued non-linearity shown in Fig. 7.29(a). If desired, the constant proportional term $K_N$ can be removed from the relationship to give the describing function in the form $G_N/K_N$ as shown in Fig. 7.28, or it can be left in the output-input curve if scaling problems do not arise. Given the output-input curve of Fig. 7.29(a), with an input $I_M \sin \omega t$, the graphical construction is as follows. A circle of radius $I_M$ is drawn to the left of the output-input curve with its centre on the input axis as shown. A distance $r$ above the input axis is then marked, and the line is extended to cut the input circle at points $P_1$ and $P_2$. Vertical lines are drawn through

$P_1$ and $P_2$. The line at distance $r$ is extended to cut the output axis. A circle of radius $r$ is now drawn with its centre at the origin of the output-input curve, to intersect the input axis at $I_1$.

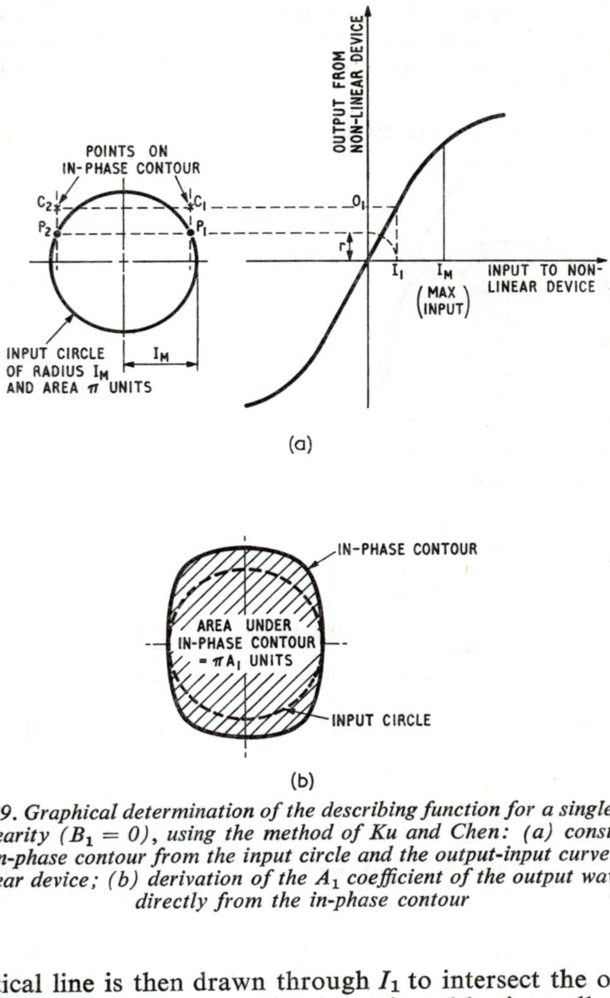


*Fig. 7.29. Graphical determination of the describing function for a single-valued non-linearity ($B_1 = 0$), using the method of Ku and Chen: (a) construction of the in-phase contour from the input circle and the output-input curve for the non-linear device; (b) derivation of the $A_1$ coefficient of the output waveforms directly from the in-phase contour*

A vertical line is then drawn through $I_1$ to intersect the output-input curve at output $O_1$, which is projected horizontally to the region of the input circle. Where the extended $O_1$ line cuts the vertical lines drawn through $P_1$ and $P_2$, points $C_1$ and $C_2$ are marked. $C_1$ and $C_2$ lie on the in-phase contour which has to be constructed to evaluate $A_1$.

Further points are constructed in like fashion until the in-phase contour can be completed, as shown in Fig. 7.29(b). The area enclosed by the in-phase contour can be determined using a planimeter, Simpson's rule, etc., and $A_1$ is then simply the ratio of the area enclosed by the in-phase contour to the area enclosed by the input circle. $A_1$ thus determined is the describing function of the element for the input $I_M$. To determine $A_1$ as a general

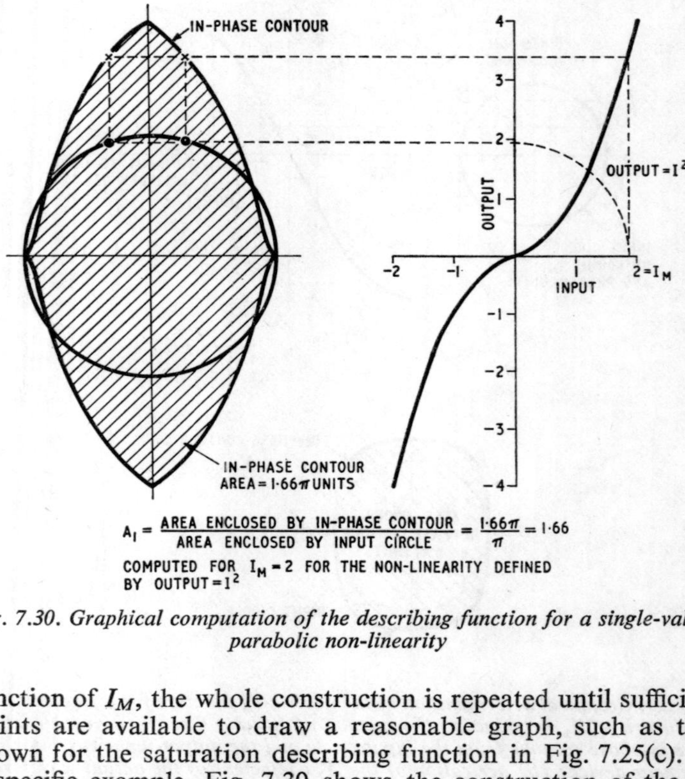


*Fig. 7.30. Graphical computation of the describing function for a single-valued parabolic non-linearity*

function of $I_M$, the whole construction is repeated until sufficient points are available to draw a reasonable graph, such as that shown for the saturation describing function in Fig. 7.25(c). As a specific example, Fig. 7.30 shows the construction of the in-phase contour for the parabolic law, output $= I^2$ for the specific case of an input of 2 sin $\omega t$. Taking the input circle area as $\pi$ units, it is found using Simpson's rule that, to the same scale, the area enclosed by the in-phase contour is $1{\cdot}66\pi$ units. For the input $I_M = 2$, $A_1$ is therefore $1{\cdot}66\pi/\pi = 1{\cdot}66$. This is the value obtained using Greif's analysis of Fig. 7.28 for $\mu = 2{\cdot}0$ and $I_M = 2{\cdot}0$. Using a planimeter, the area bounded by the in-phase contour can be evaluated very rapidly indeed, and the total time required to construct a graph relating $A_1$ to $I_M$ can therefore be quite small. The method is so simple that there is much less chance of making a mistake than when evaluating the integrals for $A_1$ mathematically. For non-linearities with many discontinuities, the discontinuities are obvious from an inspection of the output-input relationship, and can be easily translated to the in-phase contour, thus eliminating the necessity of using transcendental equations to evaluate integral limits.

Ku and Chen developed the method particularly for use with

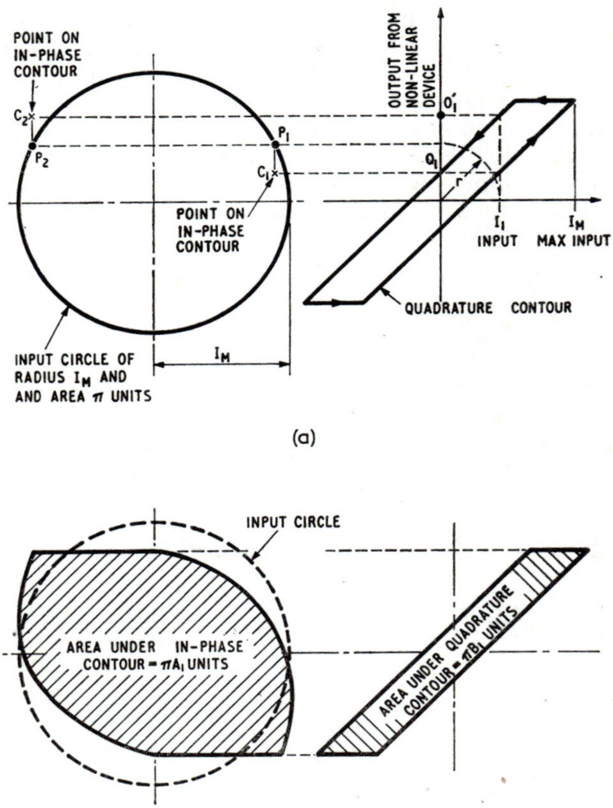


*Fig. 7.31. Graphical determination of the describing function for a non-single-valued non-linearity using the method of Ku and Chen: (a) construction of the in-phase contour from the unit circle, and the output-input relationship for the non-linear device; (b) derivation of $A_1$ and $B_1$ coefficients of the output waveform directly from the in-phase and quadrature contours*

non-single-valued non-linearities, and the method is equally suitable for continuous and discontinuous relationships. Fig. 7.31(a) shows the construction of the in-phase contour for a non-single-valued element. The input circle is drawn to the left and in line as before, and a distance $r$ above the input axis is again chosen. Intersection points $P_1$ and $P_2$ on the input circle again exist, and vertical lines are drawn through $P_1$ and $P_2$. A circle of radius $r$ is constructed with centre at the origin of the output-input graph, and this intersects the input axis at $I_1$. A vertical line is then drawn passing through $I_1$, and because the element is non-single-valued this line will intersect the output-input curve at two points corresponding to $O_1$ and $O_1'$. These two values of output are then extended to the region of the input circle, and the intersection points $C_1$ and $C_2$ with the vertical lines $P_1$ and $P_2$ are desired points on the required in-phase contour, which is shown completed in Fig. 7.31(b). With the convention adopted here, the right-hand side of the in-phase contour corresponds to the input-increasing part of the output-input locus. As before, the area enclosed by the in-phase contour is equal

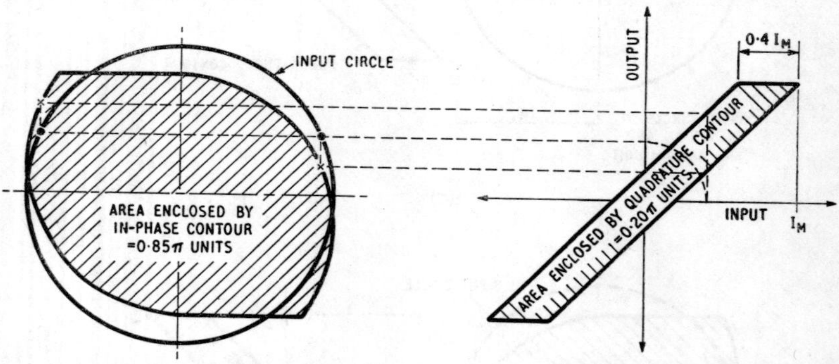


$$A_1 = \frac{\text{AREA ENCLOSED BY IN-PHASE CONTOUR}}{\text{AREA ENCLOSED BY INPUT CIRCLE}} = \frac{0{\cdot}85\pi}{\pi} = 0{\cdot}85$$

$$B_1 = \frac{\text{AREA ENCLOSED BY QUADRATURE CONTOUR}}{\text{AREA ENCLOSED BY INPUT CIRCLE}} = \frac{0{\cdot}20\pi}{\pi} = 0{\cdot}20$$

*Fig. 7.32. Graphical determination of the backlash describing function for the case of $D/I_M = 0{\cdot}20$*

to $\pi A_1$ units if the input circle area is equal to $\pi$ units. Now the quadrature contour, which is required to evaluate $B_1$, is, in fact, the locus of the output-input relationship during one cycle in the steady state and is therefore the output-input graph already drawn. This fact is a very considerable bonus, as the graph is required anyway, and now all we have to do is to evaluate the area enclosed by this quadrature contour which is equal to $\pi B_1$ units of area. Evaluating $A_1$ and $B_1$ is thus achieved by measuring the areas enclosed by the input circle, in-phase contour, and quadrature contour in whatever way is most convenient.

Fig. 7.32 shows the construction for backlash for the case where the total backlash is 0·4 $I_M$, i.e. with the terminology of Fig. 7.27(a) we have $D/I_M = 0{\cdot}20$. For an input circle area of $\pi$ units it is found, using Simpson's rule, that the area bounded by the in-phase contour is 0·85 $\pi$, giving $A_1 = 0{\cdot}85$, and the area bounded by the quadrature contour is 0·20 $\pi$, giving $B_1 = 0{\cdot}20$. The gain part of the describing function is therefore $\sqrt{A_1^2+B_1^2} = 0{\cdot}87$. Phase lag is given by $\tan^{-1}(B_1/A_1) = \tan^{-1} 0{\cdot}23 = 13°$. These two values are in reasonably close agreement with the calculated values shown in Fig. 7.27(d). From the general nature of the method, it is obvious that the relative significance of the quadrature term depends on the variation of the area enclosed by the output-input graphs as the input maximum amplitude is varied. For backlash the relative significance of the quadrature term decreases as the input is increased, since the quadrature term will become a smaller percentage of the input circle area as the input is increased. This intuitive argument based on the method of Ku and Chen confirms the detailed results of Fig. 7.27(d). Certainly, this is a useful method for the control engineer to have available in the analysis and synthesis of non-linear feedback control systems.

## 7.14 Determination of System Describing Function for Non-linear Feedback Control Systems

So far we have only been concerned with the derivation of describing functions for non-linear but frequency independent elements, and the subsequent use of these describing functions to determine the existence or otherwise of limit cycles in the feedback control system under consideration. For design work it is necessary to estimate the effect of signal amplitude on the performance of non-linear systems. The describing function obtained for an individual element can be used in a variety of ways to determine the system frequency response as the signal amplitude is varied. One major difficulty does exist, however, as it is conventional to test non-linear feedback control systems for a number of discrete values of system input amplitude, whereas the system response is more readily evaluated in terms of the signal amplitude at the input to the non-linearity, which will certainly not be a linear function of the system input. We shall discuss several techniques for overcoming this problem, but it is necessary to accept at this stage that the analysis and synthesis of non-linear feedback control systems is, in general, much more laborious than for a linear system. A system with one significant amplitude dependent non-linearity present may well require five to ten times more effort at the design stage than a linear system of the same order, unless a very specialised design

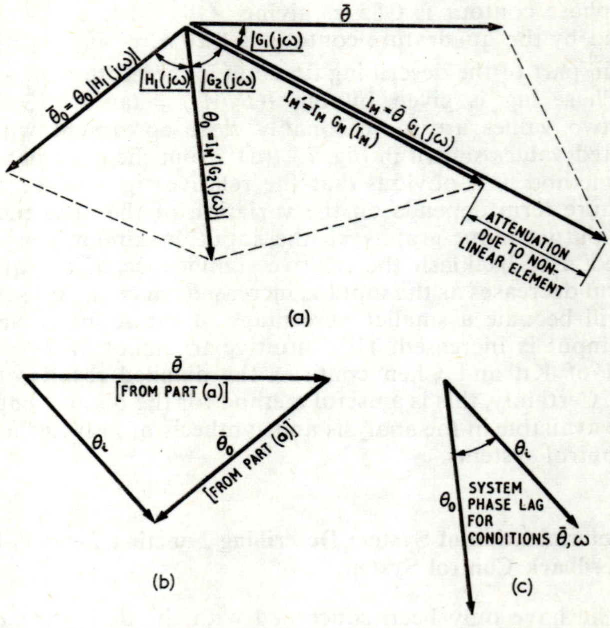


*Fig. 7.33. General method for the determination of the closed loop or system describing function for steady state conditions $\bar{\theta}$ and $\omega$ present in a feedback control system with a non-linear element: (a) vector diagram obtained by following a known signal $\bar{\theta}$ around the loop; (b) use of the comparator relationship $\theta_i = \bar{\theta}_0 + \bar{\theta}$ to determine $\theta_i$; (c) determination of system phase lag*

technique pertinent to the particular situation is already in existence.

A perfectly general method for determining the closed loop frequency response, i.e. the system-describing function, is to assume a particular value of signal amplitude and frequency at some point inside the system, and to determine the actual values of $\theta$ and $\theta_i$ graphically. For example, in the system shown in Fig. 7.19, the signal amplitude $\bar{\theta}$ is assumed, and $G_1(j\omega)$

is evaluated in modulus and phase to give the signal amplitude $I_M$ at the input to the non-linearity. Using the element describing function the fundamental output $I'_M$ from the non-linearity is determined, and $G_2(j\omega)$ is evaluated in modulus and phase to give $\theta_0$. Finally, $H_1(j\omega)$ is evaluated in modulus and phase to determine $\bar{\theta}_0$ from $\theta_0$. Fig. 7.33(a) shows the vector diagram for this process, it being assumed that the non-linearity is attenuating and does not possess a phase contribution but, as previously stated, there is no restriction to the method. In fact, if a frequency dependent non-linear element describing function is available from test data, etc., this can be incorporated very easily. [When calculating such terms as $|G_1(j\omega)|$ and $\underline{/G_1(j\omega)}$, it is a matter of choice whether the terms are obtained by direct computation using complex numbers or whether graphical measurement direct from component poles and zeros is considered more suitable.]

It is known from the comparator relationship that $\theta_i = \bar{\theta}_0 + \bar{\theta}$ and so $\theta_i$ can be determined vectorially from Fig. 7.33(b). If the phase angle between input and output is required, this can also be obtained vectorially since we can draw, from the same reference point, a vector parallel to $\theta_i$ and a vector parallel to $\theta_0$ as shown in Fig. 7.33(c), the angle between the two being the required system phase lag (if $\theta_i$ leads $\theta_0$ as indicated). If $\bar{\theta}$ is kept constant, and $\omega$ is varied, then a frequency response curve can be drawn in the usual way with $\theta_0/\theta_i$ plotted against $\omega$. However, at every point on the curve it is necessary to mark in the value of $\theta_i$ (which will be different at every frequency). This construction is repeated for a range of values of $\bar{\theta}$ and a family of responses drawn in the usual way. At selected values of frequency, subsidiary graphs are drawn relating system phase lag and amplitude ratio to $\theta_i$. Interpolation yields these quantities at specific levels of input, and hence frequency responses for constant system input can be plotted. The method is laborious but is basically straightforward, and has the advantages that the physical significance of the situation is retained and once the subsidiary graphs are available it is a simple matter to draw the frequency response for as many constant values of $|\theta_i|$ as are considered to be of interest.

As an example of the use of this general technique, consider the error saturating servo loop shown in Fig. 7.34. The basic linear servo loop suitable for controlling the control surface deflection in a guided missile was discussed in Section 2.20. Since the physical output from signal amplifier, torque motor and control valve is limited, it is found that if the servo loop error is sufficiently large the control valve will saturate and the flow will then remain constant if the error signal is increased. Which of the previously mentioned elements will actually cause the saturation in a particular servo will depend on the designer's choice of components. However, the facts as far as they are known are reasonably well fitted by the assumption of the saturation element shown in Fig. 7.34. Unity slope of this element has been chosen for convenience, and it is assumed that flow saturation takes place whenever the error $|\xi|$ exceeds the physical flow limitation $|\beta|$. The actuator time constant is $T_c$ when operating in the linear state. It is required to determine the

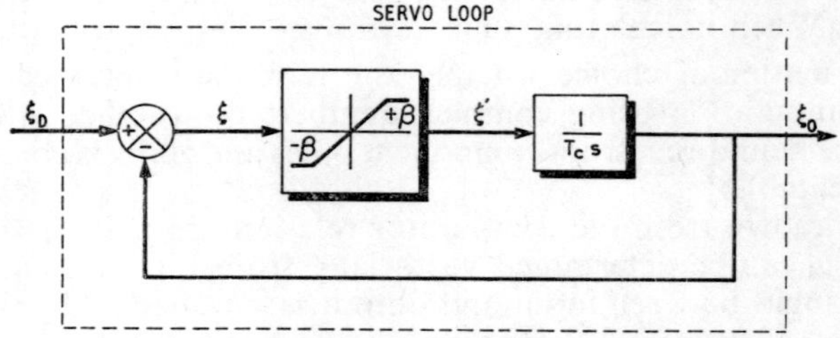


*Fig. 7.34. Block diagram of error saturating servo loop*

closed loop frequency response $|\xi_0/\xi_D|$ for various values of input $|\xi_D.|$ **In** this instance it is convenient to relate the input $|\xi_D|$ to the error amplitude causing the flow saturation $|\beta|$, and therefore the frequency response will be ultimately calibrated in terms of $|\xi_D/\beta|$.

For a given value of $\xi$, the fundamental harmonic of the flow term will be simply

$$\xi' = \xi G_N\left(\frac{\xi}{\beta}\right) \tag{7.60}$$

where $G_N(\xi/\beta)$ is, in this case, the saturation describing function shown in Fig. 7.25. Thus, knowing $\xi$ and $\beta$, $\xi$ is determined. The relationship between $\xi_0$ and $\xi'$ is simply the basic integrator, and in the steady state,

$$\xi_0 = \frac{\xi'}{T_c j\omega} \tag{7.61}$$

which is simply a phase lag of 90° and a multiplier of $|1/\omega T_c|$. Fig. 7.35(a) shows the vector diagram relating $\xi_0$ to $\xi$. Using the comparator relationship $\xi_D = \xi + \xi_0$ (vectorially, of course), $\xi_D$ is determined as shown in Fig. 7.35(b). Finally, vectors from the same point drawn parallel to $\xi_D$ and $\xi_0$ determine the system phase lag as shown in Fig. 7.35(c).

A sample subsidiary graph is shown in Fig. 7.36 for a constant excitation frequency $\omega = 1/T_c$; $\xi$ was varied in sequence such that $|\beta/\xi| = 0$, 0·2, 0·4, 0·6, 0.8, and 1·0, and the closed loop amplitude ratio, and phase lag were determined as already described. We now have a graph which relates the system steady state amplitude ratio $|\xi_0/\xi_D|$ to the relative signal ratio $|\xi/\beta|$ inside the loop. This particular graph, Fig. 7.36, may well be significant

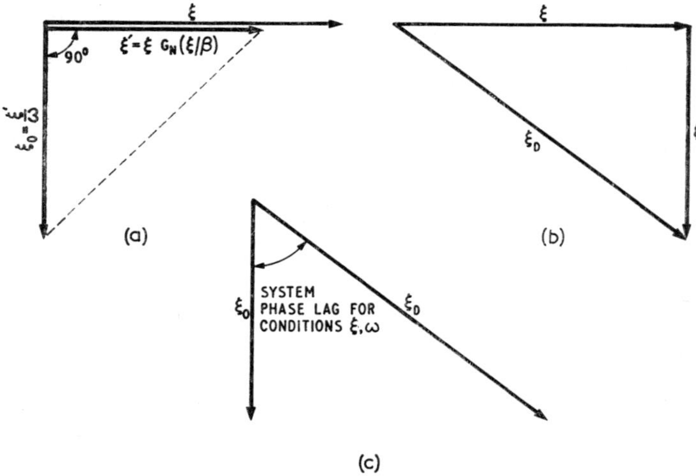


*Fig. 7.35. Method of determination of the saturating servo loop describing function relating $\xi_0$ to $\xi_D$ as a function of $\xi$ and $\beta$: (a) vector diagram; (b) use of comparator relationship $\xi_D = \xi + \xi_0$*

in its own right for some applications where the performance specification may in part be interpreted in terms of signal amplitudes in the loop. From the graphical construction the absolute value of $\xi_D$ is known and therefore a similar graph can be constructed in terms of $|\xi_D/\beta|$ as shown in Fig. 7.37, incorporating both amplitude ratio and phase lag contributions. In this example the input amplitude at which the servo loop saturates for a given frequency can be obtained using Pythagoras theorem, since in the limit $\xi/\beta = 1{\cdot}0$. From Fig. 7.35(b), we may therefore write

$$\xi_D = \sqrt{\beta^2 + \frac{\beta^2}{\omega^2 T_c^2}} \qquad (7.62)$$

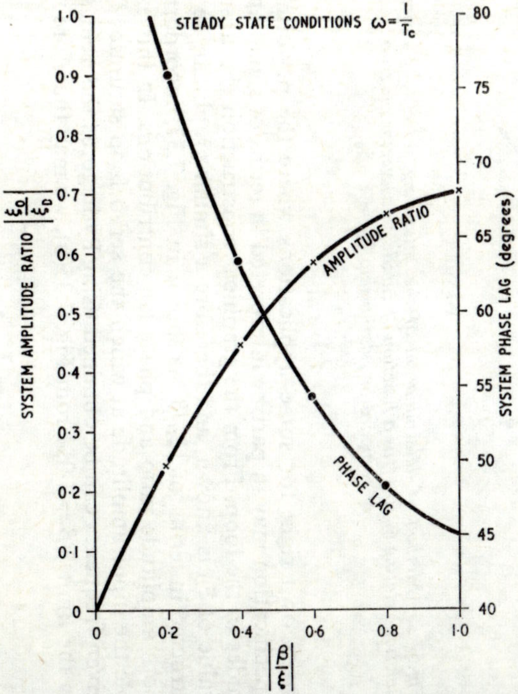


*Fig. 7.36. Describing function for error saturating servo loop in terms of the error $\xi$ for excitation frequency $\omega = 1/T_c$*

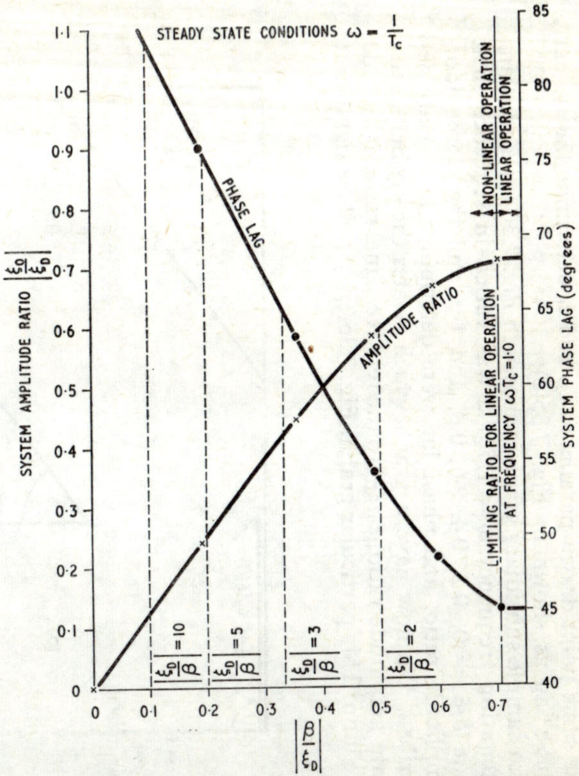


*Fig. 7.37 Describing function for error saturating servo loop in terms of the input $\xi_D$ for excitation frequency $\omega = 1/T_c$*

and hence the maximum input for linear operation at frequency $\omega$ is given by

$$\left|\frac{\xi_D}{\beta}\right| = \sqrt{1+\frac{1}{\omega^2 T_c^2}} \tag{7.63}$$

For example when $\omega = 1/T_c$, the critical value is $|\xi_D/\beta| = \sqrt{2}$, and this is shown in Fig. 7.37 as the limit of linear operation.

The values of $|\xi_D/\beta|$ of interest are 2, 3, 5, 10, and the corresponding phase lag and amplitude ratio are obtained by reading off the appropriate points as shown. This construction has been repeated for a range of input frequencies and the final results are shown in Fig. 7.38. The frequency at which the operating mode becomes non-linear can be seen to decrease as $\xi_D$ increases. It is convenient to present the phase angle and amplitude ratio contributions as separate functions of frequency: the results for $\omega = 1/T_c$ are those obtained from Fig. 7.37. In many ways this servo loop example is a severe test of the validity of using the describing function approach for predicting system frequency response, since there is only one integrator to filter the higher harmonics, and no prefilter acting before the non-linearity, which is contrary to the requirements laid down by Kochenburger in early studies on the describing function method. In fact, however, the method gives surprisingly good results, as has been previously reported by Towill and Grigg,[77] the experimental results obtained being shown as discrete points on Fig. 7.38.

Non-linear behaviour of the servo loop can be summarised by reference to Fig. 7.38, which shows that the effect, at a given frequency, is to introduce additional phase lag and system attenuation compared with the linear mode. It is therefore relatively easy to induce, say, an additional 30° system phase lag due to the non-linear effect. This additional lag can have a very significant effect on the stability of any major loop incorporating such a non-linear minor loop. For systems operating at all times in the linear mode, the bandwidth of the minor loop is commonly made three to five times faster than the bandwidth of the major loop, whilst if the minor loop saturates, this ratio may need to be increased considerably to ensure adequate stability of the major loop, as has been shown in some detail by Towill and Bryson.[78] To avoid confusion, the bandwidth ratio refers to the linear state, even if the servo loop does saturate; the bandwidth ratio chosen is then a function of the degree of saturation which may be reasonably expected in practice. Thus, all things being equal, it is generally necessary to specify a linear mode bandwidth ratio which increases with the anticipated maximum value of input to the minor loop.

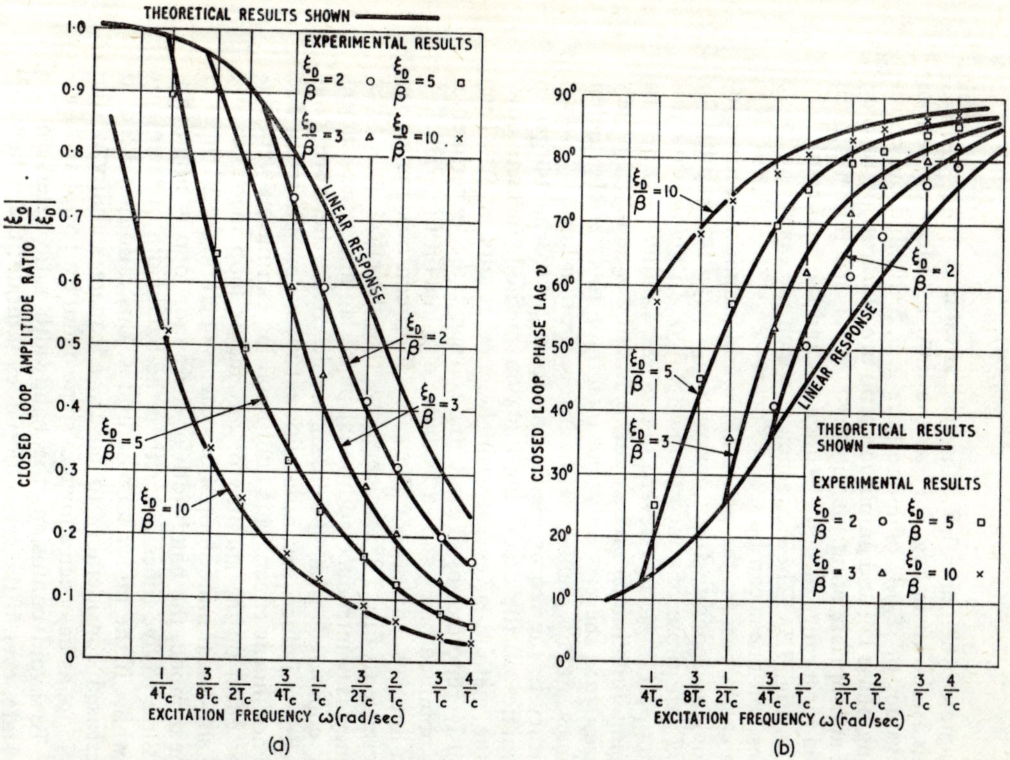


*Fig. 7.38. System describing function for error saturating servo loop in terms of the input $\xi_D$: (a) amplitude ratio component; (b) phase lag component*

It is convenient for the purposes of this chapter to illustrate a number of techniques by reference to the saturation type of non-linearity, which is very common and very important. In practice, deadzone and backlash are equally common and troublesome non-linearities. In general the same techniques are equally suitable for these other kinds of non-linearity.

## 7.15 Use of Nichols Charts to Determine System Describing Functions for Non-linear Systems

The Nichols chart may be used to determine system describing functions for non-linear systems as an alternative to the general vector method outlined in Section 7.14. It was observed in Chapter 1 that stability of a unity feedback control system could be investigated using a Nichols chart on which the broken loop response was plotted in the form of the amplitude ratio, in decibels, against the phase lag in degrees, using Cartesian coordinates. Furthermore, using standard contours superimposed on the chart, it was found that the system frequency response could be determined by interpolating between the contours to estimate, for a given frequency, the system amplitude ratio in decibels and the system phase lag in degrees. Thus for a system with broken loop transfer function $G(j\omega)$, the Nichols plot is drawn in terms of $|G(j\omega)|$ in decibels and $\underline{/G(j\omega)}$ in degrees, whilst the superimposed contours are for constant values of $|G(j\omega)/[1+G(j\omega)]|$ in decibels and $\underline{/G(j\omega)/[1+G(j\omega)]}$ in degrees.

With suitable precautions the contours may be used equally well for non-linear systems and for systems with elements in the feedback path. Let us assume that $I_M$ in Fig. 7.19 is fixed, then the describing function $G_N(I_M)$, which may have phase change as well as amplitude change, is automatically defined, as is the output fundamental harmonic from the non-linearity. Under these circumstances we may write the broken loop transfer function as

$$\frac{\bar{\theta}_0}{\bar{\theta}}(j\omega) = G_1(j\omega)\, G_N(I_M)\, G_2(j\omega)\, H_1(j\omega) \tag{7.64}$$

and if Equation 7.64 is plotted as a function of frequency on the Nichols chart then, using the contours, we obtain the relationship

$$\frac{G(j\omega)}{1+G(j\omega)} = \frac{G_1(j\omega)\, G_N(I_M)\, G_2(j\omega)\, H_1(j\omega)}{1+G_1(j\omega)\, G_N(I_M)\, G_2(j\omega)\, H_1(j\omega)} \tag{7.65}$$

The system frequency response is not given by Equation 7.65, but by the relationship

$$\frac{\theta_0}{\theta_i}(j\omega) = \frac{G_1(j\omega)\, G_N(I_M)\, G_2(j\omega)}{1+G_1(j\omega)\, G_N(I_M)\, G_2(j\omega)\, H_1(j\omega)} \tag{7.66}$$

and therefore the results obtained from the Nichols chart for the fixed value $I_M$ is modified to give

$$\frac{\theta_0}{\theta_i}(j\omega) = \frac{1}{H_1(j\omega)}\left[\frac{G_1(j\omega)\, G_N(I_M)\, G_2(j\omega)\, H_1(j\omega)}{1+G_1(j\omega)\, G_N(I_M)\, G_2(j\omega)\, H_1(j\omega)}\right]_{I_M} \tag{7.67}$$

The modification of the results obtained from the Nichols chart is straightforward, since in logarithmic form division is simply subtraction. For a range of values of $I_M$, a complete set of frequency responses $|\theta_0/\theta_i|$ and $\underline{/\theta_0/\theta_i}$ can be constructed for constant $I_M$ and constant element describing function $G_N(I_M)$. As previously explained, this is not normally the information required by the system designer, who requires to know the system frequency response as the input signal amplitude $\theta_i$ is kept constant. It is therefore necessary to calibrate these frequency responses in terms of $\theta_i$ as was done for the general graphical method described in Section 7.14. For a given value of $I_M$ and $\omega$ we may write, from Fig. 7.19,

$$\frac{\theta_0}{I_M} = G_N(I_M)\, G_2(j\omega) \tag{7.68}$$

and since

$$\frac{\theta_i}{I_M} = \frac{\theta_0}{I_M} \times \frac{\theta_i}{\theta_0} \tag{7.69}$$

we obtain the true value of $|\theta_i|$ for these particular values of $I_M$ and $\omega$ as

$$|\theta_i|_{I_M,\omega} = \frac{\theta_i}{\theta_0}(j\omega)\, G_N(I_M)\, G_2(j\omega) I_M \tag{7.70}$$

which is extremely straightforward using logarithmic plots. For a given value of $\omega$ a graph of $\theta_i$ may be constructed in terms of $\theta_i/\theta_0\,(j\omega)$, and the system amplitude ratio and phase lag corresponding to a particular value of input determined. By repeating the process for a range of frequencies a complete response for $\theta_i$ constant may be determined.

As a specific example on the determination of the system frequency response for $|\theta_i|$ fixed, let us consider the second order

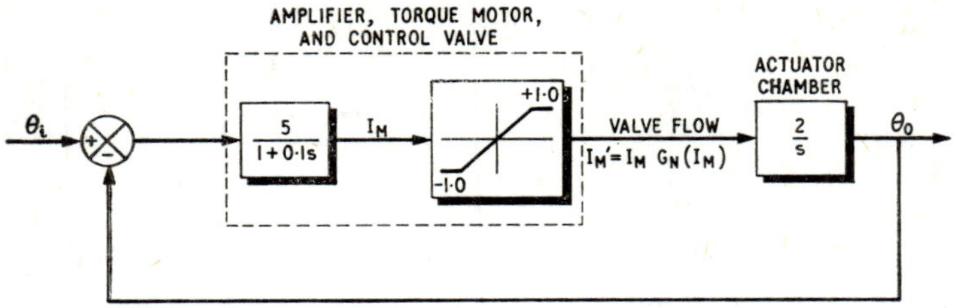


*Fig. 7.39. Simple representation of a second order flow saturating electro-hydraulic servomechanism*

system shown in Fig. 7.39, in which

$$G_1(j\omega) = \frac{5}{1+0{\cdot}1j\omega}$$

$$G_2(j\omega) = \frac{2}{j\omega}$$

and

$$H_1(j\omega) = 1$$

The system shown is a mathematical model of an electrohydraulic servo in which oil compressibility is negligible, a significant lag exists in the amplifier-torque motor combination, and the control valve saturates when the input to the control valve is unity. When studying such systems it is often helpful to evaluate as a function of frequency the value of input signal amplitude $\theta_i$ at which non-linear operation commences. This can be obtained by writing down the transfer function relating $I_M$ and $\theta_i$, substituting the limiting value for $I_M$, and evaluating $\theta_i$ as a function of frequency. *In this instance the signal leaving the comparator is the error $\theta$, and we may write*

$$\frac{\theta}{\theta_i}(j\omega) = \frac{\theta_i - \theta_0}{\theta_i} = 1 - \frac{\theta_0}{\theta_i} \tag{7.71}$$

and

$$\frac{\theta_0}{\theta_i} = \frac{G_1(j\omega)\,G_N(I_M)\,G_2(j\omega)}{1+G_1(j\omega)\,G_N(I_M)\,G_2(j\omega)} = \frac{100G_N(I_M)}{100G_N(I_M)+10j\omega-\omega^2} \tag{7.72}$$

Hence

$$\frac{\theta}{\theta_i}(j\omega) = \left[\frac{100G_N(I_M)}{100G_N(I_M)+10j\omega-\omega^2}\right]\left[\frac{10j\omega-\omega^2}{100G_N(I_M)}\right] \tag{7.73}$$

which may be expressed in terms of $\theta_0/\theta_i(j\omega)$ as

$$\frac{\theta}{\theta_i}(j\omega) = \left[\frac{\theta_0}{\theta_i}(j\omega)\right]\left[\frac{10j\omega-\omega^2}{100G_M(I_M)}\right] \tag{7.74}$$

Hence

$$\frac{I_M}{\theta_i}(j\omega) = \left[\frac{\theta_0}{\theta_i}(j\omega)\right]\left[\frac{10j\omega-\omega^2}{100G_N(I_M)}\right]\left(\frac{5}{1+0{\cdot}1j\omega}\right) \tag{7.75}$$

and, in simplest form,

$$\frac{I_M}{\theta_i}(j\omega) = \left[\frac{\theta_0}{\theta_i}(j\omega)\right]\left[\frac{j\omega}{2G_N(I_M)}\right] \tag{7.76}$$

For the limit of linear operation, $G_N(I_M) = 1{\cdot}0$, $I_M = 1{\cdot}0$ and $\theta_0/\theta_i$ is simply a second order system with undamped natural frequency; $\omega_n = 10$ rad/sec, and the damping ratio is 0·5 and can therefore be evaluated readily as a function of frequency from standard tables, pole geometry, or, in logarithmic form, by using a standard template such as that shown in Fig. 3.23(a).

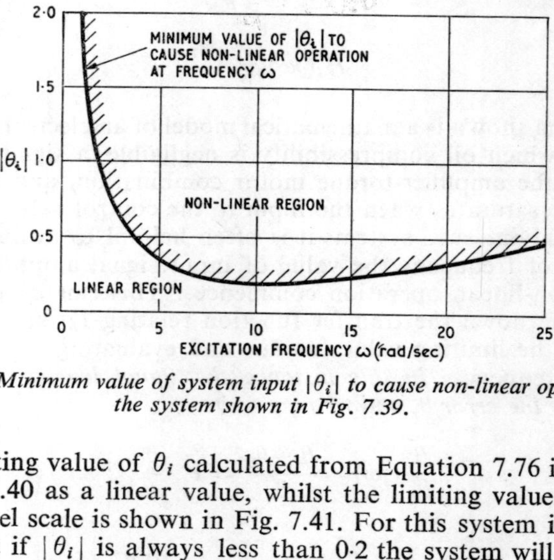


*Fig. 7.40. Minimum value of system input $|\theta_i|$ to cause non-linear operation of the system shown in Fig. 7.39.*

The limiting value of $\theta_i$ calculated from Equation 7.76 is shown in Fig. 7.40 as a linear value, whilst the limiting value of $\theta_i$ on the decibel scale is shown in Fig. 7.41. For this system it can be seen that if $|\theta_i|$ is always less than 0·2 the system will operate wholly in the linear range, whilst for larger inputs the system will be linear at both low and high frequencies but will be non-linear at intermediate frequencies. The reason for this is that at low frequencies the system is following the input signal reasonably well and therefore signal levels throughout the loop are low and saturation does not occur at the input to the non-linearity, whilst at high frequencies, although the system is following very badly, and the error is large, the attenuation due to the amplifier and torque motor lag is sufficiently large to reduce the input signal to the non-linear element below the saturation level.

The broken loop equation is, from Fig. 7.39,

$$\frac{\bar{\theta}_0}{\bar{\theta}}(j\omega) = \left(\frac{5}{1+0{\cdot}1j\omega}\right)[G_N(I_M)]\left(\frac{2}{j\omega}\right) \tag{7.77}$$

and the individual elements required in the Nichols plot are

$$\left|\frac{\bar{\theta}_0}{\bar{\theta}}\right|_{\text{db}} = 20\log_{10} 10 + 20\log_{10} G_N(I_M) - 20\log_{10}\sqrt{1+0{\cdot}1^2\omega^2} - 20\log_{10}\omega \tag{7.78}$$

$$\underline{/\bar{\theta}_0/\bar{\theta}} = -\tan^{-1} 0{\cdot}1\omega - 90° \tag{7.79}$$

From Equation 7.78 it is observed that the gain of the saturating element is required in logarithmic form. The output from the saturating element is also required at a later stage when

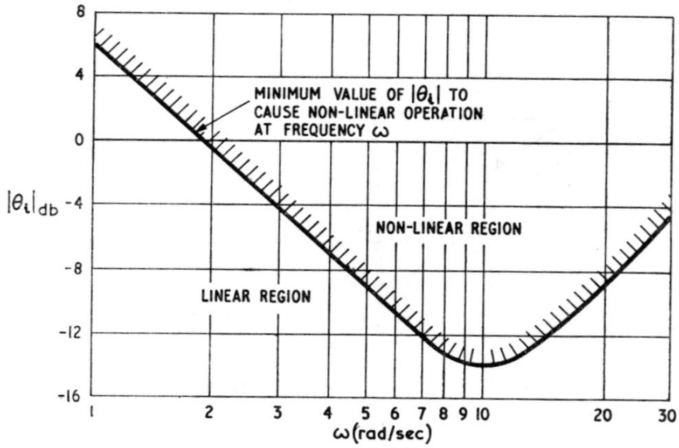


*Fig. 7.41. Minimum value of system input $|\theta_i|_{\text{db}}$ to cause non-linear operation of the system shown in Fig. 7.39*

determining $|\theta_i|$, and it is therefore convenient to tabulate all the properties inherent in a choice of $I_M$, as shown in Table 7.15.1.

The values of $I_M$ tabulated are those to be used in the determination of the system frequency response using the Nichols chart, but they have also been particularly chosen for their convenience in the application of McAllisters' technique to this particular example (Section 7.16). Fig. 7.42 shows the Nichols

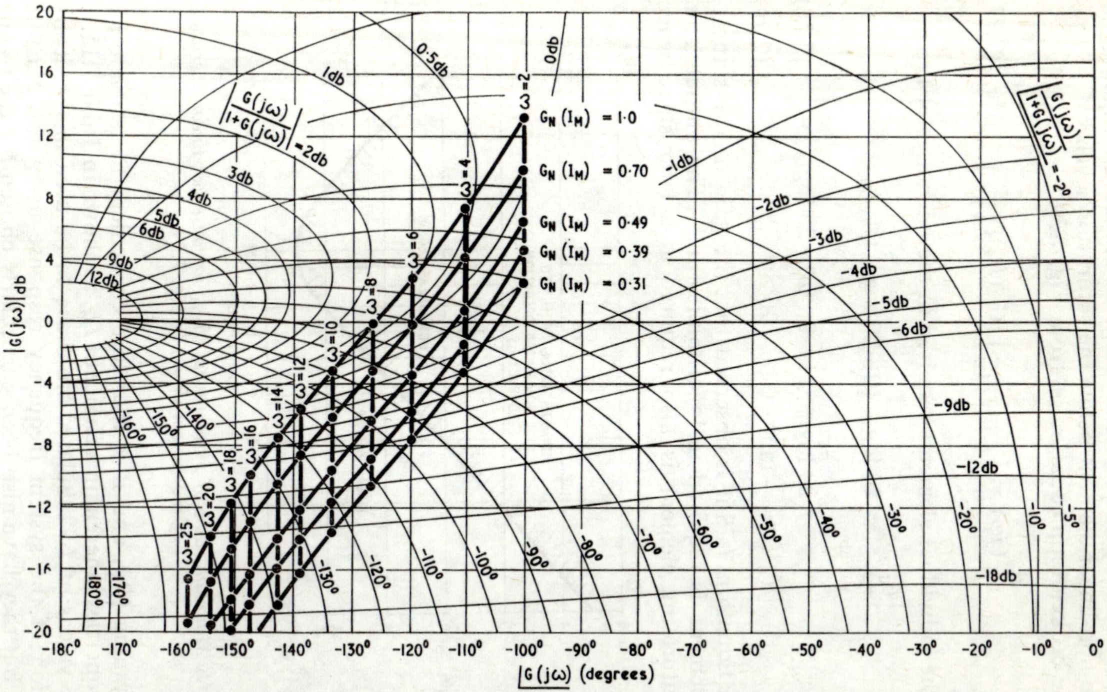


*Fig. 7.42. Nichols contours used to determine the frequency response $|\theta_0/\theta_i|$ of the saturating system of Fig. 7.39. for $|I_M|$ and $G_N(I_M)$ fixed*

**Table 7.15.1**

PROPERTIES OF THE SATURATING ELEMENT AS A FUNCTION OF THE SIGNAL AMPLITUDE $I_M$

| *Input to saturating element* | *Gain of saturating element* | *Fundamental harmonic of saturating element output* | *Logarithmic gain of saturating element* | *Logarithmic value of saturating element output* |
|---|---|---|---|---|
| $I_M$ | $G_N(I_M)$ | $I'_M = I_M G_N(I_M)$ | $20 \log_{10} G_N(I_M)$ | $20 \log_{10} I'_M$ |
| 1·00 | 1·00 | 1·00 | 0 | 0 |
| 1·72 | 0·70 | 1·20 | −3·10 | 1·58 |
| 2·53 | 0·49 | 1·24 | −6·00 | 1·87 |
| 3·22 | 0·39 | 1·25 | −8·20 | 1·94 |
| 4·08 | 0·31 | 1·27 | −10·20 | 2·08 |

plot of the broken loop Equations 7.78 and 7.79 evaluated for convenient intervals of frequency, and for the five values of $I_M$, and inherently $G_N(I_M)$, tabulated in Table 7.15.1. $G_N(I_M) = 1{\cdot}0$ is the linear response, and the total plot is exactly similar to that which would be obtained if the feedforward gain were varied, since our interpretation of the saturating element using describing function techniques is essentially a variable gain unit. Using

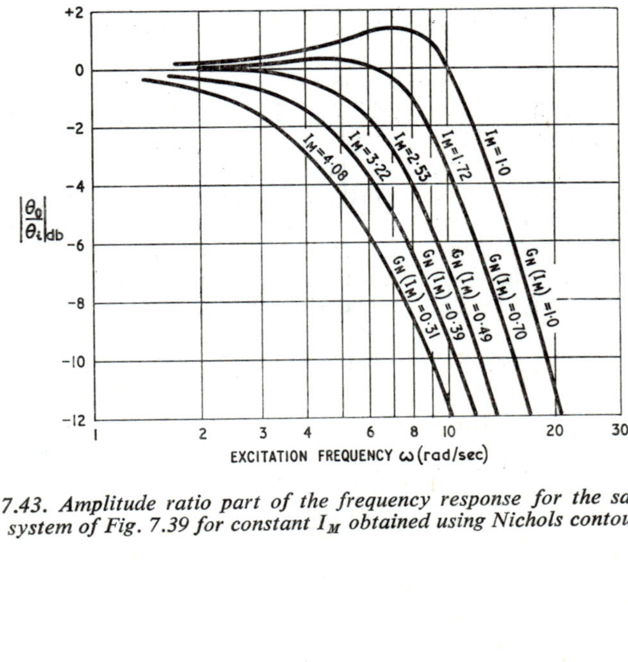


*Fig. 7.43. Amplitude ratio part of the frequency response for the saturating system of Fig. 7.39 for constant $I_M$ obtained using Nichols contours*

the constant $|\theta_0/\theta_i|$ and constant $\underline{/\theta_0/\theta_i}$ contours superimposed on the Nichols chart, the frequency response characteristics for constant $G_N(I_M)$ are constructed, and the results are shown in Fig. 7.43 for the amplitude contribution, and in Fig. 7.44 for the phase contribution. If the system is then excited such that

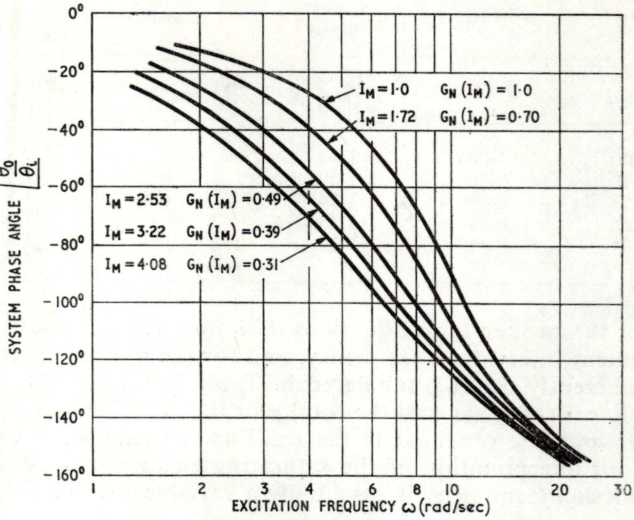


*Fig. 7.44. Phase angle part of the frequency response for the saturating system of Fig. 7.39 for constant $I_M$ obtained using Nichols contours*

$I_M$ is kept constant during the test, then the test data will approximate to that shown, and if $I_M$ is increased beyond unity there will be an increase in system phase lag and in system amplitude ratio at any particular excitation frequency.

It is now necessary, for a particular value of $I_M$ and frequency, to determine the corresponding input amplitude $|\theta_i|$. At this stage $I_M$, $G_N(I_M)$, and $|\theta_0/\theta_i|$ are already known for any point on the curves of Fig. 7.43. Therefore by using Equation 7.76 the value of $\theta_i$ may be determined directly as

$$|\theta_i| = \frac{2}{j\omega}[I_M G_N(I_M)]\left[\frac{\theta_i}{\theta_0}(j\omega)\right] \tag{7.80}$$

which in decibel form is

$$|\theta_i|_{db} = 20\log_{10}2 + 20\log_{10}[I_M G_N(I_M)]_{I_M} - 20\log_{10}\left|\frac{\theta_0}{\theta_i}\right|_{\omega} - 20\log_{10}\omega \tag{7.81}$$

For example, for $\omega = 8$ and $I_M = 1{\cdot}72$, from Table 7.15.1, $20\log_{10}[I_M G_N(I_M)] = +1{\cdot}58$, and from Fig. 7.43 for these conditions $|\theta_0/\theta_i| = 1{\cdot}20$ db. Now $20\log_{10} 8 = 18{\cdot}06$, and $20\log_{10} 2 = 6{\cdot}02$; hence $|\theta_i|_{db} = 6{\cdot}02+1{\cdot}58+1{\cdot}20-18{\cdot}16 = -9{\cdot}36$ db. In similar fashion the relationship between $|\theta_i|$ and $|\theta_0/\theta_i|$ can be determined for the values of $|\theta_0/\theta_i|$ available from Fig. 7.43 at the sample frequency $\omega = 8$ rad/sec. A curve relating $|\theta_i|$ and $|\theta_0/\theta_i|$ for $\omega = 8$ can then be drawn as shown in Fig. 7.45, and the exercise can be repeated for a reasonable number of frequencies.

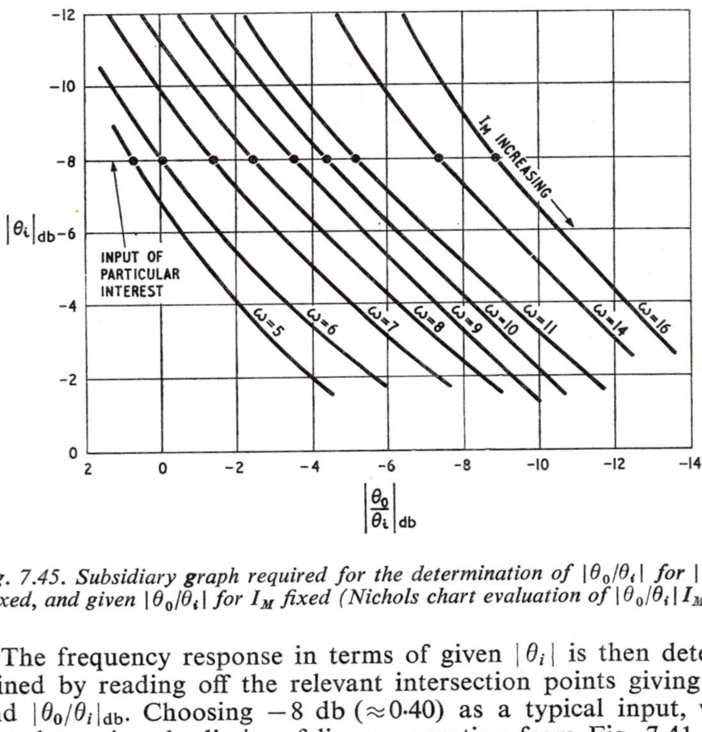


*Fig. 7.45. Subsidiary graph required for the determination of $|\theta_0/\theta_i|$ for $|\theta_i|$ fixed, and given $|\theta_0/\theta_i|$ for $I_M$ fixed (Nichols chart evaluation of $|\theta_0/\theta_i|\,I_M$)*

The frequency response in terms of given $|\theta_i|$ is then determined by reading off the relevant intersection points giving $\omega$ and $|\theta_0/\theta_i|_{db}$. Choosing $-8$ db ($\approx 0{\cdot}40$) as a typical input, we first determine the limits of linear operation from Fig. 7.41 as 4·6 rad/sec and 22 rad/sec. Thus for frequencies less than 4·6 rad/sec and greater than 22 rad/sec the system frequency response characteristics for $|\theta_i| = -8$ db will be identical to the linear system response, i.e. $I_M = 1{\cdot}0$ and $G_N(I_M) = 1{\cdot}0$ in Figs. 7.43 and 7.44. Intermediate points in the non-linear operating region are obtained directly from Fig. 7.45, and the amplitude ratio part of the response constructed in this way can be seen in Fig. 7.46. It can be seen that very considerable attenuation takes place compared with the linear response and is of the order of $-4{\cdot}5$ db at 10 rad/sec.

If system phase lag information is required, then the phase lag is known exactly at every point where the $|\theta_i| = -8$ db response cuts a constant $I_M$ line, since at that same frequency the phase lag must be the lag for the same value of $I_M$ in Fig. 7.44. Only four such points on the phase lag curve can be obtained in this way directly from Fig. 7.46, and these are the two linear operating limits and the two intersection points with the $I_M = 1{\cdot}72$ line. The latter two intersections occur at frequencies of 6 rad/sec and 12 rad/sec, and from Fig. 7.44 the phase lags at these frequencies for $I_M = 1{\cdot}72$ are 61° and 122° respectively. Additional points can be obtained in several ways. Extra system frequency responses can be determined for intermediate values of $I_M$ which will lead to more intersection points, in which case

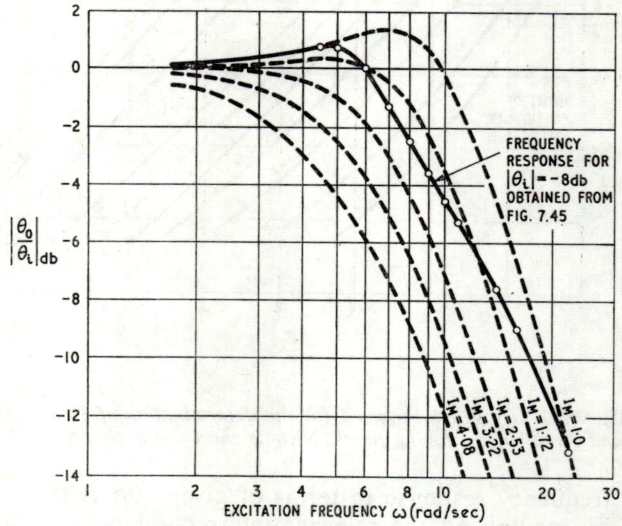


*Fig. 7.46. Amplitude ratio part of the frequency response for the saturating system of Fig. 7.39 for $|\theta_i| = -8$ db*

the problem is solved. Alternatively, the curves of Fig. 7.45 can be calibrated in terms of $I_M$ since this data is known when the curves are initially constructed. If the phase lag at, say, $\omega = 8$ rad/sec is required, then the $\omega = 8$ rad/sec curve in Fig. 7.45 is calibrated in terms of phase lag for the known values of $I_M$ using the phase lag data of Fig. 7.44. The phase lag at $\omega = 8$ rad/sec, $|\theta_i| =$

$-8$ db, is then obtained by interpolation, and is found in this case to be about 92°.

The complete phase lag plot corresponding to $|\theta_i| = -8$ db is shown in Fig. 7.47.

Probably the biggest advantage of the Nichols chart is that experimental results obtained for individual elements (which may be frequency dependent as well as amplitude dependent) can be easily incorporated into the process. The method suffers from the fact that a very considerable amount of effort must be

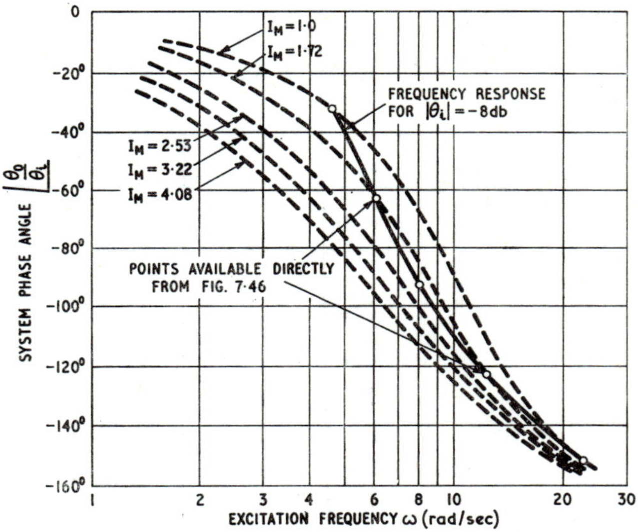


*Fig. 7.47. Phase angle part of the frequency response for the saturating system of Fig. 7.39 for $|\theta_i| = -8$ db*

expended before even one response for constant $|\theta_i|$ can be obtained, and the exact logarithmic response, not the straight-line approximation to it, must be used for all elements at all times. On the other hand the response for many different values of $|\theta_i|$ can be obtained more or less simultaneously since, having constructed Fig. 7.45, the data is readily available for a wide range of inputs.

Owing to the interpolation required when using the phase and magnitude contours the method tends to lack accuracy should this be important, and also lacks resolution for small departures into the non-linear mode.

## 7.16 Use of the Root Locus to Determine the System Describing Function For Non-linear Systems (McAllister's Method)[74]

Provided attention is restricted to the frequency domain, a root construction can be used in conjunction with Equation 7.66 with $G_N(I_M)$ as the variable. System pole-zero arrays can then be determined which are valid for a harmonic excitation with $G_N(I_M)$ constant. Frequency responses for these conditions can be determined from the system pole-zero array in the usual way. For the example evaluated using the Nichols plot in Section 7.15, Equation 7.72 may be rewritten in terms of s to yield the relationship

$$\frac{\theta_0}{\theta_i}(s) = \frac{100G_N(I_M)}{100G_N(I_M)+10s+s^2} \tag{7.82}$$

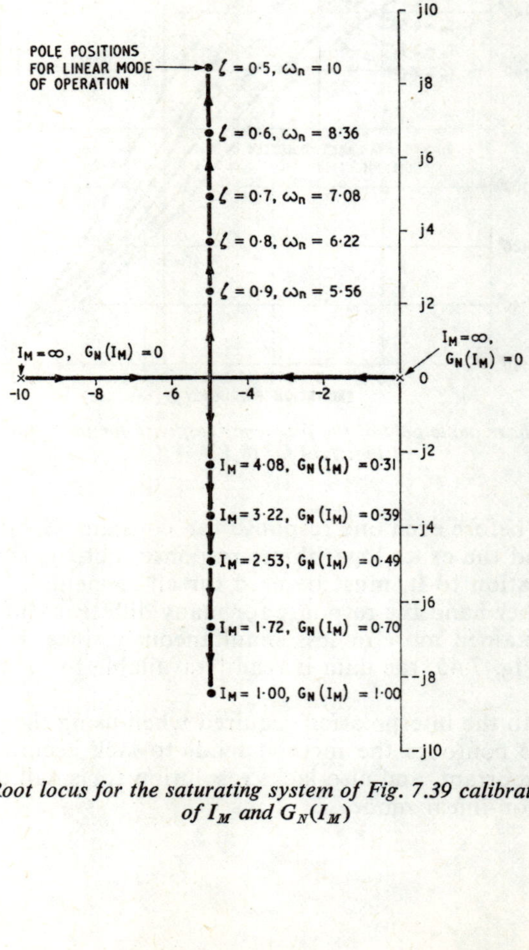


*Fig. 7.48. Root locus for the saturating system of Fig. 7.39 calibrated in terms of $I_M$ and $G_N(I_M)$*

The root locus required is therefore

$$100G_N(I_M)+10s+s^2 = 0$$

and, in conventional form,

$$\frac{100G_N(I_M)}{s(s+10)} = -1 \tag{7.83}$$

where $G_N(I_M)$ varies from 0 to 1 since 1 is the value of gain for linear operation. Clearly, the locus starting points are $s = 0$ and $s = -10$, and there will be two high-gain asymptotes at angles $\pi/2$ and $3\pi/2$. The intersection point of these asymptotes with the real axis will be midway between the open loop poles. The complete locus is shown in Fig. 7.48, and the non-linear gain $G_N(I_M)$ has been determined for damping ratios of 0·5, 0·6, 0·7, 0·8 and 0·9. Results are given in Table 7.16.1, and it can be seen that the effect of increasing $I_M$ is to decrease undamped natural frequency and to increase the damping ratio of the complex pole pair.

**Table 7.16.1**

PROPERTIES OF THE ROOT LOCUS FOR A SECOND ORDER SATURATING SYSTEM IN TERMS OF THE GAIN OF THE SATURATING ELEMENT

| *Input to saturating element* | *Gain of saturating element* | *Damping ratio of complex pole pair* | *Undamped natural frequency of complex pole pair* |
|---|---|---|---|
| $I_M$ | $G_N(I_M)$ | $\zeta$ | $\omega_n$ (rad/sec) |
| 1·00 | 1·00 | 0·50 | 10·00 |
| 1·72 | 0·70 | 0·60 | 8·36 |
| 2·53 | 0·49 | 0·70 | 7·08 |
| 3·22 | 0·39 | 0·80 | 6·22 |
| 4·08 | 0·31 | 0·90 | 5·56 |

We are now at liberty to use these poles to compute the linear or the logarithmic form of the frequency response as desired. If the logarithmic form is used, then the use of a standard template for these values of $\zeta$ will allow Fig. 7.43 to be constructed very rapidly without the interpolation necessary with the Nichols chart method, and the phase lag can also be constructed if required in a similar way using phase templates such as those to be found in Thaler and Brown.[79] These results for $I_M$ constant will be identical to Figs. 7.43 and 7.44 already obtained by the Nichols chart method and are inherently common to both techniques.

McAllister recommends the use of the logarithmic form, and having reached this stage proceeds to evaluate $|I_M/\theta_i|_{db}$ as a function of frequency and $I_M$, and plots this relationship as a separate graph. $|I_M/\theta_i|_{db}$ is obtained from the frequency responses available in Fig. 7.43 and modified according to Equation 7.76, which stated

$$\frac{I_M}{\theta_i}(j\omega) = \left[\frac{\theta_0}{\theta_i}(j\omega)\right]\left[\frac{j\omega}{2G_N(I_M)}\right]$$

for specific values of $I_M$ and $\omega$. We shall now use McAllister's method to check two points on the frequency response for $|\theta_i| = -8$db.

To determine the system frequency response for a particular input $|\theta_i|$, the procedure is to select a suitable value of $I_M$. We can then write

$$20 \log_{10}\left|\frac{I_M}{\theta_i}\right| = 20 \log_{10} I_M - 20 \log_{10} \theta_i \tag{7.84}$$

and hence for chosen $I_M$ and $|\theta_i|$, 20 $\log_{10} |I_M/\theta_i|$ is known. If the curve relating $|I_M/\theta_i|$ to $\omega$ is drawn for this value of $I_M$, then the points at which the curve cuts the required value of 20 $\log_{10}$ $|I_M/\theta_i|$ defined by Equation 7.83 will yield the frequencies at which $|\theta_i|$ and $I_M$ requirements are simultaneously satisfied. For these frequencies the phase lag and amplitude ratio are already defined by Figs. 7.43 and 7.44. As an example, Fig. 7.49 shows

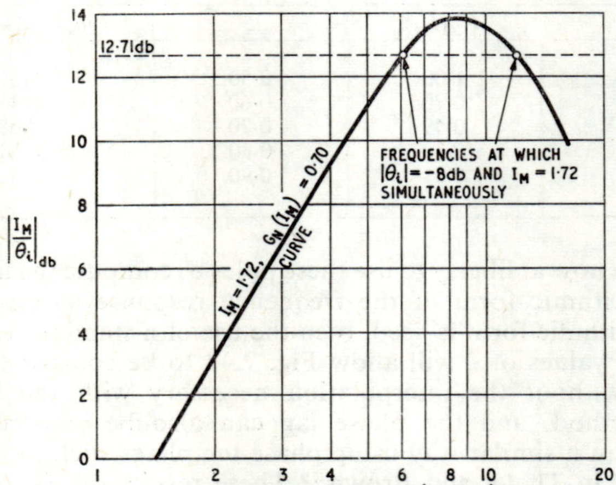


*Fig. 7.49. Subsidiary graph for the determination of the system frequency response for $|\theta_i| = -8$ db, using McAllister's method*

$|I_M/\theta_i|_{db}$ as a function of frequency for $I_M = 1{\cdot}72$. For an input $|\theta_i| = -8$ db, we may write from Equation 7.84,

$$20 \log_{10} \left| \frac{I_M}{\theta_i} \right| = 4{\cdot}71 + 8 = 12{\cdot}71 \text{ db} \qquad (7.85)$$

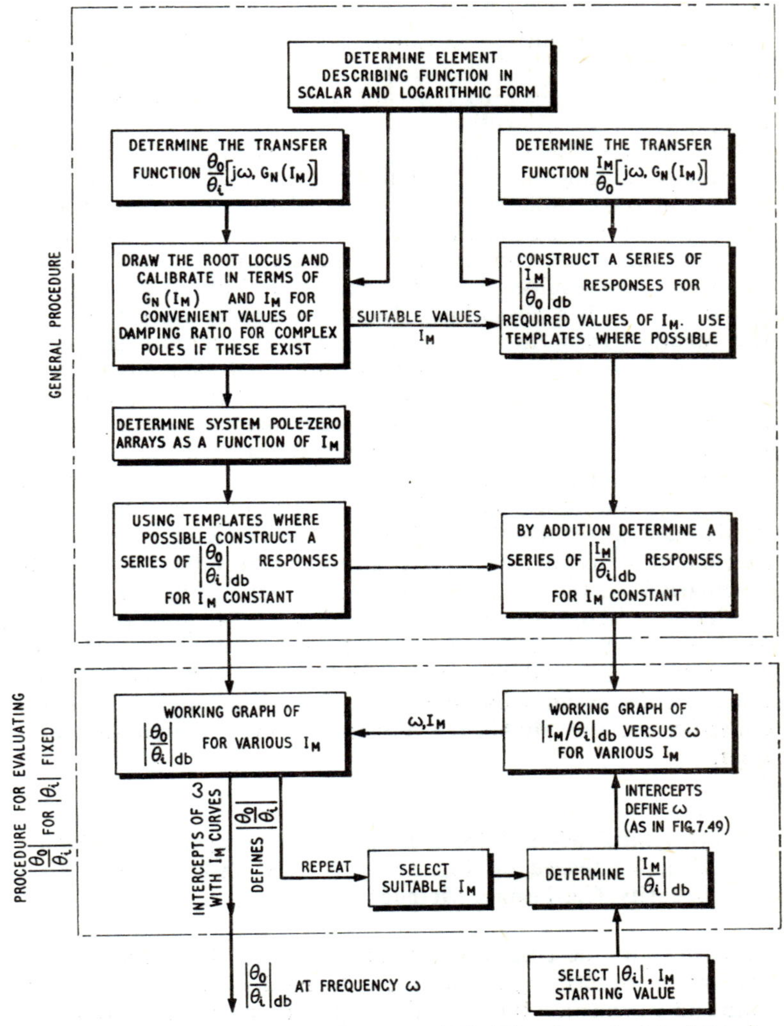


*Fig. 7.50. Sequence diagram showing use of McAllister's method to determine the system describing function (system frequency response) for $|\theta_i|$ fixed*

In Fig. 7.49, the 12·71 db line cuts the $I_M = 1{\cdot}72$ curve at frequencies of 6·1 rad/sec and 12 rad/sec. Therefore the system frequency response characteristics for $|\theta_i| = -8$ db are identical to the amplitude ratio and phase lag characteristics for $I_M =$ 1·72 at these two frequencies, and at these two frequencies only. Inspecting Fig. 7.46 we see that the $|\theta_i|$ response does indeed cut the $I_M = 1{\cdot}72$ response at frequencies of 6·1 rad/sec and 12 rad/sec, which confirms the validity of the original Nichols construction. Further points on the $|\theta_i| = -8$ db frequency response can be obtained by choosing a set of values of $I_M$ and repeating the whole process.

McAllister's method can be used for systems of any order by simply adding the logarithmic responses of all the relevant poles and zeros, and has been extended to cover systems with multiple non-linearities. Since a root locus may be drawn for systems with phase contributing non-linearities such as hysteresis, even if the resulting locus is asymmetrical about the negative real axis, there is a possibility of determining the system describing function from the poles and zeros. If accurate templates of the frequency characteristics of individual elements such as $\omega_n^2+2\zeta\omega_n s+s^2$ are used in the construction of the various curves required, the accuracy of the final response is probably better than that obtained using Nichols plots. Because of the use of the logarithmic form for speed of construction the problem of inaccurate resolution may still arise, but generally at the synthesis stage the results obtained using logarithmic scales are accurate enough anyway. McAllister's method may be conveniently summarised in the sequence diagram shown in Fig. 7.50. There is plenty of scope for the combination of Nichols plots and McAllister's method if considered desirable, since for synthesis problems where test data on individual elements are available the Nichols contours may be used as a replacement for the root locus construction, the McAllister sequence being adhered to thereafter.

## 7.17 Graphical Method for Determination of System Describing Function based on Input-output Relationship at the Non-linearity (Levinson's Method)

Probably the best-known method for the determination of the system describing function is the method commonly attributed to Levinson[80] but also developed independently by Johnson,[81] and West and Douce,[82] which is restricted to single-valued non-linearities. The method is radically different from the three methods previously outlined, and the desired input signal $|\theta_i|$ is used from the start. Fig. 7.40, which relates the saturation value of $|\theta_i|$ as a function of frequency, is therefore doubly valuable since realistic values of $|\theta_i|$ are defined and a check is available on any subsequent construction, as will be seen later. The plane used in the method is the $I_M - I_M G_N(I_M)$ plane on which is superimposed the describing function relationship for the non-linearity under consideration. It is then possible to

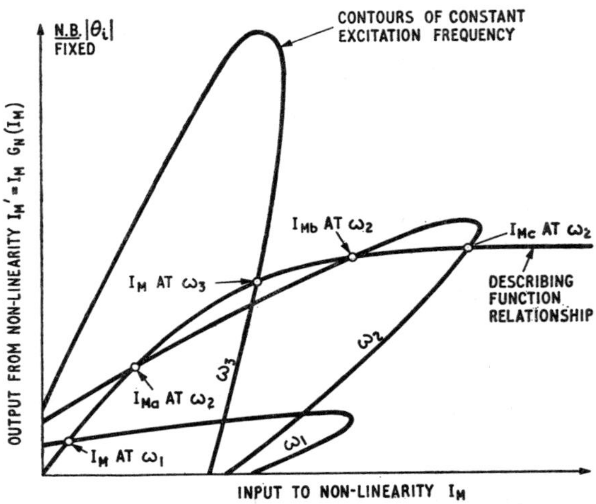


*Fig. 7.51. Typical Levinson plot for a non-linear system with jump phenomenon shown for one value of $|\theta_i|$ only*

manipulate the system equations, using signal amplitude considerations only, to obtain the equilibrium conditions for the output from the non-linearity in terms of the system input, $|\theta_i|$, the input to the non-linearity, $I_M$, and the system excitation frequency $\omega$. For $|\theta_i|$ fixed, and a particular value of $\omega$, the relevant curve may be drawn on the $I_M - I_M G_N(I_M)$ plane, and intersections with the describing function will indicate the steady state values of the input to and output from the non-linearity corresponding to the defined steady state operating conditions. The procedure is repeated for various values of $\omega$, and a frequency response for $|I_M/\theta_i|$ or $|I'_M/\theta_i|$ can be determined. Using known linear transfer functions the frequency response of interest, $|\theta_0/\theta_i|$ or $|\theta/\theta_i|$, may be obtained in a straightforward fashion. Fig. 7.51 shows a typical plot for a particular system. Jump phenomena, if existing, are clearly indicated by the multiple intersection of the describing function relationship by the constant frequency line, as shown for frequency $\omega_2$ in Fig. 7.51. The corresponding jump phenomenon is shown in the frequency domain in Fig. 7.52. Existence of such jump phenomena is also revealed using other techniques, but perhaps not so spectacularly.

As a specific example let us consider the saturating second order system already studied in Sections 7.15 and 7.16. We require

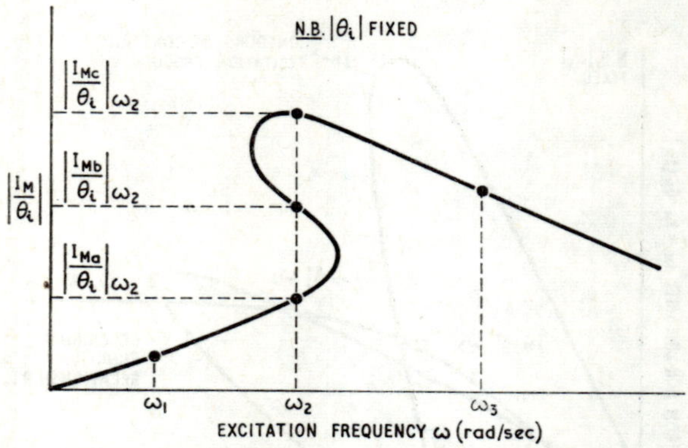


*Fig. 7.52. Frequency response $|I_M/\theta_i|$ obtained from the typical Levinson plot of Fig. 7.51*

initially $|I_M/\theta_i|$ in terms of $\omega$, but not this time in terms of $|\theta_0/\theta_i|$ as required in the Nichols and McAllister methods. Hence,

$$\frac{\theta}{\theta_i}(s) = \frac{10s+s^2}{s^2+10s+100G_N(I_M)} \tag{7.86}$$

and

$$\frac{I_M}{\theta}(s) = \frac{5}{1+0{\cdot}1s} \tag{7.87}$$

Hence

$$\frac{I_M}{\theta_i}(s) = \frac{50s}{s^2+10s+100\,G_N(I_M)} \tag{7.88}$$

Evaluating the modulus at frequency $\omega$, we have

$$\left|\frac{I_M}{\theta_i}\right|^2 = \frac{2{,}500\omega^2}{100\omega^2+[100G_N(I_M)-\omega^2]^2} \tag{7.89}$$

The procedure is now to manipulate Equation 7.89 to obtain the output from the non-linear element, $I_M G_N(I_M)$ on the left-hand side of the equation. The steps in detail are

$$2{,}500\ \omega^2\theta_i^2 = [100G_N(I_M)-\omega^2]^2 I_M^2 + 100\omega^2 I_M^2 \tag{7.90}$$

Then

$$[100G_N(I_M)-\omega^2]^2 I_M^2 = 2{,}500\omega^2\theta_i{}^2 - 100\omega^2 I_M^2 \tag{7.91}$$

and

$$[100G_N(I_M)-\omega^2] I_M = \pm 10\,\omega\,\sqrt{25\,\theta_i^2 - I_M^2} \tag{7.92}$$

Finally

$$I_M G_N(I_M) = \frac{\omega^2 I_M}{100} \pm \frac{\omega}{10}\sqrt{25\,\theta_i^2 - I_M^2} \tag{7.93}$$

The describing function relating input and output for the saturation type non-linearity may be obtained directly in the form required for plotting from Table 7.15.1 and is shown in Fig. 7.53. For consistency with the previous methods studied $|\theta_i|$

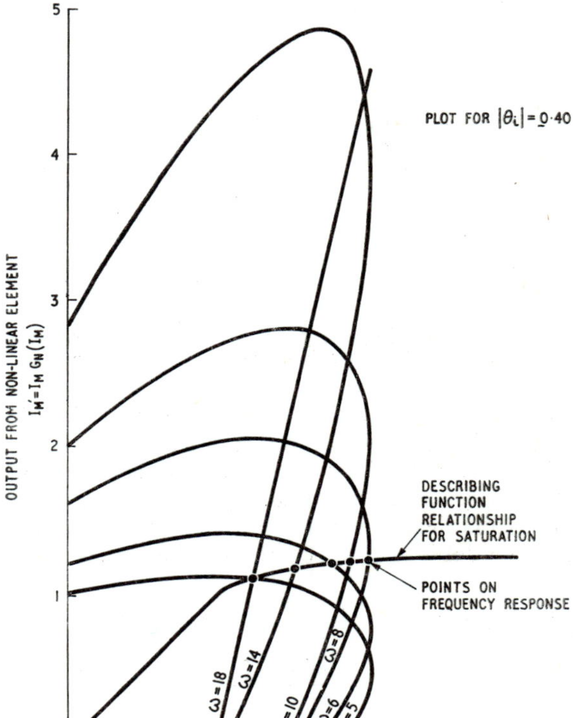


*Fig. 7.53. Levinson plot for the saturating systems of Fig. 7.39 for $|\theta_i| = 0{\cdot}40$*

is chosen to be 0·4, which is very nearly −8 db. Substituting in Equation 7.93, we obtain

$$I_M G_N(I_M) = \frac{\omega^2 I_M}{100} \pm \frac{\omega}{10}\sqrt{4-I_M^2} \qquad (7.94)$$

Equation 7.94 is a series of ellipses, and solutions for suitable values of $\omega$ are shown in Fig. 7.53. It can be seen by inspection of the intersection points with the describing function curve, that $\omega = 6{\cdot}0$ and $\omega = 18{\cdot}0$ are near the linear operating range, thus at this stage approximately confirming the response. If desired, the $\omega = 4{\cdot}6$ and $\omega = 22$ curves could be drawn to verify that the describing function curve is intersected by both at the (1,1) point which is, of course, the limit of linear operation. Using the intersection points of Fig. 7.53, the frequency response of $|I_M/\theta_i|$ shown in Fig. 7.54 may be constructed. The linear

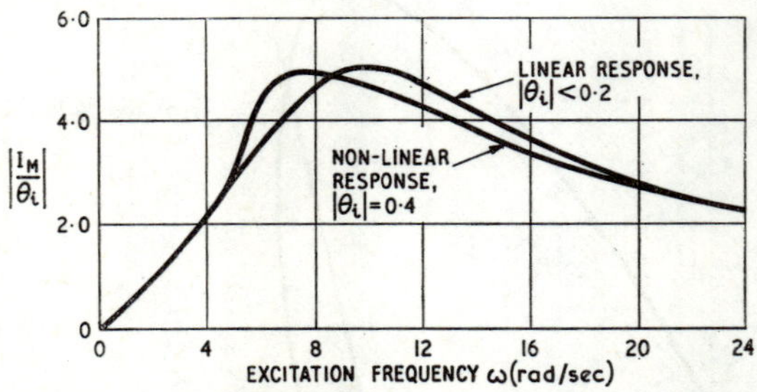


*Fig. 7.54. Amplitude ratio $|I_M/\theta_i|$ part of the frequency response for the saturating system of Fig. 7.39, obtained from the Levinson plot of Fig. 7.53*

response shown is merely Fig. 7.40 inverted with $I_M = 1{\cdot}0$, which is the limiting value for linear operation. Although Fig. 7.55 describes the behaviour at the non-linear element rather than $|\theta_0/\theta_i|$, it does so in terms of $|\theta_i|$ constant, and hence the information available at this stage is much more useful than the information available at the corresponding stage using other methods.

To obtain $|\theta_0/\theta_i|$ for $|\theta_i| = 0{\cdot}4$ we use the relationship defined from Fig. 7.39 by

$$\theta_0 = \frac{2I_M G_N(I_M)}{j\omega} \qquad (7.94)$$

for the frequencies of interest. $I_M G_N(I_M)$ can be read directly from Fig. 7.53 for the required values of $\omega$, Fig. 7.54 being completely unnecessary for this purpose. For example, if $\omega = 8$ rad/

sec, then $I_M G_N(I_M) = 1{\cdot}20$; hence $\theta_0 = 2{\cdot}40/8 = 0{\cdot}30$, giving $|\theta_0/\theta_i| = 0{\cdot}30/0{\cdot}40 = 0{\cdot}75$. The complete plot obtained in this manner is shown in Fig. 7.55. Use of the linear scale of amplitude ratio required by this method has considerably improved the resolution that may be obtained in the region of the bandwidth frequency. System bandwidth has dropped from the linear value obtained for $|\theta_i| < 0{\cdot}2$ of 12·6 rad/sec to 8·4

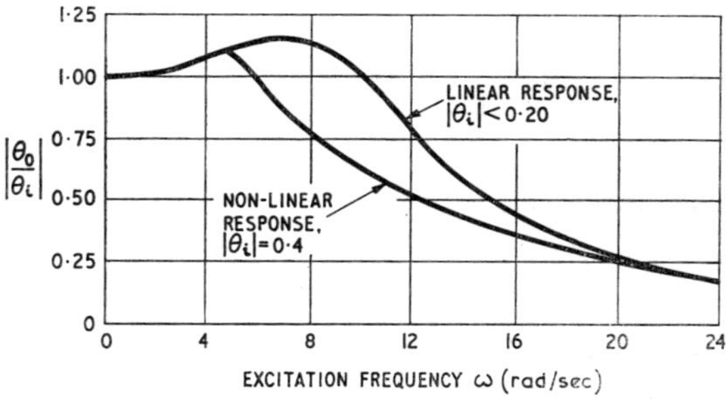


*Fig. 7.55. Amplitude ratio $|\theta_0/\theta_i|$ part of the frequency response for the saturating system of Fig. 7.39, which may be obtained at frequency ω from the non-linearity output read from the Levinson plot multiplied by $|G_2(j\omega)|$*

rad/sec for the non-linear response obtained for $|\theta_i| = 0{\cdot}40$. Higher order systems may be studied in the same way, as although the application of the technique becomes more complicated, according to Levinson the procedure does not become too unwieldy. It is probably the best technique of all for low-order systems with one single-valued non-linearity present.

## 7.18 Use of the Moving Pole-zero Concept to Estimate System Transient Response

Finnigan,[83] and Grensted,[84] suggested that provided the transient input to a non-linearity is approximately sinusoidal, considerable information about the transient behaviour of the system can be deduced with the aid of the root locus such as Fig. 7.48 calibrated in terms of the element describing function. Briefly, as crystallised in Fig. 7.56, the proposition is that the transient response can be piecewise represented between successive maxima and minima by the system poles and zeros corresponding to a particular value of equivalent gain. The equivalent gain will be constant for a segment of time between successive maxima

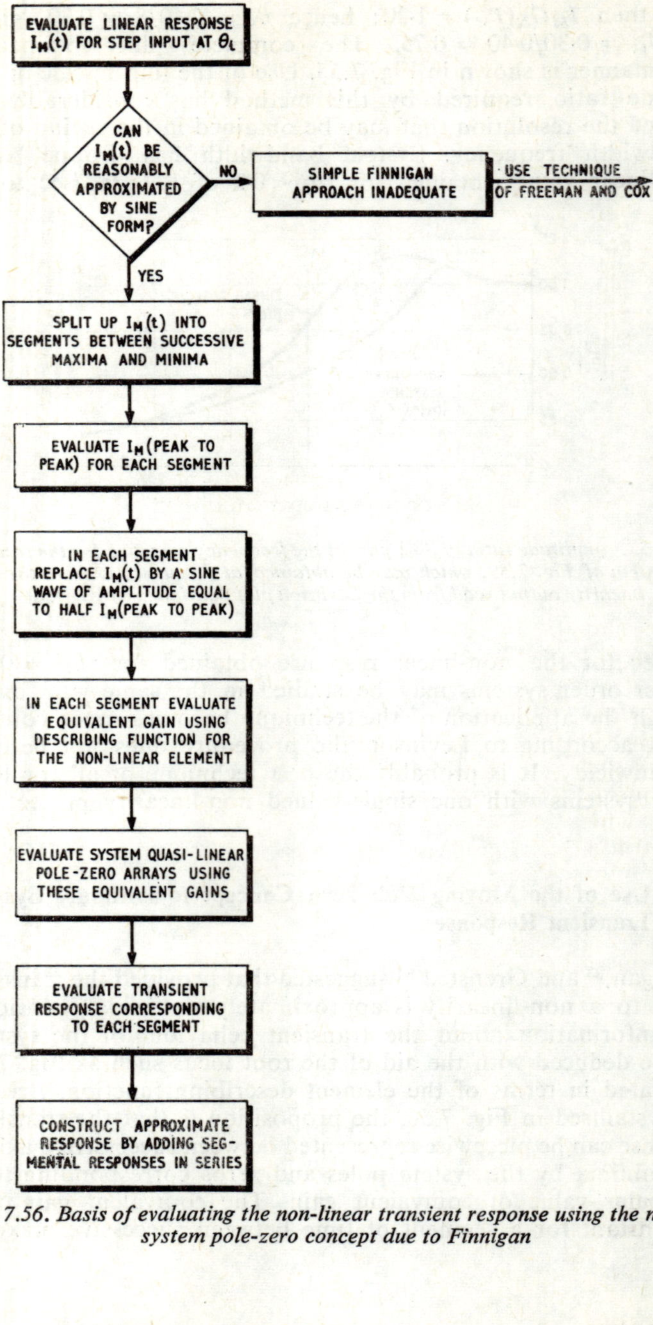


*Fig. 7.56. Basis of evaluating the non-linear transient response using the moving system pole-zero concept due to Finnigan*

and minima, but will in general be different for different segments of time. At each maximum or minimum, the absolute input to the non-linearity is known, and is used together with a knowledge of the linear response of the system, to evaluate an equivalent sinusoidal input to the element. Knowing the magnitude

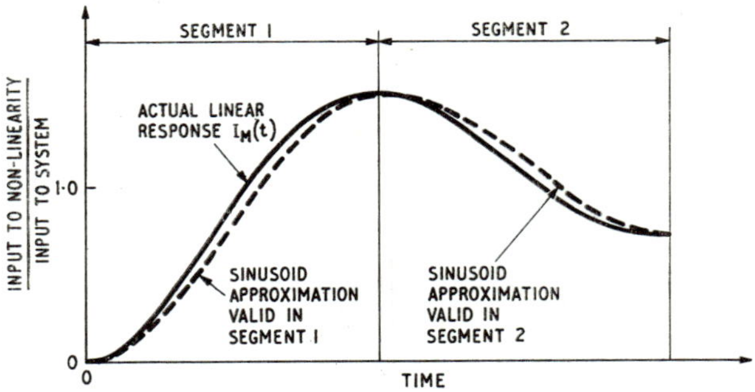


*Fig. 7.57. Time domain segmentation and replacement of transient response by equivalent sinusoid*

of the sine wave, and the characteristics of the non-linear element, it is a simple matter to replace the non-linearity by the equivalent gain, and hence determine the system pole-zero array pertinent to the segment. Fig. 7.57 shows the segmentation of a typical transient, whilst Fig. 7.58 shows a set of moving pole-zero arrays. In practice, it is the early part of the transient response which is often of interest, and it is therefore seldom necessary to evaluate the response beyond the first segment.

Consider the second order saturating system shown in Fig. 7.59, which in the linear state will have a damping ratio of $\zeta = 0{\cdot}40$. The system transfer function is given by

$$\frac{\theta_0}{\theta_i}(\mathrm{s}, \theta_i) = \frac{156{\cdot}25\ K_{\mathrm{eq}}}{\mathrm{s}^2 + 10\mathrm{s} + 156{\cdot}25 K_{\mathrm{eq}}} \qquad (7.96)$$

and the system pole-zero array is composed of two poles only with

$$\omega_n = 12{\cdot}5\sqrt{K_{\mathrm{eq}}}$$

and

$$\zeta = \frac{0{\cdot}4}{\sqrt{K_{\mathrm{eq}}}}$$

Fig. 7.60 shows the linear response at the input to the non-linear element, and it can be seen that the maximum overshoot is 1·25 times the input to the non-linearity.

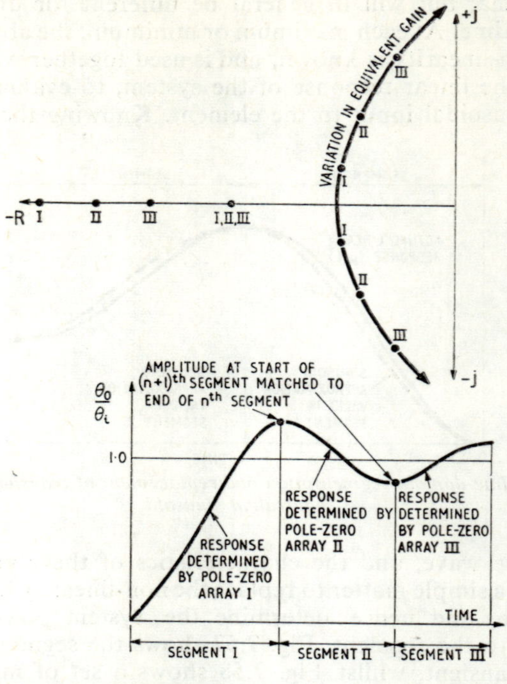


*Fig. 7.58. Moving system pole-zero arrays and corresponding segments of the transient response*

Suppose it is required to evaluate the transient response of the system when $\theta_i = 4L$. The maximum input to the non-linear element, according to Finnigan, is $1{\cdot}25 \times 4L = 5{\cdot}0L$, and this is equivalent to a sine wave of semi-amplitude $2{\cdot}5L$. The ratio of the saturating input to the actual input is thus 0·40, and the

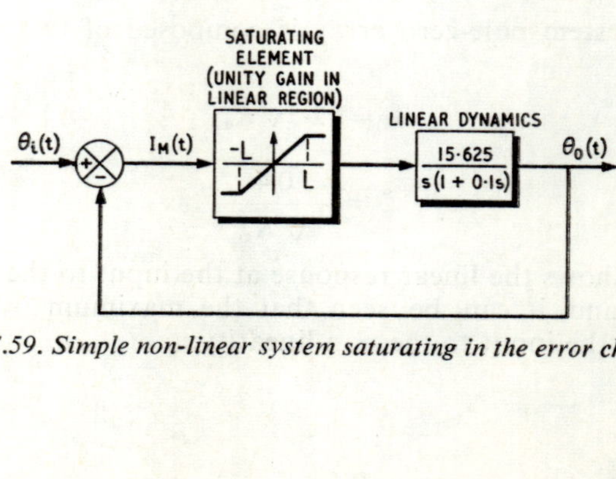


*Fig. 7.59. Simple non-linear system saturating in the error channel*

equivalent gain, from Fig. 7.25, is 0·5. Substituting for $K_{eq}$ gives $\omega_n = 8{\cdot}8$ rad/sec and $\zeta = 0{\cdot}57$. Therefore until the first peak overshoot the system behaves similarly to a linear system with these values of undamped natural frequency and damping ratio. Response is sketched in Fig. 7.61, the standard forms shown in Chapter 4 being sufficient for the purpose. Peak overshoot occurs at 4·5 sec, this also being determined by the $\zeta$ and $\omega_n$ of the equivalent system; it is *not* the time of maximum overshoot of the original linear system. At 4·5 sec, the input to the non-linearity will be well within the linear operating range, hence the

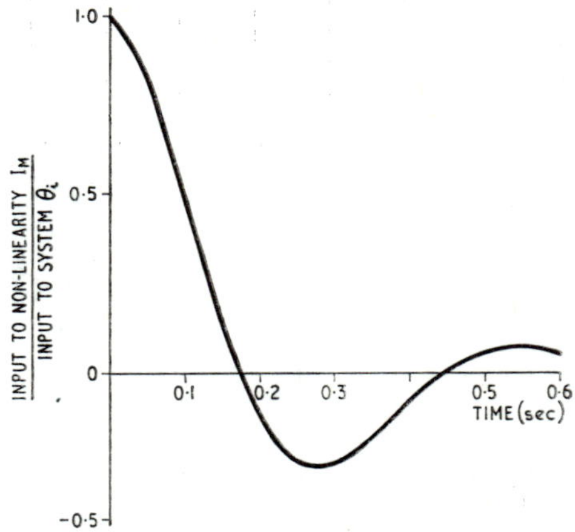


*Fig. 7.60. Signal seen by the input to non-linearity during linear transient for the system of Fig. 7.59*

remainder of the response will be identical to that for the linear system with $\zeta = 0{\cdot}4$ and $\omega_n = 12{\cdot}5$, with the initial amplitude at 4·5 sec chosen to match the terminal amplitude of the previous segment of the response. An accurate solution using numerical integration is also shown in Fig. 7.61, indicating reasonable agreement in the vicinity of the peak overshoot. As would be expected, Finnigan's technique predicts the important features of the transient responses of lightly damped systems reasonably accurately. Unfortunately it is inaccurate for heavily damped systems and it is then necessary to resort to a digital computer, or to the technique of Freeman and Cox[85], which allows for piecewise approximation over much smaller intervals, and also assumes a damped waveform rather than a pure sinusoid, both these modifications greatly increasing accuracy.

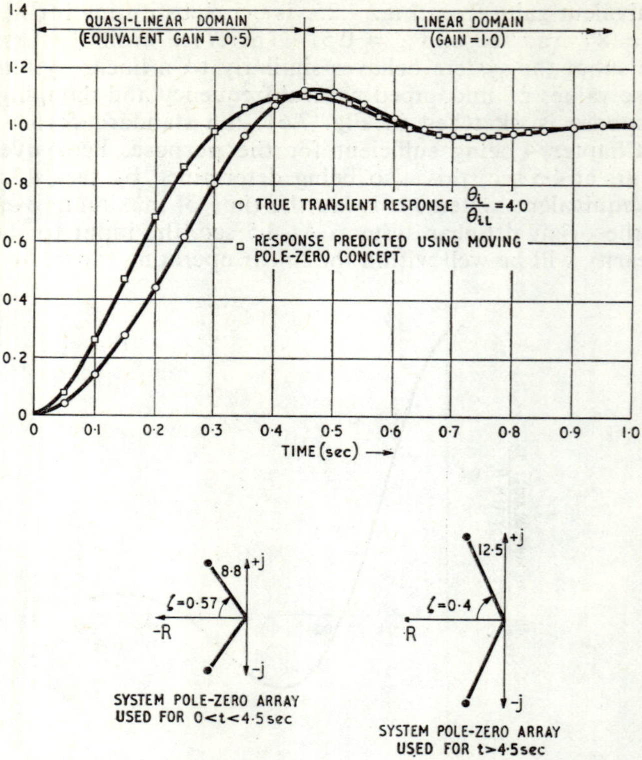


*Fig. 7.61. Transient response of the system shown in Fig. 7.59 for an input step four times that required for saturation*

## 7.19 Presentation of Performance Criteria for Non-linear Systems

Since performance criteria for non-linear systems are a function of the input amplitude, it is common at design and testing stages to plot the performance criteria of interest directly as a function of input amplitude or, in the case of small perturbations from a steady state operating point, to plot criteria as a function of the steady state parameters. Upper and lower limits can then be placed on these criteria. For example, the experimental results obtained for the electrohydraulic servo previously shown in Fig. 7.3 can be summarised for peak amplitude ratio and bandwidth as shown in Fig. 7.62. Due to the peak in $M_p$ we might suspect that there are non-linearities present which are significant at both high and low input signals. These may be saturation and deadzone respectively, or alternatively could be due to the variable gain characteristics of the electrohydraulic control value. Much more data would be needed to establish the types of non-linearity more definitely, but the information is already sufficient to attempt crude mathematical model-building in order to study the effect of the non-linear mode of operation in some larger system such as a machine tool drive or a missile autopilot.

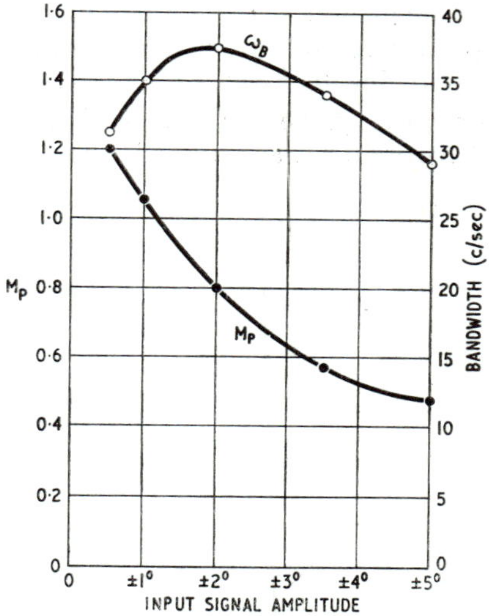


*Fig. 7.62. Variation of performance criteria for an electro-hydraulic servo as the input signal amplitude is increased*

## 7.20 Conclusion

In this chapter we have reviewed certain types of element non-linearities, and shown how various non-linear phenomena arise. Methods for the synthesis of non-linear feedback control systems should, in general, be directed at establishing broad, reliable trends before the trial design is simulated on an analogue computer. For systems with continuous non-linearities the small perturbation theory is very helpful, and meaningful small perturbation root loci and system transfer functions can be determined from which the various performance criteria may be established as for a truly linear system. The method is particularly powerful in systems with multiple cross-coupled non-linearities.

Large amplitude frequency response is best studied using the describing function technique, which becomes more exact as the order of the system is increased. For responses obtained with the input amplitude at the non-linearity fixed, a 'describing function' root locus can be drawn which yields a system describing function from which the system frequency response only can be determined. In the conventional large amplitude testing mode it is the *system input amplitude* which is kept constant, and under these circumstances the system transfer function is simply that quasi-linear transfer function which best fits the available data. Hence for system input amplitude constant the transfer function is best left in graphical form showing $|\theta_0/\theta_i|$ and $\underline{/\theta_0/\theta_i}$ as functions of $\omega$ and $|\theta_i|$. For lightly damped systems, the describing function method may be extended to predict the salient features of the transient response.

# Appendix

## Determination of Contribution of Complex Open Loop Poles in the Region of the Root Locus Real Axis Breakaway Point

Consider the point $a$ in Fig. A.1, which is on the root locus and very near the real axis breakaway point $b$ such that $ab$ is vertical. The lines joining the breakaway point to the complex open loop poles define the angles $\eta$ as shown. Owing to the incremental movement along the locus to point $a$, a phase contribution $\Delta\eta$ will arise at point $a$, both complex open loop poles contributing an amount $\Delta\eta$ of the *same* sign, giving a total contribution of $2\Delta\eta$. For the particular configuration shown with the complex open loop poles in the left-hand plane, but to the right of the assumed breakaway point, the vector rotation causing $\Delta\eta$ will be anticlockwise, i.e. a *negative* contribution at point $a$ as defined by the sign convention of Section 3.4. and Fig. 3.5. The problem is to determine $\Delta\eta$ in terms of $H$ and $V$, the horizontal and vertical displacements respectively of the complex open loop poles from the real axis breakaway point.

Consider the triangle $abc$. For a small rotation $\Delta\eta$, we may write

$$bc = R\Delta\eta \tag{A.1}$$

Angle $bac \approx \eta$, hence

$$bc = h \sin \eta \tag{A.2}$$

Eliminating $bc$

$$\Delta\eta = \frac{h \sin \eta}{R} \tag{A.3}$$

For small displacements, $ab < R$, and therefore we may consider the figure enclosing the upper complex pole, the vertical

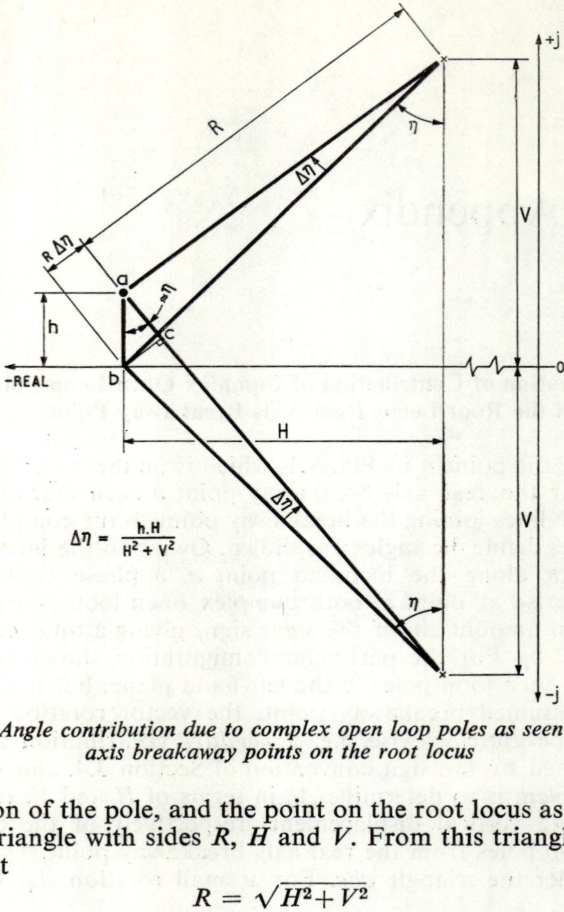


*Fig. A.1. Angle contribution due to complex open loop poles as seen near real axis breakaway points on the root locus*

projection of the pole, and the point on the root locus as a right-angled triangle with sides *R*, *H* and *V*. From this triangle it follows that

$$R = \sqrt{H^2 + V^2} \tag{A.4}$$

and

$$\sin\ \eta = \frac{H}{\sqrt{H^2 + V^2}} \tag{A.5}$$

Substituting Equations A.4 and A.5 into Equation A.3, we have the desired result

$$\Delta\eta = \frac{hH}{H^2 + V^2} \tag{A.6}$$

and the total anticlockwise phase contribution at point *a* will be

$$2\Delta\eta = \frac{2hH}{H^2 + V^2}$$

The phase contribution due to the complex open loop poles is small when $H$ is small, i.e. when, in the absence of the complex poles, the breakway point is about the same magnitude as the real part of the poles, the breakaway point is unaffected by the subsequent addition of the complex poles. For example, in Fig. 3.10, a (1,2) locus is shown for a particular open loop pole-zero configuration. Two complex open loop poles added to this configuration result in the new (1,4) locus shown in Fig. 3.45. Since the complex poles lie almost immediately above and below the breakawaypoints, to a first order approximation there is no change in these breakaway points and the root locus in the region near the real axis is therefore similar for both Figs. 3.10 and 3.45. However, in Chapter 6 a system was studied in which the presence of complex open loop poles was a significant factor in the position of the breakaway point.

EXAMPLE A.1

Determine the real axis breakaway point for the system studied in Example 6.12.2. Open loop zeros are at 0 and $-30$, and the open loop poles at $-45{\cdot}8$ and $+5{\cdot}8 \pm j25{\cdot}4$, as shown in Fig. A.2.

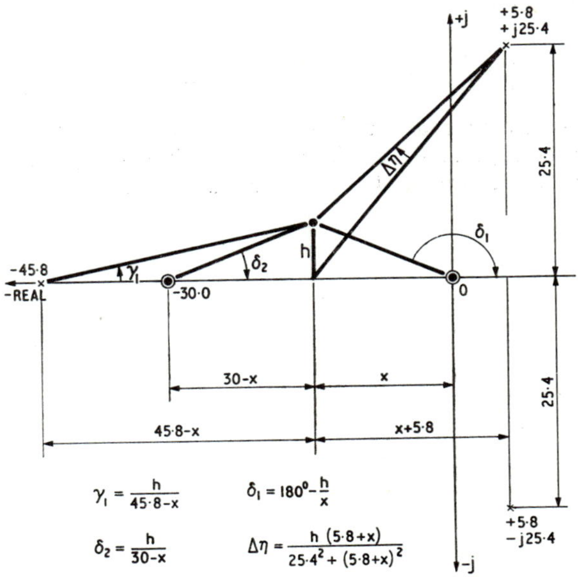


*Fig. A.2. Diagram for the computation of root locus real axis breakaway points in the presence of complex open loop poles (Example 6.12.2 and Fig. 6.46)*

The breakaway point will be in the left-hand plane at a distance $x$ from the origin. Rotation of the vectors at the open loop poles will therefore be anticlockwise and cause a negative phase contribution in the region of the real axis. The root locus angular requirement near the breakaway point is therefore, from Fig. A.2,

$$\gamma_1 - \delta_1 - \delta_2 - 2\Delta\eta = 180° \pm N360°$$

Now $V = 25{\cdot}4$ and $H = x + 5{\cdot}8$. Hence,

$$\Delta\eta = \frac{h(x+5{\cdot}8)}{(x+5{\cdot}8)^2 + 25{\cdot}4^2}$$

Reading directly from Fig. A.2,

$$\gamma_1 = \frac{h}{45{\cdot}8 - x}$$

$$\delta_1 = 180° - \frac{h}{x}$$

and

$$\delta_2 = \frac{h}{30 - x}$$

Hence the equation for iteration in $x$ becomes, on substitution and manipulation,

$$\frac{1}{x} = \left(\frac{1}{30 - x}\right) - \left(\frac{1}{45{\cdot}8 - x}\right) + \frac{2(x+5{\cdot}8)}{(x+5{\cdot}8)^2 + 25{\cdot}4^2}$$

$x = 14{\cdot}5$ gives a reasonable balance, and this value was used to construct Fig. 6.46, check points by Spirule in the vicinity confirming the estimate of the breakaway point.

# References


1. Horowitz, I. M., 'Fundamental Theory of Automatic Linear Feedback Control Systems', *I.R.E. Trans. autom. Control*, **AC-3**, 5 (1959)
2. Truxal, J. G., Reference to work by Truxal as quoted on p. 5 of Ref. 1 above
3. Towill, D. R., 'Synthesis of Optimum First Order Filters for Flat Spectra Command and Noise Signals', *Inter. J. Prod. Res.*, **3** (2) 121 (1964)
4. Coales, J. F. and Lawrence, P. J., 'The Preparation of Charts and Tables for the Optimisation of Automatic Control Systems with Random Inputs', *Int. Congr. int. Fed. autom. Control*, Moscow, 1960
5. Wiener, N., *The Extrapolation, Interpolation, and Smoothing of Stationary Time Series*, John Wiley, New York (1948)
6. Chestnut, H. and Mayer, R. W., *Servomechanisms and Regulating System Design*, John Wiley, New York, 230 (1959)
7. Towill, D. R. and Davies, P. J., 'Philosophy, Design and Development of MIARTES', *I.E.E.E. Trans. Educ.*, **E.8** (4) 140 (1965)
8. Routh, E. J., '*Advanced Part of the Dynamics of a System of Rigid Bodies*, Vol. II, 6th Edn, Macmillan, London (1930)
9. Taylor, P. L., *Servomechanisms*, Longmans, London, 304 (1960)
10. Newton, G. C., Gould, L. A. and Kaiser, J. F., *Analytical Design of Linear Feedback Controls*, John Wiley, New York, 109 (1957)
11. Chen, K., 'Quick Methods for Estimating Closed-loop Poles of Control Systems', *Trans. Am. Inst. elect. Engrs, Applics Ind.*, **80** (1957)
12. James, H. M., Nichols, N. B. and Phillips, R. S., *Theory of Servomechanisms*, McGraw-Hill, New York, 70 (1947)
13. Shinners, S. M., *Control System Design*, John Wiley, New York, 119 (1964)
14. Porter, A., *Introduction to Servomechanisms*, Methuen, London, 94 (1950)
15. Meadows, N., 'Graphical Techniques for Control Systems', *Control*, **3** (20) 133 (1960)
16. Douce, J. L., *Mathematics of Servomechanisms*, English Universities Press, London, 95 (1963)
17. Truxal, J. G., *Control System Synthesis*, McGraw-Hill, New York, 281 (1955)
18. Carrington, J. E. and Martin, H. R., 'Threshold Problems in Electrohydraulic Servomotors', *Proc. Instn mech. Engrs*, **180**, Pt. I (37) 881 (1966)
19. Whiteley, A. L., 'The Theory of Servo Systems, with Particular Reference to Stabilisation', *J. Instn elect. Engrs*, **93**, 353 (1946)

20. GRAHAM, D. and LATHROP, R. C., 'The Synthesis of Optimum Transient Response: Criteria and Standard Forms', *Trans. Am. Inst. elect. Engrs, Applics Ind.*, **72** (9) 273 (1953)
21. LATHROP, R. C., and GRAHAM, D., 'The Transient Response of Servomechanisms with Derivative and Integral Control', *Trans. Am. Inst. elect. Engrs, Applics Ind.* **73** (11) 10 (1954)
22. *Control Engineers' Handbook*, (Ed. J. G. Truxal), McGraw-Hill, New York (1955)
23. TRUXAL, J. G., *Control System Synthesis*, McGraw-Hill, New York, 161 (1955)
24. KLOEFFLER, R. C., KERCHNER, R. M. and BRENNEMAN, J. L., *Direct Current Machinery*, Macmillan, New York (1950)
25. CANFIELD, E. B., *Electromechanical Control Systems and Devices*, John Wiley, New York, 287 (1965)
26. SPEARMAN, F. R. J., 'Fundamental Aerodynamics and Control Relationships for Steering Cruciform Type Guided Missiles with Moving Control Surfaces', Royal Aircraft Establishment, Farnborough, *Technical Notes* G.W. 125 and 140
27. EVANS, W. R., 'Graphical Analysis of Control Systems', *Trans. Am. Inst. elect. Engrs*, **67,** 547 (1948), and also *Control System Dynamics*, McGraw-Hill, New York (1954)
28. DELTORO, V. and PARKER, S. R., *Principles of Control System Engineering*, McGraw-Hill, New York, 404 (1960)
29. LAWDEN, D. F., *Mathematics of Engineering Systems*, Methuen, London, 64 (1961)
30. WILCOX, R. B., *Control Engineers' Handbook* (Ed. J. G. Truxal), McGraw-Hill, New York, 3 and 55 (1958)
31. ELGRED, O. I. and STEPHENS, W., Effect of Closed-loop Transfer Function Pole and Zero Locations on the Transient Response of Linear Control Systems', *Trans. Am. Inst. elect. Engrs, Applics Ind.* **78** (42) 121 (1959)
32. COOPER, J. S., *A Discussion on the Interaction between the Servo and Autopilot Loops of a Typical Guided Weapon*, Royal Military College of Science, Fifteenth Guided Weapons Course Thesis (1965)
33. MULLIGAN, J. H., JNR., 'The Effect of Pole and Zero Locations on the Transient Response of Linear Dynamic Systems', *Proc. Inst. Radio Engrs*, **37,** 516 (1949)
34. MORGAN, M. M., 'Algebraic Function Calculations Using Potential Analogue Pairs', *Proc. Inst. Radio Engrs*, **49,** 276 (1961)
35. TOWILL, D. R., 'Frequency Response of Second Order Linear Systems with Phase Advance Signal Shaping', *Control*, **10** (95) 241 (1966)
36. JONES, F. and TOWILL, D. R., 'Control System Design Simplified Using Transfer Function Techniques', *Systems Communs*, **2** (7) 22 (1966)
37. TRUXAL, J. G., *Control System Synthesis*, McGraw-Hill, New York, 40 (1955)
38. HOROWITZ, I. M., *Synthesis of Feedback Control Systems*, Academic Press, New York, (1963)
39. TRUXAL, J. G., *Control System Synthesis*, McGraw-Hill, New York, 272 (1955)
40. HOROWITZ, I. M., 'Linear Adaptive Flight-Control Design for Re-entry Vehicles', *I.E.E.E. Trans. autom. Control*, **AC-1,** 90 (1964)
41. HOROWITZ, I. M. 'Passive Adaptive Flight-Control Design for Re-entry Vehicles', *Automatica*, **2** (3) 151 (1965)
42. TRUXAL, J. G., *Control System Synthesis*, McGraw-Hill, New York, 43 (1955)
43. TOWILL, D. R., ,Analysis and Synthesis of Feedback Compensated Third-

order Control Systems Via the Coefficient Plane', *Radio electron. Engr*, **32** (2) 119 (1966)
44. Towill, D. R., 'The Design of a Class of Third-order Control Systems Using the Coefficient Plane', Royal Military College of Science, *E.P. Tech. Note No. 19* (1965)
45. Towill, D. R., 'Step Function Response of a Linear, Zero Displacement Error, Third Order Control System', Royal Military College of Science, *E.P. Tech. Note No. 13* (1964)
46. Glaze, S. G., 'Analogue Technique and the Non-linear Jack Servomechanism', *Symposium on Recent Mechanical Engineering Developments in Automatic Control*, Institution of Mechanical Engineers, London (1960)
47. Lambert, T. H. and Davies, R. M., 'Investigation of the Response of an Hydraulic Servomechanism with an Inertia Load', *J. mech. Engng Sci.*, **5** (3) (1963)
48. Meadows, N. C., 'Resonance Peaks in Third-order, Type I Control Systems', *Proc. Instn elect. Engrs*, **112** (5) 1035 (1965)
49. Towill, D. R., 'Characteristics of Third Order Linear Zero Velocity Error Control Systems', Royal Military College of Science, *E.P. Tech. Note No. 23* (1965)
50. Truxal, J. G., *Control System Synthesis*, McGraw-Hill, New York, 326 (1955)
51. Oldenburger, R., 'Algebraic Approach to Design of Automatic Controls', *Trans. Am. Soc. mech. Engrs*, **80,** 433 (1958)
52. Lin, S. N., 'Method of Successive Approximations of Evaluating the Real and Complex Roots of Cubic and Higher Order Equations', *J. Math. Phys.*, **20,** 231 (1941)
52. Susini, A., *Filters, Amplifiers and Servomechanisms*, Heywood, London, 233 (1963)
54. Yeh, V. C. M., 'Study of Transients in Linear Feedback Systems by Conformal Mapping and the Root Locus Method', *Trans. Am. Soc. mech. Engrs*, **76,** 349 (1954)
55. Bhattacharyya, P. B., 'Root Locus Equations of the Fourth Degree', *Int. J. Control*, **1** (6) 533 (1965)
56. Chestnut, H. and Mayer, R. W., *Servmomechanisms and Regulating System Design*, Vol. I, John Wiley, New York, 413 (1957)
57. Chu, Y., 'Synthesis of Feedback Control System by Phase Angle Loci', *Trans. Am. Inst. elect. Engrs, Applics Ind.*, **71,** 330 (1952)
58. Truxal, J. G., *Control System Synthesis*, McGraw-Hill, New York, 290 (1955)
59. Thaler, G. J., Bronzino, J. D. and Kirk, D. E., 'Feedback Compensation; A Design Technique', *Trans. Am. Inst. elect. Engrs*, **57,** 300 (1961)
60. Elliott, D. W., Thaler, G. J. and Heseltine, J., 'Feedback Compensation Using Derivative Signals: Pt. I, Rouths Criterion, Root Loci; Pt. II, Mitrovic's Method', *Trans. Am. Inst. elect. Engrs*, **68,** 262 (1963)
61. Biernson, G., 'Quick Methods of Evaluating the Closed-loop Poles of Feedback Control Systems', *Trans. Am. Inst. elect. Engrs*, 53 (1953)
62. Chestnut, H., and Mayer, R. W., *Servomechanisms and Regulating System Design*, Vol. I, John Wiley, New York, 494 (1957)
63. Chen, K., Contribution to Discussion in Ref. 51
64. Towill, D. R., 'Rapid Synthesis Technique for the Duplicator Class of Feedback Control Systems', *Trans. Instn Measmt Control*, **1** (2) T 45 (1968)
65. Towill, D. R., 'Performance Advantages of Certain Non-linear Servomechanisms', *Control*, **9** (90) 673 (1965) and **10** (91) 45 (1966)
66. West, J. C., Douce, J. L. and Livesley, R. K., 'The Dual Input Describ-

ing Function and Its Use in the Analysis of Non-linear Feedback Systems', *Proc. Instn elect. Engrs*, **103,** Pt. B, 463 (1956)

67. BOGNER, I. and KAZDA, L., 'An Investigation of Switching Criteria for Higher Order Contactor Servomechanisms', *Trans. Am. Inst. elect. Engrs, Applics Ind.*, **73,** 118 (1954)
68. HARPUR, N. F., 'Some Design Considerations of Hydraulic of Jack-type', *Proceedings of Conference on Hydraulic Servomechanisms*, Institution of Mechanical Engineers, London (1953)
69. KOCHENBURGER, R. J., 'A Frequency Response Method for Analysing and Synthesising Contactor Servomechanisms', *Trans. Am. Inst. elect. Engrs*, **69,** 270 (1950)
70. GREIF, H. D., 'Describing Function Method of Servomechanism Analysis Applied to Most Commonly Encountered Non-linearities', *Trans. Am. Inst. elect. Engrs, Applics Ind.*, **72,** 243 (1953)
71. TUSTIN, A., 'The Effects of Backlash and of Speed-development Friction on the Stability of Closed-cycle Control Systems', *J. Instn elect. Engrs*, **94,** Pt. 2A, 1 (1947)
72. LOEB, J. M., 'Survey of Mathematical Methods for Non-linear Control Systems', American Society of Mechanical Engineers, *Paper No. 57-A-104* (1958)
73. CHESTNUT, H. and MAYER, R. W., *Servomechanisms and Regulating System Design*, Vol. II, John Wiley, New York, 164 (1955)
74. MCALLISTER, A. S., 'Graphical Method for Finding the Frequency Response of Non-linear Closed-loop Systems', *Trans. Am. Inst. elect. Engrs, Applics Ind.*, **80,** 268 (1961)
75. LAKSHMI-BAI, C., 'Describing-function for Dynamic Non-linear Components', *Control*, **5** (48) 105 (1962)
76. KU, Y. H. and CHEN, C. F., 'A New Method of Evaluating the Describing Function of Hysteresis-type Non-linearities', *J. Franklin Inst.* **272,** 226 (1962)
77. TOWILL, D. R. and GRIGG, R. H., 'Describing Function for an Error Saturating Minor Loop', Royal Military College of Science, *E.P. Tech. Note No. 7* (1964)
78. TOWILL, D. R. and BRYSON, L. S., 'Recent Developments in the Evaluation of Servoactuators for Trajectory Control', Royal Military College of Science, *E.P. Tech. Note No. 20* (1965)
79. THALER, G. J. and BROWN, R. G., *Analysis and Design of Feedback Control Systems*, McGraw-Hill, New York, 157 (1960)
80. LEVINSON, E., 'Some Saturation Phenomena in Servomechanisms with Emphasis on the Tachometer Stabilised System', *Trans. Am. Inst. elect. Engrs, Applics Ind.*, **72,** 1 (1953)
81. JOHNSON, E. C., 'Sinusoidal Analysis of Feedback Control Systems Containing Non-linear Elements', *Trans. Am. Inst. elect. Engrs, Applics Ind.*, **71,** 169 (1952)
82. WEST, J. C. and DOUCE, J. L., 'Frequency Response of a Certain Class of Non-linear Feedback Systems', *Br. J. appl. Phys.* **5,** 204 (1954)
83. FINNIGAN, R. E., *Transient Analysis of Non-linear Servomechanisms Using Describing Functions with Root Locus Techniques*, Ph. D. Thesis, University of Illinois (1957)
84. GRENSTED, P. E. W., 'Analysis of the Transient Response of Non-linear Control Systems,' *Trans. Am. Soc. mech. Engrs*, Paper 57, IRD-8 (1957)
85. FREEMAN, E. A. and COX, C. S., 'Dynamic Analysis of Non-linear Control Systems,' *Proc. Instn elect. Engrs*, **14** (1) 129 (1967)

# Index